SOLAR AND INTERPLANETARY DYNAMICS

INTERNATIONAL ASTRONOMICAL UNION
UNION ASTRONOMIQUE INTERNATIONALE

SYMPOSIUM No. 91

HELD IN CAMBRIDGE, MASSACHUSETTS, U.S.A.
AUGUST 27 – 31, 1979

SOLAR AND INTERPLANETARY DYNAMICS

EDITED BY

M. DRYER

Space Environment Laboratory
National Oceanic and Atmospheric Administration, Boulder, Colorado, U.S.A.

and

E. TANDBERG-HANSSEN

Marshall Space Flight Center
National Aeronautics and Space Administration, Huntsville, Alabama, U.S.A.

Cosponsored by
Scientific Committee on Solar-Terrestrial Physics
and
Committee for Space Research

D. REIDEL PUBLISHING COMPANY

DORDRECHT : HOLLAND / BOSTON : U.S.A. / LONDON : ENGLAND

Library of Congress Cataloging in Publication Data
Main entry under title:

Solar and interplanetary dynamics.

(Symposium – International Astronomical Union ; no. 91)
Includes index.
1. Solar activity–Congresses. 2. Sun–Congresses. 3. Interstellar
matter–Congresses. I. Dryer, Murray. II. Tandberg-Hanssen, Einar.
III. International Council of Scientific Unions. Special Committee on Solar-
Terrestrial Physics. IV. International Council of Scientific Unions. Committee
on Space Research. V. Series: International Astronomical Union. Symposium
no. 91.
QB524.S58   523.7   80–23953
ISBN-13: 978-90-277-1163-2   e-ISBN-13: 978-94-009-9100-2
DOI: 10.1007/ 978-94-009-9100-2

Published on behalf of
the International Astronomical Union
by
D. Reidel Publishing Company, P. O. Box 17, 3300 AA Dordrecht, Holland

Sold and distributed in the U.S.A. and Canada
by Kluwer Boston Inc.,
190 Old Derby Street, Hingham, MA 02043, U.S.A.

In all other countries, sold and distributed
by Kluwer Academic Publishers Group,
P. O. Box 322, 3300 AH Dordrecht, Holland

D. Reidel Publishing Company is a member of the Kluwer Group

TABLE OF CONTENTS

PREFACE

Informal discussions in 1977 among a number of scientists associated with solar and interplanetary physics revealed a need for a
dialogue between the two often-divergent groups. It was clear that
the latter group was dependent essentially on the sun for its *raison
d'etre*. On the other hand it was also clear that the former group
could benefit in its search for insight vis-à-vis solar activity by
looking beyond the shell of the inner corona. Needless to add that the
combined solar/interplanetary topic is relevant to astrophysics when
one considers stellar winds and binary star flows. It was felt, therefore, that a symposium was essential to bring together, for the first
time, leading solar and interplanetary physicists from the international community to discuss and record herein their own research. The
fundamental physical processes underlying our own capricious star's
activity can be understood only by the coupling of solar and interplanetary topics in an intimate observational and theoretical structure.
This book, intended for active research scientists and advanced graduate students, is an important step in this direction. The background
of solar and interplanetary dynamics is provided in Part I (The Life
History of Coronal Structures and Fields) and Part II (Coronal and
Interplanetary Responses to Long Time Scale Phenomena). The crescendo,
so to speak, comes in Part III (Solar Transient Phenomena Affecting the
Corona and Interplanetary Medium: Dynamics Deduced from Observations),
followed by Part IV which extends this subject to include Theoretical
Considerations. This theme is re-examined for short-time-scales in
Part V (Coronal and Interplanetary Responses to Short Time Scale
Phenomena: Observations) and, again, in Part VI from the viewpoint of
Theoretical Considerations. Finally, Part VII considers Future Direc-
tions followed by a Summary of the Symposium by Professor M. Kuperus.

As noted above, then, it was in 1977 when we proposed a symposium
on this topic to IAU with the support of Commissions 10, 12 and 49.
The proposal was accepted, and Symposium 91 on Solar and Interplanetary
Dynamics was formally announced together with the co-sponsorship of
SCOSTEP and COSPAR. It was held on the grounds of Harvard University
in Cambridge, Massachusetts, U.S.A., from August 27-31, 1979. The
Scientific Organizing Committee consisted of E. Tandberg-Hanssen and
M. Dryer (Co-Chairmen), V. Bumba, A. Hewish, Y. Nakagawa, R. W. Noyes,
D. E. Page, J. Rösch, D. M. Rust, M. J. Rycroft, S. F. Smerd (deceased
1978 December 21), S. I. Syrovatskii, and K. Tanaka. At this writing,

xiii

M. Dryer and E. Tandberg-Hanssen (eds.), Solar and Interplanetary Dynamics, xiii-xiv.
Copyright © 1980 by the IAU.

we were again grieved to learn of Professor Syrovatskii's death on 1979
September 26. Several of his last scientific papers, including an
invited one, are included in this volume.

The Local Organizing Committee was represented by D. M. Rust
(Chairman), A. S. Krieger, R. W. Noyes, A. J. Lazarus and K. R. Lang.
The brisk, yet relaxed, pace of the meeting was due in large measure
to the Session Chairmen, including R. W. Noyes, V. Bumba, S. W. Kahler,
Y. Nakagawa, K. Tanaka, A. Benz, and U. Anzer, to whom we owe our
thanks. We are also obliged to the 133 participants from 23 countries
for contributing the discussion, most of which we hope has been faith-
fully recorded. We also thank E. O'Neill, S. Kahler, D. F. Webb, and
R. F. Willson, Jr. for their assistance in keeping the Symposium
arrangements and the discussion record in good order. Finally, we are
indebted to C. Holladay for the inevitable re-typing of some man-
uscripts and her close attention to preparation of the discussion;
to C. L. Brown for the preparation of the Index; to the National
Aeronautics and Space Administration for some financial support; and
to Harvard University and American Science and Engineering, Inc., for
their logistical support.

                                        M. Dryer
                                        E. Tandberg-Hanssen

1980 January
Boulder, Colorado

LIST OF PARTICIPANTS

Ahluwalia, H. S., University of New Mexico; Albuquerque, N.M. 87131, U.S.A.

Anzer, U., Max-Planck Institüt für Astrophysik, München, F.R.G.

Avignon, Y., Observatoire de Paris, 92190, Meudon, France.

Baker, K., Boston University, Boston, Massachusetts 02215, U.S.A.

Bel, N., Observatoire de Paris, 92190 Meudon, France.

Benz, A. O., Swiss Federal Institute of Technology, CH-8044, Zurich, Switzerland.

Bhatnagar, A., Vedhshala Solar Observatory, Udaipur-313001, India.

Bhonsle, R. V., Physical Research Laboratory, Ahmedabad-380009, India.

Bird, M. K., University of Bonn, 5300 Bonn, F.R.G.

Bohlin, J. D., National Aeronautics and Space Admin., Washington, D.C. 20546, U.S.A.

Borrini, G., Osservatorio Astrofisico di Arcetri, 50125 Firenze, Italy.

Bratenahl, A., University of California (Riverside); Riverside, California 92521, U.S.A.

Bruzek, A., Kiepenheuer Institut für Sonnenphysik, D-7800, Freiburg, F.R.G.

Bumba, V., Astronomical Institute, Czechoslovakia Academy of Science, 251 65 Ondrejov, Czechoslovakia.

Callahan, P. S., Jet Propulsion Laboratory, Pasadena, California 91103, U.S.A.

Callebaut, D., Universiteit Antwerpen, B-2610 Wilrijk-Antwerpen, Belgium.

Cazeneuve, H., Direccion Nacional del Antartico, Buenos Aires, Argentina.

Cecchini, S., TESRE Laboratory/CNR, 40126 Bologna, Italy.

Couturier, P., Observatoire de Meudon, 92190 Meudon, France.

Crifo, F., Observatoire de Meudon, 92190 Meudon, France.

Cuperman, S., Tel-Aviv University, Ramat-Aviv, Israel.

Datlowe, D. W., Lockheed Research Laboratories, Palo Alto, California 94304, U.S.A.

Davis, J. M., American Science and Engineering, Inc.; Cambridge, Mass. 02139, U.S.A.

Degaonkar, S. S., Physical Research Laboratory, Ahmedabad-380009, India.

Dobrowolny, M., Laboratorio Plasma Spazio/CNR, 00044 Frascati, Italy.

Dryer, M., Space Environment Laboratory, NOAA; Boulder, Colorado 80303, U.S.A.

d'Uston, C., Centre d'Etude Spatiale des Rayonnements, 31029 Toulouse-Cedex, France.

Dutto, D., Fray Justo Sarmiento 1411, 1602 Florida (Buenos Aires),
   Argentina.
Emslie, A. G., The University, Glasgow G12 2QQ, Scotland, U.K.
Engvold, O., University of Oslo, Oslo 3, Norway.
Eyni, M., Ben Gurion University, Beer Sheva, Israel.
Foukal, P., Atmospheric and Environmental Research, Inc., Cambridge,
   Massachusetts 02138, U.S.A.
Frankenthal, S., Massachusetts Institute of Technology; Cambridge,
   Massachusetts 02139, U.S.A.
Gaizauskas, V., Herzberg Institute of Astrophysics, Ottawa KIA OR6,
   Canada.
Genouillac, G., Observatoire de Paris, 92190 Meudon, France.
Gergely, T. E., University of Maryland, College Park, Maryland 20742,
   U.S.A.
Godoli, G., Institute de Astronomie dell'Universite, 50125 Firenze-
   Arcetri, Italy.
Habbal, S., Harvard College Observatory, Cambridge, Massachusetts
   02138, U.S.A.
Haug, E., Universität Tübingen, D7400 Tübingen, F.R.G.
Heinemann, M., Boston College, Chestnut Hill, Massachusetts 02167,
   U.S.A.
Henoux, J.- C., Observatoire de Meudon, 92190 Meudon, France.
Howard, R. F., Hale Observatories; Pasadena, California 91011, U.S.A.
Hoyng, P., Space Research Laboratory, Utrecht, The Netherlands.
Intriligator, D. S., University of Southern California, Los Angeles,
   California 90007, U.S.A.
Ivanov, K. G., Astronomical Observatory; Varna, Bulgaria.
Jackson, B.V., University of California (San Diego); La Jolla, Califor-
   nia 92093, U.S.A.
Joselyn, J. A., Space Environment Laboratory, NOAA; Boulder, Colorado
   80303, U.S.A.
Kahler, S. W., American Science and Engineering, Inc., Cambridge,
   Massachusetts 02139, U.S.A.
Kanno, M., Center for Astrophysics; Cambridge, Massachusetts 02138,
   U.S.A.
Koomen, M. J., Naval Research Laboratory, Washington, D.C. 20375,
   U.S.A.
Krieger, A. S., American Science and Engineering, Inc., Cambridge,
   Massachusetts 02139, U.S.A.
Kuperus, M., Astronomical Institute, 3512 NL Utrecht, The Netherlands.
LaBonte, B.J., Hale Observatories; Pasadena, California 91101, U.S.A.
Lang, K., Tufts University; Medford, Massachusetts 02155, U.S.A.
Lemaire, J., Institute of Aeronomy, B1180 Brussels, Belgium.
Levine, R. H., Harvard College Observatory, Cambridge, Massachusetts
   02138, U.S.A.
Lincoln, J. V., World Data Center A for Solar-Terrestrial Physics,
   NOAA, Boulder, Colorado 80303, U.S.A.
Lockwood, G. W., Lowell Observatory, Flagstaff, Arizona 86002, U.S.A.
Low, B. C., Lau Kuei Huat Pte, Ltd., Singapore 22, Singapore.
Lundstedt, H., Institutionen for Astronomi, S-222 24 Lund, Sweden.

Mariska, J., Naval Research Laboratory, Washington, D.C. 20375, U.S.A.
Martres, M. - J., Observatoire de Paris, 92190 Meudon, France.
Mavcomichalak, H., University of Athens, Athens 144, Greece.
Maxwell, A., Harvard College Observatory, Cambridge, Massachusetts
    02138, U.S.A.
McClymont, A. N., University of California (San Diego), LaJolla,
    California 92093, U.S.A.
McIntosh, P. S., Space Environment Laboratory, NOAA, Boulder, Colorado
    80303, U.S.A.
McKenna-Lawlor, S., St. Patrick's College, Maynooth, Ct. Kildare,
    Ireland.
McLean, D. J., Division of Radiophysics, CSIRO, Epping 2121, Australia.
McWhirter, R. W. P., ARD Appleton Laboratory, Abingdon, Oxfordshire
    OX14 3DB, U.K.
Meire, R., Ghent State University, 9000 Ghent, Belgium.
Michels, D. J., Naval Research Laboratory, Washington, D.C. 20375,
    U.S.A.
Moller-Pedersen, B., Observatoire de Meudon, 92190 Meudon, France.
Moore, R. L., California Institute of Technology, Pasadena, California
    91125, U.S.A.
Mullan, D. J., University of Delaware; Newark, Delaware 19711, U.S.A.
Nakagawa, Y., Marshall Space Flight Center, NASA; Huntsville, Alabama
    35812, U.S.A.
Narayana, K. G., Boston College; Chestnut Hill, Massachusetts 02167,
    U.S.A.
Newkirk, G., Jr., High Altitude Observatory, NCAR; Boulder, Colorado
    80307, U.S.A.
Noci, G., Osservatorio Astrofisico di Arcetri, 50175 Firenze, Italy.
Noyes, R. W., Center for Astrophysics, Cambridge, Massachusetts 02138,
    U.S.A.
Orrall, F. Q., Institute for Astronomy, Honolulu, Hawaii 96822, U.S.A.
Pallavicini, R., Osservatorio Astrofisico di Arcetri, 50125 Firenze,
    Italy.
Parisi, M., Istituto di Fisica "G. Marconi" Universita, 00185 Roma,
    Italy.
Petelski, E. F., Inst. für Astrophysik and Extraterrestrische Forchung,
    D-53 Bonn, F.R.G.
Pneuman, G. W., Max-Planck Institüt für Astrophysik, 8 München 40,
    F.R.G.
Poletto, G., Osservatorio Astrofisico di Arcetri, 50125 Firenze, Italy.
Porsche, H., DFVLR Oberpfaffenhofen, D8031 Wessling, F.R.G.
Rieger, E., Max-Planck Institut für Extraterrestrische Physik, 8046
    Garching, F.R.G.
Ripken, H. W., University of Bonn, 5300 Bonn 1, F.R.G.
Rösch, J., Observatoire de Pic-du-Midi, F65200 Bagnères de Bigorre,
    France.
Rosenau, P., Technion, Israel Institute of Technology, Haifa, Israel.
Rottman, G., University of Colorado, Boulder, Colorado 80309, U.S.A.
Rust, D. M., American Science and Engineering, Inc., Cambridge,
    Massachusetts 02139, U.S.A.

Ruzdjak, V., University of Zagreb, 41000 Zagreb, Yugoslavia.
Sakurai, T., University of Tokyo, Benkyo-ku, Tokyo 113, Japan.
Sastry, Ch. V., Indian Institute of Astrophysics, Bangalore, India.
Sawant, H. S., Clark Lake Radio Observatory, Borrego Springs, California 92004, U.S.A.
Schatten, K. H., Goddard Space Flight Center, NASA; Greenbelt, Maryland 20771, U.S.A.
Scherrer, P. H., Stanford University; Stanford, California 94305, U.S.A.
Schmahl, E. J., Univeristy of Maryland, College Park, Maryland 20742, U.S.A.
Schmieder, B., Observatoire de Paris, 92190 Meudon, France.
Seidel, B. L., Jet Propulsion Laboratory; Pasadena, California 91103, U.S.A.
Sheeley, N. R., Jr., Naval Research Laboratory, Washington, D.C. 20375, U.S.A.
Sime, D. G., High Altitude Observatory, NCAR; Boulder, Colorado 80307, U.S.A.
Simon, G., Observatoire de Meudon, 92190 Meudon, France.
Simon, P., Observatoire de Meudon, 92190 Meudon, France.
Somov, B. V., P. N. Lebedeu Physical Institute, Academia Nauk, Moscow 117924, U.S.S.R.
Sotirovski, P., Observatoire de Paris, 92190 Meudon, France.
Spector, A. R., Radioastrophysical Observatory, Latvian Academy of Sciences, Riga 226524, U.S.S.R.
Steinitz, R., Ben-Gurion University, Beer-Sheva, Israel.
Steinolfson, R. S., University of Alabama (Huntsville), Huntsville, Alabama 35807, U.S.A.
Stewart, R. T., Division of Radiophysics, CSIRO; Epping 2121, Australia.
Stix, M., Kiepenheuer-Institüt für Sonnenphysik, D-7800 Freiburg, F.R.G.
Storini, M., Istituto di Fisica "G. Marconi" Universita, 00185 Roma, Italy.
Suess, S. T., Space Environment Laboratory, NOAA, Boulder, Colorado 80303, U.S.A.
Švestka, Z., Space Research Laboratory, Utrecht, The Netherlands.
Sýkora, J., Astronomical Institute of the Slovak Academy of Sciences, 059 60 Tatranska Lomnica, Czechoslovakia.
Tanaka, K., Tokyo Astronomical Observatory, Mitaka, Tokyo, Japan.
Tandberg-Hanssen, E., Marshall Space Flight Center, NASA; Huntsville, Alabama 35812, U.S.A.
Tandon, J. N., University of Delhi; Delhi, India.
Tapping, K. F., Herzberg Institute of Astrophysics, Ottawa, Ontario, KIA OR6, Canada.
Torricelli, G., Osservatorio Astrofisico di Arcetri, 50125 Firenze, Italy.
Uchida, Y., Tokyo Astronomical Observatory, Mitaka 181, Tokyo, Japan.
van Hoven, G., University of California (Irvine), Irvine, California 92717, U.S.A.
Van Tend, W., The Astronomical Institute at Utrecht; 3512 NL Utrecht, The Netherlands.
Washimi, H., Nagoya University, Toyokawa 442, Japan.

Watanabe, T., Space Environment Laboratory, NOAA, Boulder, Colorado
80303, U.S.A.
Webb, D. F., American Science and Engineering, Inc; Cambridge,
Massachusetts 02139, U.S.A.
Williams, D. J., Space Environment Laboratory, NOAA; Boulder, Colorado
80303, U.S.A
Willson, R. F., Tufts University; Medford, Massachusetts 02155, U.S.A.
Withbroe, G., Center for Astrophysics; Cambridge, Massachusetts 02138,
U.S.A.
Wu, S. T., University of Alabama (Huntsville); Huntsville, Alabama
35807, U.S.A.
Yeh, T., Metropolitan State College; Denver, Colorado 80204, U.S.A.
Zelenka, A., Swiss Federal Astronomy Observatory, 8092 Zurich,
Switzerland.

EVOLUTION OF CORONAL AND INTERPLANETARY MAGNETIC FIELDS

Randolph H. Levine
Harvard-Smithsonian Center for Astrophysics
Cambridge, Massachusetts USA

ABSTRACT. Numerous studies have provided the detailed information necessary for a substantive synthesis of the empirical relation between the magnetic field of the sun and the structure of the interplanetary field. We will point out the latest techniques and studies of the global solar magnetic field and its relation to the interplanetary field.. The potential to overcome most of the limitations of present methods of analysis exists in techniques of modelling the coronal magnetic field using observed solar data. Such empirical models are, in principle, capable of establishing the connection between a given heliospheric point and its magnetically-connected photospheric point, as well as the physical basis for the connection. We thus find ourselves at a plateau, looking back over a quarter century of empirical synthesis while anticipating a new era of detailed physical investigation on a global scale.

1. INTRODUCTION

1.1 Scientific Themes of Solar-Interplanetary Magnetism

The first synthesis of observations of the solar magnetic field was summarized by Babcock and Babcock in 1955. Using over two years of data from the original Mt. Wilson magnetograph, they were able to point out the dipolar character of the general solar field, and described many of the features of Bipolar Magnetic regions and their association with sunspots, Hale's polarity laws, and enhancements of coronal brightness. The general magnetic field of the sun, however, appeared to them to be confined to the higher latitudes; the field in the equatorial regions did not present a persistent global pattern. In the lower latitudes (between $+50^{\circ}$) the fields were seen as localized and (relatively) transient. Most of these were Bipolar Magnetic regions of varying complexity, but a very few were given the interesting (but later misleading) name "Unipolar Magnetic" (UM) regions:

1

M. Dryer and E. Tandberg-Hanssen (eds.), Solar and Interplanetary Dynamics, 1-20.
Copyright © 1980 by the IAU.

In introducing the term "unipolar," we refer to regions
which . . . show . . . almost exclusively one polarity and
which are not directly accompanied by adjacent regions of
opposite polarity; in most cases it is not at all obvious
where the magnetic lines of force emanating from the UM
region re-enter the sun.

Because we can now shed much more light on the questions that
Babcock and Babcock had to leave unanswered, their paper is an
interesting and satisfying gauge of our progress in understanding the
structure of the global solar magnetic field. At the same time,
however, we are reminded of one of the persistent themes of research in
the field of solar-interplanetary magnetism, for Babcock and Babcock
noticed that UM regions tended to last for several rotations, and found
that one particularly prominent series of geomagnetic storms in 1953
recurred in association with a large unipolar region, with a 2-3 day
phase lag. They hypothesized that the UM regions might in fact be the
"M" regions suggested by Bartels (1932) as the source of solar particle
emissions responsible for recurrent geomagnetic activity.

Another theme of scientific relevance is represented by Alfven
(1956), who criticized the then-popular view of the general magnetic
field outside the sun as being a dipole on the ground that it was an
"unfounded assumption." He pointed out the evidence for currents (in
the form of known changes in the electromagnetic field) and reminded
his readers that the flowing charges would alter the presumed structure
of the field. In the process of developing his critique, he presented
a simplified model of the interplanetary magnetic field that is
probably the first recognition of the possibility of the opening of
dipolar (or more generally potential) lines of force. The model
explicitly assumed that the open lines of force are "beamed," or
concentrated into restricted areas of the solar surface, and that other
portions of the sun are characterized by closed magnetic structures.

Perhaps the most widely referenced early insight into the
influence of the structure of the solar magnetic field on the
heliosphere is the work of Billings and Roberts (1964). Their
schematic suggestion lacks any indication of magnetic polarities,
concentrates on an invalid temperature distribution, and does not
recognize the global nature of large-scale open field regions.
However, it does recognize the role of the solar wind in extending
weaker magnetic field lines outward, and presents the issue of the
modulation of coronal temperature and density structure by the magnetic
field. Billings and Roberts' work has been cited by several authors
and has served as the basic framework for more elaborate schematic
illustrations of the basic physical processes.

A final key theme that needs to be remembered is the discovery of
the sector structure of the interplanetary magnetic field. With the
ability to measure the interplanetary field in situ near the earth,
Wilcox and Ness (1965) discovered that the averaged (over several

hours) magnetic field points predominantly toward or away from the sun for several days at a time, and that this pattern of alternating polarity sectors persists in its general form for times long compared with the time scales on which the obvious manifestations of the photospheric field (i.e., BMRs) change.

These themes, the large-scale polarity regions of the solar photosphere, the possible existence of open magnetic field lines in the corona, and the interplanetary magnetic sector structure, are the basic scientific framework for the continuing study of the way in which the solar magnetic field extends into the heliosphere.

1.2 Scope of This Review

The ability to identify the key scientific themes mentioned above is due to an ongoing synthesis of observational data and theoretical insights. Excellent summaries of this work were given by Dessler (1967) and by Wilcox (1968). The advances at the beginning of this decade, due largely to improved instrumentation and especially to the solar experiments on Skylab, are discussed in the Coronal Hole Workshop volume edited by Zirker (1977) and in a review of the work on the large-scale solar field by Howard (1977).

In this discussion, it is my intent to concentrate on the state of our present physical understanding of the global solar magnetic field and its extension into the interplanetary medium. I do not intend an exhaustive review of the literature, nor a catalog of experimental results. Rather, I wish to reflect a largely coherent understanding of the large-scale processes in the magnetic field (at least above the photosphere), and to emphasize that the period of basic discoveries which have led to this understanding must now be followed by increased theoretical efforts aimed at modelling the physical processes in realistic detail and by careful observations designed to establish a long-term, reliable data base for both empirical analysis and input to theoretical models.

The second section discusses the observational picture of the solar and the interplanetary magnetic field. The correlation between the two fields is interpreted within the framework of the understanding of the global solar magnetic field that has emerged in recent years. Section 3 points out data-based models of the global solar magnetic field that presently exist. The known limitations of each type of model are emphasized. The final section is an overview of the progress in this field, pointing out expectations and needs for the future, as well as providing recommendations for fruitful areas of theoretical and observational pursuit.

2. OBSERVATIONS OF SOLAR AND INTERPLANETARY MAGNETIC FIELDS

2.1 Whole Sun Measurements

Measurements of the mean magnetic field of the sun use integrated light from the full solar disk (the sun seen as a star) and represent the average net line of sight field strength over the disk, with the field value at each point weighted by the intensity of the magnetically sensitive line used to determine the field. In practice this means that points nearer the central portion of the disk contribute most heavily to the measured value. The values obtained also represent averages over some portion of a single day. Such measurements have been carried out since 1968 at the Crimean Astrophysical Observatory, since 1970 at Hale Observatories (Mt. Wilson), and since 1975 at the Stanford Solar Observatory. Studies based on these observations, including intercomparisons, have been published by Severny, et al. (1970), Scherrer, et al. (1977a), and Scherrer, et al. (1977b).

The physical basis of using mean field measurements is that they reflect a (weighted) average of the net solar magnetic flux over large areas of the photosphere. The properties of large-scale patterns in the photospheric field can then be investigated, although the actual net flux and the area over which it is averaged cannot be determined. The observed mean field varies slowly from day to day, with the pattern generally repeating with slow evolution after about 27 days. This is consistent with the interpretation that there are large-scale structures of consistent magnetic polarity and weak average net field strength in the photosphere. Because of the difficulty involved in interpreting mean field measurements directly in terms of photospheric structures, however, their use has been confined almost exclusively to comparisons between solar and interplanetary magnetic structure (see section 2.3, below).

2.2 Spatially Resolved Measurements

The more traditional method of studying the large-scale magnetic field of the sun has been to construct synoptic charts of spatially resolved daily magnetograms. This permits direct investigation of large-scale patterns in the photospheric magnetic field. Examples of studies using this type of data are by Babcock and Babcock (1955), Bumba and Howard (1965, 1966), Wilcox and Howard (1968), Stenflo (1972), and Svalgaard, et al. (1975), all using data from the Mt. Wilson instrument, and by Levine (1979) using measurements from Kitt Peak National Observatory.

Some properties of the large-scale photospheric field are illustrated in Figure 1. These are synoptic charts of the line of sight magnetic field (measured at Kitt Peak) averaged over eleven solar rotations. The only difference between the four panels is the grid size onto which the data are averaged. The emergence of large-scale patterns as the viewing resolution decreases is readily apparent. At

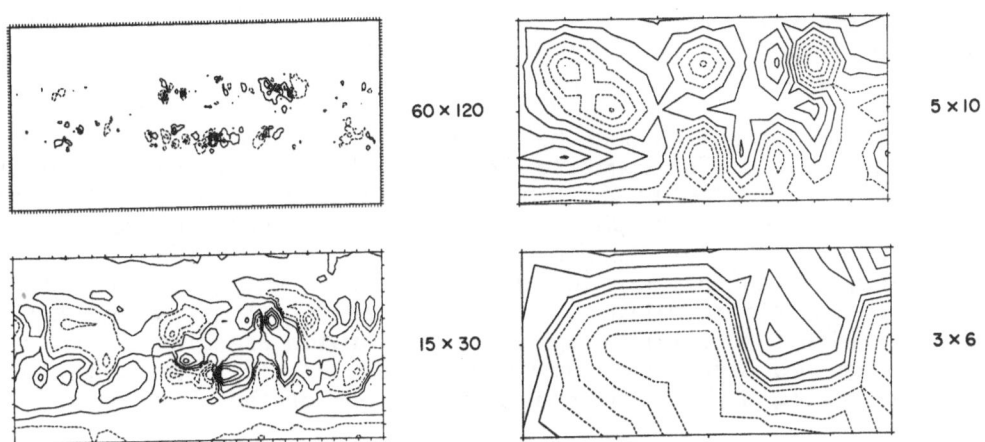

Figure 1. Contour plots of average magnetic field observed over Carrington rotations 1601-1611. The four panels each represent the same data averaged onto grids of different surface resolution. The sizes of the grids used are indicated. Contour levels (+ and -, in gauss) are (60 x 120): 5, 15, 25, 35; (15 x 30): 0, 1, 3, 5, 7; (5 x 10): 0, .4, .8, 1.2, 1.6; (3 x 6): 0, .1, .3, .5, .7 (from Levine, 1979).

its lowest resolution the pattern is remarkably similar to· the description of the large-scale field which has been inferred from interplanetary and mean field observations by Svalgaard, et al. (1974, 1975). This phenomenological model is based on the union of the observed structure of the polar caps, with uniform net polarities through most of the solar cycle, and the alternating polarity structure of the lower latitudes, which is inferred from the sector structure of the interplanetary field and the observed mean field of the sun. The result is a single large-scale neutral line which is supposed to separate the largest polarity sectors of the sun. Figure 1 shows that this model is a very good representation of the gross properties of the data, even though the intent of Svalgaard, et al. was to relate the interplanetary sector observations to coronal structure, rather than to "the complicated features found in the photosphere or in the chromosphere."

Because the sector pattern was more readily identifiable when the strong active region fields were removed from their Mt. Wilson data, Svalgaard, et al. (1975) concluded that the photospheric sectors were "not caused or controlled by the strong magnetic fields of active regions." Levine (1977a, 1977b, 1979) has studied the data from Kitt

Peak for Carrington rotations 1601-1611 and its relation to coronal holes. The conclusion of these studies is that active region flux contributes systematically to the maintenance of polarity sectors (and coronal holes, which are related to them). Thus, although the contour levels in the lower resolution panels of Figure 1 are a few gauss or less, the large-scale patterns they outline are not necessarily determined by a weak solar magnetic field component. The appearance of large-scale patterns with weak net fields might be due to the imbalance of stronger fields of opposite polarity. Because of the possibility that solar magnetic fields are concentrated into bundles of high field strength in the photosphere, however, we cannot directly make statements about the large-scale patterns being due to strong fields or weak fields; we can only examine the role of flux which is concentrated vs. flux which is more diffuse. Measured field values which differ represent either actual photospheric fields which are distributed over similar areas and have strengths which differ, or they represent fields of similar strength which cover different areas of the photosphere.

We cannot then say that only diffuse flux or only concentrated flux is responsible for the photospheric polarity pattern. By separately averaging measurements with high average |B| and low average |B|, Levine (1979) showed that evidence of the over-all pattern is present in the averages of both. Many workers (e.g. Bumba and Howard, 1965; Wilcox, 1968; Wilcox, et al. 1969; Severny, et al. 1970; Svalgaard, et al. 1975) have inferred or assumed that the existence of such patterns was due to a weak background field with a strength of a few gauss. The alternative interpretation provided by analysis of the Kitt Peak data (useful in this context because it does not saturate at high field strengths) is that the inclusion of all measured fields does bring out a sector pattern and that the visibility of the pattern is a question of spatial scale rather than of field strength. This strongly suggests that it is the imbalance of field strengths, rather than their absolute value, which establish the systematic pattern. Mean field measurements, even though sensitive to very weak net field strengths, must also include the effects of systematic imbalance of concentrated flux. A weak background field of a few gauss or less, while not excluded, is not necessary to explain the observations.

From the solar point of view we must ask what processes are responsible for the organization of fields with systematic flux imbalance extending to the largest spatial scales. The contribution of concentrated flux can occur in obvious structures like active regions, but the role of the possibly large amount of magnetic flux in Ephemeral Active Regions (Harvey and Martin, 1979; Golub, et al. 1979) cannot be ignored. The long lifetimes of magnetic polarity patterns, compared to the lifetimes of the structures which contribute their flux, reinforce the view (Svalgaard, et al. 1975; Levine, 1977a, 1977b) that the large-scale structure of the solar magnetic field is a fundamental aspect of solar magnetism. (It is interesting to note that Svalgaard, et al. concluded that the photospheric sector pattern is of fundamental importance because they found it to exist independently of the active region fields while Ievine came to the same conclusion by noting that there was a systematic component to active region evolution which maintained the photospheric sectors for times much longer than the lifetime of a typical active region.)

In order to improve the study of the influence of the sun on the earth, it will be important to know at what spatial scales the information about the solar magnetic field is propagated to the earth. Smaller features such as the active regions which make the large-scale pattern difficult to detect at high resolution are not directly sensed in the near-earth environment, while the influences of the largest magnetic sectors are known to be detectable and even dominant in the interplanetary field. It is not known in detail at what spatial scales between these extremes the magnetic pattern of the photosphere might occasionally be manifest near 1 A.U., and what processes are responsible for modulating the patterns observed. The crucial role played by the dipole and quadrupole moments of the photospheric field have been emphasized by Schulz (1973). He has shown that the presence of just these two terms is sufficient to produce a warping of the heliomagnetic equator similar to that envisioned in phenomenological models. It is the evolution of these terms, and the presence of higher order multipoles, which must be incorporated into a more satisfactory explanation of the interplanetary magnetic field (see Section 3, below).

2.3 Why Do Solar Polarity Sectors Correlate with the Interplanetary Magnetic Field?

The details of the discovery that the interplanetary magnetic field is organized into sectors of alternating polarity have been reviewed by Wilcox (1968).

By reducing the IMF and solar polarities (mean field or synoptic averages) to Bartels' chart format the similarity of the solar and interplanetary patterns can be made apparent (Scherrer, et al. 1977a; Levine, 1979). A more quantitative measure of the agreement can be obtained by cross-correlating the solar and IMF data sets. Figure 2 shows the results of such a calculation by Severny, et al. (1970) for four months in 1968. The correlation peaks at a lag of 4.5 days. This is interpreted as indicating a transit time of about 4.5 days for the solar wind to bring the solar pattern to the vicinity of the earth, where the IMF measurements are made. As discussed below, however, it is not clear exactly what pattern is being transported by the solar wind, and it is certainly not the entirety of the photospheric pattern that physically maps to the earth.

There are a series of correlation peaks in Figure 2 at 4.5 days plus multiples of 27 days, resulting from the rotation of two slowly evolving patterns. What is surprising is that the peak at 4.5+27 days is larger than the peak at 4.5 days. Subsequently, the peaks decrease slowly in magnitude, which is what would be predicted for the

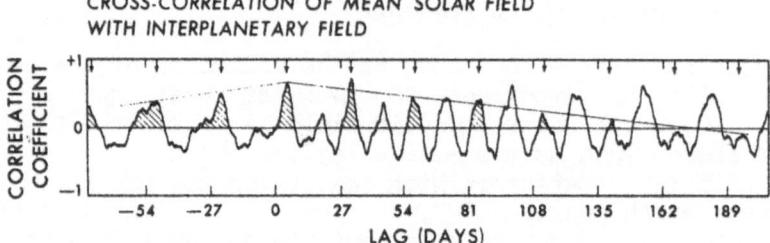

Figure 2. Cross correlation of the direction of the mean solar field with the direction of the interplanetary field during March–June, 1968 (from Severny, et al. 1970).

correlation of two slowly evolving patterns as the time lag is increased. The greater magnitude of the second peak was interpreted by Severny, et al. as indicating that there was a delay of up to one solar rotation between the appearance of a new magnetic structure in the photosphere and its manifestation in the IMF. A similar conclusion was reached by Schatten, et al. (1969). Similarly, Levine, et al. (1977) showed that smaller-scale evolutionary changes in the photospheric field took up to one rotation to be reflected in the appearance of coronal holes. Levine (1979) was able to clarify the source of this feature by cross-correlating the IMF with different spatial averages of Kitt Peak data taken in 1973–4. The increased size of the second peak (at 4.5+27 days) was found only for photospheric areas less than 20° in longitudinal extent. Averages over larger regions produced the greatest correlation at a 4.5 day lag, indicating that the evolutionary changes in the IMF are due to photospheric changes which are characteristically much smaller in scale than the largest polarity regions, and that these smaller-scale changes tend to influence the IMF after a time delay greater than several days but less than a full solar rotation.

It is instructive to inquire more deeply into the physical nature of the connection responsible for the statistical correlation of the interplanetary and solar polarity patterns. We can understand why mean field measurements correlate with the IMF by examining Figure 3. This is a contour plot from the study by Levine (1979) showing the cross-correlation of solar and interplanetary polarities where the solar polarities were determined by averaging over different size areas of the Kitt Peak synoptic magnetograms. The correlation is plotted as a function of the latitudinal and longitudinal extent of the photospheric region used to determine each day's solar polarity, and the qualitative features are the same for time lags from three to seven days. It is

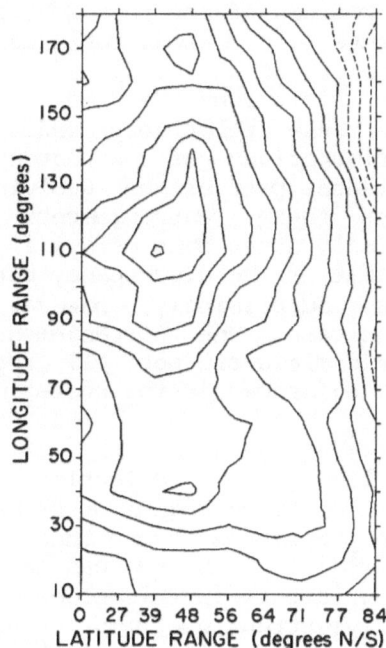

Figure 3. Contour plot of the cross correlation of solar and interplanetary polarities as a function of the size of the photospheric area used to define the solar polarity. Contour levels are from -0.09 to 0.39 in steps of 0.04 (from Levine, 1979).

clear that the maximum correlation occurs when a region of about 100° in longitudinal and latitudinal extent is used on the sun. The mean field measurements fortuitously sample the sun near this spatial scale, although the comparison cannot be exact because of the continuous weighting of the visible disk due to integrating over a varying line intensity in mean field observations vs. the sharp boundaries used to define averages of the Kitt Peak data.

Although Figure 3 helps to explain why mean field measurements are a useful representation of the solar polarity pattern, it does not explain the physical basis for the correlation with the IMF. For example, although the sectors of the IMF are generally traceable to similar sectors in the solar data, the reverse is not always true. There are large-scale polarity patterns in the photosphere which are not manifest near the earth (Levine, 1979). Some of these may be due to the latitudinal extent of the solar polarity regions and to the position of the earth with respect to the ecliptic plane, as described by Hundhausen (1977).

It has also been shown that the pattern of the photospheric field, as measured by its multipole structure, can undergo rapid large-scale rearrangements (Levine, 1977c; Altschuler, et al. 1974) and that these are accompanied by rearrangements of the corona (MacQueen and Poland, 1977). It remains to be seen whether these relatively sudden (i.e., within one solar rotation) reconfigurations of the photosphere and corona are a characteristic or fundamental part of the evolving solar cycle. It might be expected that these rearrangements would be manifest in the interplanetary field only after some delay, if at all, and that this would account for some of the discrepancy between the average photospheric and IMF patterns. Surprisingly, however, it is more nearly the reverse that is the case. The rearrangements of the multipole structure of the photospheric field are not all represented in the photospheric averages, but are evidenced in the evolution of the interplanetary sector structure.

These difficulties are most likely a reflection of the shortcomings of choosing solar polarities by averaging or by mean field measurements. Comparisons using such procedures assume that the entire extent of each large-scale polarity region on the sun can be mapped in interplanetary space. The controlling factors in this process, though, are global and depend on the relative strength of most or all of the large-scale regions on the sun (including the polar regions). It is then not surprising that the analysis of the field in terms of spherical harmonics does show a varying strength of the strongest multipoles which more closely matches the interplanetary pattern during this time of two sectors initially, four sectors briefly, then two sectors again (Levine, 1977c). The interplanetary field is a reflection of the relative strengths as well as of the polarities of large regions of the solar photosphere.

Additionally, explicit empirical models of the sources of interplanetary magnetic flux have supported the "nozzle" theory of the origin of the field, in that the footpoints of open lines of magnetic force tend to concentrate in relatively small portions of photospheric polarity regions (Levine, et al. 1977; Levine 1977a, 1977b, 1978; Burlaga, et al. 1978). Similarly, the phenomenological model of Svalgaard, et al. (1974) explicitly recognizes and attempts to account for magnetic arcades and helmet streamers. These closed structures are directly above the large-scale photospheric neutral lines and form a geometrical barrier to the direct connection of photospheric and interplanetary polarity boundaries. Direct evidence that most of the sun is covered by closed magnetic structures comes from analysis of x-ray observations of the corona (Vaiana, 1976; Vaiana and Rosner, 1978).

Because most of the area of photospheric polarity regions is not likely to connect to interplanetary space, the use of the extent of these regions as tracers of the extent of interplanetary magnetic field sectors is a very simple approximation at best. That this procedure is successful at all is due to the fact that the open field lines concentrated in the photosphere diverge rapidly into the corona where they then tend to reflect the extent (and strength) of the region in which they are rooted. The longitudinal extent of this mapping is largely preserved at 1 A.U., although the latitudinal structure is substantially altered by electromagnetic and dynamical effects (Schulz, 1973; Svalgaard, et al. 1975; Seuss, et al. 1977; Hundhausen, 1977).

3. PHYSICAL MODELS OF THE SOLAR CONNECTION TO THE INTERPLANETARY MEDIUM

As outlined above in Section 2.3, the difficulty with studying the source of the interplanetary magnetic field by observing the sun with little or no spatial resolution is that the physical basis for the correlation of the patterns is never clear in detail. Thus, for example, when four large-scale sectors are observed in the photosphere at a time when there are only two interplanetary sectors the statistical measures cannot help us to decipher the actual mapping.

An ideal procedure would be to establish a connection between each point in the heliosphere and a point on the solar surface using a fully magnetohydrodynamic model of a rotating, magnetized sun and its ionized corona. In practice this is well beyond present capability, for both physical and computational reasons. The important priorities for constructing useful, but admittedly incomplete, models of the extension of the solar magnetic field into the heliosphere are (1) that such models reflect the distribution of magnetic flux at the solar surface in detail, (2) that such models attempt to provide a realistic description of the inner corona, where the influence of closed structures is greatest, and (3) that such models take account of the expansion of the corona and its influence on the magnetic field, i.e., that they include the appropriate mhd effects in each portion of the corona.

Steps toward these goals have been taken and have resulted in substantial progress in understanding the solar-interplanetary connection. Some of the methods now in use or under development are discussed below in Sections 3.1 to 3.4.

3.1 Potential Field Models and Source Surfaces

The potential magnetic field extrapolation technique was developed for the sun by Schatten, et al. (1969) and by Altschuler and Newkirk (1969), and resulting calculated distributions of the large-scale coronal magnetic field were compared with observations of coronal structures with varying degrees of success (Schatten, 1968; Newkirk and Altschuler, 1970; Smith and Schatten, 1970; Smerd and Dulk, 1971; Uchida, et al. 1973). The method is based on a spherical harmonic analysis of solar magnetic field observations taken on a synoptic basis. The basic assumptions are that the field above the photosphere is potential (current-free) and that the line-of-sight component of the model field at the photosphere agrees with the observed line-of-sight field as given by synoptic magnetograms. The presence of currents in the low corona, as well as all short time scale evolution or transient effects, are ignored.

If the harmonic extrapolation of a potential field were not modified, however, all magnetic structures would be closed (except for a finite set of open footpoints with measure zero). So a further provision of all these models is the ability to study the effect of open structures by including a source surface in the mathematical model (Chapman and Bartels, 1940; Schatten, et al. 1969; Altschuler and Newkirk, 1969). The source surface in such models is spherical, concentric to the solar surface, and required to be equipotential, so that the magnetic field has no angular components there and is thus perpendicular to the source surface. In this way the complicated effects of plasma-field interactions that create open structures can be emulated by varying one parameter, the radius of the source surface. Although the source surface is a useful and mathematically simple device, it is not meant to be an exact representation of physical processes and care must be exercised in interpreting its use.

More recently, Altschuler, et al. (1976) have greatly improved this technique by using higher resolution synoptic magnetograms and by increasing the highest principal polynomial index used in the expansion from 9 to 90. An equivalent technique, using Fourier decomposition, was developed by Adams and Pneuman (1976). These newer modelling procedures have been used to investigate the coronal geometry associated with radio storms (Jackson and Levine, 1979; Gergely and Kundu, 1979), the detailed geometry and evolution of coronal holes (Levine, et al. 1977, Levine, 1977a, 1977b), the relation of open magnetic structures in the low corona to solar wind flow (Burlaga, et al. 1978; Levine, 1978), the properties of possibly open field lines within active regions (Svestka, et al. 1977), and the relation of recurrent energetic particle phenomena to coronal structure (Roelof and Levine, 1978). These studies have demonstrated the strong relation of open magnetic structures to active regions, and have emphasized that the sources of magnetic flux in the ecliptic plane can originate over a broad range of photospheric latitudes. This implies that almost all of the solar wind flow in the low corona is significantly non-radial.

An example of the difficulty of applying spherical source surface techniques to the study of solar structures is that investigations of the corona at eclipse (e.g., Altschuler and Newkirk, 1969) or of structures associated with active regions or active region interconnections (Howard and Svestka, 1977; Gergely and Kundu, 1979) usually find the best agreement for a source surface at a distance of about 2.5 solar radii from the center of the sun. Studies of interplanetary structure (Schatten, et al. 1969; Burlaga, et al. 1978; Levine, 1978) and of coronal holes (Levine, 1977a, 1977b), however, have used a source surface radius of about 1.5 solar radii to fit the observations. It is likely that both of these determinations are correct, in the sense that the height at which field lines become open over active regions is higher than it is over coronal holes. This can be expected by noting that the radius at which the solar wind flow speed reaches the local Alfven velocity is defined by a surface where the plasma density times the square of the magnetic field strength is constant. Because both the density and the field strength are larger at a given radius over active regions than over coronal holes, the Alfven radius will be closer to the sun over coronal holes. Attempts to account for this in self-consistent models are discussed in the next two sections.

3.2 Models Including the Effects of Current Sheets

In addition to the difficulty mentioned above, Schatten (1971) noted that both polar plumes and streamers have a significant (and systematic) non-radial orientation. The use of a spherical source surface at which the field is required to be radial cannot produce these effects. Because the plasma pressure is much less than the magnetic field pressure out to many solar radii, Schatten argued that currents could only exist in the inner corona at places where the field was weak, i.e., near neutral sheets. In contrast, source surface models ignore all currents, except for an implied current distribution on the source surface itself.

By using an ingenious trick dependent on the symmetry of the electrostatic Maxwell stress tensor, Schatten was able to devise a relatively simple procedure to account for sheet currents in a self-consistent manner while retaining the boundary condition of observed photospheric fields. His model uses a spherical, heliocentric shell to define the direction of field lines in a potential model with no source surface. At the height of this shell field lines are not necessarily radial; in fact, if the potential field lines were allowed to extend beyond the shell they would all close. However, Schatten's model specifically requires that no closed structures exist above the shell. This opening of field lines is due to an implied distribution of currents, much as in a source surface model, except that in Schatten's model the currents flow on neutral sheets in the region outside the specified shell, and the field distribution can be calculated there. As in a source surface model, there is one free parameter, the radius of the shell, and all closed structures must lie below the shell. The necessity of using shells as low as 1.6 solar radii, however, results in a poor representation of the observed structure directly over large closed regions such as helmet streamers.

Yeh and Pneuman (1977) developed a current sheet model in which the positions and strengths of neutral-line current sheets are determined in an iterative solution to the pressure balance equation, including the effects of coronal expansion. While generalizable in principle, this technique is so numerically intricate that it will be difficult to use it in contexts much more complicated than the example

of a photospheric dipole. In that case, however, it agrees very well
with a full mhd solution.

3.3 Non-Spherical Source Surface Models

Schulz, et al. (1978) have recently produced a close simulation of
mhd behavior by introducing a source surface that is non-spherical.
Arguing that an isogauss surface of a potential field solution will
have the desired property of being closest to the sun over regions of
low average B, they calculate an analytic solution for the case of a
photospheric dipole in which the source surface is defined by a
constant value of B in a purely dipole field.

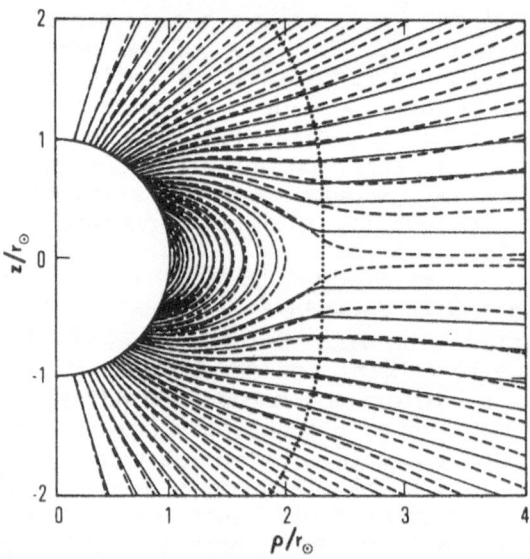

Figure 4. Configuration of magnetic field lines due to an
internal solar dipole in the mhd solution of Pneuman and Kopp
(1971, dashed lines) and in the non-spherical source surface
model of Schulz, et al. (1978, solid lines). The position of
the source surface is dotted and has an equatorial radius of
2.3 solar radii and a polar radius of 2.9 solar radii (from
Schulz, et al. 1978).

Their example, shown in Figure 4, demonstrates the variation of
the source surface radius and also a close agreement with the mhd
solution of Pneuman and Kopp (1971) for an isothermal corona over a
photospheric dipole. The field lines are deflected systematically
equatorward from their equivalent positions in a spherical source

surface solution. The exterior solution, produced by extrapolating the field outward normal to the source surface (rather than radially) in a rotating heliosphere, gives a field value at earth that is a few gammas for a one gauss dipole at the sun. This value is not highly dependent on latitude because of the deflection of flux from the polar regions toward the equator. Another important feature of the illustrative model of Schulz, et al. is that the (equatorial) current sheet is highly localized compared with the current distribution implied in a spherical source surface model.

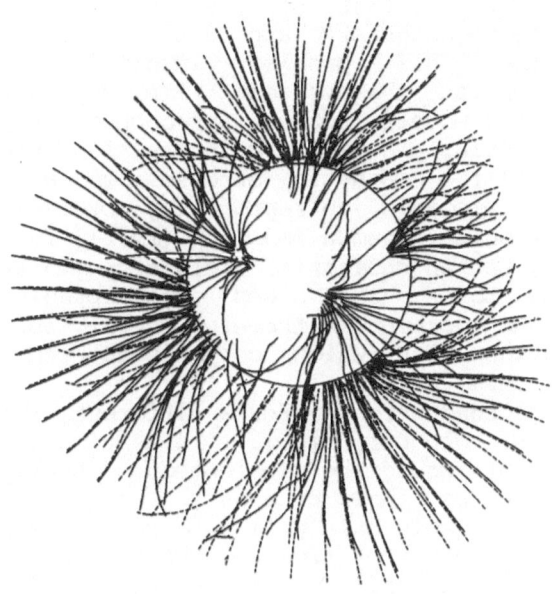

Figure 5. Configuration of magnetic field lines in a non-spherical source surface model being developed by Levine, Schulz, and Frazier. The photospheric data represents Carrington rotation 1601 and the source surface was chosen as an isogauss of the potential solution with no source surface.

It is possible to extend the formalism of Schulz, et al. (1978) for a non-spherical source surface to the analysis of actual photospheric data. This is being studied by Levine, Schulz, and Frazier. The advantages of this type of model are the expected ability to simulate more closely an exact mhd solution, and the provision for specifying any continuous distribution of radius as the source surface. Preliminary results are shown in Figure 5. The source surface was

chosen as an isogauss of the potential field model for Carrington
rotation 1601 and field lines have been plotted inward from the source
surface. The highly non-spherical nature of the source surface is
apparent, as is the presence of non-radially oriented coronal
structures. Further work on this model, including the exterior
solution for the interplanetary field and its evolution, is in
progress.

4. OVERVIEW

Although I have attempted to present a coherent view of the
structure of the sun's magnetic field and how it might be understood in
a degree of detail matching that of the best observations, I have
chosen a basically magnetostatic approach. In Section 3 I alluded to
the fact that the full problem involved magnetohydrodynamic
considerations which at the present time can only be simulated in
realistically detailed models. Other approaches to the dynamically
coupled problem of solar plasma-field interactions place more emphasis
on accounting for all the magnetic stresses to a known degree of
accuracy and less emphasis on reproducing observed magnetic fields.
The exact mhd solution of Pneuman and Kopp (1971), for example, uses an
isothermal corona, has an equatorial field strength at the photosphere
half that at the poles, and gives unacceptable flow speeds.
Nevertheless, it has proven extremely useful as the only fully
nonlinear calculation with which comparisons can be made. An
improvement over this calculation, either in terms of physics (e.g.,
inclusion of thermal conduction and/or a different photospheric field
distribution) or of geometry (e.g., relaxation of azimuthal and/or
north-south symmetry) would be welcomed. Analyses using mhd
approximations, such as those of Suess, et al. (1977) or Nerney and
Suess (1975), are applicable only well into the interplanetary medium
but have the great advantage of treating the problem self-consistently.
Regretfully, there is not sufficient space here to discuss these and
other efforts in more detail.

Realistic expectations of progress in studying the structure and
evolution of the solar magnetic field are contingent on continuing
efforts in three major areas: (1) The synoptic observation of the solar
magnetic field with sufficient spatial resolution to discriminate the
major sources of magnetic flux. (2) Continued efforts to produce
realistic theoretical models of the actual field structure in the low
corona and its extension into the interplanetary medium. (3) The
development of more mhd models for physically interesting test cases.

REFERENCES

Adams, J. and Pneuman, G.W.: 1976, Solar Phys., 46, 185.
Alfven, H.: 1956, Tellus, 8, 1.
Altschuler, M.D. and Newkirk, G., Jr.: 1969, Solar Phys., 9, 131.

Altschuler, M.D., Trotter, D.E., Newkirk, G., Jr. and Howard, R.: 1974, Solar Phys., 39, 3.
Altschuler, M.D., Levine, R.H., Stix, M. and Harvey, J.W.: 1976, Solar Phys., 51, 345.
Babcock, H.W. and Babcock, H.D.: 1955, Astrophys. J., 121, 349.
Bartels, J.: 1932, Terr. Mag. Atm. Elec., 37, 48.
Billings, D.E. and Roberts, W.O.: 1964, Astrophysica Norvegica, 9, 147.
Bumba, V. and Howard, R.: 1965, Astrophys. J., 141, 1502.
Bumba, V. and Howard, R.: 1966, Astrophys. J., 143, 592.
Burlaga, L.F., Behannon, K.H., Hansen, S.F. and Pneuman, G.W.: 1978, J. Geophys. Res., 83, 4177.
Chapman, S. and Bartels, J.: 1940, Geomagnetism, London, Oxford Univ. Press.
Dessler, A.J.: 1967, Revs. Geophys., 5, 1.
Gergely, T. and Kundu, M.: 1979, submitted for publication.
Golub, L., Davis, J.M. and Krieger, A.S.: 1979, Astrophys. J. Lett., 229, L145.
Harvey, K.L. and Martin, S.F.: 1979, Solar Phys., 64, 93.
Howard, R.: 1977, Ann. Rev. Astron. Astrophys., 15, 153.
Howard, R. and Svestka, Z.: 1977, Solar Phys., 54, 65.
Hundhausen, A.J.: 1977, in Zirker, J.B. (ed.), Coronal Holes and High Speed Solar Wind Streams, Boulder, Colo. Assoc. Univ. Press, ch. 7.
Jackson, B.V. and Levine, R.H.: 1979, presentation to Skylab Solar Workshop on Active Regions.
Levine, R.H.: 1977a, Astrophys. J., 218, 291.
Levine, R.H.: 1977b, in Zirker, J.B. (ed.), Coronal Holes and High Speed Solar Wind Streams, Boulder, Colo. Assoc. Univ. Press, ch. 4.
Levine, R.H.: 1977c, Solar Phys., 54, 327.
Levine, R.H., Altschuler, M.D. and Harvey, J.W.: 1977, J. Geophys. Res., 82, 1061.
Levine, R.H.: 1978, J. Geophys. Res., 83, 4193.
Levine, R.H.: 1979, Solar Phys., 62, 277.
MacQueen, R. and Poland, A.I.: 1977, Solar Phys., 55, 143.
Nerney, S.F. and Suess, S.T.: 1975, Astrophys. J., 196, 837.
Newkirk, G., Jr. and Altschuler, M.D.: 1970, Solar Phys., 13, 131.
Pneuman, G.W. and Kopp, R.A.: 1971, Solar Phys., 18, 258.
Roelof, E.C. and Levine, R.H.: 1978, EOS, 58, 1204.
Schatten, K.H.: 1968, Nature, 220, 1211.
Schatten, K.H., Wilcox, J.M. and Ness, N.F.: 1969, Solar Phys., 9, 442.
Schatten, K.H.: 1971, Cosmic Electrodynamics, 2, 232.
Scherrer, P.H., Wilcox, J.M., Kotov, V., Severny, A.B. and Howard, R.: 1977a, Solar Phys., 52, 3.
Scherrer, P.H., Wilcox, J.M., Svalgaard, L., Duvall, T.L., Jr., Dittmer, P.H. and Gustafson, E.K.: 1977b, Solar Phys., 54, 355.
Schulz, M.: 1973, Astrophys. Space Sci., 24, 371.
Schulz, M., Frazier, E.N. and Boucher, D.J., Jr.: 1978, Solar Phys., 60, 83.
Severny, A., Wilcox, J.M., Scherrer, P.H. and Colburn, D.S.: 1970, Solar Phys., 15, 3.
Smerd, S.F. and Dulk, G.A.: 1971, in Howard (ed.), IAU Symposium No. 43, Solar Magnetic Fields, 616.

Smith, S. and Schatten, K.H.: 1970, Nature, 226, 1130.
Stenflo, J.O.: 1972, Solar Phys., 23, 301.
Suess, S.T., Richter, A.K., Winge, C.R. and Nerney, S.F.: 1977,
    Astrophys. J., 217, 296.
Svalgaard, L., Wilcox, J.M. and Duvall, T.L.: 1974, Solar Phys., 37,
    157.
Svalgaard, L., Wilcox, J.M., Scherrer, P.H. and Howard, R.: 1975, Solar
    Phys., 45, 83.
Svestka, Z., Solodyna, C.V., Howard, R. and Levine, R.H.: 1977, Solar
    Phys., 55, 359.
Uchida, Y., Altschuler, M.D. and Newkirk, G.A., Jr.: 1973, Solar Phys.,
    28, 495.
Vaiana, G.S.: 1976, Phil. Trans. Roy. Soc. Lond. A, 281, 365.
Vaiana, G.S. and Rosner, R.: 1978, Ann. Rev. Astron. Astrophys., 16,
    393.
Wilcox, J.M. and Ness, N.F.: 1965, J. Geophys. Res., 70, 5793.
Wilcox, J.M.: 1968, Space Sci. Revs., 8, 258.
Wilcox, J.M. and Howard, R.: 1968, Solar Phys., 5, 564.
Wilcox, J.M., Severny, A. and Colburn, D.S.: 1969, Nature, 224, 353.
Yeh, T. and Pneuman, G.W.: 1977, Solar Phys., 54, 419.
Zirker, J.B. (ed.): 1977, Coronal Holes and High Speed Solar Wind
    Streams, Boulder, Colo. Assoc. Univ. Press.

DISCUSSION

Scherrer: You have suggested that the field pattern seen in low
resolution observation of the photospheric field and in the interplanetary field has its origin in active regions. Is this consistent
with the many year lifetime of the low resolution structure?

Levine: I should clarify that the active regions contribute to the
pattern but they do not constitute its only source. As I mentioned,
separate averages of concentrated and diffuse flux (essentially including and excluding the active region) both show the photospheric
pattern. The data I used to draw this conclusion do not yet cover a
time period of years. When that data becomes available, your question
will be an important one to investigate. However, even for the time
coverage I was able to study there were large-scale structures which
lasted much longer than active regions, indicating that some active
regions had to have a systematic tendency to reinforce the pattern
rather than to destroy it.

Stix: Are there methods available to compute coronal fields with
current sheets, which do not require knowledge about the location of
these sheets? I mean methods which yield these sheets (e.g., its
latitude) together with the surrounding potential field?

Levine: The technique of Schatten which I mentioned results in a
model with implied current sheets at all neutral lines outside the
shell he uses to define which field lines will be open. The work of Yeh
and Pneuman allows specification of the location of current sheets, but
only the simplest geometries are tractable. Source surfaces also contain

current sheets, although it is only on the non-spherical source surfaces that the current system is well localized. Of course, if you are only interested in the latitude of possible current sheets the positions of neutral surfaces in any potential model will give a good first approximation.

McIntosh: The Stanford solar magnetic-field data can be formatted into synoptic charts of large-scale patterns of neutral lines. These agree closely with the Hα synoptic charts of chromospheric neutral lines. Doesn't this agreement indicate that solar mean field measurements detect solar sector boundaries that originate on, or near, the solar surface?

Levine: The Stanford instrument operates in a mode which produces magnetograms with three arc minute resolution (which are used to make the synoptic charts you mention) and in a mean field mode (this is the data I referred to in my talk). Synoptic charts of their resolved data do agree well with the nuetral lines of the Hα charts, which is to be expected because they both refer to fields near the surface. The mean field measurements also indicate a photospheric pattern. However, because the photospheric neutral line is covered by arcades of coronal loops (visible in X-ray or EUV emission) there can be no direct connection between the photospheric neutral line (sector boundary) and the interplanetary sector boundary. The interplanetary sectors spread like funnels from isolated areas within photospheric sectors. Their boundaries in the outer corona will have longitudes typically near those of the photospheric neutral lines, but their exact position is determined by the relative strengths of the fields in photospheric sectors and by MHD effects in the corona.

Moore: What is your opinion on whether some open field lines may be rooted in the umbras of sunspots?

Levine: The resolution of the numerical techniques used in extrapolating the coronal field is not sufficient to answer this question unambiguously. However, the models and the X-ray data (Švestka, et al. Solar Phys., 55, 359, 1977) are not inconsistent with this possibility. Note, however, that the possibly open structures in active regions tend to be elongated so that only a portion of the field lines could be rooted in a sunspot umbra. Further, if you consider the electrostatics of magnetic multipoles, you will recall that the highest field lines (those which are candidates for opening in the solar analogy) are rooted in the strongest fields.

Pneuman: Referring to the Schulz et al., computation, I think that the model is attractive in that it does incorporate a non-spherical source surface. However, I worry about the physical basis for choosing the surface as an iso-gauss contour. I would think a more appropriate assumption might be to relate the source surface to, perhaps, the Alfven speed which depends upon the solar wind velocity as well as the magnetic field.

Levine: Schulz et al., used an iso-gauss of a dipole field as a non-spherical source surface in order to obtain an analytical solution

in a test case which had an appropriate quantitative behavior. This
is perfectly acceptable. I agree, however, that other definitions
of the source surface should also be explored in realistic cases. This
is one goal of our program.

SEARCH FOR GIANT CELLS IN THE SOLAR CONVECTION ZONE

B. J. LaBonte and R. Howard
Hale Observatories*, Carnegie Institution of Washington

The Mount Wilson Observatory has obtained daily full disk digital magnetograms of the Sun since 1966, with 12 to 17 arcsecond resolution. As
each magnetogram is taken, the position of the Doppler line shift compensator
is also recorded, thus giving a full disk map of the longitudinal velocity.
This entire dataset is currently being rereduced on a uniform basis (Howard
et al., 1980), and daily arrays of residual velocities are being formed by
removing large scale patterns, e.g., Earth's motions, solar rotation,
limbshift. Data from the years 1972 through 1978 are used here.

For this study, we have searched for large scale persistent east-west
horizontal flows. The procedure is to construct synoptic velocity maps,
with the amplitude at each point given by the average of the daily velocities
at that point, weighted by the sines of their central meridian distances. This
reinforces longitudinal flows and cancels vertical and meridional flows.
Since a synoptic (e.g., Carrington) longitude is visible for many days,
random solar velocity noise (supergranules, 5 minute oscillations) is also
reduced.

Figure 1 shows two such synoptic maps, for Carrington rotations 1603
and 1606. The velocity pattern during rotation 1603 is random, but during
rotation 1606 stripes of alternating velocity sign are seen. A stripe of one
sign extends from about 60° north to 60° south latitude, with a constant width
of about 30° in longitude. The latitude-averaged velocity amplitude is about
10 m s^{-1}. A pair of stripes, east-flowing on the east and west-flowing on the
west, appears to be a single outflowing velocity cell. This pattern is similar
to that deduced from theory for the largest scale solar convection cells (e.g.,
Gilman, 1979). Velocity stripes of this type are seen about $\frac{1}{2}$ the time from
1972 to mid-1974, but only occasionally after that.

Unfortunately these stripes appear to be an instrumental artifact, rather
than a solar phenomenon. Stripes of larger amplitude, about 30 to 60 m s^{-1},
can occur on a single observation, and leak into the synoptic maps due to

* The Hale Observatories are operated jointly by the Carnegie Institution of
Washington and the California Institute of Technology.

M. Dryer and E. Tandberg-Hanssen (eds.), Solar and Interplanetary Dynamics, 21-23.
Copyright © 1980 by the IAU.

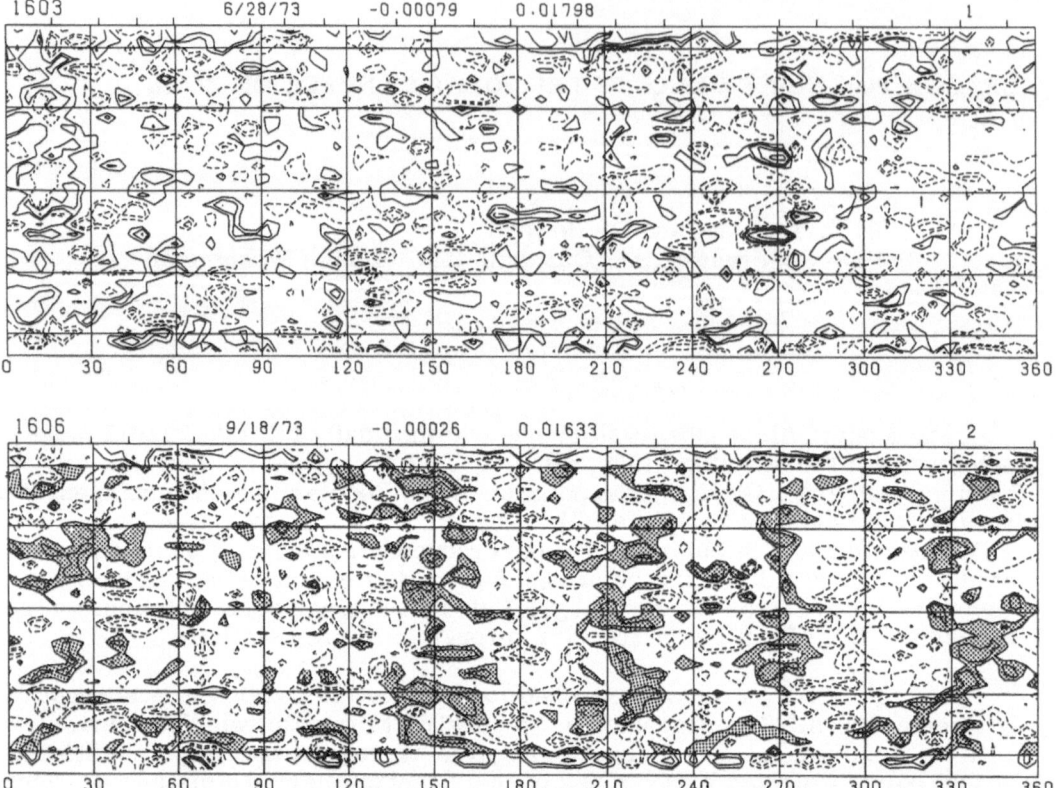

Fig. 1.   Synoptic maps of the observed longitudinal velocities on the solar surface, for Carrington rotations 1603 and 1606. Solid contours display eastward flow, dashed contours westward flow. Contour levels are 10, 20, 40 and 80 m s^{-1}. Eastward velocities have been shaded on the lower map to show the large-scale organization of velocities into stripes.

---

incomplete sampling or shifts in the daily stripe locations on the solar disk. The cause of the velocity stripes in the daily observations is not yet certain, but is either optical fringes in the spectrum or mechanical problems in the exit slits of the magnetograph.

Because these instrumental stripes are not always present, we can still search for persistent solar motions of this type. Power spectra and auto-correlation functions of the latitude averaged longitudinal velocity have been computed for intervals from 3 months to $3\frac{1}{2}$ years. Synoptic rotation rates from 0.8 to 1.1 times the Carrington rate have been searched. Figure 2 shows one example of the results, power spectra of the velocity for the two halves of the dataset. The only significant features are the broad peaks in the spectra near 3, 6, and 9 cycles (cells) per rotation, which are caused by the instrumental stripes. This result is true of all the analyses thus far. We can set upper limits on the amplitude of a true solar cell-like velocity pattern which are quite small. A broad spectral feature (> 1 cycle per

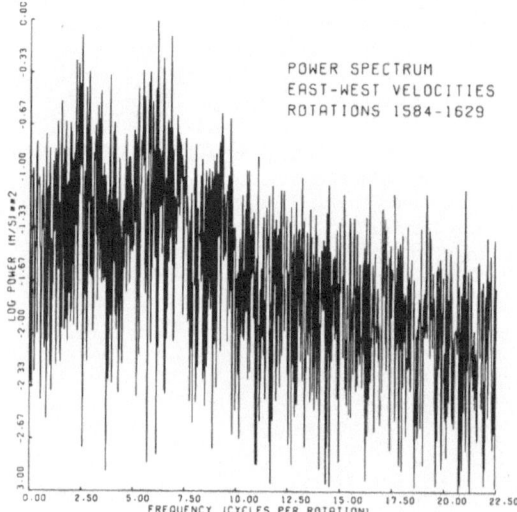 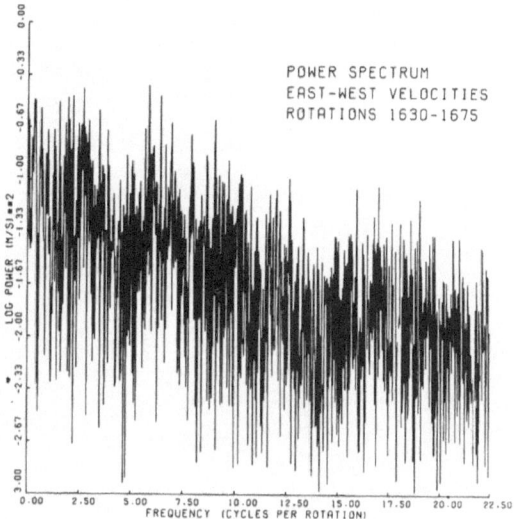

Fig. 2.   Power spectra of the latitude-averaged longitudinal velocity, for
Carrington rotations 1584 to 1629, and 1630 to 1675.   Broad peaks at 3, 6,
and 9 cycles per rotation in the first spectrum are caused by the instrumental
velocity stripes.   Their amplitude is much reduced in the second spectrum,
for a time interval during which stripes were rare.   The integrated amplitude
(RMS) of the instrumental stripes is only $\simeq 5$ m s^{-1}, and any solar features
must be weaker.

---

rotation) must have an amplitude $\leqslant 5$ m s^{-1} for lifetimes 1 year, and
$\leqslant 10$ m s^{-1} for lifetimes of a few months.   A narrow spectral feature (< 1 cycle
per rotation) must be $\leqslant 2$ m s^{-1} for lifetimes 1 year, and $\leqslant 5$ m s^{-1} for lifetimes of a few months.   These limits apply at all frequencies between
2 and 20 cycles per rotation.

REFERENCES

Gilman, P. A.: 1979, Astrophys. J., 231, pp. 284–292.
Howard, R., Boyden, J. E., and LaBonte, B. J.: 1980, 66, pp. 167.
    Solar Phys.

DISCUSSION

   Bratenahl: Yoshimura's global convection cells, I believe, have a
retrograde motion.   Did you research for that possibility?
   LaBonte: Yes.   By analyzing the data while using a variety of
rotation rates, we have allowed for the possibility that the convective
cells slip with respect to the surface or the core.

DYNAMICS OF LARGE-SCALE MAGNETIC FIELD EVOLUTION DURING SOLAR CYCLE 20

Patrick S. McIntosh
NOAA Space Environment Laboratory
Boulder, Colorado 80303

Abstract

The evolution of large-scale solar magnetic fields has been studied for a complete solar cycle using the atlas of H-alpha synoptic charts for 1964-1974. The results include: a unique magnetic pattern coinciding with major coronal holes; variations in the rate of solar rotation through the solar cycle; discovery of convergence and divergence among long-lived magnetic patterns; periodic discontinuities in the organization of large-scale magnetic fields; and a new cause for coronal transients.

Large-scale solar magnetic fields are clearly outlined by filaments and filament channels in ordinary solar patrol photographs taken with H-alpha filters. Synoptic charts based on H-alpha observations (McIntosh, 1979) provide a new perspective on the large-scale magnetic patterns by emphasizing the neutral lines in the radial component of these fields, whereas the locations of these lines on longitudinal magnetograms appear as diffuse zones where there is an absence of signal. The H-alpha structures, therefore, permit a more accurate mapping of these lines. The magnetic patterns are more complete than on conventional magnetograms because of the interconnections between filaments and active regions formed by the filament channel and the transient filament-like features. The result is a sun laced by only a few, very long neutral lines, one of which encircles the sun like a stitch on a baseball, underlying both polar-crown filament-systems and crossing the solar equator at two longitudes. Because the polar-crown neutral lines are part of this single, fundamental boundary in the magnetic fields, there exists a gap in the polar crown in each hemisphere. There may be more than one gap in each hemisphere for short periods of time, but one of those gaps dominates in size and lifetime.

The polar-crown gap (PCG) and other long-lived features in the large-scale solar magnetic fields are best studied by dividing synoptic charts into narrow zones of latitude and assembling time series showing single zones. The areas of negative polarity are shaded to emphasize the patterns. The latitude limits for this initial study of

25

M. Dryer and E. Tandberg-Hanssen (eds.), Solar and Interplanetary Dynamics, 25-28

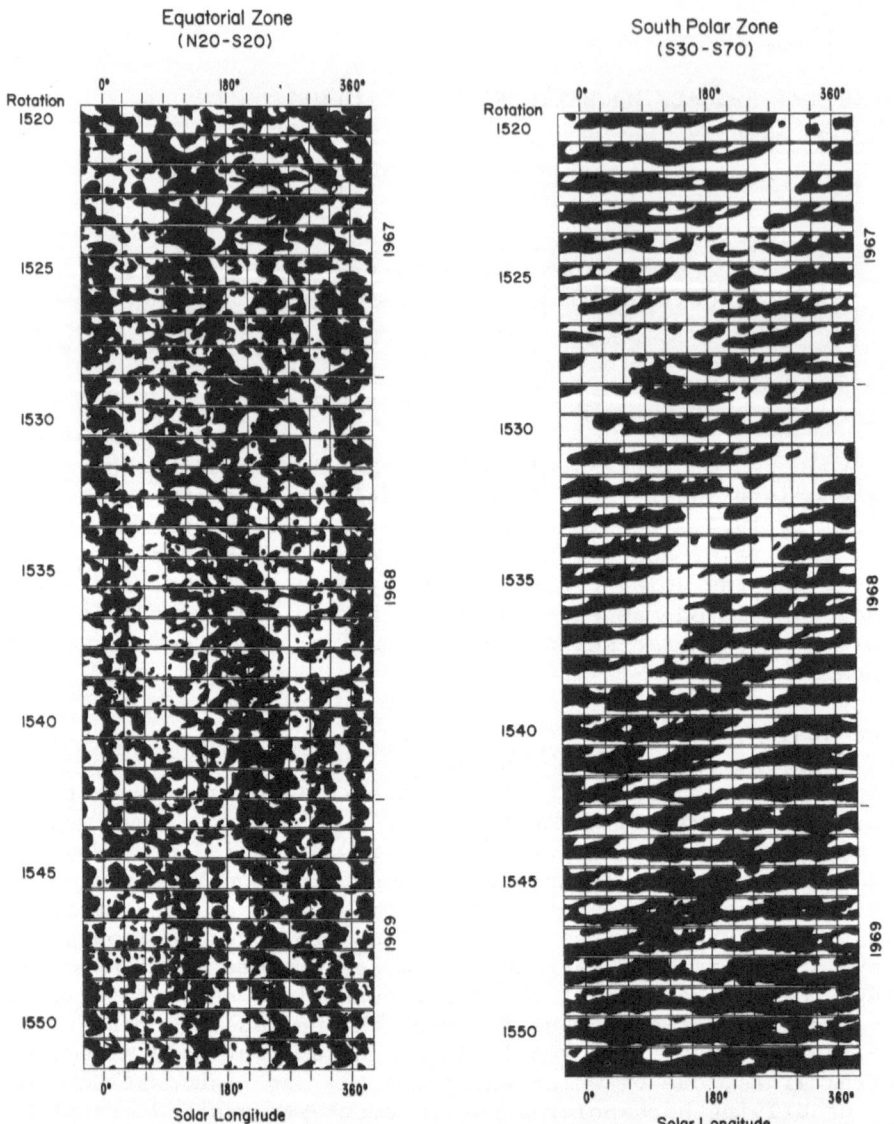

Figure 1. Large-scale patterns of solar magnetic polarity for the equatorial and southern high-latitude zones during the rise to solar maximum of Solar Cycle 20 (Carrington Rotations 1520-1551). Each strip is a zone from an H-alpha synoptic chart in McIntosh (1979). Negative polarity is black, positive polarity is white.

Solar Cycle 20 are: N20-S20, N30-N70 and S30-S70. Figure 1 shows the equatorial and south polar zones for 1967-1969, the period of rapid rise to solar maximum. Figures for the northern polar zone and for the other portions of Solar Cycle 20 will appear elsewhere.

The polarity of the south polar cap and of the southern PCG was positive at the beginning of the solar cycle. The PCG appears in the sequence of south polar zones in Figure 1 as a white diagonal, indicating the slow rate of solar rotation at this latitude. Midway through the figure the white diagonal terminates and a black diagonal emerges. This signifies the polarity reversal that took place at high latitudes in mid-1968. The southern PCG divided into two during 1969. Identical behavior, with polarities reversed, occurred in the northern polar zone for this same period.

The PCG could be identified in both hemispheres at the start of Solar Cycle 20 and both continued without interruption, and without dividing, until the time of polarity reversal in 1968, a period of almost four years.

These PCGs can be directly correlated with minima in the white-light K-corona observations in 1967. The PCGs of the 1972-1974 period coincide with the XUV and X-ray coronal holes (McIntosh et. al., 1976) observed by OSO-7 and Skylab. The PCGs of the current solar cycle coincide with the major coronal holes inferred from infrared observations of the helium line.

The slopes of the diagonals formed by the PCG indicate the rate of rotation of the PCG relative to Carrington coordinates. The slopes of both northern and southern PCG were constant from 1964 to the end of 1966 at a 29-day period, then decreased (decelerated) in 1967-1968 to a 32-day period. The slopes of features in the equatorial zone show no systematic variation during the same interval, implying that the degree of shear across latitude increased as the solar cycle built to maximum.

Large-scale magnetic-field patterns in the equatorial zone persist as diagonals in the zonal series, indicating that solar activity produces only minor and gradual changes in the large-scale patterns. An individual feature has active regions preferably near its leading (western) boundary.

The large-scale patterns drift with respect to one another, producing convergence and divergence among the patterns. The merger of two patterns forces the disappearance of large filaments that marked their adjacent boundaries. Filament disruption is closely associated with coronal transients; therefore, large-scale magnetic-field mergers represent a major category, or source, of coronal transients.

Convergence between adjacent features in the polar magnetic patterns closes the PCGs in 1968 at the time of polarity reversal.

Divergence opens new PCGs a few rotations later. Therefore, the birth
and death of coronal holes is caused by large-scale divergence and
convergence, respectively.

The phenomenon of convergence and divergence implies that we are
observing the presence of large-scale circulation superposed on the
mean differential solar rotation.

Zonal time series such as Figure 1 were combined so that the 10-
year period could be viewed as a single figure in each latitude interval. This continuous view of the magnetic-field evolution revealed
periodic changes in the patterns that escaped notice in viewing shorter
intervals. Discontinuities in the equatorial patterns occurred every
2-1/2 years, marked by the periodic emergence of an especially large
and long-lived negative-polarity feature following the discontinuities.
More distinct and more frequent discontinuities occurred in the northern polar patterns, at roughly yearly intervals. Discontinuities
occurred in the southern polar zone at the same time, but with less
reorganization of the patterns. The yearly discontinuities may be
present in the equatorial patterns, but cannot be proven at this time.

REFERENCES

McIntosh, P. S., Krieger, A. S., Nolte, J. T. and Vaiana, G.: 1976,
    Solar Phys., 49, pp. 57-77.

McIntosh, P. S.: 1979, UAG Report 70, World Data Center A for Solar-
    Terrestrial Physics, Boulder, Colo. USA.

DISCUSSION

Moore: Do the polar cap gaps rotate faster or slower than the
photospheric plasma at the same latitude, and if so, by how much?
  McIntosh: No detailed direct comparison has yet been made. The
average rotation rate for the photospheric plasma at polar-crown gap
latitudes is approximately 32 days. The observed motion of the polar-
crown gaps varies from 29 to 35½ days.

A TWO-LEVEL SOLAR DYNAMO BASED ON SOLAR ACTIVITY, CONVECTION, AND DIFFERENTIAL ROTATION

A. Bratenahl and P. J. Baum
IGPP, University of California, Riverside, CA 92521
W. M. Adams
The Aerospace Corporation, El Segundo, CA 92957

In orthodox dynamo theory (Stix, 1976), the two basic processes, generation of toroidal from poloidal field and conversion of toroidal into reversed poloidal field, are both located in the high β regime convection zone. Generation requires that regime, since its function demands it be driven by mechanical forces. But the function and therefore the operating requirements of conversion are entirely different, and there seems to be no a priori reason, other than historical tradition coupled with failure to recognize those differences, for the assumption that conversion must also operate there. Conversion transforms the topological structure of generated flux by altering the field line connectivity, so that the principal task performed is reconnection. Reconnection is a spontaneous process which must compress and accelerate plasma if any is present. Obviously it must perform much more work in the high β convection zone than in the low β solar atmosphere. It seems natural, therefore, to expect the reconnection aspect of conversion to be located there, where the least work needs to be performed. To transfer the generated flux there, we may add to conversion another spontaneous process: eruption of bipolar structure (Parker, 1955). To transfer the reconnected flux back down, we add to generation another mechanically driven process called topological pumping (Drobyshevski and Yuferev, 1974). Topological pumping depends on the diamagnetic effect of eddy-motion (Wiess, 1966), the kind possessed by supergranulation: 3-dimensional arrangement of isolated rising plumes, surrounded by a continuous network of descending sheet-like flow. In the two-level dynamo presented here, conversion may be observed directly, since we expect it to express itself in terms of all forms of solar activity: sunspots, flares, faculae, filaments, coronal structures including coronal holes, etc., and their organization and evolution in a "solar meteorology". It is clearly important to investigate a model that thus unites the two disciplines of solar activity and dynamo theory. Each strengthens the other and brings a greater unity to solar physics.

Although not all flux tubes above the photosphere get there by eruption from below (reconnection produces some), we assume that subduction from above accounts for all the flux tubes below. An important relation is thereby established between erupted flux, and previously subducted

29

M. Dryer and E. Tandberg-Hanssen (eds.), Solar and Interplanetary Dynamics, 29-32.
Copyright © 1980 by the IAU.

flux. We expect the subduction process to produce flux fibers containing a characteristic quantity of flux and a characteristic initial mass-per-unit length. Horizontal field, lying in the 600 km temperature minimum, ionization fraction $n_i/n_H = 10^{-4}$ (Giovanelli, 1977), loses its buoyancy by soaking up a sufficient mass of gas. During the 20 hr life of a super-granule, the characteristic quantity of flux will have soaked up the characteristically sufficient mass-per-unit length. This mixture of field and gas becomes frozen-in quickly and permanently as it is pulled below and becomes fully ionized. The effect of this is that the convection zone is maintained as a two-phase mixture of magnetized fibers confined to downflows in a sea of unmagnetized plasma.

The stretching action of differential rotation will eventually pro-duce such a large increase in the fiber's field-to-mass-density ratio that the convective downflow can no longer prevent eruption (Parker, 1955). Zwaan (1978) suggests that the ubiquitous occurrence of facular points ($\gtrsim 1500$ gauss, $\sim 10^{17}$ maxwells) is indicative of their preexistence as flux fibers before eruption and not some process operating at the time they appear. (In the model presented here, preexistence begins at subduction.) Zwaan further suggests that large numbers of fibers become packed into bundles and that sunspots result from their eruption. He avoided the term "flux rope", implying systematic twisting, since there is no obvious mechanism to produce it and observations do not support it.

Conversion must somehow reorient flux bundles from a westward-equa-torward (eastward-poleward) tilt to a westward-poleward (eastward-equa-torward) tilt in both hemispheres at once. This topological change in bundle-connectivity is not possible without invoking reconnection on a massive scale. We propose that it is best carried out in the solar atmosphere through an extended series of small elementary steps (Figure 1). Portions of neighboring bundles are erupted, reconnected; and then, while one of the new interconnections is subducted, the other expands out into the corona. The subducted interconnected link contributes to the required rotation in a stepwise reorgnization of the whole set of bundles. But since this series of steps is governed somewhat by chance, the process is untidy. For instance, the general field and its reversal is much less obvious than Hale's polarity law.

Consider two neighboring bipolar systems, Nos. 1 and 2, representing erupted portions of two independent bundles. Let No. 2 lie equatorward of No. 1. Reconnection leads to the sharing of flux of the two preceding (p) and following (f) polarities so as to create four flux cells with inter-connections as follows: p_1f_1, p_2f_2, p_1f_2, p_2f_1 (Sweet, 1958; Bratenahl and Baum, 1976; Baum and Bratenahl, 1980). Cells with like indices we call parents, those with unlike indices, daughters. Conservation of flux in the transfer to the daughters requires that each of them receive the same amount, and the two parents must contribute equally. The daughter interconnecting the shorter p-to-f distance must lie underneath her sister, exposing weak horizontal field to subductive activity. As the topological pumping proceeds, more and more flux is transferred to both she and her sister, but when no more is available under the conservation

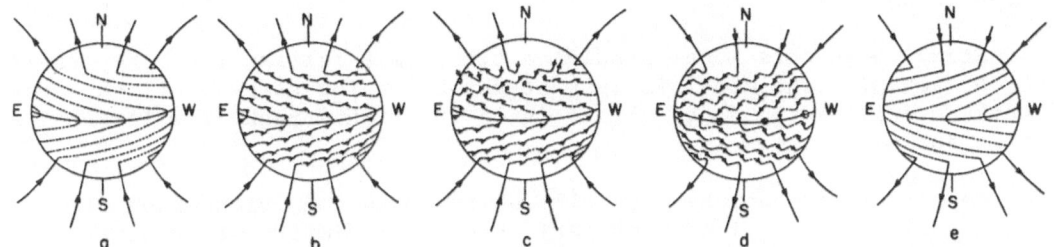

Fig. 1.   Schematic representation of the rotation (from Adams, 1977).

rules, she disappears below, while her sister rises into the corona as
an expanding loop system. The subducted daughter now forms a new link between the affected bundles and with an orientation in the convection zone
corresponding to that which she had in the atmosphere. If that orientation
corresponds to a rotation in the appropriate direction, the cycle is advanced; if not, it is retarded. The appropriate rotation can only be
accomplished by the daughter named p_1f_2, which means that No. 2 must be
somewhat westward of No. 1, its poleward neighbor. It may be demonstrated
that the combination of differential rotation and the proper motion of
polarity regions within bipolar structures, on the average, gives the
necessary statistical advantage to steps that advance the cycle.   The
cumulative effect of thousands of such elementary steps reverses both the
poloidal and toroidal components and thus ensures an efficient dynamo
with little need for turbulent dissipation to get rid of unwanted flux.

The coronal loops undergo further reconnection processes; the net
effect of which is the cancellation of the original polar fields and the
merging of loops across the equator (Babcock, 1961). In this way, consistency is maintained with the corresponding changes in the convection zone.

This two-level dynamo model, if it has merit, could soon find con-
firmation in observations since nearly every step is linked to some
aspect of solar activity, although subduction may present difficulties.

REFERENCES

Adams, W. M.: 1977, Big Bear Observatory, Caltech, BBSO No. 0163 (unpub).
Babcock, H. W.: 1961, Astrophys. J. 133, pp572-587.
Baum, P. J. and Bratenahl, A.   1980, Solar Phys. 67, 245.
Bratenahl, A. and Baum, P. J.: 1976, Geophys. J. Roy. Astron. Soc. 46,
     pp259-293.
Drobyshevski, E. M. and Yuferev, V. S.: 1974, J. Fluid Mech. 65, pp33-44.
Giovanelli, R. G.: 1977, Solar Phys. 52, pp315-325.
Parker, E. N.: 1955, Astrophys. J. 121, pp491-507.
Stix, M.: 1976, in V. Bumba and J. Kleczek (eds.) "Basic Mechanisms of
     Solar Activity", IAU Symp. 71, pp367-388.
Sweet, P. A.: 1958, Nuovo Cimento Suppl. 8, Ser. X, pp188-196.
Wiess, N. O.: 1966, Proc. Roy. Soc. A. 293, pp310-328.
Zwaan, C.: 1978, Solar Phys. 60, pp213-240.

DISCUSSION

Stix: Does this model predict a time scale for the field reversal?

Bratenahl: The model is just at the beginning stage. All the real work is still to be done. If it is on the right track it will set its time scale.

Moore: Is the north-south tilt observed in the orientation of most active-region bipolar magnetic systems an important property for the operation of your model?

Bratenahl: No, not at all; the tilt can be either way providing it is not too large. The process is advanced by the statistically preferred reconnection geometry: preceding spot polward to follower spot equatorward.

Newkirk: How does the model which you are advancing, with the two field amplification located just below the photosphere, overcome the problem produced by field ropes being bouyed to the surface in a time short compared to any reasonable amplification time?

Bratenahl: We suppose amplification takes place near the base of super granulation ($\gtrsim$ 20,000 Km depth). The flux fibers and flux bundles are always in downflows. Bouyancy does not lead to eruption until it is sufficient to overcome downward drag forces.

LOCATION OF COMPACT MICROWAVE SOURCES WITH RESPECT TO CONCENTRATIONS OF MAGNETIC FIELD IN ACTIVE SOLAR REGIONS

V. Gaizauskas and K.F. Tapping
Herzberg Institute of Astrophysics, Ottawa, Canada

1. INTRODUCTION

The slowly varying (S-) component of solar microwave emission originates in the vicinity of sunspots and chromospheric plages. A significant portion of this component is produced by bright, compact sources which are much smaller than the active regions in which they occur (Lang, 1974; Kundu *et al.*, 1977; Donati Falchi *et al.*, 1978; Lang and Willson, 1979). The combined radio-optical investigation reported here was initiated to explore, in regions at various stages of evolution, both the variability of compact microwave sources and their association with specific optical features.

2. OBSERVING PROCEDURE

The 46m radio telescope at the Algonquin Radio Observatory[1] is used at a wavelength of 2.8 cm in real-time collaboration with the multiple Hα photoheliograph of the Ottawa River Solar Observatory[2]. Both telescopes scan the sun in rasters in order to locate and record the regions of interest. Microwave sources are designated compact when they produce no detectable broadening of the 2.7' antenna beam, an indication that their angular size must be less than 20". Subsequent to the scanning procedure, the radio observer measures the position of maximum intensity of each microwave source to within 20 arc-sec or less depending upon the brightness distribution within the source. The optical raster scans provide a series of photographs on a scale of 15 arc-sec/mm such that 102 overlapping images cover the entire solar disk at one wavelength. At each solar position, the tunable Hα filter in the photoheliograph is cycled through 5 wavelength steps in the range Hα±1.0A. An optical scan of the entire disk on this scale at 5 wavelengths requires 25 min while the corresponding radio scan requires 45 min. Filtergrams of the whole disk in Hα are also secured on a small scale in order to assist in establishing the spatial correspondence between radio and optical features.

From September 1977 to July 1979, joint optical and radio observations were made successfully on a total of 20 days; in all, 28 active

M. Dryer and E. Tandberg-Hanssen (eds.), Solar and Interplanetary Dynamics, 33-36.
Copyright © 1980 by the IAU.

regions with compact microwave sources have been examined. This sample
includes active regions at various stages of their development.
Although the analysis of these data is incomplete, examples of microwave
emission from two regions at extreme positions on the evolutionary scale
merit special attention.

3. EMERGING FLUX REGION: McMATH 15467

This small plage region developed from pre-existing elements of
the chromospheric network which were well isolated from larger active
features. On 11 and 12 August 1978, the bipolar elements of this region
were interconnected by a complex system of fine arches; by 13 August,
the region had produced a miniature but conspicuous arch-filament system
in which flare-like activity was evident on a micro-scale. Pores were
first visible on 14 August when four of them, each 1" or less, formed
a cluster about 5 x 10^3 km across at the leading edge of the widening
plage. Microwaves were undetectable from this region at λ2.8 cm during
radio searches made on 12 and 13 August; a compact, steadily-emitting
source of microwaves was located in the region on 14 August.

4. REGION NEAR MAXIMUM DEVELOPMENT: McMATH 14943

This magnetically complex region was oriented with its "neutral
line" principally E-W rather than N-S. Between 11 and 15 September 1977
the sunspots in this region exhibited strong proper motions: a spreading
apart of all spots and a pronounced rotation by the large leader of this
complex, a spot of f polarity. The microwave emission from this region
consisted of two components: a diffuse plateau at an antenna temperature
of 10^4 K distributed over the whole region and a single compact source
with a peak antenna temperature of 1.1 x 10^5 K. The absence of beam-
broadening for this source implies a brightness temperature of at least
10^7 K. The compact source was located in the leading plage which
partially obscured a rapidly decaying spot to the large leader spot of
f polarity. On 13 September, the source pulsated at a period of 2.5s
throughout the entire observing period of 5½h. The measured peak-to-
peak fluctuations in brightness temperature were at least 2.5 x 10^5 K.
There was no evidence from optical data to suggest that the pulsations
were excited by flare processes.

5. DISCUSSION

In 11 of the 28 observed regions, the compact microwave sources
varied in intensity with time-scales from minutes to several hours by
as much as 50%, neglecting obviously impulsive events. The brightness
temperatures of the sources at λ2.8 cm were in the range 3 x 10^4 K to
more than 10^7 K; for six sources, the brightness temperatures exceeded
2 x 10^6 K. Because these values were maintained for hours, they do not

describe impulsive events. Such high values cannot be explained in
terms of thermal emissive processes.

The presence of just a single compact microwave source in a large,
magnetically complex region suggests a uniqueness to the immediate
environment of the source which is also difficult to explain in terms
of thermal models. Alternative mechanisms, including plasma oscillations
and gyrosynchrotron emission, require a continuous supply of high energy
electrons. The source would otherwise decay in minutes. Electron
acceleration is most likely at particular points rather than generally
over an active region. Microwave emission driven by a supply of high
energy electrons would consequently arise only at those points.

[1]The Algonquin Radio Observatory is operated as a national
facility by the National Research Council Canada.

[2]The Ottawa River Solar Observatory is operated by the National
Research Council Canada.

REFERENCES

Donati Falchi, A., Felli, M., Pampaloni, P., and Tofani, G., 1978,
    Solar Phys. 56, 335.
Kundu, M.R., Alissandrakis, C.E., Bregman, J.D., and Hin, A.C., 1977,
    Astrophysics J. 213, 278.
Lang, K.R., 1974, *Solar Phys.* 36, 351.
Lang, K.R., and Willson, R.F., 1979, *Nature* 278, 24.

DISCUSSION

Pallavicini: I would like to point out that high-resolution
observations at 2.8 cm from Stanford (Pallavicini et al. 1979, Ap. J.,
229, 375) have shown that all bright compact radio sources with a
life-time of the order of several days occur directly above the umbra
of large sunspots, where the photospheric magnetic field has an
intensity in excess of about 2000 gauss. Comparison of the radio data
with simultaneous X-ray and EUV observations, together with extra-
polations of the photospheric magnetic field to coronal levels in the
current-free approximation, indicate that the bright radio sources can
be interpreted as produced by thermal cyclotron emission in the strong
sunspot magnetic field (Pallavicini, 1979, IAU Symposium No. 86). Are
you referring to the same type of compact radio sources, or are your
sources of much shorter life-time?
Gaizauskas: All the sources we examined were long-lived. Our
results, as well as some of those in the cited references, indicate
however that not all long-lived compact microwave sources occur
directly over sunspot umbrae. Some of our sources had brightness
temperatures well in excess of thermal values. In order for thermal

cyclotron processes to be significant in these cases (especially the
emerging region McMath 15467), we would have to postulate unacceptably
high magnetic field strengths in the corona. Furthermore, although
the very large region McMath 14943 contained many spots and many areas
with large magnetic field gradients, it had just a single, very bright,
compact microwave source. A thermal model would predict multiple
sources.

RADIO OBSERVATIONS OF CORONAL HOLES

K.V. Sheridan and G.A. Dulk†
Division of Radiophysics, CSIRO, Sydney, Australia
†Also Department of Astro-Geophysics, University of Colorado,
Boulder, U.S.A.

ABSTRACT

Coronal holes have been observed on several occasions with the
80 and 160 MHz radioheliograph at Culgoora. At 160 MHz the holes
invariably appear as areas of low brightness, either on the disk or at
the limb. At 80 MHz holes on the limb always appear less bright than
their surroundings but on the disk they frequently appear brighter.

The simplest interpretation is that the coronal temperature in
holes near the 80 MHz critical density (8×10^7 cm^{-3}) is higher than in
normal quiet regions, but that the density at this level is lower.

1. INTRODUCTION

In the absence of solar activity, radio pictures of the Sun at
metre wavelengths show several kinds of structure: bright regions
(which are usually coronal streamers and are not related to the active
regions and coronal condensations so prominent at centimetre and
decimetre wavelengths and in soft X-rays); intermediate "quiet regions"
(which correspond to the moderately bright loop-like structure seen in
soft X-rays); and dark regions (which usually correspond to coronal
holes). In this report we shall concentrate on the coronal holes and,
in particular, their change in appearance with wavelength, using
observations made with the 80 and 160 MHz radioheliograph at Culgoora.

2. OBSERVATIONS

A comparison of "quiet Sun" maps at 80 and 160 MHz (e.g. Figs. 1
and 2) shows that, as would be expected, the corona is less extensive
at 160 MHz than at 80 MHz. The average brightness temperature in the
central region tends to be fairly uniform at a value $\lesssim 10^6$ K; at the
limb the brightness falls away quite steeply to about half the central
value and then more slowly to a low intensity that blends into the sky

37

M. Dryer and E. Tandberg-Hanssen (eds.), Solar and Interplanetary Dynamics, 37-43.

background.   Bulges and indentations in the contours near the limb are common.

Coronal Holes on the Solar Disk

Observations of the well-known coronal hole of the Skylab period, CH1, have been reported by Dulk et al. (1977). The hole was clearly visible on 160 MHz and higher-frequency maps as a depression with boundaries similar to, but not identical with, those seen on X-ray and EUV maps. However, at 80 MHz there was no indication of depressed brightness in the hole; in fact there was a brightness enhancement slightly to the west of the hole.

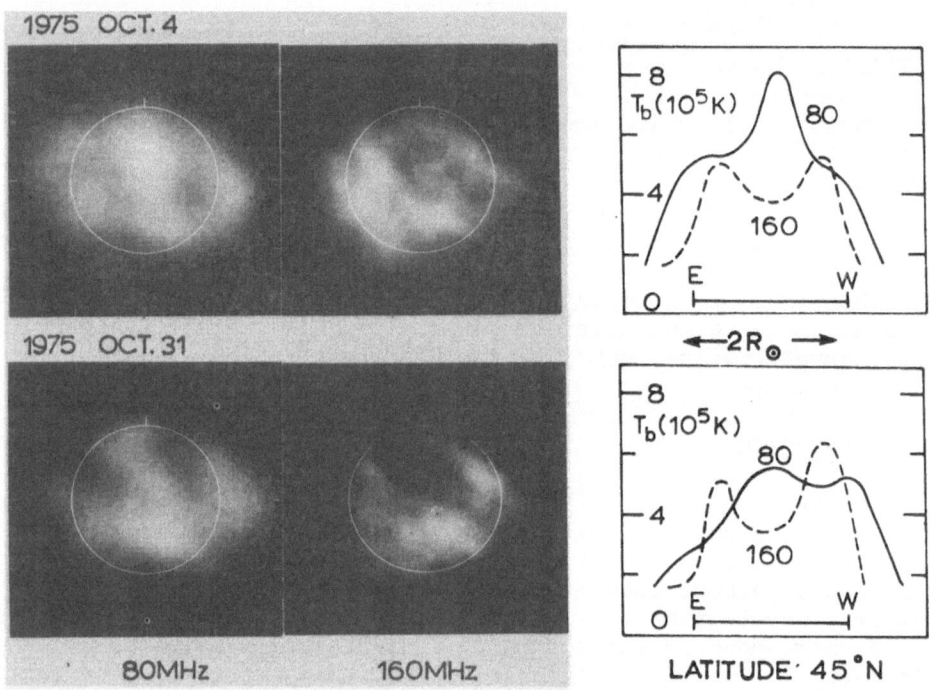

Fig. 1 - The large northern coronal hole of 1975 shown when it was near
   disk centre on two rotations, 1975 October 4 (top) and 1975 October 31
   (bottom). The hole appears about 20% darker than its surroundings at
   160 MHz and about 30% brighter at 80 MHz. The graphs at the right
   show the brightness temperature vs. position in the east/west
   direction at a latitude of 45°N. The scale of T_b assumes flux densities
   of 1.5 SFU (80 MHz) and 4.5 SFU (160 MHz).

Figure 1 shows radio pictures of another, very large, coronal hole which was observed on two successive solar rotations in 1975. The hole is clearly visible on the solar disk as a dark region at 160 MHz while a bright region is visible at the corresponding place at 80 MHz. The

line drawings in Figure 1 show the observed brightnesses at the two
frequencies on east-west scans across the Sun at latitude 45°N., the
location of the deepest part of the hole. From these scans it is clear
that the 80 MHz enhancements correspond very closely to the 160 MHz
depressions. We note that the same hole appeared as a depression in
brightness at 3.8 GHz (Shibazaki et al., 1977).

Coronal holes on the disk may not always appear as enhancements at
80 MHz, as indicated by Figure 1(b) of Dulk and Sheridan (1974). A
low-brightness region was observed on FeXV maps obtained by the NASA-
Goddard experiment on OSO-7 (Solar-Geophysical Data, 1972). The same
region was seen as a slight depression both at 80 and 160 MHz.

Coronal Holes on the Limb

When coronal holes are on the limb their brightness temperature is
less than that of their surroundings at both 80 and 160 MHz (and
probably at all other wavelengths as well). Figure 2 compares radio

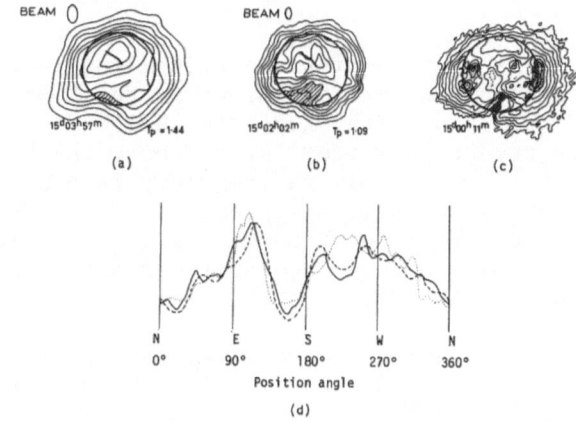

Fig. 2 - (a),(b),(c): Contour maps of coronal brightness on 1972 July
    15: (a) observed at 80 MHz; (b) observed at 160 MHz; (c) the 284 Å
    line of FeXV. The circles represent the visible disk. The radio
    contours represent 0.1,0.2,...0.9 of T_p (units of 10^6 K), the peak
    brightness temperature of the map; the hatching emphasizes regions on
    the disk which are of unusually low brightness. On the EUV map the
    brightest regions, above contour 640, are stippled; on the disk, the
    darkest regions, below contour 10, are filled in. The dark line
    segment on the radio maps shows the position of a quiescent filament.
    (d): Scan of coronal brightness as a function of position angle at
    fixed heights above the limb. Full lines: 160 MHz at 1.2 R_Θ; dashed
    lines: 80 MHz at 1.45 R_Θ; dotted lines: K-corona at 1.57 R_Θ. The
    radio observations were made at the times of maps (a) and (b) and the
    K-corona observations (from the High Altitude Observatory in Hawaii)
    were made at about 18^h00^m UT on the previous day. The intensity
    scale is arbitrary for all data (after Dulk and Sheridan (1974) and
    Sheridan (1978)).

observations at 80 and 160 MHz with FeXV brightness observations on the
same day, 1972 July 15.   The outer contours in each case show general
agreement in shape and extent.   The darker regions in the south-central
region on each map show the presence of a coronal hole near the limb.
In Figure 2(d) we show scans made around the limb at fixed heights:
R = 1.2, 1.45 and 1.57 $R_\odot$ at 160 MHz, 80 MHz and white light respectively;
the white light data were obtained with the K-coronameter operated by
HAO at Mauna Loa, Hawaii (Hansen et al., 1969).   The coronal hole at
PA = 160° can be seen on each scan.   The brightness depression at 80 MHz
is similar to that observed at 160 MHz and in white light.

Other examples of coronal holes near the limb have been given by:
(i) Dulk et al. (1977), where CH1 was observed to be prominent near the
northern solar limb at both 1.4 and 10.7 GHz;   (ii) Sheridan (1978),
where CH1 became visible at 80 MHz as it rotated to the limb;   and
(iii) Shibasaki et al. (1977) where a north polar hole was visible near
the limb at 3.8 GHz.

3.   INTERPRETATION

The radio brightness of the quiet Sun arises entirely from thermal
bremsstrahlung due to electron-ion collisions in the ionized plasma.
The electron density decreases with increasing height, and radiation at
a given frequency can escape only from layers above the "plasma level",
the height at which the electron plasma frequency is equal to the wave
frequency and the index of refraction approaches zero.   The radiation
observed at metre wavelengths is emitted mainly from the layers not far
above the corresponding plasma level and so the diameter of the Sun at
160 MHz is smaller than at 80 MHz.

It is interesting to consider how and why the observed brightness
temperature varies across the Sun.   But first we briefly review the
emission theory.

The brightness temperature due to thermal bremsstrahlung is

$$T_b = \int_0^\infty T(\tau_\nu) \exp(-\tau_\nu) \, d\tau_\nu , \qquad (1)$$

where the optical depth at distance s from the observer is

$$\tau_\nu = \int_0^s \kappa_\nu \, ds , \qquad (2)$$

and the integral is along the (possibly refracted and scattered) ray
path from the observer to the point s, perhaps including a reflection
at the plasma level.   The absorption coefficient κ_ν in (2) is given,
for a fully ionized plasma with 10% helium, by

$$\kappa_\nu = 0.0108 \, n_e^2 \, T^{-3/2} \, \nu^{-2} \, \mu_\nu^{-1} \left[17.63 + \ln(T^{3/2}/\nu)\right] cm^{-1} , \qquad (3)$$

where n_e is the electron density (cm^{-3}), T the electron temperature (K), ν the frequency (Hz) and μ_ν the index of refraction.

For an optically thin medium at constant temperature, T_c, (1) to (3) reduce to

$$T_b = T_c \, \tau_c \, , \tag{4}$$

where

$$\tau_c = C \, \nu^{-2} \, T_c^{-3/2} \int_0^\infty n_e^2 \, ds \, , \tag{5}$$

and the proportionality factor C depends only weakly on ν and T. For an optically thick medium we have

$$T_b = T_c \, . \tag{6}$$

We now consider possible reasons for the observed brightnesses of coronal holes and normal quiet regions, first at 80 MHz, then at 160 MHz.

At 80 MHz, the plasma level occurs at a density of $8 \times 10^7 \, cm^{-3}$; this density probably occurs in the corona rather than the transition region, even within coronal holes. Our calculations and those of others, starting with Smerd (1950), indicate that the corona above the plasma level, near disk centre, is optically thick, typically $\tau_c \approx 5$. Thus from (6) the measured brightness temperature equals the coronal electron temperature. The enhancements in brightness temperature over coronal holes on the disk therefore indicate that the electron temperature is higher there than in the surrounding quiet areas.

An alternative, although seemingly unlikely, interpretation could be that the corona at 80 MHz is not optically thick, which would make the brightness dependent on the density gradient near the plasma level. If this were the case, the higher brightness in a hole would imply a lower density gradient there than in normal quiet regions.

We noted earlier that the 80 MHz enhancement near CH1 was displaced to the west of the hole. This observation, if indeed the enhancement was related to the hole, would indicate an asymmetry in the emergence of magnetic flux from the hole.

Sufficiently far above the limb the corona at 80 MHz is optically thin. The brightness is then proportional to $\int n_e^2 \, ds$ by (4) and (5). Because n_e is lower at a given height in holes than in normal quiet regions, the 80 MHz brightness is also lower. The indentations in the 80 MHz contours and the dips in the limb scans of Figure 2 reflect this effect.

At 160 MHz the situation is more complicated. The plasma level corresponds to the height where $n_e = 3.2 \times 10^8$ K, and this can occur either in the corona (for quiet regions) or in the upper transition region (for holes). Near disk centre, in quiet regions, the optical depth of the corona at 160 MHz is large, so, as with 80 MHz, $T_b \approx T_c$. In holes the corona has an optical depth of order unity, and thus the coronal material makes a substantial contribution to the observed brightness; however, the material of the transition region also contributes to the observed brightness, but because of its lower temperature its contribution is diminished. Thus coronal holes on the disk appear dark on 160 MHz pictures.

Above the limb, where the ray path is entirely in the corona and the corona is optically thin, the situation at 160 MHz is similar to that at 80 MHz: the brightness is proportional to $\int n_e^2 \, ds$ and is lower at a given height if the line of sight passes through a hole than if it passes through a normal quiet region.

4. CONCLUSIONS

We have illustrated the features of coronal holes and quiet regions as they appear at 80 and 160 MHz. Our major result is the strong evidence that coronal holes on the disk at 80 MHz are sometimes brighter than surrounding quiet regions. This possibility had been raised by Dulk et al. (1977), but the 80 MHz enhancement in their data was displaced from the centre of the hole. At the limb, holes always appear as low-brightness indentations on quiet Sun maps.

We have qualitatively examined the reasons for the high brightness at 80 MHz and found that the coronal electron temperature is probably higher in the hole than in quiet regions, at least near the height where $n_e \approx 10^8$ cm^{-3}. We have not analysed the data quantitatively, partly because of observational uncertainties, and partly because of the more serious problem of the discrepancy between radio and other wavelengths as to the density-temperature structure of the transition region and low corona (Dulk et al., 1977; Chiuderi-Drago et al., 1977; Dulk, 1977; Chambe, 1978; Trottet and Lantos, 1978; Chiuderi-Drago, 1979). Progress must be made on the latter problem before we can be confident of any densities or temperatures that we might derive from radio observations alone.

References

Chambe, G.: 1978, *Astron. Astrophys.* 70, p. 255.
Chiuderi-Drago, F.: 1979, "EUV and radio spectrum of coronal holes" - *Solar Phys,* 65, p. 237.
Chiuderi-Drago, F.C., Avignon, Y., and Thomas, R.J.: 1977, *Solar Phys.* 51, p. 43.
Dulk, G.A.: 1977, *Proc. of the Nov. 7-10, 1977 OSO-8 Workshop,* LASP, Univ. of Colo., p. 77.
Dulk, G.A., and Sheridan, K.V.: 1974, *Solar Phys.* 36, p. 191.
Dulk, G.A., Sheridan, K.V., Smerd, S.F., and Withbroe, G.L.: 1977, *Solar Phys.* 52, p. 349.

Hansen, R.T., Garcia, C.J., Hansen, S.F., and Loomis, H.G.: 1969,
    Solar Phys. 7, p. 417.
Sheridan, K.V.: 1978, *Proc. Astron. Soc. Aust.* 3, p. 185.
Shibasaki, K., Ishiguro, M., Enome, S., and Tanaka, H.: 1977,
    "Contributed Papers to the Study of Travelling Interplanetary
    Phenomena/1977" (Ed. M.A. Shea et al.) AFGL-TR-77-0309 Spec.
    Rep. No. 209.
Smerd, S.F.: 1950, *Aust. J. Sci. Res. A* 3, p. 34.
Solar-Geophysical Data: 1972, IER-FB337, Part 1, U.S. Dept. of
    Commerce, Boulder, Colo.
Trottet, G., and Lantos, P.: 1978, *Astron. Astrophys.* 70, p. 245.

DISCUSSION

Newkirk: What is the quantitative temperature enhancement required
at the 80 MHz level?
Sheridan: The observed 80 MHz enhancement (see Fig. 1) is about
20-30% brighter than the surrounding regions.

Moore: What is the height of formation of the 80 MHz emission in
coronal holes? The recent Lyα/whitelight rocket observations from
Harvard indicate that the temperature in coronal holes does increase
with height by about a factor of two over a few solar radii, which
would be consistent with your results.
Sheridan: The 80 MHz critical density (8×10^7 cm^{-3}) level would
occur in the corona, even inside the hole, and so the result you
quote would be in agreement with our observations.

A MODEL FOR THE NORTH CORONAL HOLE OBSERVED AT THE 1973 ECLIPSE,
BETWEEN 1.3 AND 3.2 $R_\odot$

Francoise Crifo and Jean-Pierre Picat
D.A.P.H.E. - Observatoire de Meudon
92 190 - Meudon, France

1. THE OBSERVATIONS

At the 1973 eclipse, S. Koutchmy (Institut d'Astrophysique, Paris) obtained several pictures of the white-light corona, using polarizers and a radially-compensated filter. These pictures provide a very good opportunity for studying the large coronal hole at the north polar cap; this hole has been extensively studied during the Skylab period. On the plates of Koutchmy, we could record reliable intensities between 1.3 and 3.2 $R_\odot$. The absolute calibration was made using the stars observed in the field at the same time. This method allows a direct comparison of well-exposed objects on a same plate and must therefore be highly reliable (see Koutchmy et al., 1978). It is well-known that the absolute calibration of eclipse plates is a difficult problem.

The northern hole was very dark and from the synoptic maps and the X-ray pictures, one can conclude that probably no high-latitude streamers were projected over the hole in the plane of the sky. Intensities in the radial and tangential directions of polarization were recorded in the darkest part of the hole between the visible plumes.

2. DENSITY MODEL

The hole is considered as consisting of a weak background over which plumes are superimposed. We estimate that our intensity represents this background, which we assume to be homogeneous and constant with heliographic latitude. Thus a spherical inversion, with the formulae of Van de Hulst (1950) can be performed. As the density calculated at the highest points in the corona by this method is strongly dependent on the gradient adopted for $(I_t - I_r)$, we decided to extend our data by those obtained by Munro and Jackson (1977) from the Skylab coronograph in the same hole, between 2 and 5 $R_\odot$. Although these two curves exhibit exactly the same gradient, the data of Munro and Jackson lie 1.51 times higher than ours; so we divided them by this factor. This choice may be of course questionable but the absolute calibration

45

of Koutchmy seems to be very good; the intensities measured by Lilliequist (1977) from his eclipse pictures are close to ours and below those of Munro and Jackson in the region where they overlap.

The intensities are best fitted by the following expressions:

$$(I_t - I_r)/10^{10} \cdot \overline{B}_{\odot} = 500.r^{-7.565} + 120.r^{-4.245}$$

$$(I_t + I_r)/10^{10} \cdot B_{\odot} = 700.r^{-4.174} + 208.r^{-2.455}$$

where $\overline{B}_{\odot}$ is the mean solar disk brightness and r the distance from the sun's center.

The density derived is plotted on Fig. 1, together with the one determined by Munro and Jackson along the polar axis in their axysymmetric geometry, the density calculated by Koutchmy (1976) in the same hole from a different plate he took at the same eclipse, and the Van de Hulst minimum model.

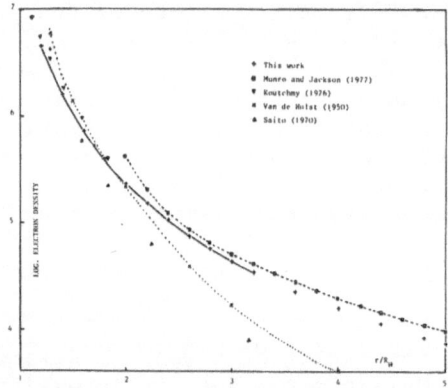

Figure 1.   Electron density versus height

3.   SOLAR WIND SPEED

Munro and Jackson determined an analytical representation for the shape of the hole's boundary that fits extremely well the eclipse pictures, and even other eclipses (1966 for example). So the surface of the hole is known and the wind speed may be calculated from the equation of mass conservation:

$$Ne(r) \cdot V(r) \cdot S(r) = K.$$

Munro and Jackson calculate the constant K at 1AU:

$$K = <Ne \cdot V>_{1AU} \cdot S(1AU)$$

and take a mean value: $<Ne \cdot V>_{1AU} = 3.10^8 \ cm^{-2} \ s^{-1}$, from the work of Feldman et al. (1976), similar to that given by Kopp and Orrall (1977).

The wind speed obtained that way is shown on Figure 2. It is larger than the current values: $V \sim 100$ km/s at 1.3 $R_\odot$ and 400 km/s at 3.2 $R_\odot$ and if the density obtained from our data extended by the reduced intensities of Munro and Jackson is correct above 3.2 $R_\odot$, the speed reaches 750 km/s at 5 $R_\odot$. From Hundhausen (1977) it appears that this speed is observed at 1AU for the fast streams leaving coronal holes; usual wind models show that the speed about doubles between 5 $R_\odot$ and 1AU.

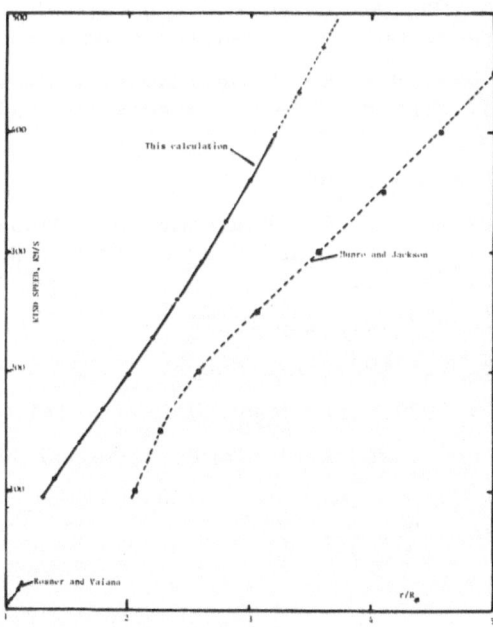

Figure 2. Wind speed versus height

However:

a)   It can be reasonably compared with the result of Rosner and Vaiana
(1977), who also obtain a very strong acceleration in the transition
region.

b)   The calculated speed is strongly subordinated to the value adopted
for the proton flux at 1AU. Although the work of Feldman et al. (1976)
seems rather convincing, the data published in "Solar Geophysical Data"
for June and July 1973 from the Pioneer VIII and IX satellites suggest
a value about 5 times lower. Using it would reduce the wind speed by
this factor of 5, and thus bring it to a more realistic value (if not
even too low). Anyway, the real flux over the poles is still unknown.

c)   Electron density has been calculated here between the polar plumes.
If plumes are inside the coronal hole, they contribute significantly to
its mass budget (see Ahmad and Withbroe, 1977), and the mass ejected by
the interplume region should be smaller. The mass flux at 1AU of course
does not take such inhomogeneities into account.

BIBLIOGRAPHY

Ahmad, I. A. and Withbroe, G. L.: 1977, *Sol. Phys.*, *53*, 397.

Feldman, W. C., Ashbridge, J. R., Bame, S. J. and Goslin, J. T.,:
        1976, *J. Geophy. Res. 81*, 5054.

Hundahusen, A. J.: 1977, in: *Coronal Holes and High Speed Wind
        Streams*, ed. by J. Zirker, Colorado Associated University Press.

Kopp, R. A. and Orrall, F. Q.: 1977, in *Coronal Holes and High Speed
        Wind Streams*, ed. by J. Zirker, Colorado Associated University
        Press.

Koutchmy, S.: 1977, *Sol. Phys.*, *51*, 399.

Koutchmy, S., Lamy, P., Stellmacher, G., Koutchmy, O., Dzubenko, N.,
        Ivanchuk, V., Popov O., Rubo, G., and Vsekhsvjatsky, S.: 1978,
        Astron. and Astrophys., *69*, 35.

Lilliequist, C.: 1977, *N.C.A.R./T.N.*, 128 + STR.

Munro, R. H. and Jackson, B. V.: 1977, *Astrophys. J.*, *213*, 874.

Rosner, R. and Viana, G. S.: 1977, *Astrophys. J.*, *216*, 141.

Saito, K.: 1970, *Annals of the Tokyo Astron. Obs., Second Series*,
        Vol. XII.

Van de Hulst, H. C.: 1950, *B.A.N.*, *11*, 135.

DISCUSSION

   Steinitz: The flux is variable; how to evaluate it will be shown
in the afternoon session.
   Crifo: I am very excited to know about this!

ON THE POSSIBILITY OF IDENTIFYING CORONAL HOLES ON SYNOPTIC MAPS OF THE GREEN CORONA

V. Letfus
Astronomical Institute of the Czechoslovak Academy
of Sciences, 251 65 Ondřejov, Czechoslovakia

L. Kulčár, J. Sýkora
Astronomical Institute of the Slovak Academy of Sciences,
Skalnaté Pleso, 059 60 Tatranská Lomnica, Czechoslovakia

The coronal holes of the Skylab period are treated to identify them with the low-brightness regions in our synoptic maps of the λ 530.3 nm emission corona. Possibilities and difficulties of this identification are discussed.

INTRODUCTION

The coronal holes were identified as a very important feature on the Sun, namely for their responses in geoactivity. Most of the physical and morphological studies of them are, of course, related to the period of the manned Skylab missions, when excelent pictures of the Sun in X-ray and XUV were taken. On acquiring a fairly good understanding of the physical substance of the holes, attempts appeared to extend into the past the period in which the position and properties of the holes could be identified and studied from other type of observations. Broussard et al. (1978) have used X-ray and XUV images obtained from rockets, OSO-6, OSO-7, and some others, to investigate coronal holes and their solar wind associations through-.out the whole sunspot cycle 20. Waldmeier (1975) has reported that it is possible to distinguish coronal holes on heliographic maps, constructed from daily observations of the green coronal line 530.3 nm. This is possible because the intensity of this line decreases strongly in regions of low density and temperature (Waldmeier, 1971), which are properties the coronal holes have been shown to posses. In this connection Waldmeier (1975) pointed out that for such regions on coronal synoptic maps he already used the term "löcher", which stands for holes in German, more than twenty years ago (Waldmeier, 1957). Guldbrandsen (1975) maintains that the known facts on solar sources of high-speed plasma streams, presented during the last 4 decades, almost inevitably seem to lead to the conclusion that the solar M-regions (Bartels, 1932) should be identified with the central portion of magnetically open solar regions, or coronal holes. In ad-

M. Dryer and E. Tandberg-Hanssen (eds.), Solar and Interplanetary Dynamics, 49-53.

dition, it is to be said that in several papers of Guldbrandsen the
low green-line intensity regions are found to have very similar geophysical consequences as they were ascribed to the hypothetical
M-regions.

RESULTS

     In the light of the above-mentioned facts, but in contradis-
tinction to the general conclusions of Waldmeier and Guldbrandsen,
we will try to show the possible identity of the low green-line intensity regions and coronal holes, in details. For this purpose the
synoptic tables of the green corona (prepared as described by Sýkora, 1971) were visualized for the period of the Skylab missions and
then the positions of the coronal holes, taken from Bohlin and Rubenstein (1975), were drawn into them (see Figure 1). Isophotes are in
absolute coronal units, the regions of intensity lower than the surroundings are hatched (of course, they are not necessarily the lowest
intensity regions on a given map), and coronal holes are cross-
-hatched and, at the same time, the generally accepted numbering of
the Skylab-period holes is given. In Figure 1 the following can be
seen: (1) Parts of polar coronal holes reaching up to our maps (they
only cover ±60° within equator) certainly fit in well with the low

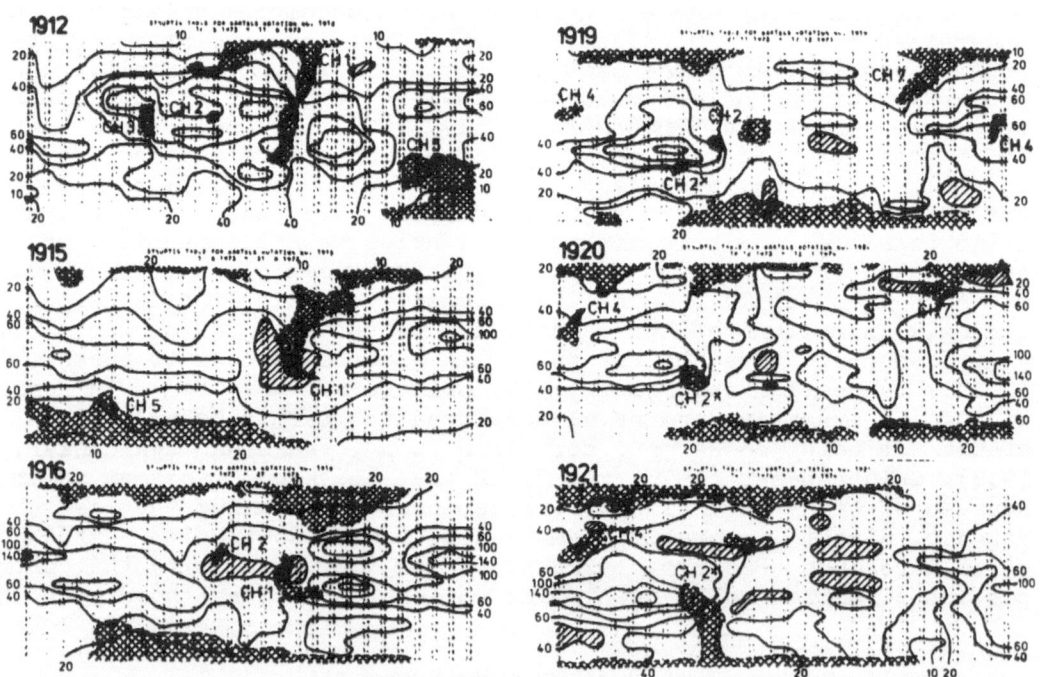

Figure 1. Synoptic charts of the green corona for indicated Bartels
rotations. Details see in the text.

coronal intensity regions towards the poles. (2) The low-latitude
holes are somewhat more complicated. In some cases they clearly agree
in position with the low corona (CH1, CH2 in Bartels rotations Nos.
1915, 1916 and 1919). Then there are cases in which both these features became displaced mutually for about 10-30° as, for example,
CH1 shows in rot. No.1912, CH2^X in rot. Nos.1919 and 1921 and CH4
in rot. No.1921. Two sources of this displacement are very probable.
Firstly, the coronal data are plotted as if they were measured at
12 00 U.T., but really they could have be observed at any hour of
the given day. Secondly, the position of the green corona features
are derived from limb observations, there is then uncertainty in the
position of the low corona regions along the line of sight and perhaps the large-scale changes of position during some days, when the
region was observed as a hole on the disk, are also possible. As
Timothy et al. (1975) showed, the geometry and position of the holes
must also be thoroughly determined. Owing to the said causes the found
differences in position of both features are quite possible. (3) The
last case are "small active region coronal holes" - SARCH - which
are common phenomenon on the Sun (Bohlin and Sheeley, 1978; Nolte

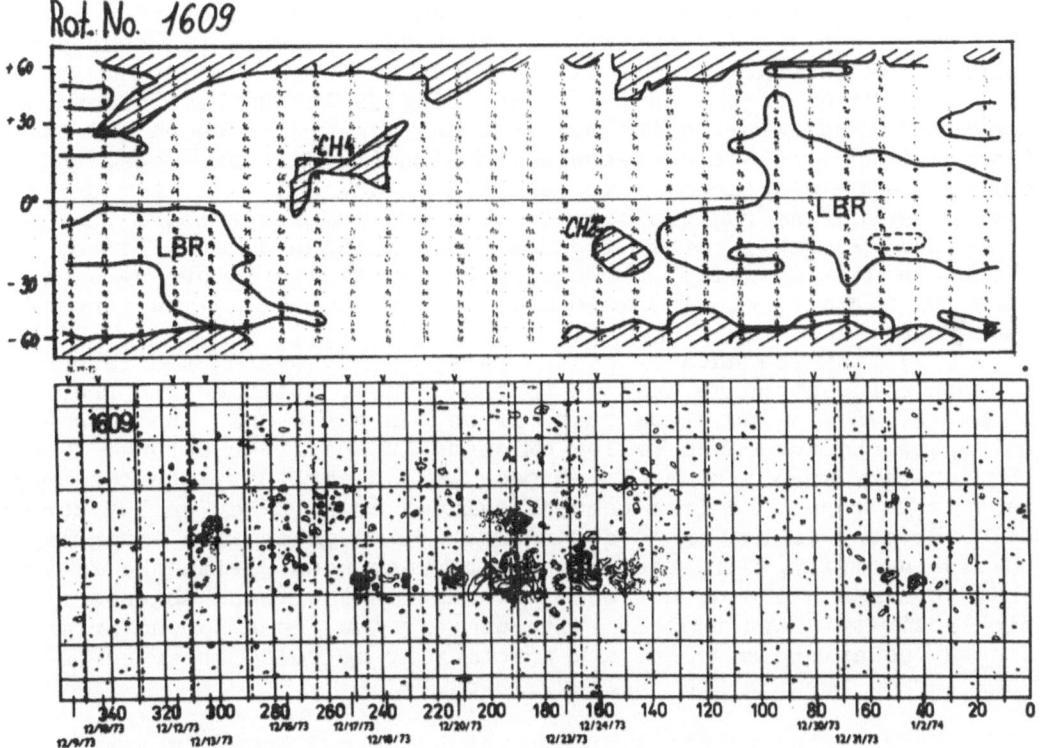

Figure 2. Green corona (above) and photospheric magnetic field (below) maps for Carrington rot. No.1609. In the coronal map the holes
are hatched and low-brightness regions are indicated by "LBR". The
holes CH4 and CH2^X are localized in clear area of higher coronal
activity (compare lower magnetic map).

et al., 1978). These holes are CH4, CH2^X, shown in Figure 2 (where
photospheric magnetic field and coronal maps for Carrington rot.
No.1609 can be compared). The holes CH4 and CH2^X are surrounded by
or are in the vicinity of active regions. The corona in places where
the holes are located does not show or shows only little decrease
of brightness. There are many other regions in the interval from
320° to 130° longitude (and also outside it) which are almost totally void of disk and coronal activity, but there are no holes observed in these regions. We are of opinion that low green line intensity
in SARCH really exists, but it is not possible to reveal it from
limb corona observations because the disturbing influence of the
adjacent active regions (once more the line of sight is working)
which fact is strengthened by the dimensions of these holes usually
not beeing very large.

CONCLUSIONS

        This short communication may be summarized as follows:
        (i) The coronal holes·, at least those of the Skylab period, are
characterized by low green corona intensities, in accordance with
physically motivated expectations. This fact is more pronounced with
polar holes and large low-latitude holes.
        (ii) Owing to the spatial uncertainties in the limb corona ob-
servations and owing to the possible changes of the shape and position of the hole in time between its limb and disk passages, the
mutual 10-30° displacement in position of the holes and low brightness green corona regions seem to be quite comprehensible.
        (iii) Probably rather frequent low-latitude coronal holes, ad-
jacent to disk activity, can hardly be identified in our coronal
maps for validity of (ii) in combination with small dimensions of
such holes.
        (iv) From the data presented here, but also from the inspection
of much more extensive data on coronal holes, published in Broussard
et al. (1978), we conclude that, if we can state that the coronal
holes are characterized by low emission in the green coronal line,
the contrary is not true. In no case we can say that each low-brightness green corona regions becomes evident as an X-ray or EUV coronal
hole. Because of the known dependence of the green line intensity
on disk activity, this result is in correspondence with the facts
that "coronal holes do not seem to exist in the total absence of
disk activity", and that in spite of very low green corona brightness at solar minima there are practically no holes observed in
this period (Bohlin and Sheeley, 1978).

        Although the relation of coronal holes and low-brightness green
corona regions seem to be somewhat one-sided, we are still attempting
to exploit the green corona synoptic maps, which are now in our pos-
session, covering the last three cycles, to clarify these questions
in greater details.

REFERENCES

Bartels, J.: 1932, Terr. Magn. 37, 1.
Bohlin, J.D. and Rubenstein, D.M.: 1975, Report UAG-51, "Synoptic
  Maps of Solar Coronal Hole Boundaries ... from the Manned Skylab
  Missions", World Data Center A for Solar-Terrestrial Physics,
  NOAA, Boulder, Colo.
Bohlin, J.D. and Sheeley, Jr., N.R.: 1978, Solar Phys. 56, 125.
Broussard, R.M., Sheeley, Jr., N.R., Tousey, R., and Underwood, J.H.:
  1978, Solar Phys. 56, 161.
Guldbrandsen, A.: 1975, Planet. Space Sci. 23, 143.
Nolte, J.T., Davis, J.M., Gerassimenko, M., Krieger, A.S., Solodyna,
  C.V., and Golub, L.: Solar Phys. 60, 143.
Sýkora, J.: 1971, Solar Phys. 18, 72.
Timothy, A.F., Krieger, A.S., and Vaiana, G.S.: 1978, Solar Phys.
  42, 135.
Waldmeier, M.: 1957, Die Sonnenkorona, Vol. II, Verlag Birkhäuser,
  Basel.
Waldmeier, M.: 1971, in C.J. Macris (ed.), "Physics of the Solar
  Corona", D. Reidel Publ. Co., Dordrecht, p. 130.
Waldmeier, M.: 1975, Solar Phys. 40, 351.

SOLAR OBSERVATIONS WITH A NEW EARTH-ORBITING CORONAGRAPH

N. R. Sheeley, Jr., R. A. Howard, D. J. Michels & M. J. Koomen
E. O. Hulburt Center for Space Research
U. S. Naval Research Laboratory
Washington, D. C.   20375

Abstract: Since March 28, 1979, the Solwind coronagraph has been
observing the Sun's white light corona (2.6 - 10.0 $R_\odot$) routinely with
a spatial resolution of approximately 1.25 arc min and a repetition
rate of 10 minutes during the one-hour sunlit portion of each 97-minute
satellite orbital period. These are the first satellite observations
of the outer corona near the peak of a sunspot cycle when coronal
transients and high-latitude streamers are common.

Satellite observations of the Sun's white light corona began on
October 4, 1971 when the Naval Research Laboratory's coronagraph on
OSO-7 started its 2.7-year interval of operation (Howard et al., 1975;
Koomen et al., 1975). The High Altitude Observatory continued coronal
observations during May 1973-February 1974 with its coronagraph on the
Skylab Apollo Telescope Mount (MacQueen et al.,1974). This paper presents
a sample of observations obtained with a new NRL coronagraph called
Solwind that is currently operating on the U. S. Department of Defense
Space Test Program Satellite P78-1 in a noon-midnight polar orbit.

In Figure 1, coronal images obtained with the OSO-7, Skylab, and
Solwind instruments are compared at the same spatial scale. A small
white area and an NRL He II 304 Å spectroheliogram have been placed
in the centers of the OSO-7 image (left) and the Skylab image (center),
respectively, to indicate the size of the solar disk. The OSO-7 and
Solwind images have similar fields of view ranging from the edges of
their occulting disks at approximately 2.6 $R_\odot$ to their outer limits of
10.0 $R_\odot$. The Skylab image shows a corresponding field of view from
1.5 $R_\odot$ to 6.0 $R_\odot$. However, the OSO-7 and Solwind images have a spatial
resolution of only 1.25 arc min, whereas the Skylab image has a resolution of 8 arc sec (MacQueen et al., 1974). Perhaps the greatest advantage
of the Solwind instrument over its predecessors is its relatively larger
duty cycle which is limited only by the periodic passage of the satellite
into the Earth's shadow for approximately 30-minute intervals.

Figure 1 also illustrates the changing character of the outer corona
during the sunspot cycle (cf. Van de Hulst, 1953). The OSO-7 and Skylab

55

M. Dryer and E. Tandberg-Hanssen (eds.), Solar and Interplanetary Dynamics, 55-59.
Copyright © 1980 by the IAU.

images were obtained during the declining phase of the previous cycle, and show streamers that are confined primarily to low latitudes. The Solwind image was obtained during the present approach to sunspot maximum, and shows streamers in more-or-less all directions. In particular a narrow, bright streamer is visible 15° east of north and a faint one is visible 10° west of north. Such high-latitude streamers were not observed during the OSO-7 and Skylab missions.

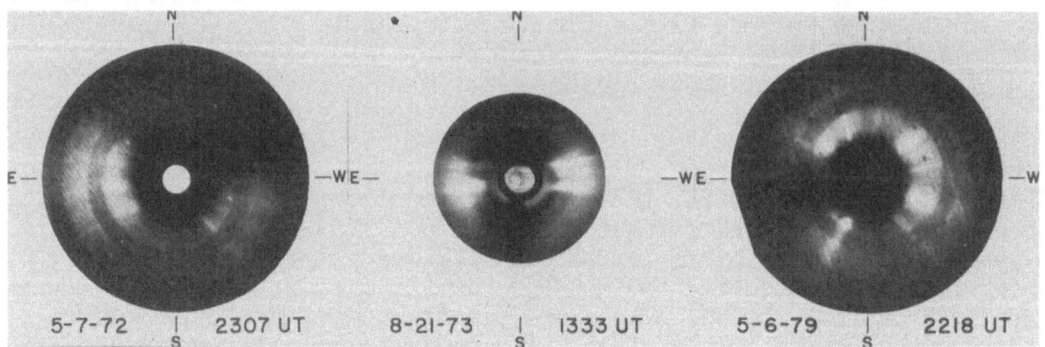

Fig. 1: OSO-7 (left), Skylab (center), and Solwind (right) coronal images illustrating the fields of view, spatial resolutions, and coronal conditions obtained with each instrument. The size of the solar disk is indicated by the white area and He II 304 Å image in the centers of the OSO-7 and Skylab images, respectively.

Figure 2 shows a sample of Solwind coronal images during May 6-9, 1979. Here, the outer limit of the field of view has been masked down to approximately 8.4 $R_\odot$. The narrow streamer 15° east of north seems to persist (with some changes) from May 6 to May 9. This suggests that the intervening frames will show the evolution of such high-latitude streamers relatively free from the effects of overlap and solar rotation that hindered previous studies of low-latitude streamers.

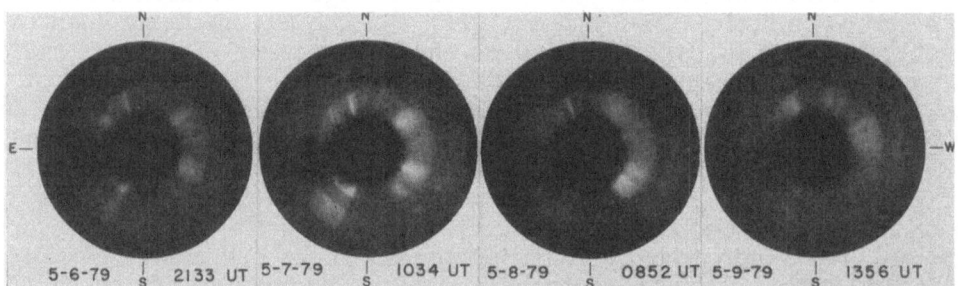

Fig. 2: A sample of Solwind coronal images during May 6-9, 1979 illustrating the daily changes in the pattern of coronal streamers.

Figure 2 also shows some major coronal changes from day to day. A structure in the southeast is relatively bright at 1034 UT on May 7, but

hardly visible at 0852 UT on May 8. A similar change occurred in the
southwest between May 8 and May 9. Subtle, but definite, changes also
are visible in the northeast during May 6-7 and northwest during May 7-8.
Intervening images show that all of these changes were associated with
coronal transients. The May 7-8 transients are shown in the next figure,
and the May 8 transient is described elsewhere (Michels et al. 1979a, b).

Figure 3 shows four "difference" images formed by subtracting pairs
of images obtained at the indicated times. In the southeast the 8.4 $R_\odot$
mask has been cut to show the structure out to 10.0 $R_\odot$. In the northwest,
a mass ejection is clearly visible during the interval 1805-2214 UT on
May 7. Unlike the "loop" transients which were so common during Skylab
and which we observed in the southwest on May 8, this transient does not
show a well-defined, continuous, curved, advancing front. Instead, this
transient consists of several outward-extending, spiky structures. Also
during this event the isolated streamer 15° east of north seems to have
moved or been pushed slightly away from the transient.

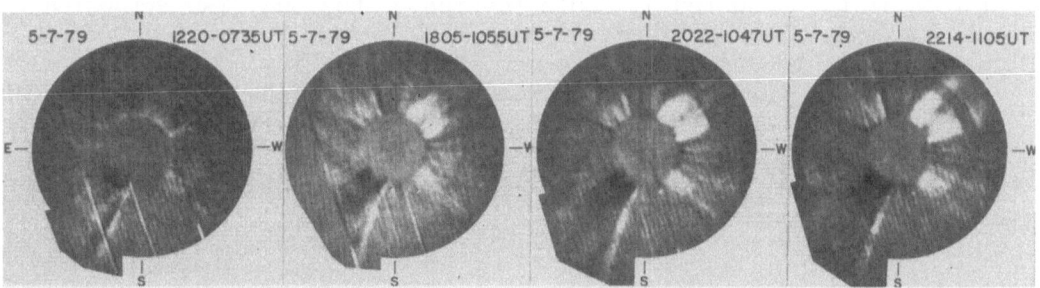

Fig. 3: "Difference" images formed by subtracting pairs of images at the
specified times. The dark pattern in the southeast and the bright
pattern in the northwest indicate coronal mass ejections that were in
progress during the first and second halves of May 7, respectively.

Figure 3 also shows evidence for the outward expansion of a looplike
transient in the southeast quadrant during the interval 0735-1220 UT. The
"1220-0735" difference image shows a dark loop near 3 $R_\odot$ and a faint
bright structure near 7 $R_\odot$ as if the transient had expanded outward during
this interval. In the "1805-1055" difference image the entire pattern is
dark, indicating that the transient faded by 1805 UT. Near the southward
boundary of this transient, a very long streamer seems to have been displaced and bent southward. Such transient-induced streamer displacements
were observed during the Skylab mission (Hildner et al., 1975) ,but are
shown much more dramatically in these Solwind difference images.

This effort to orbit a small coronagraph capable of nearly continuous
solar monitoring has received substantial assistance from several sources.
The NASA Office of Solar Physics provided spare coronagraph and solar
pointing flight hardware from its OSO-7 program, and also supplied ground
station support that has allowed quick access to the Solwind data. The
Department of Defense Space Test Program provided integration and launch

support and the Office of Naval Research provided financial support. At NRL, F. Harlow, D. Roberts, R. Chaimson, and R. Seal provided the technical and engineering support that helped to make this project a success.

REFERENCES

Hildner, E., Gosling, J. T., MacQueen, R. M., Munro, R. H., Poland, A.I., and Ross, C. L.: 1975, Solar Phys. 42, 163.
Howard, R. A., Koomen, M. J., Michels, D. J., Tousey, R., Detwiler, C. R., Roberts, D. E., Seal, R. T., and Whitney, J. D.; Hansen, R. T. and S. F., Garcia, C. J., and Yasukawa, E.: 1975, Report UAG-48, World Data Center A for Solar-Terrestrial Physics, NOAA, Boulder, CO.
Koomen, M. J., Detwiler, C. R., Brueckner, G. E., Cooper, H. W., and Tousey, R.: 1975, Applied Optics 14, 743.
MacQueen, R. M., Eddy, J. A., Gosling, J. T., Hildner, E., Munro, R. H., Newkirk, Jr., G. A., Poland, A. I., and Ross, C. L.: 1974, Astrophys. J. Letters 187, L85.
Michels, D. J., Howard, R. A., Koomen, M. J., and Sheeley, Jr., N. R.: 1979a, in M. Kundu (ed.), Radio Physics of the Sun, IAU Symposium No. 86, College Park, Maryland, August 7-10, 1979.
Michels, D. J., Howard, R. A., Koomen, M. J., Sheeley, Jr., N. R. and Rompolt, B.: 1979b, in M. Dryer and E. Tandberg-Hanssen (eds.) Solar and Interplanetary Dynamics, I.A.U. Symposium No. 91, Cambridge, MA, August 27-31, 1979.
Van de Hulst, H. C.: 1953, in G. P. Kuiper (ed.) The Sun, University of Chicago Press, Chicago, Ill. p. 207.

DISCUSSION

Webb: Have you developed sufficient statistics to be able to confirm the predictions of the Skylab/HAO group of the increased frequency of transients during solar maximum?
Sheeley: No. We have been observing routinely since March 28, 1979, but we have received less than 10% of these data so far.

Kahler: Can you compare the instrumental sensitivities of the P78-1 coronograph with those on Skylab and the OSO-7?
Sheeley: I have not made this comparison, but my co-authors say that the sensitivities of the Solwind and OSO-7 instruments are comparable, but that the Solwind instrument has a lower noise level than the OSO-7 instrument. Eventually we will probably compare the sensitivities of these coronographs with that of the Skylab coronograph and the planned SMM coronograph.

Bhonsle: (1) It is possible to delineate the time evolution of coronal transient from your observations? If so, with what time resolution? (2) What is the frequency of occurrence and average life-time of a coronal transient?
Sheeley: (1) Yes. On Thursday Donald Michels will describe the detailed solution of the May 8, 1979 transient with 10-minute temporal

resolution during the 60-minute sunlit portions of 2 or 3 consecutive 96 minute satellite orbital periods.

(2) We have not yet seen enough of our data to determine the frequency of occurrence of transients at this phase of the sunspot cycle. However, during the interval May 6-9, 1979 we have seen a significant part of our data and in this case we observed at least 4 transients. At a propagation speed of 500 km/sec, it takes about 3 hours for a transient to cross our field of view from 2.6 $R_\odot$ to 10.0 $R_\odot$. The remaining time for the coronal intensity to fade varies from event to event.

Bird: The coronograph images seem to exhibit a decrease in intensity at all position angles about halfway out in the field of view. Is this an instrumental effect?

Sheeley: Yes. Our instrument contains two eccentric, but nearly circular, analyzers for the linear polarization that is expected for Thomson-scattered radiation from electrons in the plane of the sky.

Dryer: Thanks to the excellent (and gratefully received) prompt reporting by the NRL team, we have been able to ascertain the possible interplanetary signature of your 8 May 1979 transient (in the South-west) at Venus (0.78 AU, about 110° west of the sun-earth line on this date). The preliminary identification was possible with the cooperation of John Wolfe, PI for the Pioneer-Venus solar wind plasma analyzer of the NASA-Ames instrument.

Sheeley: Also, it will be interesting to see if the Helios II space-craft (which was suitably located at 0.3 AU from the sun) detected an interplanetary signature from this event.

Jackson: When will you have a list of transients from your instrument?

Sheeley: We have a small list now and are adding to it day by day as we receive and look at our data.

X-RAY STRUCTURES ASSOCIATED WITH DISAPPEARING Hα FILAMENTS IN ACTIVE REGIONS

S. W. Kahler
American Science and Engineering, Inc.
Cambridge, Massachusetts 02139 USA

Several studies using data from Skylab instruments have been carried out to determine the spatial and temporal relationships between disappearing Hα filaments and the associated coronal emission features. Webb et al. (1976) studied 30 transient coronal X-ray enhancements which could be associated with the disappearances of Hα filaments outside active regions. They found that in the early phase of the transient X-ray brightening, emitting structures appeared at or near the filament location with shape and size resembling the filament. Sheeley et al. (1975) examined a long-lived X-ray enhancement of expanding loops associated with an active region filament which disappeared. Rust and Webb (1977) found a good statistical correlation in time and position between large scale (length > 60,000 km) active region X-ray enhancements and Hα filament activity, in particular, events of an eruptive nature.

The purpose of the present study is to examine in detail the relationship between active region disappearing Hα filaments and the associated coronal X-ray structures observed both before the disappearance event and afterwards. The results presented here constitute a "first order" overview of the events chosen for study.

The events chosen for study were first selected from a list of active region X-ray transients observed in the images from the AS&E X-ray telescope on Skylab which were the basis of the Rust and Webb (1977) study. Additional events were selected from a list compiled by D. Webb of sudden disappearances of filaments during the Skylab period. Only those events for which an active region filament disappearance could easily be seen in the NOAA Hα patrol films were used. There was a total of 14 events in 8 different active regions.

The event of 29 August 1973 shown in Figure 1 is one of the events studied. Meudon filaments F7 and F8, which are overlain by X-ray loops, remained after the filament disappearance of F6 at ~1830 UT. There were no obvious X-ray loops overlying the latter filament.

M. Dryer and E. Tandberg-Hanssen (eds.), Solar and Interplanetary Dynamics, 61-65.
Copyright © 1980 by the IAU.

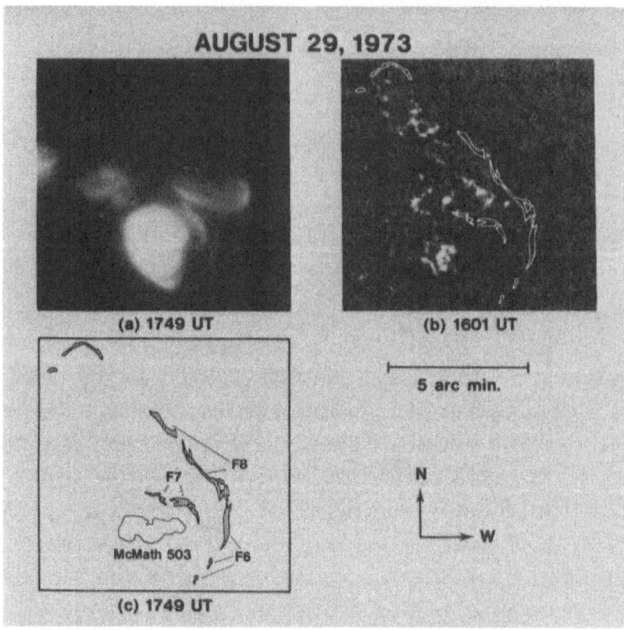

Figure 1. (a) The X-ray structures of McMath 497 and 503 prior to the
filament disappearance event of ~ 1830 UT on 29 August. (b) The outline
of the Hα filament structure on the aligned daily KPNO magnetogram.
(c) Schematic showing Meudon F7 and F8, which remained, and F6, which
disappeared. Note that X-ray loops overlie F7 and F8, but not F6.

The event of 3 September 1973 illustrated in Figure 2 shows that we can
not in general associate disappearing filaments with a lack of visible
associated X-ray loops. In this case a loop lies along both the western
part of the filament which disappears as well as the eastern portion
which survives. No new X-ray emission feature appears at the site of
the remaining filament section, but a new X-ray loop appears along the
site of the disappearing filament.

In the first part of the study the 14 events were analyzed for their pre-
event spatial associations with X-ray structures. The sections of the
filaments that later disappeared were considered separately from the
filaments that remained after the events. In each case the associated
X-ray structure was classified into three categories: (1) a distinct loop
or set of loops; (2) an X-ray cloud with no resolved loop structures; and
(3) no obvious X-ray emission observed in photographic images of the
thin filter (2 - 32, 44 - 54 Å) 64 sec exposure. In some cases a disappearing or remaining filament was covered by structures of different

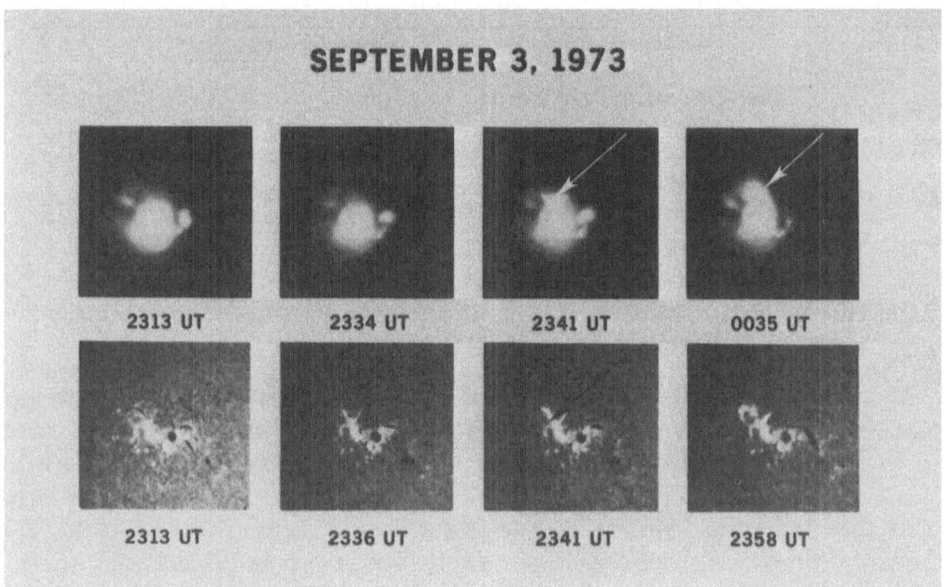

Figure 2 (from Rust and Webb, 1977). Aligned X-ray and Hα images of McMath 510 before and during the disappearance of the western part of the Hα filament on 3 September. A curved loop lies along the entire section of the Hα filament shown at 2313 UT. In the event an X-ray loop (indicated by an arrow) is present at the site of the disappearing filament, but no new X-ray emission is apparent at the remaining filament. Each frame is 15 arc min square.

categories, for example, partially by an X-ray cloud and partially by no emission. In these cases the filament was considered to be half of one category and half of the other. The results of the X-ray associations with the 14 disappearing filaments and the 11 remaining filaments are shown in Table I. It can be seen that within the limited statistics of the study there is little difference between the X-ray features associated with disappearing filaments and those associated with remaining filaments in the same active regions.

The second part of the study has been to examine the X-ray images during a period of 3 hours following the onsets of the filament disappearances. These existed for 10 of the 14 events of the study. In 8 of the 10 cases an X-ray loop structure of some kind brightened spatially close to the site of the disappearing filament, either along or over the magnetic inversion line. In one of the cases where no new X-ray feature was seen, the disappearing filament had completely reformed prior to the time of the first X-ray image of the event, so an X-ray structure may well have

TABLE I: ASSOCIATION OF X-RAY FEATURES AND Hα FILAMENTS

		Loop	Cloud	No X-Ray
Pre-Event	Disappearing Filament	7-1/2	4-1/2	2
	Remaining Filament	4	4-1/2	2-1/2
Event	Disappearing Filament	8	0	2
	Remaining Filament	0	0	9

brightened and then faded in that time. In none of the 9 cases in which a portion of the active region filament remained was there any new X-ray emission associated with that part of the filament.

The results indicate that there is no distinction between disappearing and remaining active region filaments in terms of their pre-event associated X-ray emission features. The presence or absence of pre-existing overlying or parallel X-ray loops does not appear to influence the stability of the filament against a disappearance. On the other hand, X-ray brightenings were associated in a nearly one-to-one correspondence with disappearing portions of the filaments. The pre-event result is valid only for a characteristic time scale of at least an hour. Martin and Ramsey (1972) found a statistical pattern of filament activity beginning within an hour prior to flares of class 1 or larger. One might then ask whether there is a simultaneous pattern of changes in X-ray features associated with such filament activity. That question will be the subject of future studies.

This work has been supported by NASA contract NAS8-27758.

REFERENCES

Martin, S. F. and Ramsey, H. E.: 1972, in (P. S. McIntosh and M. Dryer, eds.) Solar Activity Observations and Predictions, MIT Press, Cambridge, Massachusetts.

Rust, D. M. and Webb, D. F.: 1977, Solar Phys. 54, 403.

Webb, D. F., Krieger, A. S., and Rust, D. M.: 1976, Solar Phys. 48, 159.

Sheeley, N. R., Jr., Bohlin, J. D., Brueckner, G. E., Purcell, J. D., Scherrer, V. E., Tousey, R., Smith, J. B., Jr., Speich, D. M., Tandberg-Hanssen, E., Wilson, R. M., DeLoach, A. C., Hoover, R. B., and McGuire, J. P.: 1975, Solar Phys. 45, 377.

DISCUSSION

VanHoven: My question has two parts: (a) Do you mean to distinguish between disappearing and erupting filaments in these cases? and (b) Are the X-ray brightenings different from subflares?

Kahler: (a) No, I cannot distinguish between the two using the NOAA Hα patrol films. (b) Most of the brightenings of this study can be considered to be X-ray subflares or flares.

Pneuman: You seem to be saying that the X-ray emission occurs after the filament disappears. Is there any evidence from your observations that X-ray emission might occur also <u>before</u> the eruption of the prominence?

Kahler: I have not yet examined the data with that question in mind. You should ask David Webb about the results of his studies.

Webb: In the paper by Webb, Krieger and Rust we noted that in every case where we had images during or before the onset of a filament disappearance, we observed a compact brightening at a location where the filament was doing something interesting (e.g., bend or kink in filament, location where it first showed high velocity, motion, etc.).

Uchida: What are the time scales (rise time, duration, fading time scale, etc.) for the X-ray emitting objects appearing after the disappearance of the dark filament?

Kahler: The study was not really designed to answer that question, but I can give you a general idea from this and other studies. At first there may be small loop brightenings, manifested as flares, in which loops are bright with lifetimes on the order of 1-3 x 10 min. There are also often larger scale loops observed later which constitute the long decay events lasting for perhaps hours.

THE ORIGIN OF INTERPLANETARY SECTORS

Kenneth H. Schatten
Goddard Space Flight Center
NASA
Greenbelt, MD   20771

1.   EARLY CORONAL MAGNETIC MODELS

The coronal magnetic models of Altschuler and Newkirk (1969), Schatten, Wilcox and Ness (1969), and Schatten (1971) that allowed calculations of the coronal magnetic field from the observed photospheric magnetic field shed light on the origin of sectors. Figure 1 from Schatten's (1971) "Current Sheet Model" is a schematic representation of these similar models. There are three distinct regions in these models where different physical phenomena occur. The photosphere, where the magnetic fields are governed by the detailed motions and currents in the plasma is considered a boundary condition for the model. Above the photosphere, the plasma density diminishes very rapidly with only moderate decreases in the magnetic energy density. This results in the middle region where the magnetic energy density is greater than plasma energy density and hence controls the configuration. One may then utilize the force-free condition, $\underline{j} \times \underline{B} = 0$, and in fact make the more restrictive assumption that this region is current free. The magnetic field in this region can be derived from a solution to the Laplace equation.

Substantially farther out in the corona the total magnetic energy density diminishes to a value less than the plasma energy density, and the magnetic field can no longer structure the solar wind flow. The magnetic field has, however, become oriented very much in the radial direction, as suggested by Davis (1965). Thus, before the total magnetic energy density falls below the plasma energy density, a region is reached where the transverse magnetic energy density does so. It is this component of the magnetic field that interacts with the outward flowing plasma. On the "source surface", another boundary condition applies – transverse magnetic fields are transported away from the sun by the radially flowing plasma. Thus the magnetic field passing through the surface boundary is oriented in the radial direction (the 1969 models), serving as a source for the interplanetary magnetic field. In the current sheet model the fields calculated were in a Maxwell-stress equilibrium rather than requiring coronal forces to

M. Dryer and E. Tandberg-Hanssen (eds.), Solar and Interplanetary Dynamics, 67-72.
Copyright © 1980 by the IAU.

artificially bend the field into a purely radial configuration. The
fields were constrained only to be "open" rather than radial. This
conformed better to the theoretical MHD treatments of Pneuman and Kopp
(1970).

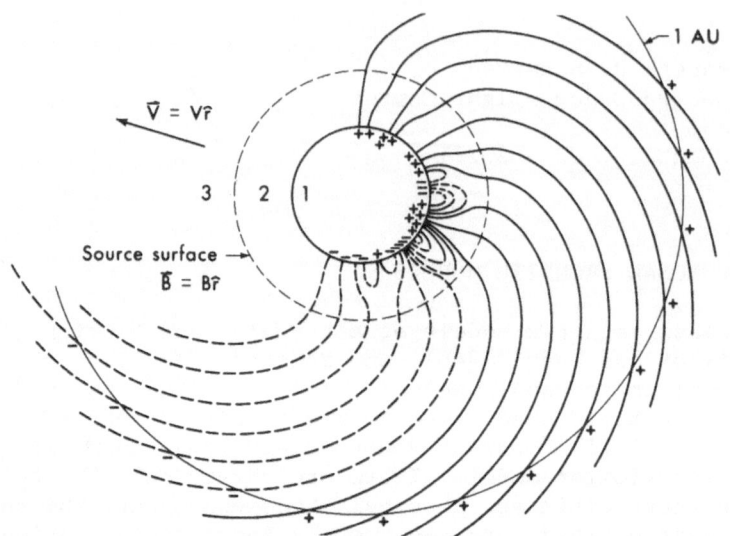

Figure 1.   Concepts related to the current sheet model
of sector boundaries.

2.   CORONAL HOLES

Early Mariner 2 and IMP observations revealed that geomagnetic
activity was related to streams of abnormally fast solar wind. The
solar source of such streams (particle streams before 1960 or high-
speed solar wind streams since) remained an intriguing mystery until
recently. An important step in this development was the discovery
around 1970 of coronal holes-sharply defined regions of very low
emission, in X-ray and ultra-violet images of the sun. The appearance
of holes in both X-rays and white light suggests that they are regions
of open magnetic field lines that diverge rapidly with increasing
altitude; this suggestion is strengthened by comparisons of the
observed locations of holes with the coronal magnetic field models
described.

3.   THE ORIGIN OF SECTORS

It is the purpose of this section to clear up some misconceptions
about interplanetary magnetic field sectors. Ness and Wilcox (1964)
first related interplanetary sectors to solar features during the time
of a quiet inactive sun. At this time a more or less one-to-one mapping

existed between solar features and the interplanetary sector structure. Subsequently, however, the situation grew more complicated and the previously described coronal models became necessary to solve the configuration of the magnetic field patterns involved.

These models tell us that the quiet IMF sector structure is effectively, a low-bandpass filtered version of the photospheric field. This gives rise, often to a 2 or 4 sector structure pattern in the ecliptic plane and a unidirected high solar latitude field. Superposed on this are transients-filaments kinks, bottles, holes, etc. in the IMF.

Thus the sector structure arises in the corona and does not exist on the sun, either in the photosphere or below the surface. Suitably tailored analyses (see Wilcox and Gonzales, 1971, and Wilcox et al., 1970) give correlations for certain time periods suggesting sector structure exists within the sun. The description is partly correct, partly not. Firstly, suitable low-bandpass averaging shows the "average fields" are calculable, however, their influence is often misinterpreted. The average field value in a region, although calculable, is not physically present - thus has little meaning. Secondly, considering the size-scale of the features and the time it takes to transport magnetic flux on the sun, these "average fields" certainly persist for a while. Thirdly, there are no significant individual solar features found, despite numerous investigative correlations, which statistically correlate with these "average fields". Shapiro (1965), for example, finds "the principal source of power at harmonics of the 27-day fundamental in... solar indices is due to the periodic but non-sinusoidal nature of the variation ... not due to a tendency for equidistant spacing of active regions on the sun."

Thus, if one ascribes a physical reality to these "average fields" on the sun, one would make the same error ascribing reality of the main solar dipole field to the photospheric fields at low latitudes. That is, suitably tailored averaging (over the northern and southern hemisperes) could show a north-south average dipole over all latitudes, nevertheless the sun's polar fields appear to be limited to the higher latitudes and few would argue for their reality at all latitudes despite a tailored averaging process. Thus, interplanetary sector structure should be confined to studies of the outer corona, interplanetary space and objects therein, not the sun itself. Solar structure does exist but not in the form of "Solar Sector Boundaries". Coronal holes or open field line boundaries, which are very different from the projection of the IMF current sheet onto the solar surface, may play some role.

REFERENCES

Altschuler, M.D., and G. Newkirk, Jr., Magnetic Fields and the Structure of the Solar Corona, *Solar Phys.*, *9*, 131, 1969.

Davis, L., Jr., Mariner II Observations Relevant to Solar Fields, *Stellar and Solar Magnetic Fields, 202,* 1965.

Ness, N. F., and J. M. Wilcox, Solar Origin of the Interplanetary Magnetic Field, *Phys. Rev. Lett., 13,* 461, 1964.

Pneuman, G. W. and R. A. Kopp, Coronal Streamers III: Energy Transport in Streamer and Interstreamer Regions, *Solar Phys., 13,* 176, 1970.

Schatten, K. H., Current Sheet Magnetic Model for the Solar Corona, *Cosmic Electrodynamics, 11,* 4, 1971.

Schatten, K. H., J. M. Wilcox and N. F. Ness, A Model of Coronal and Interplanetary Magnetic Fields, *Solar Phys., 9,* 442-455, 1969.

Shapiro, R., Comparison of Power Spectrums of Artificial Time Series with Spectrum of the Solar Plage, *J. Geophys. Rev., 70,* 3581, 1965.

Wilcox, J. M. and W. Gonzales, A Rotating Solar Magnetic "Dipole" Observed from 1926 to 1968, *Science, 174,* 820-821, 1971.

Wilcox, J. M., K. H. Schatten, A. S. Tanenbaum, and R. Howard, Photospheric Magnetic Field Rotation: Rigid and Differential, *Solar Phys., 14,* 255-262, 1970.

DISCUSSION

Ahluwalia: In your historical introduction you stated that no one knew about the sector structure in the interplanetary magnetic field (IMF) before it was actually observed. I wish to set the historical record straight by drawing your attention to the fact that Ahluwalia and Dessler (1962) predicted the existence of the sector structure in IMF. In fact, John Wilcox acknowledged this in his paper announcing the discovery of the sector structure in IMF!

Schatten: Yes, you are quite right! At the 1st solar wind conference, where a fuller review was given, I also referred to your work. Only due to the brevity of this talk was your work omitted. As I recall, you predicted the sector structure on the basis of the IMF originating from sunspots. So, like all of us, you were partly right, partly wrong. Thank you for the correction.

McIntosh: I must continue to disagree with you on two points: (1) the Hα synoptic charts clearly show a major neutral line, on most solar rotations, that divides the sun into two hemispheres in longitude, revealing the fundamental magnetic dipole in the east-west direction; (2) the sunspot, flare and X-ray emissivity statistics show an unmistakable longitude dependence. Active longitudes are clearly discernable in Hα synoptic charts as well. Confusion as to their reality has arisen from taking data over too long a time interval so that the drifts of long-lived active zones with respect to heliographic longitude produced an appearance of random occurrence. Certainly the active longitudes are more clearly defined with the larger and less-frequent categories of activities.

Schatten: I understand that not everyone will agree with this view as other views have been forcefully supported with analyses which appear to yield an opposite picture.

With regard to your question (1): Yes, during certain time intervals there is an east-west dipole, but at others there is not. For example, the June 1954 solar eclipse showed an almost exact north-south symmetry-denying an east-west dipole! In any case, the corona responds to the large-scale photospheric field and if it is east-west, then there is a two-sector structure at 1AU.

With regard to (2): I believe for "active region longitudes" to have <u>real</u> <u>physical</u> <u>importance</u>, they must be long-lived, and continue to arise at a specific longitude in a particular rotating (Ω = constant) coordinate system. Otherwise, they can be attributed to just relatively short persistence effects. So that for a few months a particular longitude is active. As an example, Jysiter's Red Spot exists at a specific longitude (in one coordinate system) but over a long-time it wanders in longitude - more than 360°! Thus it has a similarity to "active longitudes" in so far as it exists at one longitude for only a limited period. Thus, I can agree with you, in part, as follows. Large-scale solar fields have a long persistence due to the divergence B condition and the problems of changing the net flux in a short amount of time. This large-scale structure would to a large extent relate to the IMF sector structure as shown by the fine work of Wilcox and colleagues. It may then, also relate, for moderately long time intervals, to active regions, but this hasn't to my knowledge been demonstrated. In the absence of this, I have yet to understand "active longitudes".

Bhonsle: Is the presence of coronal hole a necessary condition for the existence of interplanetary magnetic field (IMF) sector structure? If so, what happens to the sector structure in the absence of coronal holes?

Schatten: The solar wind depletes the corona of material, and cools it. Thus, there is generally a close relation between the IMF sector structure and coronal holes. This is, however, a <u>quasi-stationary</u> view. If conditions are <u>changing</u>, the above will not be <u>universally</u> true! For example, a new sector may take time to deplete the coronal material, and no hole would be immediately apparent; thus, a 1 to 1 relation need not always exist.

Further, sector structure would generally exist even if no coronal holes were visible. However, some interplanetary field structures may arise from dynamic events - blast waves, bottles, bubbles, etc. That is, interplanetary magnetic fields are sometimes more complex than the simple sector viewpoint.

Tandon: Would you comment on the correlation of the sector structure with coronal holes, extending from pole to equator, in the absence of high latitude solar wind observations?

Schatten: I believe these features (coronal holes) occur through depletion of coronal plasma on long-lived "open" field lines, as calculated in the coronal theories reviewed by Levine. Thus, the "high latitude" spacecraft observations should show a general high speed wind with a uni-directed sector structure above about 20° latitude (as suggested in Schatten, 1971, Reviews of Geophys. and Space Phys.) except when the polar fields are reversing; with perhaps occasional variability due to solar activity (blast waves, bottles, etc.).

CORONAL STRUCTURE AND SOLAR WIND

J. N. Tandon
Department of Physics & Astrophysics
University of Delhi
Delhi-110007, India

ABSTRACT

Recent observations of large scale coronal structures and solar wind have been studied. The intercorrelation of the two have been qualitatively explained through the focussing of solar-ion streams taking account of the local and general solar magnetic fields. This explains the association of coronal holes with weak, diverging open magnetic field lines and envisages the transfer of hydromagnetic wave energy from nearby active centers to account for the enhanced outflow of solar wind associated with coronal holes.

1. CORONAL STRUCTURES AND SOLAR WIND

During the past one decade, we gained some new information regarding the solar wind and their association with coronal structure. The coronal structures are primarily responsible for filling the entire interplanetary space with fast and slow solar wind. Variation of solar wind speed by several hundred kilometer over its mean value of 450 km/s have been observed (Gosling et al., 1976). In addition, Feldman et al. (1976) have revealed several instances of high speed streams of velocity $\geq$ 650 km/s during March 1971 through July 1974. None of the existing theory predicts speeds in excess of 650 km/s at 1 A.U. with proton flux of $(3.3 \pm 0.5) \times 10^8 \ cm^{-2} \ s^{-1}$. These streams have different characteristics than the usual solar wind; eg. $T_p^{\perp}/T_p^{||}$ = 2.4 instead of normal 0.5; also $T_p > T_e$ instead of $T_e > T_p$. Thus there is a significant gap between theory and observation and input of some additional energy source is required.

Further, information collected through X-ray and EUV measurements by OSO-7 and Skylab ATM mission and by 1.5 $R_\odot$ K-corona observations have revealed the existence of two prominent coronal structural features; vis.

M. Dryer and E. Tandberg-Hanssen (eds.), Solar and Interplanetary Dynamics, 73-78.
Copyright © 1980 by the IAU.

(i)   The Active Centres: They constitute large bright areas of high
density and temperature in corona similar to coronal condensations overlying strong bipolar or more complex magnetic field regions of the photosphere. Many loop-like structures are seen invariably connecting regions
of opposite magnetic polarity in the photosphere. Hansen et al. (1976)
showed that the bright coronal regions of 1.5 $R_\odot$ K-corona, being identical to active centres, inhibit geomagnetic activity on +3 day of its
CMP-consistent with the "cone of avoidance" model of Allen (1944).

(ii) The Coronal Holes: The large coronal structures of greatly reduced
X-ray, EUV and metric radio emission are usually referred to as coronal
holes (CH). The density and temperature above CH are respectively lower
by a factor of 3 and 2 than the average quiet sun, while the thickness
of the chromospheric-coronal transition zone increased by a factor of
3 in holes over that in quiet sun. K-corona observations also reveal
similar properties of the holes. Most important and striking aspect
of these holes is their apparent occurrence in weak open diverging unipolar magnetic regions. These holes are seen to occur both at the centre
and the pole.   The polar holes, though similar to equatorial ones, are
somewhat larger and long lived structures.

CMP of CH are now known to be followed by increase in geomagnetic
activity (K_p maximum at +3 day).   In general near equatorial
$CH(\leq |40^\circ|)$ are associated with enhanced solar wind speed at 1 A.U. in
the ecliptic plane with the interplanetary magnetic polarity in stream
agreeing with solar polarity beneath the CH and that they are responsible for M-storms.   (Altschuler et al., 1972; Munro and Withbroe, 1972;
Bell and Noci, 1973, 1976; Krieger et al., 1973; Vaiana et al., 1973;
Neupert and Pizzo, 1974; Timothy et al., 1975; Hansen et al., 1976;
Nolte et al., 1976; Rickett et al., 1976; Sheeley et al., 1976).   In
fact both equatorial CH and M-regions are known to develop in areas
empty of spots.

Earlier, Tandon (1958, 1963, 1966) proposed that M-regions have
coronal origin and can be placed in two groups; vis. (i) the disturbed
M-region and (ii) the quiet M-region. The basic properties of these
M-regions, as envisaged through various possible identification of
M-regions with variety of solar features, are respectively identical
with the active centres and the CH.

2.   FOCUSSING OF SOLAR-ION STREAMS

Considering the effect of "frozen in" magnetic feild on the
focussing of solar plasma streams, especially in the neighborhood of
the solar atmosphere, Tandon (1958, 1963, 1966) has obtained the
following conditions for the focussing of streams:

$$i_c \geq i_c = (2N\Psi + q^2 + 2M)^{\frac{1}{2}} \tag{1}$$

that is, the stream will get focussed if the total current in the stream
of unit length exceeds the critical value i_c. Here N, Ψ, q, and M are
respectively the total number of particles per unit length of the stream,
the mean kinetic energy of the particles due to transverse component of
velocity, the total charge per unit length of the stream and the
polarized electric field energy per unit length of the stream. Further,
M involves a vector product of frozen-in field H and the velocity V and
hence will be zero when the condition (1) may be satisfied and hence
the stream may get focussed. Further it is evident from the condition
(1) that in the neighbourhood of the active centre of the sun, M will
be relatively large and hence the magnetic field will disperse the solar
plasma particles from the stream and the stream may not get focussed,
i.e., there will be an increase in the non-radial dispersion of the
stream, referred to as "magnetic dispersion". However, there is a likelihood for the stream to get focussed in the neighbourhood of coronal
regions where the magnetic field of the active centres decreases
considerably.

       Let us now consider the bearing of condition (1) on coronal
structures. The magnetic field associated with active centre will give
rise to large magnetic dispersion which may continue to a distance of
about one solar radius or so. Afterwards the focussing is likely to
set in since the magnitude of the magnetic feild at such large distances
decreases considerably on account of low density. Such a process of
magnetic dispersion and subsequent focussing will lead the particles to
remain away from the active centres or, in other words, the streams
form a 'cone of avoidance'. This process is roughly represented in
Figure 1. The edges of the cone on projection will appear to come from
the adjacent areas of the active centres and that the corona overlying
these active regions should possess large density and temperature due
to associated large magnetic field forming coronal condensations.
Further, the width of the cone is expected to be about 30°-40° near the
maximum development of the active region and zero at the time of
extinction of the spot (see Figure 2).

       During the decaying phase the magnetic field of the active region
is known to fall off rather sharply ejecting large magnetic field in the
form of detached plasma clouds in accordance with the plasma ring
experiment of Alfven et al. (1960). These detached plasma clouds are
likely to form coronal regions of low density and temperature with open
field line structures overlying the decaying spot – the CH. At the photospheric and low chromospheric level features like granulation, super
granulation and oscillations will become indistinguishable with the
neighbouring regions outside the spots. This indicates that the
mechanical input passing up through the photosphere may be substantially
the same inside and outside the holes. The radiative and conductive
losses will be less due to low density and temperature in the holes.
This results in an excess energy which goes into creating and
accelerating a solar wind that emanates primarily from the regions of
the holes. Further, extrapolating the observations of plasma ring

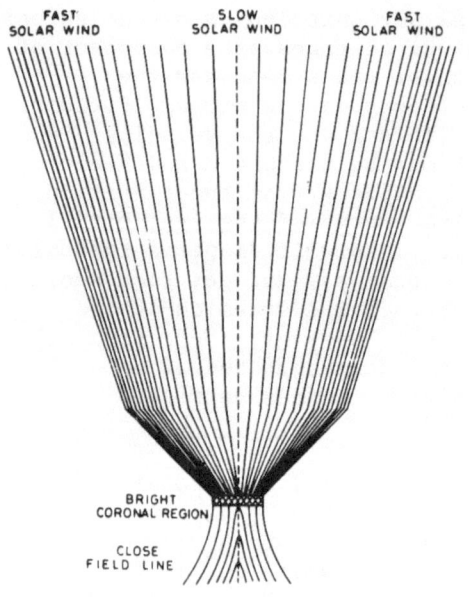

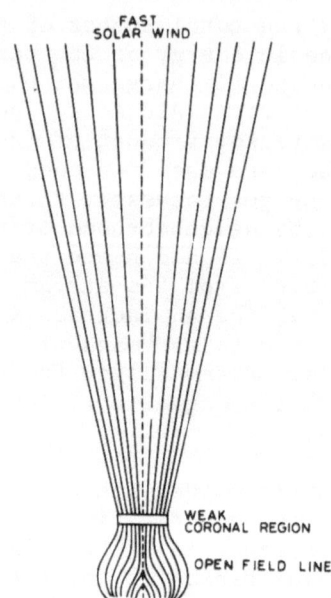

Figure 1. Figure illustrating
   qualitatively the effect of
   magnetic field on the focussing of solar ion stream and
   formation of 'cone of avoidance.'

Figure 2. Structure of proposed
   solar beam at the time of
   extinction of solar active region.

experiment to incorporate the effect of large conductivity and magnetic
field of neighbouring regions one should expect that the detached
plasma clouds at the time of vanishing of parent active region should
form closed field lines either with local field of another active
region as shown in Figure 3 or with other weak field region including
the general magnetic field of the sun. In the former event the coronal
structure above the parent active region should disappear with the
decay of active centre. During this phase, a large amount of energy in
the form of Alfven waves will be imparted to this coronal structure
enhancing considerably the solar wind speed at times > 650 km/s and
enhanced proton temperature such that $T_p > T_e$ even. This thus throws
light on the mechanism for explaining the observations of Feldman
et al. (1976). On the other hand, in the latter case, there will be
comparatively low feedback of Alfven wave energy creating high speed
solar winds conceivably dwarfing the rather modest solar wind, to form
a solar wind with variable wind velocity.

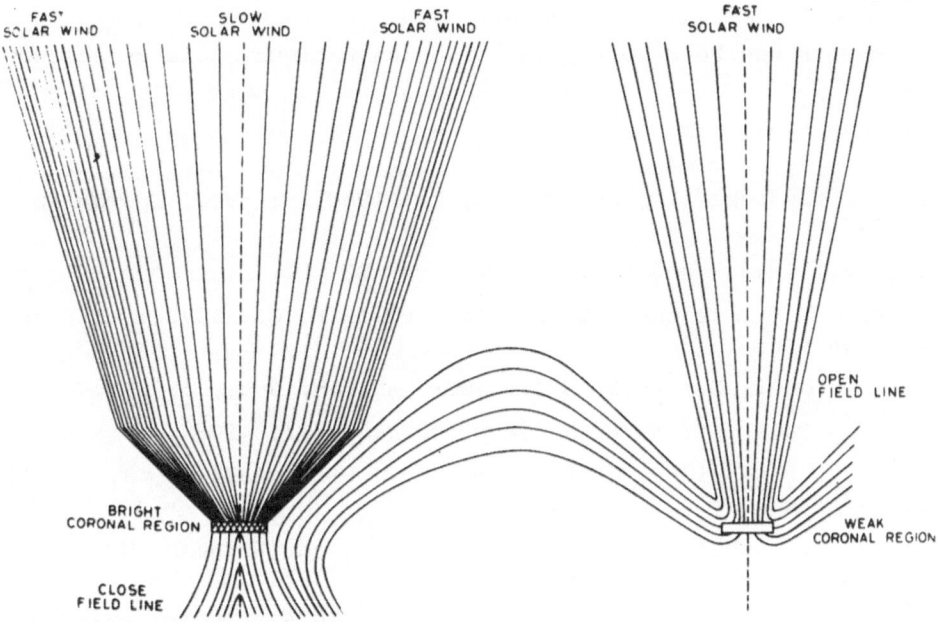

Figure 3.   Structure of proposed solar bean when the activity of solar
            region (right) has died out.

During the high sunspot activity period the dispersed streams from
the neighbouring active solar regions might collide with the solar plasma
stream of decaying region and are likely to get disrupted into small
clouds of space charge due to electrostatic instabilities. As we go
towards the low sunspot activity period such interaction of solar plasma
streams would be small and there is then a likelihood for the existence
of a stable stream. By the time we reach the sunspot minimum period the
frequency of these stable streams decreases since it is proportional to
the number of parent active centers apart from their dependence on the
frequency of disruption. Since frequency of disruption is also
dependent on the number of active centres, one should expect maximum
frequency for these stable streams about one or two years before the
sunspot minimum as expected.

At times it may happen that a particular stream escapes collision
for a considerably longer period. Such a possibility can occur only
near the sunspot minimum. In this case the stream may last as long as
3-4 years and the projection of such a stream on the sun's surface
should coincide with low coronal line regions - the equatorial CH.
Converse of this may not be true. Further, it is also clear from
Figures 1-3 that the CH will, in general, be associated with open field
line structure.

ACKNOWLEDGMENT

This work has been supported by Indian National Science Academy.

REFERENCE

Alfven, H., Lindberg, L., and Mitlid, P.: 1960, *J. Nucl. Energy Pt. C*, *1*, 116.

Allen, C. W.: 1944, *Mon. Not. R.A.S.*, *104*, 13.

Altschuler, M. D., Trotter, D. E. and Orrall, F. Q.: 1972, *Solar Phys.*, *26*, 354.

Bell, B. and Noci, G.: 1973, *Bull. Amer. Astron. Soc.*, *5*, 269.

Bell, B. and Noci, G.: 1976, *J. Geophys. Res.*, *81*, 4508.

Feldman, W. C., Asbridge, J. R., Bame, S. J., and Gosling, J. T.: 1976, *J. Geophys. Res.*, *81*, 5054.

Gosling, J. T., Asbridge, J. R., Bane, S. J., and Feldman, W. C.: 1976, *J. Geophys. Res.*, *81*, 5061.

Hansen, R. T., Hansen, S. F. and Sawyer, C.: 1976, *Planet. Space Sci.*, *24*, 381.

Krieger, A. S., Timothy, A. F. and Roelof, E. C.: 1973, *Solar Phys.*, *29*, 505.

Munro, R. H. and Withbroe, G. L.: 1972, *Astrophys. J.*, *176*, 511.

Neupert, W. M. and Pizzo, V.: 1974, *J. Geophys. Res.*, *79*, 3701.

Nolte, J. T., Krieger, A. S., Timothy, A. F., Gold, R. E., Roelof, E. C., Vaiana, G., Lazarus, A. J., Sullivan, J. D. and McIntosh, P. S.: 1976, *Solar Phys.*, *46*, 303.

Rickett, B. J., Sime, D. G., Sheeley, N. R., Crocket, W. P. and Tousey, R.: 1976, *J. Geophys. Res.*, *81*, 3845.

Sheeley, N. R., Harvey, J. W. and Feldman, W. C.: 1976, *Solar Phys.*, *49*, 271.

Tandon, J. N.: 1958, *Some Problems in Astrophysics*, Thesis, Delhi University.

Tandon, J. N.: 1963, *Indian J. Met. Geophys.*, *14*, 302.

Tandon, J. N.: 1966, *Bull. NAT. Inst. Sci. India*, *No. 33*, 15.

Timothy, A. F., Krieger, A. S., Vaiana, G. S.: 1976, *Solar Phys.*, *42*, 135.

Vaiana, C. S., Krieger, A. S., and Timothy, A. F.: 1973, *Solar Phys.*, *32*, 81.

SOLAR POLAR FIELD REVERSALS AND SECULAR VARIATION OF COSMIC RAY INTENSITY

H. S. Ahluwalia
Department of Physics and Astronomy
The University of New Mexico
Albuquerque, New Mexico 87131

The profile of the well-known 11-year variation of the cosmic ray intensity appears to depend upon the emerging solar polar magnetic field regime in a very characteristic manner. During the solar activity cycle 19, the cosmic ray intensity takes about seven years to recover to its solar activity minimum level. But during the solar activity cycle 20, the recovery takes place in only about two years. It appears that these characteristic recovery modes are obtainable every other solar activity cycle. We are led to suggest two model configurations for the heliosphere. We believe that an "open" heliosphere model applies to solar activity cycles 18 and 20. A "closed" heliosphere model is obtainable during solar activity cycles 17 and 19. Our results are discussed.

1. INTRODUCTION

The variation of the annual mean cosmic ray intensity, at a given site, with the sunspot activity cycle has been known for many years (Forbush, 1966). It is referred to as the eleven-year variation of the cosmic ray intensity in the literature. Perhaps several causes contribute to the observed modulation of the cosmic ray intensity. Over the last three decades many brave hypotheses have been suggested to identify and to describe the contributory causes. These attempts have only helped us in developing some interesting "insights", but have not yet resulted in a satisfactory theory. Somehow the lower energy cosmic rays are prevented from reaching the earth, as the solar activity increases. The "obstruction" gradually disappears with the decline in the solar activity. There is probably a general agreement, however, that a major underlying physical process which contributes to the "obstruction" is the scattering of the cosmic rays by the inhomogenieties.in the interplanetary magnetic field. This idea was first suggested by Morrison (1956) and developed by Parker (1958). The present status of this approach to the problem is given by Forman (1975) and by Quenby (1977).

M. Dryer and E. Tandberg-Hanssen (eds.), Solar and Interplanetary Dynamics, 79-86.

2. ELEVEN YEAR VARIATION OVER FOUR SOLAR ACTIVITY CYCLES

Figure 1(a) shows a plot of the annual mean Zurich sunspot numbers over the period 1937-78. The data are represented by crosses (x). The epochs of the minima and the maxima in the solar activity are indicated by arrows. The data cover four solar activity cycles; namely, 17, 18, 19, and 20.

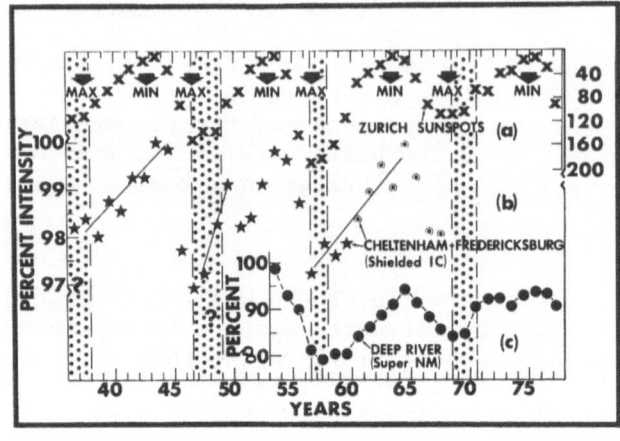

Figure 1(b) depicts the annual mean muon intensity measured with a shielded ion-chamber at Cheltenham-Fredericksburg by Forbush's group (Lange and Forbush, 1957; Beach and Forbush, 1969). The data cover the period 1937-68 and are indicated by: * and ⊙. One can recognize the 11-year varia-

Figure 1(a), (b), (c)

tion of the cosmic ray intensity. Note that the intensity of the muons is the highest when the sunspot activity is the lowest and vice versa. Solid lines are eye-fit estimates to the muon data for the periods 1938-45, 1947-50, and 1957-64. These lines represent recovery of the cosmic ray intensity during solar activity cycles 17, 18, and 19 respectively. One can see that the recovery of the cosmic ray intensity occurs more rapidly in 1947-50 period than during the other two epochs. The significance of this will become clear as you read on!

Figure 1(c) gives a plot of the annual mean hourly counting rate of the neutron monitors at Ottawa for 1954-59 and at Deep River for 1960-78. The counting rates of the two monitors are normalized for the month of December 1959. The neutron monitors respond to lower energy cosmic rays (median primary energy of response is ∿7 GeV) and the shielded ion-chambers respond to higher energy cosmic rays (median primary energy of response is ∿50 GeV). So a comparison of the amplitudes of the 11-year variations recorded by the two detectors gives us an estimate of the energy dependence of the solar modulation processes. I might also mention that the median primary energy of response of the neutron monitor at Deep River is slightly higher than that of the Ottawa neutron monitor.

The epochs of the solar polar field reversals for 1957-58 (Babcock, 1959) and 1969-71 (Howard, 1974) are indicated by the vertical shaded areas. The shaded areas with question marks (?) are our estimate of the earlier epochs of solar polar field reversals during 1947-49 and 1937-38. These estimates are made from the study of the recovery modes of the 11-year variation of the cosmic ray intensity. The basis of these predictions will become clear from the discussion presented in the following section.

I wish to draw the attention of the reader to the following salient features present in the data that are summarized in the Figures 1(a), (b), and (c).

(1) Eleven year variation of the cosmic ray intensity is clearly seen in the data of the neutron monitors as well as that of the shielded ion-chamber. As expected, the amplitude of the effect is larger in the neutron monitor data than in the muon data. The amplitude for the neutron data is nearly 7 times larger than muons in the 1954-58 and about 3 times larger in 1965-68 periods.

(2) Cosmic ray muon intensity recovers completely near each solar activity minimum. This is not true of the neutron monitor data. The latter intensity is still depressed about 6% below 1954 level in 1965 as well as in 1976. A part of this difference is undoubtedly due to the fact that the neutron monitor at Ottawa has a slightly lower median primary energy of response than does the super neutron monitor at Deep River. This does not account for the entire difference, however.

(3) A very remarkable effect manifests itself in the recovery mode of the neutron intensity as well as in the recovery of the muon intensity. Neutrons take about 7 years to recover to a level nearly 6% below the 1954 level in the solar activity cycle 19 and only about 2 years to attain the same level in the solar activity cycle 20. In both cases the recovery follows the epochs of the solar polar field reversals. The latter recovery has been called anomalous by some research workers.

3. MODELS OF THE HELIOSPHERE

Long recovery times characterize recovery by diffusion whereas short recovery times imply almost a direct connection to the source. We also note here that after solar polar field reversal in 1969-71 the solar dipole is oriented in a direction opposite to that of the geomagnetic dipole. The large-scale interplanetary magnetic field, away from the ecliptic plane, corresponds to this new orientation (Smith and Tsurutani, 1978). Keeping these facts in mind, if we now recall the rapid recovery of muon intensity in 1947-49 period compared to the recovery of muon in 1937-45 and 1957-65 periods, we realize that the solar polar field configuration after the reversal in 1947-49 must be the same as that obtainable after the reversal in 1969-71; if indeed the anomalous recovery is to be ascribed to a particular configuration of the solar polar field.

The above line of reasoning demands that the most recent solar polar field reversal must connect the heliosphere to the interstellar medium in such a manner as to make it easier for the interstellar cosmic rays to enter the heliosphere in more or less unrestricted manner. This leads us to suggest an open heliosphere model shown in Figure 2(b). The model is inspired by the work of Alfven (1954), Dungey (1961, 1963), and Levy et al (1964), who have invoked similar models to understand the responses

of the comets and the magnetosphere
to the super-sonic solar wind.

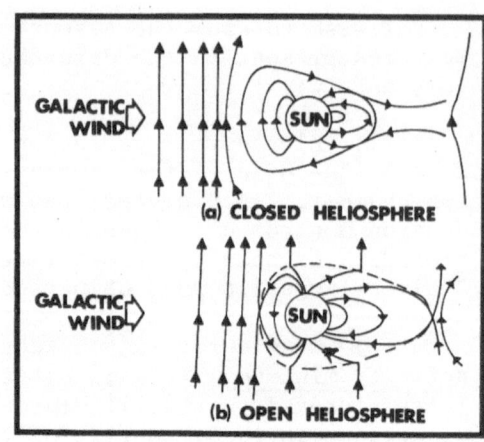

Figure 2(a), (b)

The main rationale for the indicated shape of the heliosphere is
that our solar system is moving with
a velocity of about 20 km/sec, with
respect to the interstellar medium,
in the direction of the constellation
Hercules. Reliable parameters per-
taining to the local interstellar
medium are not available at present.
Sparse data (Heiles, 1976) seem to indicate that magnetosonic mach number
may not exceed unity. So a bow shock
may not form. An important requirement for the models is that there must
exist a significant component of the interstellar magnetic field oriented in the manner shown in Figure 2(b). If this is granted then the
"anomalous" recovery of the cosmic ray intensity, during even solar activity cycles, may be readily understood. The given magnetic configuration of the heliosphere enables the interstellar cosmic rays to reach
locations at fairly low heliolatitudes. This inference also provides a
very natural explanation for the results reported by McKibben et al
(1979). They find large latitudinal gradients for the anomalous helium
as well as for the protons. Also our inference provides a strong support for the assumption made by Fisk et al (1974) that the anomalous
components are of interstellar origin. In this scheme of things gradient
and curvature drifts (Jokipii et al, 1977) play only a secondary role in
the transport of the interstellar cosmic rays.

Quite naturally one expects that during odd solar activity cycles a
closed heliosphere configuration, shown in Figure 2(a), must be obtainable in the nature. Diffusion plays a major role in the transport of
the interstellar cosmic rays to earth. So we expect recovery for the
11-year variation of the cosmic ray intensity to occur over a longer
period, as is indeed observed.

The fact that the neutron intensity in 1965 and 1976 is below 1954
level might be due to the fact that there is some residual modulation of
the lower energy cosmic rays still present during these years. We note
that the solar activity during 1954 was extremely low. The point is that
if our ideas are correct then we expect the cosmic ray particle density
gradient to undergo characteristic time variations over a Hale cycle.
We intend to examine this question further in our study of the observed
long-term changes in the parameters of the solar anisotropy of cosmic
rays (Ahluwalia, 1977a, b).

It is quite interesting to note that heliosphere is not really
"closed" at any time due to the presence of the neutral points on the

heliopause. The observed modulation is therefore likely to be different in different regions of the heliosphere. For example, one would expect to observe very dynamic modulation of cosmic rays at high heliolatitudes. We have to wait for the solar polar missions to make this discovery.

These models may also have some implications for the solar wind and the origin of the coronal holes. During odd solar activity cycles the closed heliosphere configuration permits an almost radial expansion for the solar wind. On the other hand, non-radial expansion of the solar wind is perhaps to be expected during the open heliosphere regime, if the polar wind is deflected towards lower heliolatitudes, at the heliopause. More work is clearly necessary to examine this question in detail. One may also note here that the connection of the solar polar field lines with the interstellar magnetic field might encourage the formation of the polar coronal holes. Moreover, the charged particles precipitating in the solar polar regions might constitute an additional nontrivial source of energy needed to heat the solar wind in the outer solar corona (Zirker, 1977). This question also needs to be examined in more detail. We expect to do this in the future.

ACKNOWLEDGEMENTS

My travel to this symposium was supported by the National Science Foundation Grant # ATM78-10727.

REFERENCES

Ahluwalia, H.S.: 1977a, 15th Intern. Cosmic Ray Conf., Plovdiv. Conference Papers (Bulgarian Academy of Sciences), 11, 298.
Ahluwalia, H.S.: 1977b, 15th Intern. Cosmic Ray Conf., Plovdiv. Conference Papers (Bulgarian Academy of Sciences), 11, 304.
Alfven, H.: 1957, Tellus, 9, 92.
Babcock, H.D., 1959, Astrophys. J., 130, 364.
Beach, L., and Forbush, S.E.: 1969, Cosmic-ray Results, Carnegie Inst. Wash., Publ. 175, Vol. 22.
Dungey, J.W.: 1961, Phys. Rev. Letters, 6, 47.
Dungey, J.W.: 1963, Geophysics: The Earth's Environment. Lectures given at Les Houches during 1962 Summer School of Theoretical Physics of the Univ. of Grenoble, Gordon and Breach Science Publishers, New York.
Fisk, L.A., Koslovsky, B., and Ramaty, R.: 1974, Astrophys. J., 190, L39.
Forbush, S.E.: 1967, J. Geophys. Res., 72, 4937.
Forbush, S.E.: 1966, in S. Flugge (ed), Handbuch der Physik, Springer-Verlag, New York, 49/1, 159.
Forman, M.A.: 1975, Rapporteur Paper at the 14th Intern. Cosmic Ray Conf., Munich, Conference Papers (Max-Planck-Institut fur Estraterrestrische Physik), 11, 3820.
Howard, R.: 1974, Solar Phys., 38, 283.

Heiles, C.: 1976, Ann. Rev. Astron. Astrophys., 14, 1.

Jokippi, J.R., Levy, E.H., and Hubbard, W.B.: 1977, Astrophys. J., 213, 861.

Knapp, G.R., Tremaine, S.D., and Gunn, J.E.: 1978, Astron. J., 83, 1585.

Lange, I., and Forbush, S.E.: 1957, Cosmic-ray Results, Carnegie Inst. Wash., Publ. 175, Vol. 20.

Levy, R.H., Petschek, H.E., and Siscoe, G.L.: 1964, A.I.A.A. Jour., 2, 2065.

McKibben, R.B., Pyle, K.R., and Simpson, J.A.: 1979, Astrophys. J., 277, L147.

Morrison, P.: 1956, Phys. Rev., 101, 1397.

Nagashima, K.: 1977, Rapporteur paper at the 15th Intern. Cosmic Ray Conf., Plovdiv. Conference papers (Bulgarian Academy of Sciences), 10, 380.

Parker, E.N.: 1958, Phys. Rev., 110, 1445.

Quenby, J.J.: 1977, Rapporteur paper at the 15th Intern. Cosmic Ray Conf., Plovdiv, Conference Papers (Bulgarian Academy of Sciences), 10, 364.

Schatten, K.H., and Wilcox, J.M.: 1969, J. Geophys. Res., 74, 4157.

Smith, E.J., and Tsurutani, B.T.: 1978, Astrophys. J., 83, 717.

Wagner, W.J.: 1976, Astrophys. J., 206, 583.

Zirker, J.B.: 1977, Rev. Geophys. and Space Phys., 15, 257.

DISCUSSION

Datlowe: The magnetic configuration shown in Figure 2 is of a type, like the earth, in which the magnetic field dominates the plasma pressure. But the interplanetary situation has the solar wind plasma dominating the magnetic field. Is this figure a useful model for understanding cosmic ray propagation?

Ahluwalia: Let me answer the first part of your question first. I wish to emphasize that I do not suggest that the interplanetary magnetic field dominates the solar wind, inside the heliosphere. Far from it! Figures 2(a), (b) are drawn for the case of a non-rotating sun. In the real world the field lines would be bent into Archimedian spirals due to the rotation of the sun. However, in the open heliosphere model, the field at the heliopause may be strong enough to deflect the polar solar wind towards lower helioaltitudes. Since the magnetic field is frozen into the wind, it would follow the wind. In this manner we are able to understand the result of Wagner (1975) that much of the interplanetary magnetic field sampled at earth, during 1972-73 period, originates at high heliolatitudes. This inference is probably valid for much of the even solar activity cycles when an open heliosphere regime is obtainable.

Now let me attempt to answer the second part of your question. As I said in my talk the models do explain the recovery modes of the 11-year cosmic ray intensity variation, during odd and even solar activity cycles. They also explain the presence of the anomalous components of the energetic particles measured by McKibben et al. (1979), during solar activity cycle 20. Schatten and Wilcox (1969) have invoked a similar idea, but with a spherical heliosphere, to explain the 20-year

wave in the solar diurnal variation of cosmic rays reported by Forbush
(1967). Schatten-Wilcox model is invoked by Nagashima (1977) to
explain a variety of energetic particle measurements reported in the
literature. Our models explain everything that Schatten-Wilcox model
does. But they are more general. They imply that the traditional
concept of a spherical symmetry, so often invoked by the theoreticians
to explain the observed solar modulations, is at best obtainable only
during odd solar activity cycles. The models also imply that one must
drastically revise the concept of a modulation region surrounding the
sun. During even cycles the modulation is much more dynamic. I am
satisfied that we are on the right track. However, one can see that the
models are in a skeletal form. In the future we have to build-in more
details by confronting the models with a variety of observations. At
this point in time I am quite optimistic!

Stix: High altitude solar prominences and faculae also indicate
polar field reversals. Using them one could thus infer the epochs of
the two earlier reversals of your Figure 2.
   Ahluwalia: I thank you for your suggestion. My motive in coming to
this meeting was to invite comments from my solar colleagues that would
help me refine my hypothesis. I will certainly compare the two results.

Newkirk: Your model appears to require that you identify the increase of galactic cosmic ray flux after each sunspot maximum as the
result of diffusion in the heliosphere with a characteristic time of
about 5 years. This would require an unusually low diffusion
coefficient. Would it not be simpler to view the recovery of galactic
cosmic ray flux as the response to the evolution of the heliosphere
over the $\sim$ 6 years of declining activity?
   Ahluwalia: Gordon, the models that I have attempted to describe are
much more dynamic than you seem to think. We are just now beginning
to appreciate how solar modulation of cosmic rays comes about. Probably
the following causes contribute to the observed modulation:
   (a) Solar active regions. Details of how they contribute are still
        obscure.
   (b) Fast streams and stream interaction regions.
   (c) Solar flare initiated shocks in the interplanetary medium.
   (d) Large scale organization of the interplanetary magnetic field.
   (e) State of magnetic connection of the heliosphere with the inter-
        stellar medium.
I must emphasize that it is not yet clear what fraction of the observed
modulation is contributed by each of the above causes. We can not rule
out the possibility that there might exist another set of contributing
mechanisms which are unidentified yet. The point is that during the
period when (a), (b) and (c) do not contribute, the cosmic ray
intensity at earth must attempt to rise to the level obtainable in the
interstellar medium. Under closed-heliosphere regime this is brought
about, primarily, by cosmic ray entry into the heliosphere through
neutral points. Under open-heliosphere regime, rise in intensity level
is brought about by 'unrestricted' flow of interstellar cosmic rays into
the heliosphere. Since all solar disturbances propagate radially out-

wards, <u>recovery is much more rapid under the open-heliosphere regime</u>,
since the interstellar cosmic rays have almost direct access to the
'depletion' region surrounding the earth. This is how the 'anomalous
recovery' of the eleven-year variation of cosmic ray intensity comes
about in EVEN solar activity cycles. Contributions from (a), (b) and
(c) described above are controlled by the level of activity in each
solar cycle but recovery mode at earth depends upon which of the two
heliospheric regimes is obtainable, at a given point in time.

Levine: For the recovery of the mean counting rates beginning about
1947, you draw a line through the data indicating a much shorter time
scale than for the even-numbered solar cycles. Because your conclusions
are strongly dependent on this faster recovery, can you explain why
you ignored the points after 1950, which would give a slope the same
as for the other cycles?

Ahluwalia: The line representing the recovery of the 11-year cosmic
ray variation, during solar activity cycle 18, is obtained by joining
three consecutive data points
after cosmic ray intensity
minimum in 1947. As you point
out there is a depression in
the cosmic ray intensity for
the period 1951-52. This is
probably due to the Forbush
decreases. They occur all the
time. They are seen more
clearly in the monthly averages
of the data. Figure 3 shows a
plot of the monthly mean
intensity of cosmic rays
recorded by the neutron
monitors at Huancayo and Chimax,
for the solar activity cycles 18,
and 19. The reported epochs of
the solar polar field reversals
are indicated by vertical shaded
areas.   The features discussed in reference to the neutron
monitor data at Deep River (Figure 2(c)) are also seen in these data.
But you also see sharp, temporary depressions of the intensity due
to the Forbush decreases, during both cycles. Some of these depressions
are quite large. <u>But the point is that when the Forbush event is over,</u>
<u>the long-term recovery is resumed uninterrupted.</u>   This is true of muon
data also for 1952-54 period. The eye-fit line joining these data points
is <u>parallel</u> to the line that I have drawn for the recovery during
1947-1950.

I take this opportunity to point out that the large fluctuations of
the cosmic ray intensity observed after 1971 is referred to as a "mini"
solar activity cycle. Fast streams play a dominant role in producing
the observed modulation. Probably a similar situation is available in
1951-52 period. We intend to investigate this further.

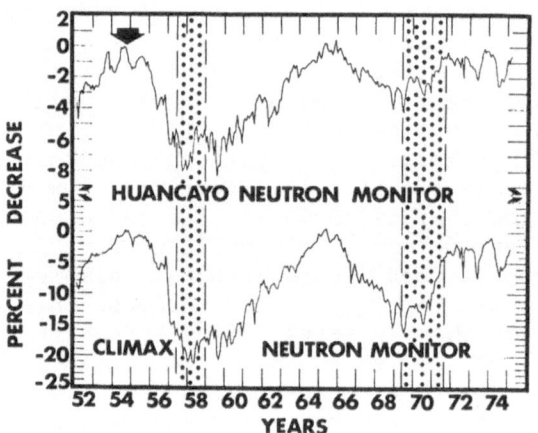

Figure 3

THE CORONAL RESPONSES TO THE LARGE-SCALE AND LONG-TERM PHENOMENA
OF THE LOWER LAYERS OF THE SUN

J. Sýkora
Astronomical Institute of the Slovak Academy of Sciences,
Skalnaté Pleso, 059 60 Tatranská Lomnica, Czechoslovakia

Based on the assumption, generally accepted over the past decade,
that all the forms of solar and interplanetary activity are responses to the magnetic fields generated initially in the subphotosphere,
some characteristics of the large-scale and long-term behaviour of
the solar corona during the last three solar cycles are presented.

Most part of our report is based on the statistical and graphical processing of the abundant observational material pertinent to
530.3 nm Fe XIV coronal line. Apart from these, some white-light corona and EUV observations were used to demonstrate the long-term
phenomena. Namely, the longitudinal and latitudinal distributions
of the corona brightness are shown and, in this connection, active
longitudes and coronal rotation are discussed. Some solar cycle properties and N-S asymmetry are compared with similar features found
for sunspots and solar flares. Large-scale characteristics and long-
term evolution of coronal holes are briefly summarized.

1. INTRODUCTION

Although our symposium is focused primarily on short-term phenomena on the Sun and their effects in the interplanetary medium,
the Scientific Organizing Committee felt that it was also necessary
to describe the recent understanding of the "background" solar and
interplanetary medium characteristics. We will attempt to do this
for the solar corona.

It is certainly correct to follow idea, generally accepted over
the past decade, that all the forms of solar and interplanetary activity are consequences of magnetic fields, generated initially in
the subphotosphere, and that it would be difficult and misleading
to study either short-lived or long-term coronal phenomena apart
from their connection with photospheric magnetic fields. Several
fundamental introductory papers giving reasons for this point of
view were presented at the last solar symposia (e.g., Newkirk, 1971;

87

M. Dryer and E. Tandberg-Hanssen (eds.), Solar and Interplanetary Dynamics, 87-104.
Copyright © 1980 by the IAU.

Wilcox, 1971; Altschuler, 1974; Pneuman, 1974; Bumba, 1976). Therefore, taking into account the number of papers referred to in the
mentioned summarizing papers, speaking of coronal responses to long-
term phenomena on the Sun means, above all, to discuss responses
to long-term and large-scale phenomena of the photospheric magnetic
fields. Of course, in analysing the distribution of coronal brightness through the solar cycle, N-S asymmetry and coronal rotation we
will also frequently make comparisons with the distribution of sunspots and solar flares, the association of which with magnetic fields
is indisputable.

It is clear that responses both to short- and long-term phenomena do not stop in the solar corona. They are transported through
and certainly transformed by this layer and proceed into the interplanetary medium to the upper and lower layers of the Earth's atmosphere and beyond. Viewed in this way, the study of the large-scale
behaviour of the solar corona should prove a good tool to understand
disturbances in terrestrial magnetism, the ionosphere and possibly
the influences on life in the troposphere. But this topic is certainly closer to the speaker who will follow me.

During the last few years the observational material enabling
us to study the individual physical processes in the corona, on the
one hand, and statistical large-scale manifestations of coronal activity, on the other hand, has markedly increased. This is not only
as a result of space research, but also thanks to the systematic,
routine ground-based observations.

Our report is primarily based on the statistical and graphical
analysis of the abundant observational material obtained by the network of corona stations, headed by Pic du Midi, which observe the
monochromatic corona, particularly in the green (530.3 nm) and red
(637.4 nm) lines. A detailed description of the data used and of the
process of their homogenization is given in (Sýkora, 1971a, 1971b).
These data refer to the height of about 43 500 km above the photosphere and their resolution on the Sun's surface is quite small -
- one day in heliographic longitude and 5° in heliographic latitude
- but for studying the large-scale phenomena this even seems to be
an advantage. The green corona data are used in our study because
they are fairly complete over sufficiently long period of time and
as we shall see further on, they allow us to draw some conclusions
on the corona as a whole (i.e. on the corona observed in the X-ray,
optical and radio intervals).

We should also mention what is meant by large-scale and long-
-term phenomena in this paper. Large-scale means that the dimensions
of the investigated features are greater than the scale of one active region and long-term refers to features with a duration of one
solar rotation to one solar cycle, nevertheless, the last three solar
cycles will be compared.

2. LONGITUDINAL DISTRIBUTION OF THE GREEN CORONAL EMISSION

Most of the solar magnetic field features we are interested in association with long-term coronal responses were obtained from an all-round study of the synoptic charts constructed from daily low-resolution magnetograms of the Mount Wilson Obsevatory. The charts were partially published as the "Atlas of Solar Magnetic Fields 1959-1966" by Howard et al. (1967). The analysis of the separate polarities allowed to Bumba and Howard (1969) to postulate certain new patterns in the large-scale field distribution - sections, rows and streams. These are features, the width of which is some tens of degrees in latitude, and they clearly show that the activity distribution on the Sun's disk is not regular, but, on the contrary, during long periods - as long as several years - the activity in certain longitudinal intervals is higher in comparison with other heliographic longitudes. A very well pronounced property of Bumba's and Howard's formations is their recurrence period which is close to 27 days, in higher latitudes 28-29 days (see also Bumba, 1976).

Similar regularities, in most studies called "active longitudes", have been found by many authors for different types of solar phenomena. But the understanding and definition of active longitudes differed from author to author and, so far, only a few attempts have been made to identify the "active longitudes" of the various phenomena.

Also we can comment only qualitative similarity (response) of the green corona longitudinal distribution to that of the magnetic field. Examples of four times repeated 27-day sequences of daily data on green corona brightness, prepared in a way similar to that described in (Sýkora, 1971b), are shown in Figure 1. Originally, the data from the period 1947-1976 were processed separately for every 5° in the interval of ±60° of solar latitude. In this figure only data for every 10° of the northern hemisphere are presented. Variations of activity with the solar cycle were subtracted. It can be clearly seen that the activity in the last three solar cycles is not distributed randomly, but on the contrary, this distribution holds almost all the characters of the active longitudes as they were postulated in the past and reminds many properties (width, duration, inclination) found by Bumba and Howard for the background magnetic fields and perhaps even more so of the sector structure of the interplanetary magnetic field as it is seen for example in Figure 10 of Bumba's paper in IAU Symp. No.71 (Bumba, 1976). The different inclinations seen in our Figure 1 are mainly connected with coronal rotation and differential rotation and we shall discuss them in Section 4.

From the point of view of solar-terrestrial relations, in recent years the coronal holes and, generally speaking, the low-brightness coronal regions seem to be very important. It is clear, that Figure 1 shows the distribution of these features just as well, if the bright area of the figure is analysed.

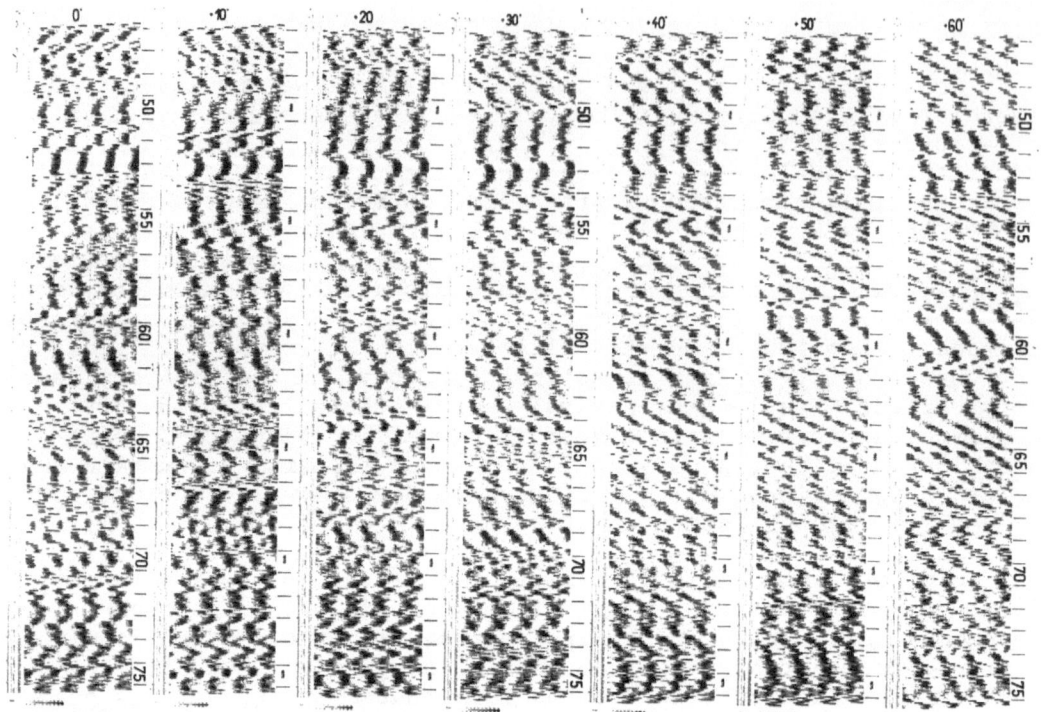

Figure 1. Four times repeated green corona brightness observed in
discrete latitudes is visualized in succession of Bartels rotations
over the last two and a half solar cycles.

    The green corona synoptic tables we prepared for the period
1947-1970 have been extensively used by Antonucci and her co-workers
for an all-round study of the large-scale distribution of the coronal emission. She found (Antonucci, 1974) that enhancements of the
green corona show an organized pattern within solar magnetic sectors.
Antonucci and Svalgaard (1976) found correlation peaks amplitude in
excess of 0.1 between corresponding northern and southern latitude
zones. They conclude: "The very fact that any correlation exists
between corresponding latitude zones in opposite hemispheres at a
lag significantly different from zero for these 24 yr time series,
strongly suggest that the green line corona includes a component
which is organized on a very large scale." The high correlation
between northern and southern high-latitude emissions at a 15-day
time lag is explained as a feature of the two-sector solar magnetic
structure, while four sectors are associated with 6- and 24-day peaks.

    As an interesting response to the long-term distribution of the
photospheric magnetic fields some years ago we have found (Bumba and
Sýkora, 1973; Bumba and Sýkora, 1974) that the maxima of the coronal
emission coincide in position with the negative (at least during
the declining phase of the 19th cycle and during the increasing phase

of the recent cycle of solar activity) and the minima with the positive polarity of the photospheric magnetic field. Stenflo's (1972)
assumption that the green line intensity should not be related so
much to the polarity pattern of the field, but more directly to the
field strength is based on the comparison of the very low resolution
butterfly diagram of the green line corona for the period 1959-1970
with a similar diagram of magnetic field.

3. LATITUDINAL DISTRIBUTION OF THE GREEN CORONAL EMISSION

The latitude-time variation of corona brightness can be illustrated in various ways. Charts of isophotes were used, for example,
by Trellis (1957), Waldmeier (1957a), Stenflo (1972), etc. All these
diagrams seem to be rather smoothed, revealing only little of the
details. In Figure 2 we present similar diagrams for the last three
solar cycles. The coronal data, as published in the Quarterly Bul-
letin on Solar Activity, are expressed in physical units, i.e. in
millionths of the energy radiated from the centre of the Sun's disk
in 0.1 nm strip of the spectrum near the 530.3 nm coronal emission

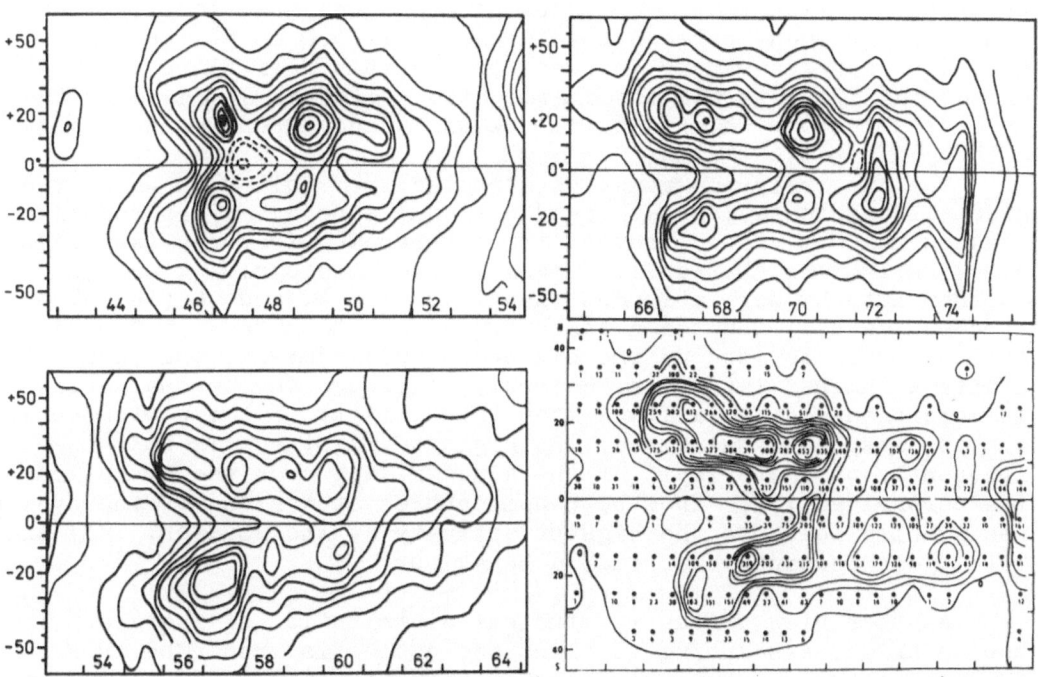

Figure 2. Latitude distribution of the intensity of the green line
(λ 530.3 nm) corona in the last three solar cycles. The lowest contour level is 15 absolute coronal intensity units and the contours
are given with step of 10 units. In the lower right part a similar
distribution of the flare occurrence is shown.

line. The homogenized data of various corona stations (Sýkora, 1971b)
were averaged over the intervals of six Bartels rotations for each
5° of solar latitude. We omitted the data for the polar regions. In
the lower right-hand part of the figure the diagram of occurrence of
flares in the 20th cycle (Knoška et al., 1978) is shown for comparison with the coronal data.

The following characteristics can be seen in Figure 2: 1) There
have place remarkable differences - as large as one or one and a half
year - in the beginnings of the cycle in both hemispheres. 2) The
fact that activity during cycle consists of several large impulses,
on an average two years apart, seems to be realistic. In the previous
cycle the impulses in 1967, 1970, 1972 and 1974 were well expressed.
The cores of impulses gradually shift in latitude from about 25° at
the beginning to about 10° at the end of the cycle. There is no evident correlation between their time occurrence in both hemispheres.
3) Because the intensity of the green line increases with coronal
temperature, Stenflo (1972) found that the latitude distribution of
the green line intensity follows closely that of sunspots, faculae
and strength of magnetic fields. On the right-hand side of Figure 2
a comparison can be made with flare occurrence for the 20th cycle.
We should mention that the flares were not weighted for importance.
The good agreement between both diagrams is broken at the beginning
of 1971, when the strong decrease of the number of flares till the
end of the cycle is not associated with a similar decrease in the
coronal emission. This fact will be discussed in analysing Figure 6.

The butterfly diagrams, constructed from points in which the
green line intensities, everaged over the single Bartels rotations,
reached maximum values, reveal two interesting features (see Figure
3). Firstly, the "butterfly wings" in the 19th cycle extend 10°
further into higher latitudes than in cycles 20 and 18 (see also Figure 9). We have adopted a tentative explanation that in the higher
cycles (characterized by the higher sunspot numbers) the activity
starts at higher latitudes. Secondly, although the coronal activity
is more extended in latitude than the activity of the photospheric
and chromospheric phenomena, the highest brightness is only in-
frequently characteristic of the equator. Not even the summation of
the coronal activity from both hemispheres at the end of the cycles
necessarily leads to the highest activity at the equator. Only the
20th cycle perhaps shows such a behaviour.

Another view of the latitudinal distribution of the green line
corona brightness during the last two cycles is presented in Figure
4. Here, the distribution of the average radiance at 5° increments
of latitude, is shown for a succession of Bartels six-rotations. During the solar minima of 1954, 1964 and 1976 (curves Nos. 1, 1 and
28) the brightness is practically uniform over all latitudes between
±60° and reaches 10-15 absolute coronal units. Responses to the running solar cycles are as follows: At the beginning the activity
starts (namely in the 19th cycle) at latitudes of 30-35°, then it

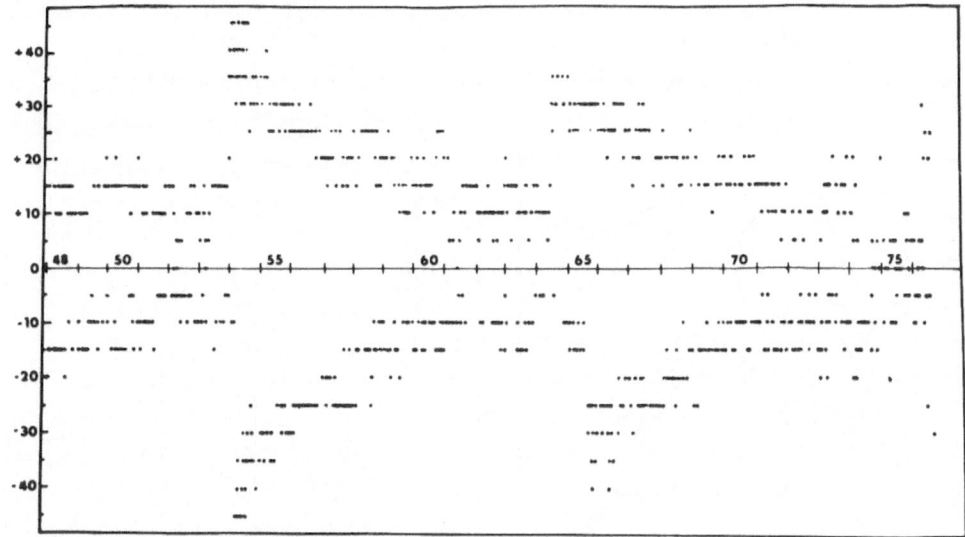

Figure 3. Butterfly diagrams of the green corona intensity for the last two and a half solar cycles.

grows rapidly and forms two very well expressed peaks at about ±20- -25° in the cycles maxima. The brightness in this period is about ten times higher than that in the solar minima. This fact is in severe contradiction with the maximum-minimum ratio in brightness of the K-corona as published by Hansen et al. (1969a) in their Table II (data from four sources) where this ratio is found to be 2.0 - 2.5. The disagreement certainly indicates a much higher sensitivity of the green line intensity to complexity of magnetic fields on the solar disk. Towards the cycle minima the activity regularly migrates to the equator, to latitudes of ±5-10° and sometimes (as in 20th cycle) the mutual overlapping of the activity from both hemispheres results in the impression that the brightness maximum is located at the equator.

Hansen et al. (1969a) state that the arithmetic averages, as shown in our Figure 4, tend to obscure the fact that the coronal features occur in discrete latitude zones. In Figure 5, adapted from their paper on the K-corona, it is seen clearly that the statistical standard deviation of the intensity from six-months mean, except of brightness maxima well associated with sunspot and plage regions, reveals the high-latitude or polar zone of activity at 60-70° as they were perhaps firstly pointed out by Trellis (1957). Hansen et al. show that these high-latitude zones steadily migrated to the poles over period 1964-1967 which fact also is known from the movement of prominences and filaments as well as from theory and measurements of solar magnetic fields. We did not attempt to find these polar zones of activity in our green-line data because they are limited by ±60° in latitude.

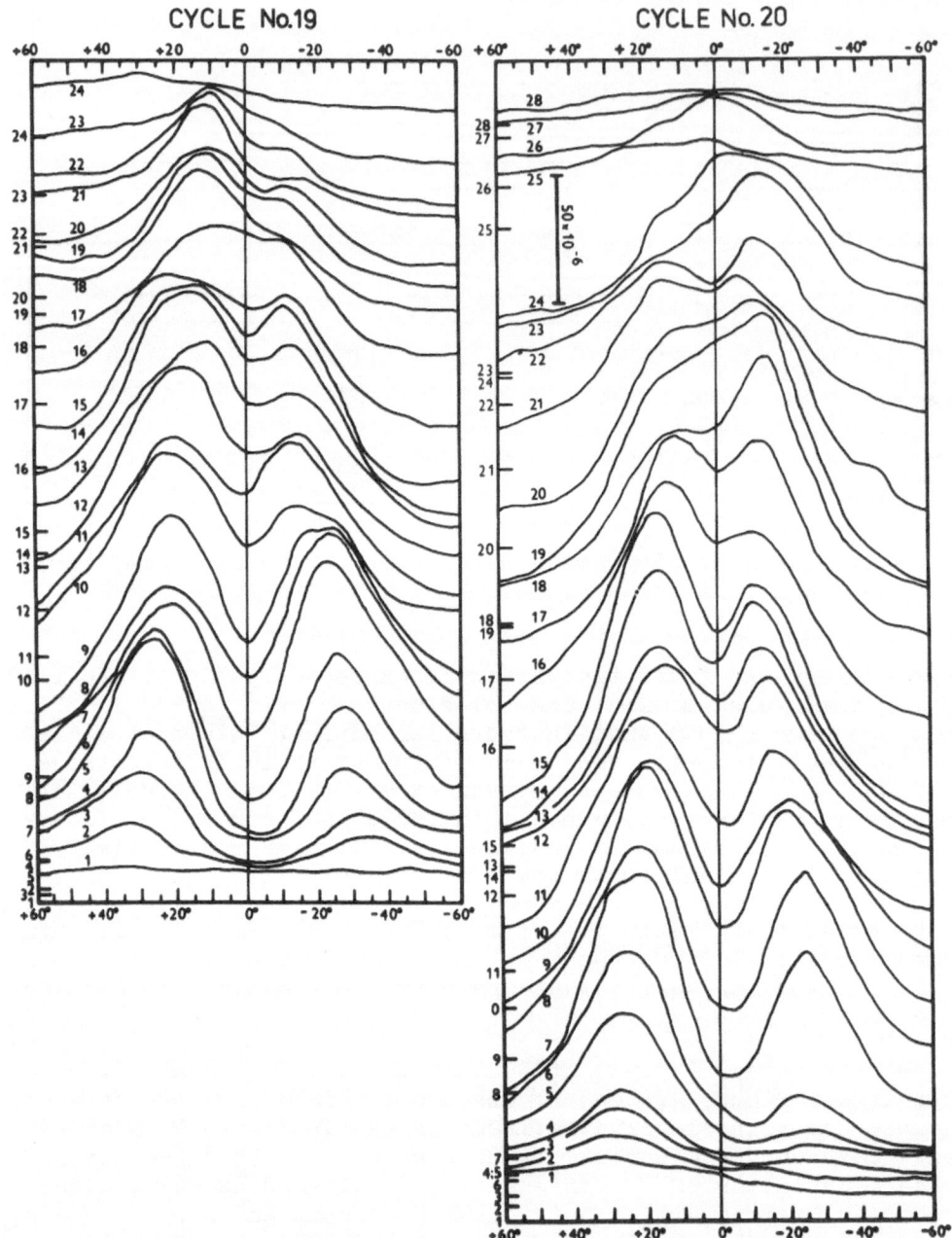

Figure 4. Average brightness distribution of the green corona for
Bartels six-rotations (denoted by numbers 1 to 24 and 1 to 28).
Number 1 in the cycle No.19 denotes six-rotation 1652 (beginning on
Feb. 25, 1954) - 1657 (ending on Aug. 5, 1954), and number 1 in the
cycle No.20 denotes six-rotation 1790 (beginning on May 9, 1964) -
- 1795 (ending on Oct. 17, 1964). Zero points of the curves and the
scale in coronal units are given.

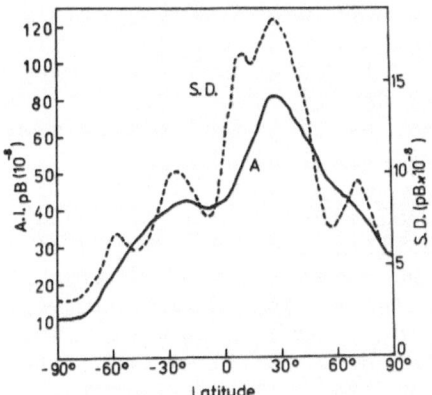

Figure 5. Comparison of the latitude distribution of average K-corona brightness (A) and standard deviation (S.D.) from this mean for the period July – December 1966. – 1.125 R_o (from Hansen et al., 1969).

4. CORONAL ROTATION

     Much work has been done in the study of the coronal rotation and its latitude-dependence. The results of the past decade seem to be convincing enough with regard to the variety of the analysed data and with respect to quite good agreement among them. Determination of rotation rates is possible because of the existence of the regions (enhanced or quiet, bright or dark) which reappear for several rotations with sufficient stability.

     Hansen et al. (1969b) applied an autocorrelation analysis to the limb observations of the white-light corona during 1964-1967. At low latitudes the corona was found to rotate at the same rate as the sunspots, but at higher latitudes it was consistently faster than the underlying photosphere - at the same time at latitudes 60--65° it amounts to 12.8°/day (period of 28.13 days). There were differences as large as 3-4% in the rate at specific latitudes from year to year and between the two hemispheres. No obvious relation of these changes to increasing solar activity in the studied period has been found.

     To our knowledge the paper of Hansen et al. (1969b) is the only one in which the coronal rotation was studied at different height levels using the same observational data. In 1967 a nearly constant rotation was found for heights ranging from 1.125 to 2.0 R_O. For 1966 there was a more complicated pattern of height dependence, with the rate generally decreasing with height at low latitudes and increasing at high latitudes.

     A graphical study of the green-corona synoptic tables, divided into six latitude zones 20° in width and covering the period 1947-

1968 showed only little latitude dependence of the rotation rate
(Sýkora, 1971b). These rates were derived from the recurrence of
the long-lived large-scale active features and the average values
of the rotation period are 27.80 days for the +7.5° zone and 28.23
days for the +47.5° zone. If one was to judge just by the slopes of
the strips of activity seen in Figure 1 of the present paper, one
should notice the rather complicated pattern of these slopes. If we
neglect the changes of the coronal structures, especially those occuring in the longitudinal direction, then the slopes of the strips
of activity are indicators of the rotation rate at given time and
latitude. Generally, two tendencies of the slopes can be seen. At
lower latitudes (from +10° to about +30°) almost vertical formations,
which are only occasionally interrupted by formations characterized
by sufficiently larger slopes, dominate. The large slopes are more
pronounced in higher latitude zones (+40° to +60°). But at these
latitudes the vertical formations are still visible and perhaps they
are indicators of "almost rigid" coronal rotation, discussed below.

This very qualitative picture of coronal rotation can be well
fitted to Bumba's results on regularities in the large-scale distribution of the solar magnetic fields (Bumba, 1976), especially to
the features, the recurrence period of which is 27 and 28-29 days.

The green corona synoptic tables (Sýkora, 1975), mentioned above,
were later processed by autocorrelation techniques, and Antonucci
and Svalgaard (1974) and Antonucci and Dodero (1977) have published
some interesting results. They deduce that the short-lived coronal
features (observed in green line 530.3 nm of Fe XIV), associated
with the activity, rotate differentially, while features with a
lifetime of the order of about one year or more and characteristic
preferably of the years of declining solar activity do not show
differential rotation, but instead all rotate in ≈27 days, independent of latitude. These results were also extended to the decreasing
phase of the last cycle (Antonucci and Dodero, 1977, 1979).

An effect, similar to that just mentioned, was found earlier
in an analysis of the photospheric magnetic fields by Bumba and Howard (1969) and also by Wilcox et al. (1970). The rotation curve of
the photospheric magnetic field polarity, computed for period 1959-
-1970 by Stenflo (1974) is in good agreement with the average corona
rotation curve (for comparison see Figure 6 in Antonucci and Dodero,
1977). In addition, Stenflo (1977) has found that the foot points
of the large-scale rigidly rotating pattern of the corona are very
likely associated with the long-lived component of the photospheric
magnetic field. Thus the rotational behaviour of the photospheric
magnetic fields and corona are the same, averaging over a solar
cycle and, hence, Antonucci and Dodero conclude, that "at present,
on analogy of the corona, a variation of the degree of differential
rotation of large-scale photospheric magnetic fields through a solar
cycle, can not be excluded."

Another noticable result, presented in the lower part of Figure
4 in the paper of Antonucci and Dodero, is very good coincidence of
the rotational characteristics of the EUV (Fe XV, 28.4 nm) coronal
holes detected by OSO-7 spectroheliograph from May 1972 to October
1973 and analysed by Wagner (1975) with a quiet corona rotation curve. Hence, the rotation characteristics of the corona and coronal
holes are very probably the same during low activity periods, and
low emission regions probably coincide with coronal holes (see also
Letfus et al., this Symposium).

The results on coronal rotation discussed so far have been criticized by Timothy et al. (1975) in the sense that they may heavily
be dependent on statistics in which the changes in the coronal structure, particularly those occuring in the longitudinal direction, introduce uncertainties into the analysis of the individual features.
It is possible to agree with this opinion, but anyway the result
obtained by Timothy et al. is in quite good agreement with the criticized results. They studied the rotation rate of the elongated coronal hole CH1 of the Skylab mission. During five successive rotations an "almost rigid" rotation of this hole has been observed.

Figure 7 in Bohlin (1977) clearly shows that also the rotation
of other holes of the Skylab period (except of unusual case of CH5,
which was more likely due to intrinsic evolution rather than rotation) confirm the results of Timothy et al. But because features 2^X
and 4 are too low in latitude Bohlin calls for further observations
to prove the validity of the "almost rigid" rotation equation
$\omega \approx (13.3 \pm 0.1) - 0.4 \sin^2 \varphi.$

In summarizing the coronal rotation data one should take into
account that the results were obtained from the analysis of quite
different data - K-corona, emission corona and coronal holes. The
results are sufficiently similar to conclude that at least long-lived
large-scale coronal features display a substantially smaller differential rotation than derived from a number of photospheric and
chromospheric phenomena. The fact that the results are consistent
with the hypothesis on a rigidly rotating "subsurface source", mentioned in Bumba and Howard (1969), and in some other papers on large-
scale solar magnetic field structure and "active longitudes", is
very important.

Of course, the situation is not so simple if we go into details.
It is sufficient to look once more at Figure 1 to realize how complicated the rotation rate can be within the cycle and in dependence
on latitude. Perhaps it depends on the actual distribution and phase
of the development of the large-scale magnetic structures with a
lifetime of about 10-20 rotations. Sometimes we see symptoms of rigid
rotation, at other times we can not be sure about this. The physical
causes of such behaviour are quite obscured.

5. SOLAR CYCLE CHARACTERISTICS

Perhaps we can start this Section by presenting the runs of two global characteristics - the sunspot number and the 530.3 nm coronal line intensity averaged over ±60° and six Bartels rotations (see Figure 6). The correlation between these two phenomena through the solar cycles does not seem very good. First of all, the relatively equal heights of the coronal cycles are accompanied by the sizable differences in the heights of sunspots cycles. One can obtain the impression that during the cycle the corona reaches some degree of saturation (in the sense of particles), without any further possibility to increase its brightness (19th cycle). The conspicuous decrease of the number of the sunspots after 1970 is accompanied by only a small decrease of the coronal intensity. At the same time two sharp coronal maxima in 1972 and 1974 correspond to the presence of the well-known active complexes in these years which also produced huge proton flares. At the same time, complicated, closed magnetic structures, characteristic of the regions with large flares, serve as a trap for the coronal plasma particles, a long-term enhancement of plasma concentration takes place and this results in long-term increase of coronal brightness in these regions. The number of sunspots does not seem to be important. Similar facts were analysed and quantitatively well expressed by Xanthakis (1969). He found that, for the long-time variation of the coronal intensity, the area of sunspots and faculae and the number of proton flares are far more important characteristics than the relative sunspot number. Frequency of occurrence of fast green corona events during the 20th cycle is also found to be rather higher than it was in cycle 19 (Demastus et al., 1973).

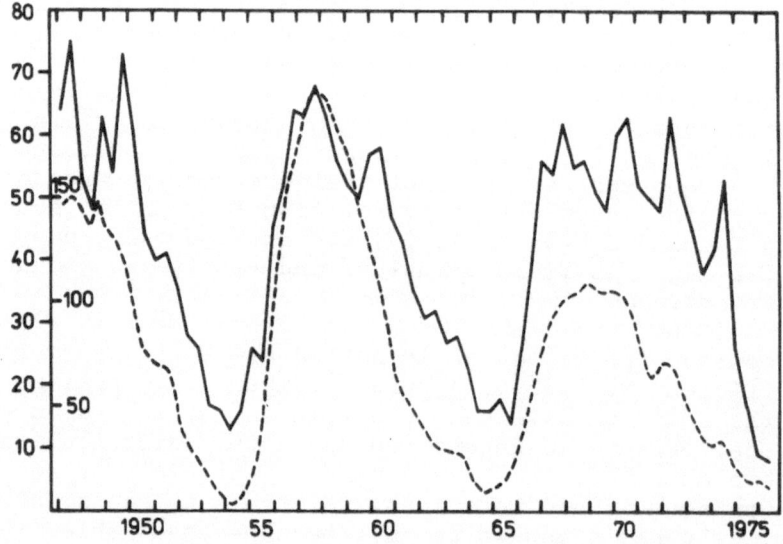

Figure 6. 11-year cycle variations of the green corona (solid line - - left hand scale in absolute coronal units) and Wolf numbers (dashed line - right hand scale).

North-south asymmetry of the solar activity certainly belongs
to the long-term properties of the Sun, because during long time
intervals, months or years, the solar activity remains asymmetrical.
It even seems (Waldmeier, 1957b, 1971) that the asymmetry is not
connected with the 11-year cycle, but perhaps it is ruled by a cycle
with a longer period.

The N-S asymmetry of the green corona, defined as N-S/N+S, was
mostly positive during the last three cycles, as it was for spots,
faculae, prominences, the white light corona, etc. The most outstanding asymmetry (see Figure 7) is seen in the period of 1958-1970
with a maximum in 1966 when, according to Dodson and Hedeman (1969),
the N-S asymmetry in sunspots was the highest during the past 100 yrs

An interesting result was published by Waldmeier (1971) for the
period 1959-1969). During this period of outstanding positive asymmetry of the green corona, he found the intensity of the red emission
line (637.4 nm) to be asymmetric in the opposite sense. Applying the
theory of dielectric recombination to the found green-red line intensity ratio, from such a behaviour it follows that over the most active hemisphere the corona is really denser and hotter.

In the last decade the question of the existence and reality
of two maxima of activity during the 11-year solar cycle was raised.

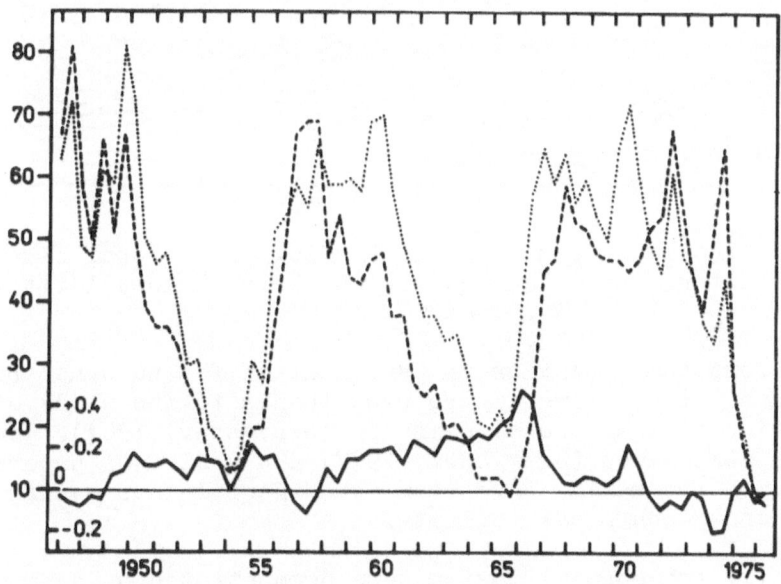

Figure 7. 11-year cycle variation of the green corona intensity,
drawn separately for northern (dotted line) and southern (dashed
line) hemispheres (left hand scale in absolute coronal units). The
N-S/N+S asymmetry is drawn by the full line (right hand scale).

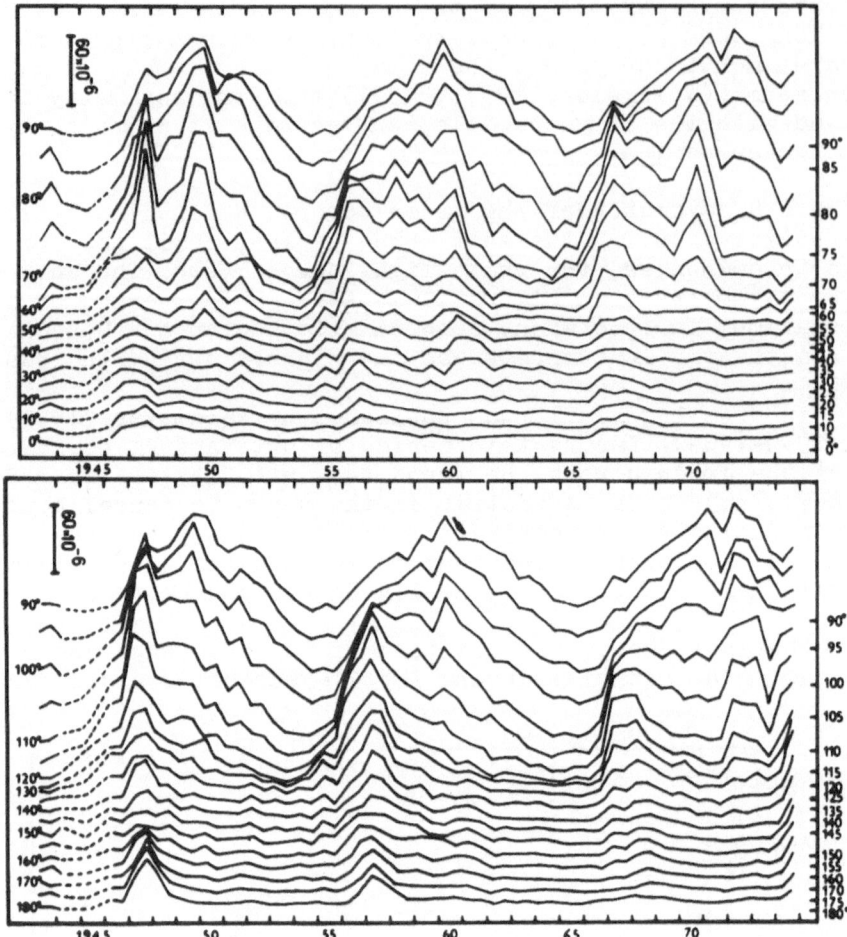

Figure 8. The time dependence of the 530.3 nm coronal line intensity
for different northern (above) and southern (below) latitudes.

In our report this question is pertinent if for no other reason,
then because it was first introduced thanks to the study of the time
distribution of the coronal emission (Gnevyshev, 1967). Recently it
was again summarized (Gnevyshev, 1977) and a list of papers was given in which two maxima have been found for various manifestations
of solar and geophysical activity.

    Usually two coronal maxima were demonstrated on data of 19th
cycle. Some papers declaring the reality of this feature in 20th
cycle showed to be premature (Pathak, 1972; Cuperman and Sternlieb,
1972). These authors only had data up to 1971 at their disposal and
they evaluated the activities in 1967 and 1969-1970 as the two maxima.
But it is generally known and it can clearly be seen in our Figures

2, 6 and also 8 (constructed from the data of Pic du Midi only,to
avoid any possible influence of the non-linear relation of the photo-
metric scales of different observatories) that there were two subsequent coronal maxima in 1972 and 1974. Because cycle No.18 - the
first for which the coronal data are sufficiently complete - does
not seem to be very expressive in the sense of the existence of two
maxima either, we suggest one should be more careful before declaring
this feature to be a "basic feature of the 11-year cycle". From Figure 8, in combination with Figure 3 it follows that: (1) The curves
sometimes are not so smooth as they are frequently presented. (2) The
activity of the green corona seems to follow the development of the
features known as "complexes of activity", the lifetime of which is
1-2 years. (3) This process is relatively independent in both hemispheres.

    What we only can mention as some sort of secondary activity at
low latitudes just before the solar cycle minima is reality shown
in Figure 9, where the largest half-yearly values of the green corona in the whole latitudinal interval (±60°) are marked separately
for the northern and southern hemispheres. The regular displacement
of activity to the equator seems to have been disrupted in 1949,
1961 and 1971. Beyond these years the activity remained at the same
or was even shifted to higher latitudes. This may be, but is not necessarily connected with a real increase of the activity level.

6. CORONAL HOLES

    In dealing with coronal responses to long-time phenomena one
should definitely summarize briefly our state-of-knowledge on coro-

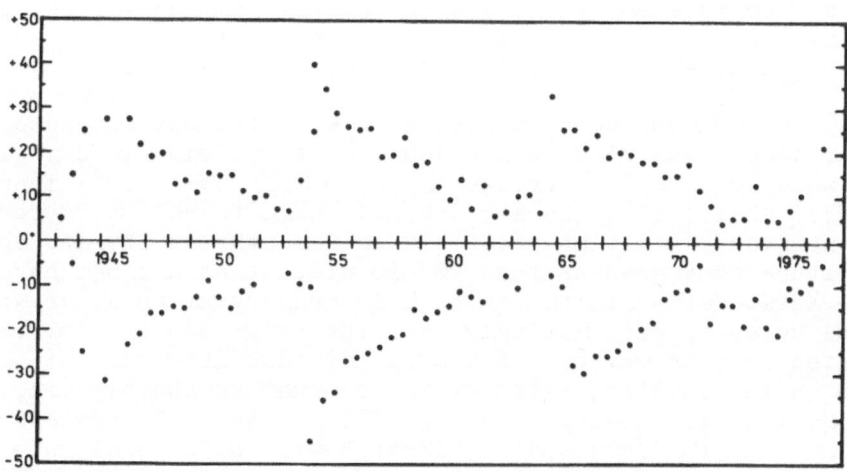

Figure 9. Maximum values of half-yearly averages of the green coronal line intensity.

nal holes. During the last five years a large number of papers appeared which clearly indicate that the coronal holes belong to the fundamental features of the Sun. From the point of view of our introductory paper their following properties are of principal importance:

(1) Coronal holes are predominantly magnetic features. The photospheric magnetic fields underlying a hole are weak, predominantly unipolar, with an open configuration of the field lines. During the Skylab mission low-latitude holes tended to develop in magnetic cells of the same polarity as the polar cap in that same hemisphere (Bohlin and Sheeley, 1978).

(2) Generally, coronal holes are divided into two groups: polar and low-latitude coronal holes. The first are more extended in size and more stable. For the Skylab period Bohlin (1977) has reported an average of 18 to 19% of the Sun's total surface covered by coronal holes, of which 15% was contained in the polar caps. Altschuler et al. (1972) have found 13% of the Sun's surface covered by lower--latitude holes in November 1966. From a study of coronal holes in the period of 1963-1974 Broussard et al. (1978) were able to draw a picture of the coronal holes variation with the sunspot cycle. The polar coronal holes, prominent at solar minimum, decreased in area as the solar activity increased and were small or absent at maximum phase. During the maximum coronal holes occurred poleward of the sunspot belts and in the equatorial region between them. The observed equatorial holes were small and persisted for one or two solar rotations only. The general conclusion is that the coronal holes wax and wane during the cycle in consistence with most long-term features observed in emission- and K-coronas.

(3) According to Bohlin (1977) the lifetime of holes typically exceeded 5 or more rotations and frequently reached values of 8-10 rotations. In 1974-1975 one hole or its remnants may have lasted 20 rotations (indications see, for example, in Sheeley et al., 1976).

(4) As we have shown in Section 2 a well-known characteristic of solar corona activity is its tendency to persist in certain preferred zones on the Sun, frequently over many solar rotations. Bohlin (1977) claims that the same result seems to hold true for coronal holes. However, on a time-scale of the 8 months of Skylab, there is no longitude zone greater than 40° in width that did not have a hole at some latitude, north or south. This means that on a long-term average the holes appear uniformly over the solar sphere. The result formulated in this way probably does not take into account the difference in the rotation rates of the corona and photosphere, pointed out in Section 4. Sheeley et al. (1976) adopted a better approach to demonstrate the longitudinal distribution of coronal holes. On a sequence of 27-day Bartels rotations for the interval 1973-1975 the arrangement of the holes into two long-term strips is clearly seen, nevertheless, the statement, similar to Bohlin's, that over this 3-year interval some hole occurred at each longitude, is also valid.

(5) Evolution of the coronal holes through the solar cycle has been discussed in some details by Bohlin and Sheeley (1978) who suppose that the holes form as a result of the emergence of two or more new BMR's which are sufficiently close together to prevent their respective proceding and following magnetic fields (of like sign) from achieving flux balance. This was the case of all four major holes observed during the manned Skylab missions. Large-scale shifts in boundary locations were found to explain most of the evolution of coronal holes (Nolte et al., 1978a). The authors conclude that coronal holes evolve by magnetic field lines opening when the holes are groving, and by fields closing as the holes shrink. The disappearence of the large holes near solar minimum and perhaps of the holes in general was hypothesized by filling the area of the holes with many small-scale, magnetically closed features, known as X-ray bright points (Nolte et al., 1978b).

(6) Finally we should mention a fact important, let us say, from the practical point of view. From observations of coronal holes, solar wind streams and geomagnetic disturbances during 1973-1976, Sheeley et al. (1976) conclude, that "the results leave little doubt that coronal holes are related to high-speed streams and their recurrent geomagnetic disturbances". Similarly, "during 1963-1974 whenever XUV or X-ray images were available, nearly all recurrent solar wind streams of speed $\geqslant 500$ kms^{-1} were found associated with coronal holes at less than 40° latitude" (Broussard et al., 1978). On these two quotations it is appropriate to hand word over to the next speaker.

REFERENCES

Altschuler, M.D.: 1974, in G. Newkirk, Jr. (ed.), "Coronal Disturbances", IAU Symp. 57, 3.
Antonucci, E.: 1974, Solar Phys. 34, 471.
Antonucci, E. and Dodero, M.A.: 1977, Solar Phys. 53, 179.
Antonucci, E. and Svalgaard, L.: 1974, Solar Phys. 34, 3.
Antonucci, E. and Svalgaard, L.: 1976, Solar Phys. 36, 155.
Bohlin, J.D.: 1977, Solar Phys. 51, 377.
Bohlin, J.D.and Sheeley, Jr., N.R.: 1978, Solar Phys. 56, 125.
Broussard, R.M., Sheeley, Jr., N.R., Tousey, R., and Underwood, J.H.: 1978, Solar Phys. 56, 161.
Bumba, V.: 1976, in V. Bumba and J. Kleczek (eds.), "Basic Mechanisms of Solar Activity", IAU Symp. 71, 47.
Bumba, V. and Howard, R.: 1969, Solar Phys. 7, 28.
Bumba, V. and Sýkora, J.: 1973, in M.J. Rycroft and S.K. Runcorn (eds.), COSPAR Space Research, Vol. XIII, Berlin, p. 1973.
Bumba, V. and Sýkora, J.: 1974, in G. Newkirk, Jr. (ed.), "Coronal Disturbances", IAU Symp. 57, 73.
Cuperman, S. and Sternlieb, A.: 1972, Solar Phys. 25, 493.
Demastus, H.L., Wagner, W.J., and Robinson, R.D.: 1973, Solar Phys. 31, 449.
Dodson, H.W. and Hedeman, E.R.: 1969, IQSY 4, 3.

Gnevyshev, M.N.: 1967, Solar Phys. 1, 107.

Gnevyshev, M.N.: 1977, Solar Phys. 51, 175.

Hansen, R.T., Garcia, Ch.J., Hansen, S.F., and Loomis, H.G.: 1969a, Solar Phys. 7, 417.

Hansen, R.T., Hansen, S.F., and Loomis, H.G.: 1969b, Solar Phys. 10, 135.

Howard, R., Bumba, V., and Smith, S.F.: 1967, Carnegie Inst. of Washington Publ. No. 626, Washington.

Knoška, Š., Křivský, L., and Sýkora, J.: 1978, paper presented at IX. Consultation on Solar Physics, Sept. 24 - Oct. 1, 1978, Wroclaw, in press.

Letfus, V., Kulčár, L., and Sýkora, J.: 1980, this Symposium.

Newkirk, Jr., G.: 1971, in R. Howard (ed.), "Solar Magnetic Fields", IAU Symp. 43, 547.

Nolte, J.T., Davis, J.M., Gerassimenko, M., Krieger, A.S., and Solodyna, C.V.: 1978a, Solar Phys. 60, 143.

Nolte, J.T., Gerassimenko, M., Krieger, A.S., and Solodyna, C.V.: 1978b, Solar Phys. 56, 153.

Pathak, P.N.: 1972, Solar Phys. 25, 489.

Pneuman, G.W.: 1974, in G. Newkirk, Jr. (ed.), "Coronal Disturbances", IAU Symp. 57, 35.

Sheeley, Jr., N.R., Harvey, J.W., and Feldman, W.C.: 1976, Solar Phys. 49, 271.

Stenflo, J.O.: 1972, Solar Phys. 23, 307.

Stenflo, J.O.: 1974, Solar Phys. 36, 495.

Stenflo, J.O.: 1977, Astron. Astrophys. 61, 797.

Sýkora, J.: 1971a, Bull. Astron. Inst. Czech. 22, 12.

Sýkora, J.: 1971b, Solar Phys. 18, 72.

Sýkora, J.: 1975, Contr. Astron. Obs. Skalnaté Pleso 5, 5.

Timothy, A.F., Krieger, A.S., and Vaiana, G.S.: 1975, Solar Phys. 42, 135.

Trellis, M.: 1957, Suppl. Ann. Astrophys. No.5.

Wagner, W.F.: 1975, Astrophys. J. 198, L141.

Waldmeier, M.: 1957a, Die Sonnenkorona, Verlag Birkhäuser, Basel.

Waldmeier, M.: 1957b, Z. Astrophys. 43, 149.

Waldmeier, M.: 1971, Solar Phys. 20, 332.

Wilcox, J.M.: 1971, in R. Howard (ed.), "Solar Magnetic Fields", IAU Symp. 43, 744.

Wilcox, J.M., Schatten, K.H., Tanenbaum, A.S., and Howard, R.: 1970, Solar Phys. 14, 255.

Xanthakis, J.: 1969, Solar Phys. 10, 168.

INTERPLANETARY RESPONSE TO SOLAR LONG TIME-SCALE PHENOMENA

C. D'Uston and J.M. Bosqued
Centre d'Etude Spatiale des Rayonnements
9, avenue Colonel Roche - B.P. 4346
31029 TOULOUSE CEDEX - France

In this paper, we briefly review the experimental knowledge gained in the recent years on the interplanetary response to solar long-time scale phenomena such as the coronal
magnetic structure and its evolution. Observational evidence
that solar wind flow in the outer corona comes from the unipolar diverging magnetic regions of the photosphere is discussed
along with relations to coronal holes. High-speed solar wind
streams observed within the boundary of interplanetary magnetic
sectors are associated with these structures. Their boundaries
appear as very narrow velocity shears.
    The value of the maximum velocity increase is related to
the amount of divergence of the field lines at the base of
the corona which is less in large unipolar regions and causes
the expansion to be faster according to theoretical models.
Radial variations and solar cycle modulations of these structures are also presented.

1. INTRODUCTION

Sunspots are traditionally a marker for solar activity. Their
variable number is used to keep track of the solar cycle. In order to
understand the 11-year solar cycle effects on the relations between the
Sun and the interplanetary medium, it is important to determine the
magnetic structure of the base of the corona and its variations. Changes
in the properties of the interplanetary medium result from these. This
paper does not intend to describe all the long time scale variable
phenomena in the interplanetary medium ; its goals are to review the
periodic variations due to solar rotation and to present the changes
which have been observed during the cycle of activity of about 11 years.
    There are areas of the solar surface where the magnetic field lines
are connected to the interplanetary medium. These open structures are
surrounded by regions where the magnetic field has different polarities
linked by closed loops of the magnetic field lines. In these areas where
the magnetic field is open, the divergent geometry of the field lines
acts on the coronal expansion of the solar wind. Pneuman and Kopp (1971)

M. Dryer and E. Tandberg-Hanssen (eds.), Solar and Interplanetary Dynamics, 105-125.

have shown that the expansion speed is greater where the divergence is
less. When such regions pass in front of an observer the plasma flow
which is detected is changed and temporal variations are observed with
a 27-day period of recurrence if the lifetime of the source structure
is greater than the time for one solar rotation. The size and the position
of these areas of the corona evolve with time and general trends can be
established during the solar cycle.

These long term variations occuring during a solar cycle have
received increasing interest in recent years (see Nolte et al. 1978 and
references therein). Progress has been made with identification of the
coronal holes as those parts of the solar structure which act as sources
of high-speed streams (Krieger et al. 1973). In this paper we intend to
give a brief review of the experimental evidence gained in the past
years concerning the evolution of the properties of the solar wind and
of the interplanetary magnetic field during solar cycle 20 (1965-1976)
and the beginning of the cycle 21.

2. RECURRENT HIGH SPEED SOLAR WIND STREAMS

2.1.Coronal holes

Coronal holes appear as dark areas on the white light photographs
taken during solar eclipses and also on photographs made in soft X rays.
Since they have been recognised as sources of interplanetary fast streams,
these structures habe been the object of regular observations, summarized
recently by Bohlin (1977) and Broussard et al. (1978). Several empirical
properties can be deduced from observations (Zirker 1977). First the
coronal holes are seen as regions where the intensity of the coronal
emission is lower : dark areas on the white light photographs of the
corona taken during total solar eclipse, or on the photographs of the
corona taken in XUV and EUV (Broussard et al. 1977, Bohlin 1977). The
simplest explanation of the general darkness of coronal holes is that
the electron density is reduced and/or that temperature is lower.

Coronal holes are found in large unipolar magnetic regions. At the
photospheric level the magnetic field intensity is weak in the coronal
holes by comparison with surrounding areas (Vaiana et al. 1973, Bohlin
1977). At an altitude of about 2 R_s (solar radii) the magnetic field
intensity of the coronal holes is of the order of 2-10 Gauss larger than
in quiet background areas where it is of the order of one Gauss (Svalgaard
et al. 1978). It is noted that coronal holes are quasi-permanent features
of the poles of the Sun. Their polarities are constant through 11 years
and reverse at the time of the maximum of the solar cycle. We are
presently observing a positive polarity (away from the sun) in the
northern hemisphere polar hole, (negative in the southern polar hole),
and a reversal is expected around 1981, the forecast maximum year for
solar cycle 21. Coronal holes which are present at lower, high, or
mid-latitudes are often extensions of the polar caps. Lifetimes of coronal
holes are found from one to more than 10 solar rotations (the average is

about 5 solar rotations). In 1973-1974 a coronal hole was observed to last as long as 16 solar rotations. It is during the period of decreasing solar activity that the largest and the most stable coronal holes are observed. Another characteristic feature of coronal holes is their apparent rigid rotation rate with a 27.2 day synodic period. They are comparatively unaffected by the usual differential rotation rate of the surface of the Sun even when they extend from the polar cap through a large heliolatitude range across the equator (Timothy et al. 1975).

Extensive investigations in the past years have demonstrated clearly that coronal holes are the sources of most of the high-speed streams in the interplanetary medium ; notably solar wind high velocity flows are often observed to be recurrent during several solar rotations with only slight variations in amplitude and structure. Intercomparisons between coronal hole data, solar wind velocity, interplanetary magnetic field variations and geomagnetic activity enforce the idea of the relation between high-speed flows and coronal holes first presented by Krieger et al. (1973, 1974). However it is important to note that not all coronal holes are associated with fast streams in the interplanetary medium and that some long lived high-speed streams are not correlated with coronal holes. Nevertheless the latter high-speed streams seem to be associated with diverging unipolar coronal magnetic fields areas (Levine 1978, Burlaga et al. 1978).

2.2. High-speed streams

In the literature there are numerous examples illustrating the association between high-speed streams and coronal holes (Nolte and Roelof 1977, Burlaga et al. 1978 a, 1978 b, Levine 1978, Schwenn et al. 1978). Figure 1 is an illustrative example of the observation made at 1 AU by IMP 7 and IMP 8 during 1973. The velocity profile (upper panel) shows three high velocity streams. Each begins by a velocity increase which corresponds to an increase in the interplanetary magnetic field intensity (second panel from the top) measured by the HEOS 1 and HEOS 2 magnetometer. The magnetic polarity is also shown. The panel at the bottom is a map of coronal holes at the photospheric level during solar Carrington rotation 1607. A clear association can be seen between the central meridian passage of coronal holes and observations of high velocity streams when taking into account the delay time needed for the solar wind to reach 1 AU. Several characteristic features can be seen from this display.

2.2.1. High-speed solar wind streams are observed within a single magnetic sector. These come from a source position located within a unipolar magnetic region. Figure 2 of Gosling et al. (1976) shows the position of high-speed streams relative to interplanetary magnetic sectors measured during many successive solar rotations. Most of the high velocity streams are located inside unipolar magnetic sectors. Sometimes several streams can be seen in the same sector as shown in the right hand side of Figure 1, and they are related to different coronal holes of the same coronal unipolar magnetic region. Inside a

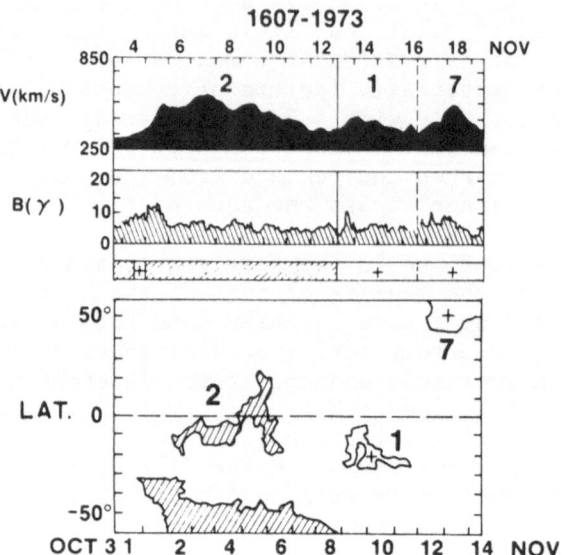

Figure 1. The top panel shows three interplanetary streams observed at 1 AU together with the magnetic field intensities and polarities in these streams. The bottom panel shows a map of the coronal holes from which the streams probably originated. The photospheric magnetic field polarities in the coronal holes agree with the polarities in streams at 1 AU. (from Burlaga et al. 1978a).

magnetic sector there is a tendency for the streams to be present near the leading edge of the sector and sometimes across this boundary due to the interaction of the fast flow with the slow ambient plasma during the corotation. Figure 2 indicates also that the high-speed streams have variable lifetimes. Some are present for only one solar rotation while others are recurrent during many solar rotations as from the end of 1973 through 1974.

2.2.2, Most of the high-speed solar wind streams are associated with equatorial or low latitude coronal holes. This has been established through studying the correspondence between the magnetic polarity inside high-speed streams and inside coronal holes deduced from magnetic observations of the Sun and also through comparing the longitudinal position of the sources of the streams with the position of the holes.

The first kind of association can be explained by the example of Figure 3 where observations made during solar roation 1609 are presented. We note that high-speed solar wind streams are well associated with low latitude or equatorial coronal holes. It remains to be explained why there is a negative polarity sector between streams n° 7 and n° 4 and why stream

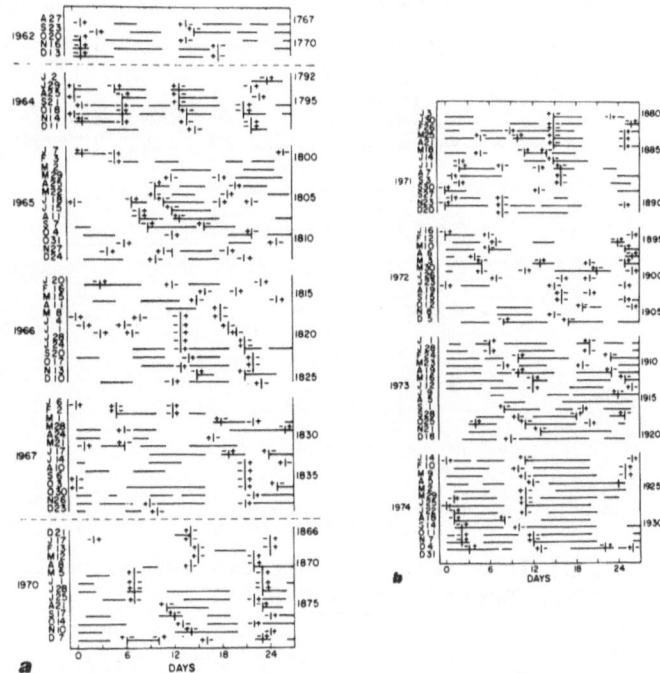

Figure 2. High-speed solar wind streams (horizontal bars) and interplanetary magnetic field sector boundaries (vertical lines) plotted according to 27-day Bartels rotations. Note that there are no reliable stream data for 1963, 1968, or 1969. The plus and minus refer to the field polarities on either side of the sector boundaries. The numbers on the right refer to the Bartels rotation numbers. Note that almost all streams are unipolar. Two series of recurrent streams dominate the late 1973 and 1974 period. (from Gosling et al. 1976)

n° 2* lasts for such a long time. Burlaga et al. (1978a) computed the magnetic field using the assumption that the field is force-free in this range of altitude and forcing the field lines to be radial at the outer boundary. The photospheric feet of the open field lines are mapped in Figure 4. For rotation 1609 several coronal holes are also present on this Figure along with areas which were not reported as coronal holes but with diverging open magnetic structures and whose polarities explain the observations in the interplanetary medium. For example a negative polarity area on the solar equator at a heliographic longitude of 320° can be correlated to the negative interplanetary sector observed between December 17 and 19. Also the negative polarity areas seen close to the solar equator and located between heliographic longitudes of 40° and 100° can explain the extension of the interplanetary magnetic sector after January 3. In order to be sure that this is not merely a fortuitous coincidence it is useful to locate the source of the high-speed streams near the base of the corona. A simple way to do this is to assume that

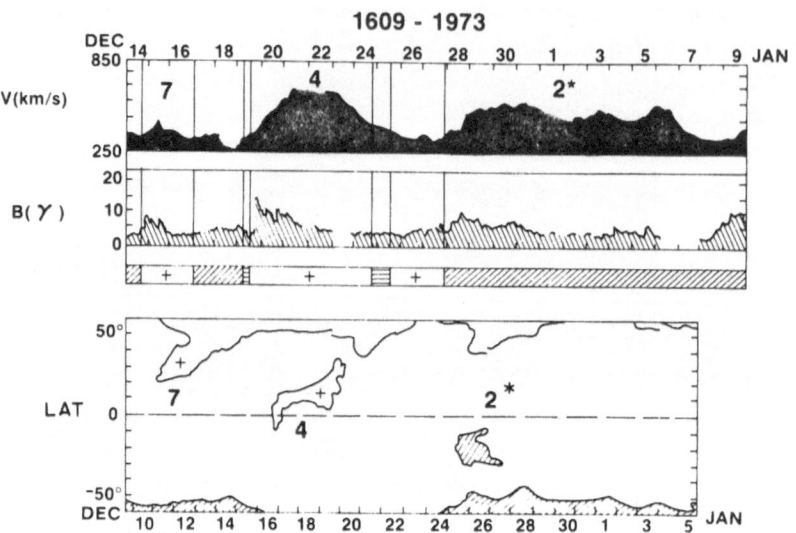

Figure 3. Interplanetary streams and magnetic fields for CR
1609 (from Burlaga et al. 1978a)

the plasma propagated radially from the Sun at the constant velocity
observed at 1 AU. Figure 5 is from Nolte et al. (1976) who used this
method for rotation 1609 (note that time is running from right to left).
The first point of each day is indicated by a heavy dot and alternate
days are indicated. It indicates that fast stream n° 7 is connected to
a place at about 340° heliolongitude which corresponds to coronal hole
n° 7. In the same way stream n° 4 is consequently correlated to thecoro-
nal hole between longitudes 210° and 270°. Observation of a fast stream
with a negative magnetic polarity during all of the remaining time of
rotation 1609 can also be interpreted with less certainty as corresponding to coronal hole n° 2 and to negative polarity areas open to the
interplanetary magnetic field as computed by Burlaga et al. (1978a). Thus
it seems that cases which give rise to high-speed streams in the solar
wind are more fundamentally related to an open magnetic field structure
than to coronal holes. Thus at this time of the solar cycle the solar
wind behavior around 1 AU has been shown to depend mainly on the equatorial and low latitude structure of the corona.

2.2.3, High-speed wind streams have very thin boundaries. In the
preceeding discussion concerning the interplanetary magnetic polarity
in relation to the associated coronal sources, it could be seen that
several adjacent regions of the Sun were competing. During solar rotation
1608 a positive polarity zone was observed on December 9 and was no
longer registered during rotation 1609. From one rotation to the next
the positive area and also the negative area at 60° of heliolongitude
were displaced southward by about 10° (Figure 4), the negative polarity
area coming closer to the equator and the positive open structure decaying as suggested by the presence of a positive polarity coronal hole

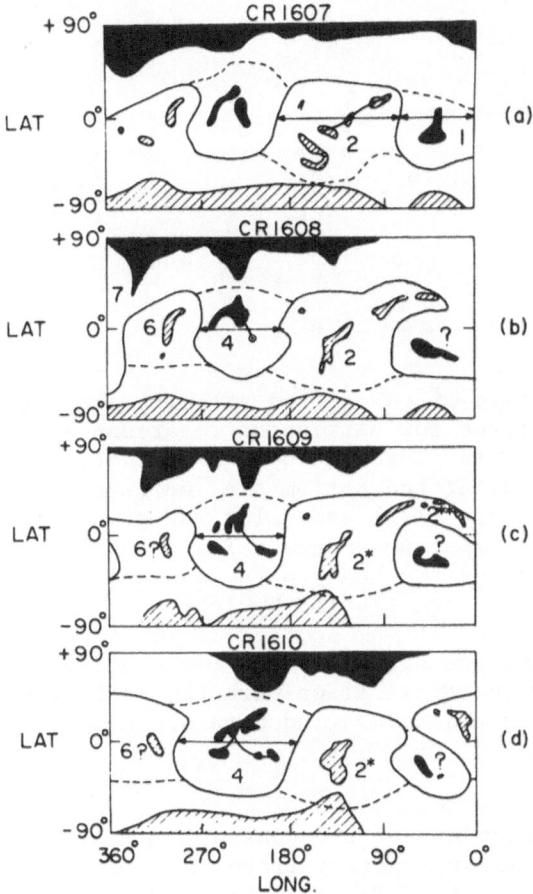

Figure 4. Map of the foot-prints of open magnetic field lines, their polarities (solid areas have positive polarity), the null lines (solid curves), and the saddle lines (dashed curves) for CR 1607-1610. (from Burlaga et al. 1978a)

at this place two rotations earlier. This could explain why an inter-planetary flow with positive magnetic field became a flow with negative magnetic field during rotation 1609.

Schwenn et al. (1979) investigated the boundaries of the streams observed by HELIOS 1. In the front edge of a stream, the sharp velocity increase can be weakened by interacting with the slow ambient wind ; in certain circumstances it can begin to strengthen again when the spiral angle of magnetic field lines becomes larger and it can form a corotating shock structure. The boundary layers in front of a stream are thinner at 0.3 AU than at 1 AU and are more faithful images of the source boundaries. The angular extent of this limit is less than 5°.

The trailing edge of a fast stream does not interact with the slow ambient flow and thus it is possible to map back directly the source position using the extrapolated quasi-radial hypervelocity approximation (Nolte et al. 1973). Figure 6 from Nolte et al. (1977) gives a proof of

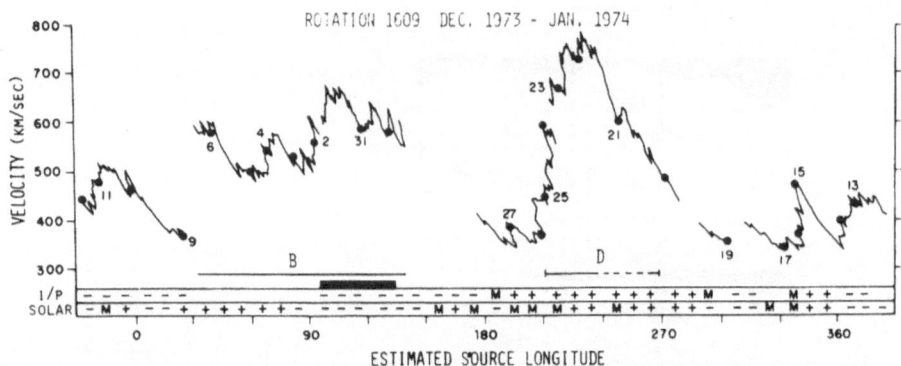

Figure 5. Hourly averages of solar wind velocity plotted against
estimated source longitude for Carrington rotations 1609. The
first point of each day is indicated by a heavy dot, and alternate days are indicated. At the bottom are indicated the 10°
averages of interplanetary (I/P) and solar magnetic polarity.
Above these polarity strips, near-equatorial coronal hole
locations are indicated by the heavy bars, and the high-speed
solar wind stream sources are labeled by letters. The estimated
extent in longitude of the stream sources is marked by the
horizontal lines, which are dashed during times of rapidly
rising velocity. Short vertical lines mark the estimated edges
of the sources whenever the solar wind data are complete
(from Nolte et al. 1976)

the small width of the eastern boundary of the source. Apparent westward
position changes indicated that interplanetary travel of the stream
does not leave its structure unaffected. Taking into account interplanetary interaction effects and velocity related altitude change in the
source position (Nolte et al. 1977), the angular width of the eastern
boundary of the source is again of the order of 5°.

Using the observations made by IMP at 1 AU and by HELIOS near
0.3 AU at various heliocentric latitudes, Schwenn et al. (1978) have
shown that the latitudinal boundary of high-speed wind streams was less
than 10° wide. Thus with an average velocity increase of 300 km/s in
streams, we are led to a velocity gradient of the order of 30 km/s/deg.
Schwenn et al. (1979) related that in some cases, the observations made
by HELIOS 1 and HELIOS 2 were completely different when these spacecraft
were more than 5° apart in latitude. However, due to the longitude
separation of the spacecraft, the measurements were not made simultaneously and thus temporal variations could have occured. Nevertheless
in other circumstances when the latitudinal and longitudinal separation
was very small, the observations suggest that the boundary could be very
strongly inclined relative to the solar meridian plane.

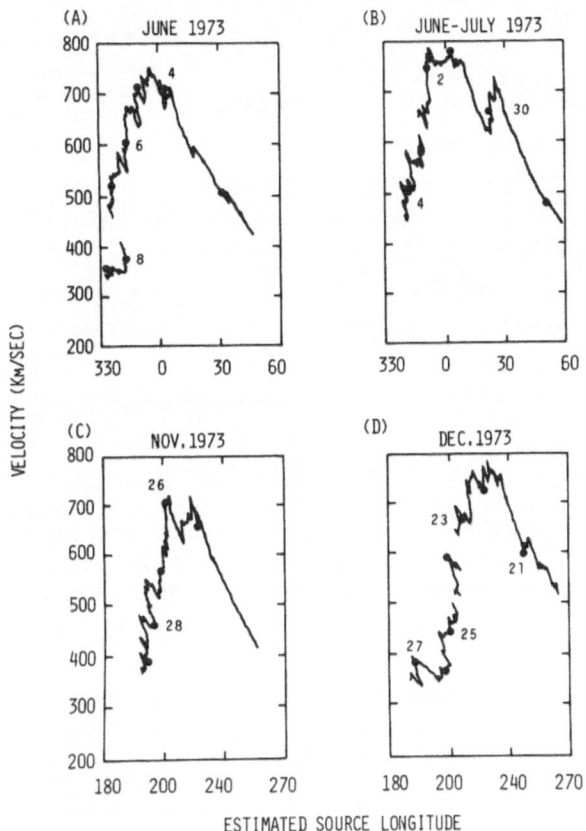

Figure 6. Same as Figure 5 for selected high-speed streams.
Note the near vertical drop, or dwell in longitude, for two
or three days at the eastern edge of each streams. (from Nolte
et al. 1977)

2.2.4, In high-speed solar wind streams maximum velocity increases
depend on the size of the associated coronal hole. Nolte et al. (1976)
found a clear linear relation as shown on Figure 7. This relationship
supports the theory that the divergence rate of the open magnetic field
lines modulates the coronal expansion rate in the interplanetary medium.
Magnetic field lines are expected to diverge more rapidly near the edges
of coronal holes than in the central part. Thus the larger the hole, the
more radial the magnetic field lines of the central part, or the less
divergent the magnetic structure (Figure 8). Pneuman and Kopp (1971) and
Pneuman (1973) investigated theoretically the flow resulting from coronal
expansion in open and in closed magnetic configuration. They have shown
that in open, divergent geometry the expansion rate is increased. Several
theoretical models refined this approach (Kopp and Holzer 1976, Steinolfson
and Tandberg-Hanssen 1977, Suess 1976, Pneuman 1976, Kopp and Orrall 1976).
For a review see Suess (1979).

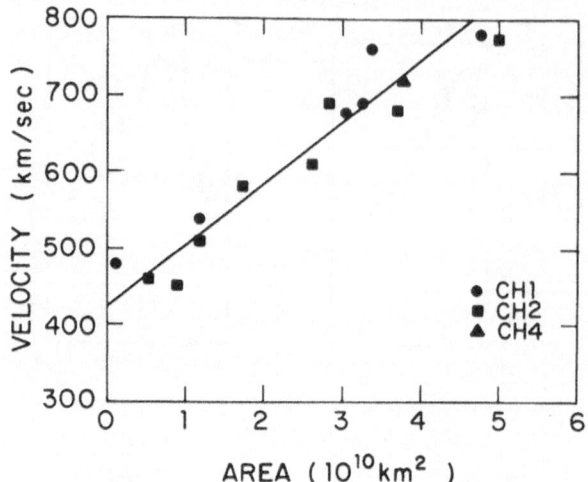

Figure 7. Maximum solar wind velocity vs. area of the associated coronal hole within 10° of the ecliptic plane. The least squares straight line is also shown. (from Nolte et al. 1976)

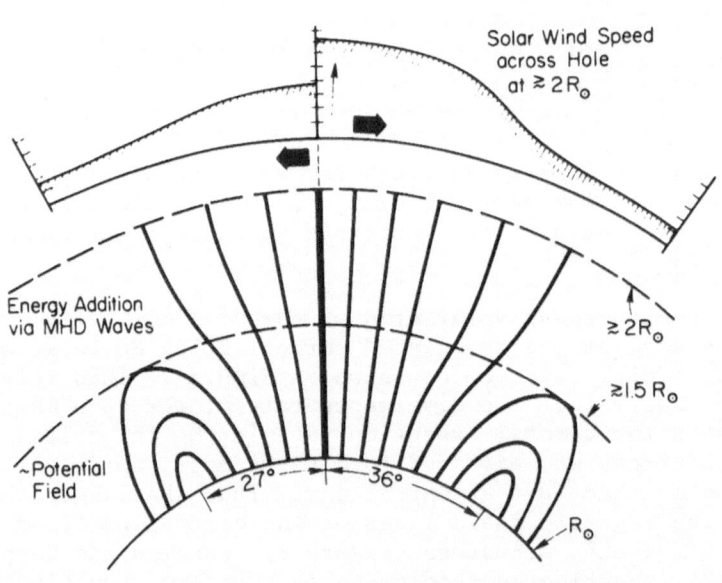

Figure 8. A hypothetical example of the influence of the size of coronal holes is illustrated on this schematic cross section of two coronal holes differing only in width. (from Suess,1979)

Generally the solar poles are extended coronal holes and it is logical to expect some higher velocity in the high-latitude solar wind. Statistical investigations done using interplanetary scintillation measurements of radio sources allow a determination to be made of the latitudinal gradient of the solar wind speed which was found to be 2.1 km/s/deg. (Coles and Rickett 1976). This value of the latitudinal gradient could be the effect of the diverging structure of the solar polar magnetic field. The divergence of flow tubes takes place in the base of the corona up to an altitude of about 3 R_s (Munro and Jackson 1977). In the case of a polar coronal hole observed in June-July 1973, this divergence is such that it could be deduced that 8 % of the total solar surface of the northern hemisphere was the source of plasma for about 60 % of the flow at an altitude of 3 R_s in the same hemisphere.

During March 1975 HELIOS reached perihelion at 0.31 AU and registered a fast stream with a longitudinal extent of about 30°. The associated coronal hole was identified as an equatorward extension of the southern polar coronal hole. Its longitudinal extent at the latitude of the spacecraft was not more than 10°. Thus the flow diverged by a factor of 3 between the solar surface and 0.31 AU (Burlaga et al. 1978b). This value is of the same order as the one which was established directly by Munro and Jackson (1977) : 2.9 between the solar surface and 5 R_s. Bohlin (1976) estimated that 15 % to 20 % of the solar surface was occupied by coronal holes during 1973-1974 and Bame et al. (1977) proposed the hypothesis that all the solar wind came from them. Following the same idea, several authors suggest that the basic state of the solar wind is the one observed in high velocity flows.

2.3. Radial variation

A large amount of data were obtained from interplanetary spacecraft between 0.3 AU and 15 AU so that it is possible to determine the characteristic behavior of fast streams with increasing distance from the Sun. The radial evolution of high speed wind streams has been described by Collard and Wolfe (1974), Gosling et al. (1976), Hundhausen and Gosling (1976), and others ; they showed that the interplanetary medium acts like a low pass filter. The rapid fluctuations are smoothed and large scale gradients are strengthened and finally give the sawtooth like velocity profile of the solar wind in terms of temporal evolution. Due to solar rotation, the flow undergoes an asymmetrical evolution. The leading edge interacts with the slower and denser plasma. Starting near the Sun, the high pressure in the stream tends to widen the range of longitudes in which the velocity increase takes place. Then, because of the strong convergence of the fast and slow streams, the velocity and density gradients increase again and can result in the formation of a forward front shock and of a reverse shock. The momentum transfer through the interface reduces the velocity increase further from the Sun (Gosling et al. 1978).

Figure 9 gives an example of observations of a fast stream in two positions : at PIONEER 11 at 1.5 AU and at PIONEER 10 at 4.4 AU (Smith and Wolfe, 1977). The first spacecraft registered a velocity increase of

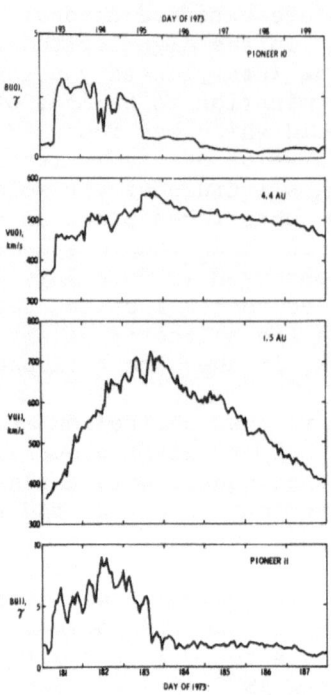

Figure 9. Evolution of a high speed stream between Pioneer 11
and Pioneer 10. The two upper panels show the field and velocity
at Pioneer 10, B (10) and V (10), and the two lower panels
show V (11) and B (11) corresponding to the same stream at
Pioneer 11.

350 km/s within 2 1/2 days during which the magnetic field maintained an
enhanced value characterizing the interaction region. During the same
period (June-July 1973) IMP 7 and IMP 8 were radially aligned with PIONEER
11 and show that at 1 AU the velocity profile is steeper in the leading
part with a twofold increase (Intriligator 1977). The velocity at 1 AU
reaches about 800 km/s, slightly higher than the value of 760 km/s
observed at PIONEER 11. Furthermore this higher value was observed for a
longer time at 1 AU. This high speed solar wind stream is related to an
equatorial coronal hole seen during Carrington rotation 1602 (Bohlin and
Sheeley 1978). This coronal hole is located between longitudes 5° and 25°
and the resulting fast stream at 1 AU is observed between longitudes
-23° and 45° (stream B on figure 6) which shows again the divergence of
the flow tube between the Sun and 1 AU. Applying the constant velocity
radial propagation assumption to the data of PIONEER 10 at 4.4 AU it is
found that the longitudinal width of the source of the stream is of about
70°. This value is the same order of magnitude as the one found from IMP
data so it is concluded that the divergence takes place close to the Sun

before 1 AU and that the flow expands spherically further out in the ecliptic plane. Following its propagation to 4.4 AU, where it was observed by PIONEER 10, this stream undergoes important modifications. First the velocity increase takes place in several abrupt steps and secondly the maximum velocity seen in the stream at 4.4 AU is much weaker ($\sim$ 560 km/s). It is clear that a forward shock front appeared at the leading edge of the stream between 1.5 AU and 4.4 AU. Smith and Wolfe (1977) interpret this fact by suggesting that after they are formed, the shocks tend to mask the positive velocity gradients and eventually to eliminate the original large velocity difference associated with a stream. The interaction region here again is indicated by the high value of the observed magnetic field and probably ends as a discontinuity, which could be a reverse shock. This would tend to erode the velocity peak in the stream and then to propagate through the decreasing velocity portion. At larger radial distances a sawtooth like velocity profile develops. Another example is given on Figure 10 which shows the evolution of some high speed streams between 1 AU (IMP 7 and IMP 8) and 11 AU to 15 AU (PIONEER 10) (Collard, 1979) ; global characteristics of velocity profiles are observed to persist as far out as 12 AU, but they are much more attenuated at large heliocentric distances. The velocity fluctuations are smoothed and the average velocity in the stream is of the same order at large distances as at 1 AU but the standard deviation of the velocity histogram is much less at a larger radial position. (Collard 1979). Several discontinuities are then likely to appear as a result of the possible interaction between shocks. VOYAGER spacecraft will provide more details about the behavior of the solar wind at large radial distances.

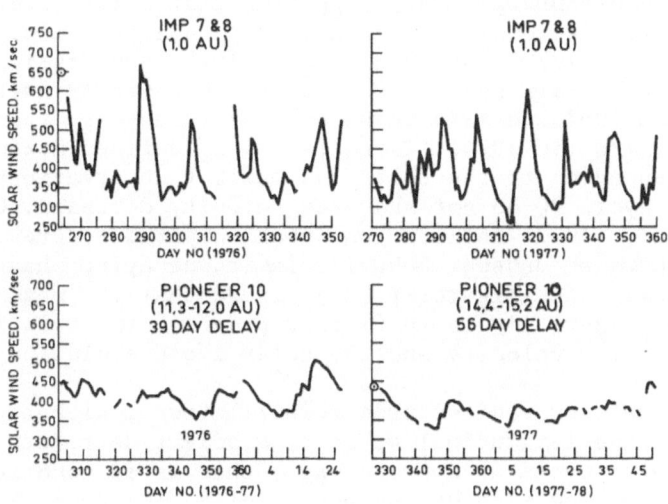

Figure 10. Comparisons of IMP and PIONEER solar wind speed measurements at the beginning solar cycle 21 (courtesy of Collard, ARC NASA, 1979)

3. SOLAR CYCLE VARIATION

     Measurements of the solar wind registered since the sixties cover
the entire solar cycle 20 and the beginning of cycle 21. Broussard et al.
(1978) summarized the evolution of coronal holes during a solar cycle.
In particular, they observed that during minimum solar activity, there
were large polar coronal holes and that their area was decreasing with
increasing solar activity. During the increase of activity, small sized
polar coronal holes are seen to grow high latitude extensions toward the
polar side of the sunspot belts. When the polar coronal holes have almost
completely vanished during solar maximum, coronal holes can be observed
north (south) of the northern (southern) sunspot belt and also at the
equator between them. They have relatively short lifetimes. During the
decaying phase of the solar cycle, equatorial holes tend to join the
higher latitude ones giving birth to a stable disruption of the sunspot
belts, and polar coronal holes appear again.

     Comparing this evolution with that of the high speed streams, one
concludes that small low latitude coronal holes are related to relatively
narrow fast streams with velocities not more than 600 km/s (Bame et al.
1976) while large coronal holes even at mid latitudes give rise to
extended streams with velocities greater than 700 km/s. Thus the long
time scale changes of the corona modulates the solar wind behavior at
1 AU (Zirker 1977, Sheeley and Harvey, 1978, Sheeley et al. 1976)

3.1. Solar wind variations

     Several authors investigated the long term variations of solar wind
parameters at 1 AU. Hirshberg (1969) suggested that solar wind velocity
should increase with increasing solar activity. Analyses made by Gosling
et al. (1971), Wolfe (1972), Diodato et al. (1974) and Gosling et al.
(1976) failed to ascertain any clear tendency when yearly averaged values
were used. Between 1962 and 1974 these average values are from 398 km/s
(in 1965) to 521 km/s (in 1974). There is a well defined maximum in 1968
(473 km/s) corresponding to the solar maximum but observations made at
the end of solar cycle 20 do not show any velocity decrease. On the other
hand Gosling et al. (1977) noticed that when compared to previous cycles,
solar cycle 20 shows an unusual behavior in its decaying phase. Looking
at the fast streams, Intriligator (1977) argued that the maximum value
observed in 1968 (Figure 11) is sufficient proof of the relation-ship
between the solar wind velocity and the solar cycle evolution.

     From interplanetary radio-source scintillations, Rickett and Coles
(1979) found that the latitudinal velocity gradient decreases when the
maximum of the solar cycle is approaching ; this is in good accordance
with the decrease of the size of the polar holes. Nolte et al. (1977)
noticed that in 1976, the variations of the solar wind velocity were

very similar to those in 1965 during the previous solar minimum. Diodato
et al. (1974) gave evidence of a variation of the plasma density which
decreases by 40 % between minimum and maximum solar activity and this
was observed whatever velocity range was considered. This density variation
was related to the latitudinal effect suggested by Hundhausen et al.
(1971). During the minimum of solar activity the zones where the magnetic
field is divergent are mainly located at the poles so that the flow of
plasma converges into the equatorial plane of the sun at large distance.
Thus at this time and in this plane there flows a slow and dense plasma
whereas during the maximum of solar activity the flow probably comes
from small coronal holes or narrow areas of divergent magnetic field at
low latitudes and close to the equator. No clear temperature variations
with solar activity were observed in the last solar cycle (Diodato et al.
1974). Feldman et al. (1978) have investigated the variation of the
relative abundance of α particles and have shown that between 1971 and
1977 the relative abundance changed from 5 % to 3 % in the slow stream
while it stayed at a constant value of about 5 % in the fast streams.
Ogilvie and Hirshberg (1974) examined the observations between 1962 and

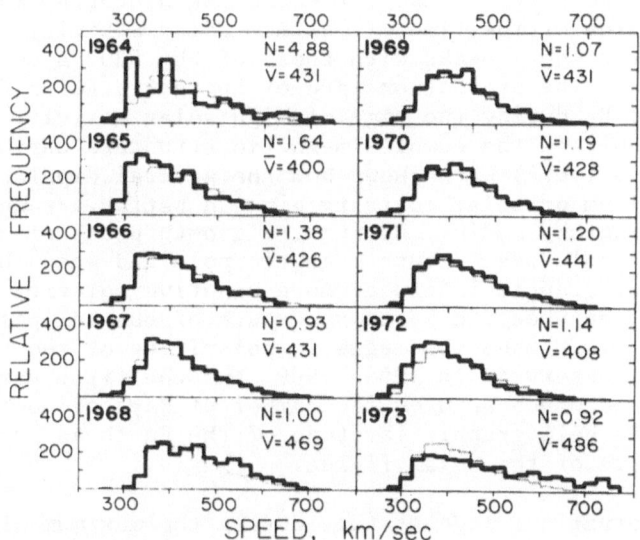

Figure 11. The dark lines indicate relative frequency of occu-
rence of the solar wind streaming speed in intervals of 25 km/
sec. for each year from 1964 through 1973. The light lines
indicate the frequency histogram of the speed for the entire
(parent) population composed of all the data from mid 1964
through 1973. The histograms have been normalized (there is the
same area under each histogram) as indicated by the N's in the
figure. The yearly average solar wind speeds are shown by the
V̄s. (from Intriligator 1977)

1972 and found a variation of the relative abundance from 3-4 % at the time of minimum activity to more than 5 % at time of maximum activity, although with large error bars (about 1 %). These authors have also shown that this variation could not be attributed either to a variation of the solar wind expansion or to a variation of the number of transient phenomena in the corona.

3.2. Magnetic field variations

The interplanetary magnetic field was also measured during the period from 1964 to 1974 by several spacecraft. King (1976) and Mariani et al. (1975) conclude that the magnetic field intensity does not show any long term variations during this period of time. However King (1976) noticed that the magnitude and the azimuthal component of the positive magnetic fields went through a variation different from that of the negative fields. Siscoe et al. (1978) revealed that the histogram of occurence of B_z (normal to the ecliptic plane) changed with the solar cycle phase. The mean value of $|B_z|$ is always 0 but the mean value of the absolute value of B_z is larger during the time of the maximum of solar activity. Figure 12 shows the variation of $|B_z|$, of the related standard deviation and of the inverse of the slope of the corresponding histogram and reveals two peaks on each side of the time of maximum solar activity. Siscoe et al. (1978) associated these peaks with those of the 5303 Å coronal emission and with those of the production rate of energetic solar flares. The larger value of B_z during the time of high solar activity can be interpreted as a result of the complex magnetic structure during this period. Svalgaard et al. (1974) have shown how the general dipolar magnetic field during the minimum of solar activity gives a better arranged appearance to the interplanetary medium. During the growth phase of solar cycle 20 the solar north pole was a south magnetic pole and when the Earth reached its maximum heliographic latitude, more negative polarity magnetic fields were observed than when the Earth was south of the ecliptic plane (Figure 13). Then after 1970 when the magnetic polarities of the sun were reversed, the inverse was observed. In 1968, 1969, and the first part of 1970 there was no clear dependence between the number of days of observed negative polarity and the heliographic latitude of the Earth as expected from the complex structure of the solar fields.

These observations suggest that during the maximum of the solar activity, the interplanetary magnetic field is irregular and this explains why there is a minimum in the flux of cosmic rays at that time (Winkler and Bedijn, 1976). The irregular structure of the interplanetary magnetic field during the sunspot maximum increases the number of scattering centers for the energetic particles and, more importantly, the magnetic field lines in the heliosphere are likely to be more tightly and more strongly closed near the sunspot maximum making it more difficult for the cosmic rays to penetrate deep into the inner heliosphere (Yoshimura, 1977).

A delay of more than one year was observed between the sunspot

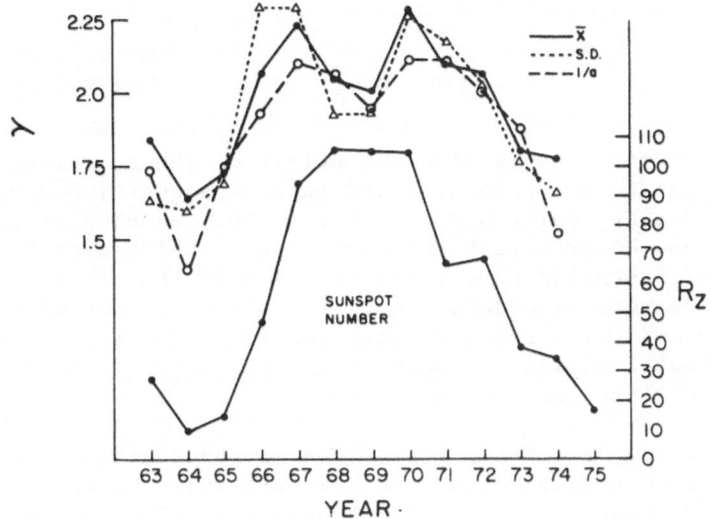

Figure 12. The top data field shows the yearly determined
values of $\bar{x} \equiv$ average B_z, S.D. $\equiv$ the associated standard
deviation, and $1/a$, the slope parameter which is defined in the
text. The bottom data field shows the yearly averaged sunspot
number, R_z. (From Siscoe et al. 1978).

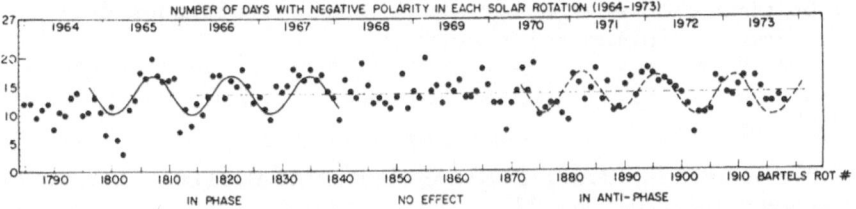

Figure 13. Number of days with negative polarity observed in
the inferred interplanetary magnetic field for each Bartels
27-day rotation since 1964. Before sunspot maximum the amount
of negative polarity observed varies in phase with the heliographic latitude of the Earth, but after maximum the two
variations are out of phase (from Svalgaard et al. 1974).

minimum and the maximum of the flux of cosmic rays (Forbush, 1954) and
indicates that the heliopause could be as far away as 100 to 200 AU. This
estimate is done by considering that the solar wind carries the image of
the coronal magnetic structure and consequently that the properties change

at the heliopause late after the coronal modifications (Yoshimura, 1977).

4. CONCLUSION

In the light of these results the origin of the so called "quiet" solar wind remains to be explained, and it can be questioned how likely it is that the "quiet" solar wind represents the fundamental regime of coronal expansion. Nevertheless the fast streams which seem to be the normal regime of expansion have velocities ($\sim$ 800 km/s) that cannot be reproduced by theoretical models, since the latter do not give velocities larger than 500 km/s. The role of the magnetic field divergence seems to be predominant but additional processes should occur at an important rate, probably through Alfven wave dissipation.

The solar cycle associated variations of the solar wind are not established with certitude yet due to the particular aspect of solar cycle 20 and also because data have not been collected for more than one and one half solar cycles. In the future, continuous observations of the interplanetary medium and further investigations of the measurements should make it clear if there is a strong relationship between the mean solar wind velocity and the solar cycle phase. Other questions remain unsolved which could be answered by extended programs of investigations. Among these are the question of the life history of coronal holes, the problem of the detailed behavior of the plasma processes involved between 2 R_S and 5 R_S, the problem of the flow over the solar poles and its variations. The last point will certainly become clearer after 1986 when the solar polar mission spacecraft pass over the solar poles at time of minimum solar activity.

REFERENCES

Bame, S.J., Asbridge, J.R., Feldman, W.C. and Gosling, J.T. : 1976, *Astrophys. J.* 207, p. 977.
Bame, S.J., Asbridge, J.R., Feldman, W.C. and Gosling, J.T. : 1977, *J. Geophys. Res.* 82, p. 1487.
Bohlin, J.D. : 1976, in D.J. Williams (ed.), *Physics of Solar Planetary Environment,* Proc. Internat. Symp. on Solar-Terrestrial Physics, Boulder, Colorado, Vol. 1, A.G.U., p. 114.
Bohlin, J.D. : 1977, *Solar Phys.* 51, p. 377.
Bohlin, J.D., and Sheeley, N.R., Jr : 1978, *Solar Phys.* 56, p. 125.
Broussard, R.M., Underwood, J.H., Tousey, R., and Sheeley, N.R. : 1977, *Bull. Amer. Astron. Soc.* 8, p. 557.
Broussard, R.M., Sheeley, N.R., Jr, Tousey, R. and Underwood, T.H. : 1978 *Solar Phys.* 56, p. 161.
Burlaga, L.F. : 1979, *Space Sci. Rev.* 23, p. 201.
Burlaga, L.F., Behannon, K.W., Hansen, S.F., Pneuman, G.W., Feldman, W.C. 1978a, *J. Geophys. Res.* 83, p. 4177.

Burlaga, L.F., Ness, N.F., Mariani, F., Bavassano, B., Villante, U., Rosenbauer, H., Schwenn, R., and Harvey, J. : 1978b, *J. Geophys. Res.* 83, p. 5167.

Coles, W.A. and Rickett, B.J. : 1976, *J. Geophys. Res.* 81, p. 4797.

Collard, H.R. : 1979, *private communication*.

Collard, H.R. and Wolfe, J.H. : 1974, in C.T. Russell (ed.) *Solar Wind Three*, Geophys. and Planetary Phys., U. of California, Los Angeles, p. 281.

Diodato, L., Moreno, G. and Signorini, C. : 1974, *J. Geophys. Res.*, 79, p. 5095.

Feldman, W.C., Asbridge, J.R., Bame, S.J. and Gosling, J.T. : *J. Geophys. Res.* 83, p. 2177.

Forbush, S.E. : 1954, *J. Geophys. Res.* 59, p. 525.

Gosling, J.T., Hansen, R.T. and Bame, S.J. : 1971, *J. Geophys. Res.* 76, p. 1811.

Gosling, J.T., Asbridge, J.R., Bame, S.J. and Feldman, W.C. : 1976, *J. Geophys. Res.* 81, p. 5061.

Gosling, J.T., Asbridge, J.R. and Bame, S.J. : 1977, *J. Geophys. Res.* 82, p. 3311.

Gosling, J.T., Asbridge, J.R., Bame, S.J. and Feldman, W.C. : 1978, *J. Geophys. Res.* 83, p. 1401.

Hirshberg, J. : 1969, *J. Geophys. Res.* 74, p. 5814.

Hundhausen, A.J., Bame, S.J. and Montgomery, M.D. : 1971, *J. Geophys. Res.* 76, p. 5145.

Hundhausen, A.J. and Gosling, J.T. : 1976, *J. Geophys. Res.* 81, p. 1436.

Intriligator, D.S. : 1977, in M.A. SHEA et al. (eds.), *Study of Traveling Interplanetary Phenomena*, D. Reidel Publ. Co, Dordrecht, Holland, p. 195.

King, J.H. : 1976, *J. Geophys. Res.* 81, p. 653.

Kopp, R.A. and Holzer, T.E. : 1976, *Solar Phys.* 49, p. 43.

Kopp, R.A. and Orrall, F.Q. : 1977, *Astron. Astrophys.* 53, p. 363.

Krieger, A.S., Timothy, A.F., and Roelof, E.C. : 1973, *Solar Phys.* 29, p. 505.

Krieger, A.S., Timothy, A.F., Vaiana, G.S., Lazarus, A.J. and Sullivan, J.D. : 1974, in C.T. Russell (ed.), *Solar Wind Three*, Geophys. and Planetary Phys. U. of California, Los Angeles, p. 132.

Levine, R.H. : 1978, *J. Geophys. Res.*, 83, p. 4193.

Mariani, F., Diodato, L. and Moreno, G. : 1975, *Solar Phys.*, 45, p. 241.

Munro, R.H. and Jackson, B.V. : 1977, *Astrophys. J.* 213, p. 874.

Nolte, J.T. and Roelof, E.C. : 1973, *Solar Phys.* 33, p. 241.

Nolte, J.T., Krieger, A.S., Timothy, A.F., Gold, R.E., Roelof, E.C., Vaiana, G., Lazarus, A.J., Sullivan, J.D., Mc Intosh, P.S. : 1976, *Solar Phys.* 46, p. 303.

Nolte, J.T. and Roelof, E.C. : 1977, *J. Geophys. Res.* 82, p. 2175.

Nolte, J.T., Krieger, A.S., Roelof, E.C. Gold, R.E. : 1977, *Solar Phys.* 51, p. 459.

Nolte, J.T., Davis, J.M., Gerassimenko, M., Krieger, A.S., and Solodyna, C.V. : 1978, *Solar Phys.* 60, p. 143.

Ogilvie, K.W. and Hirshberg, J. : 1974, *J. Geophys. Res.* 79, p. 4595.

Pneuman, G.W. : 1973, *Solar Phys.* 28, p. 247.

Pneuman, G.W. : 1976, *J. Geophys. Res.* 81, p. 5049.

Pneuman, G.W. and Kopp, R.A. : 1971, *Solar Phys.* 18, p. 258.

Rickett, B.J. and Coles, W.A. : 1979, *submitted to Nature*.

Schwenn, R., Montgomery, M.D., Rosenbauer, H. Miggenrieder, H. Mulhauser, K.H., Bame, S.J., Feldman, W.C. and Hansen, R.T. : 1976, *J. Geophys. Res.*, 83, p. 1011.

Schwenn, R., Mulhauser, K.H. and Rosenbauer, H. : 1979, to be published in *Lecture notes in Physics*, Springer-Verlag, Berlin, W. Germany.

Sheeley, Jr. N.R., Harvey, J.W. and Feldman, W.C. : 1976, *Solar Phys.* 49, p. 271.

Sheeley, Jr., N.R. and Harvey, J.W. : 1978, *Solar Phys.* 59, p. 159.

Siscoe, G.L., Crooker, N.U. and Cristopher, L. : 1978, *Solar Phys.* 56, p. 449.

Smith, E.J. and Wolfe, J.H. : 1977, in M.A. Shea et al. (eds), *Study of Traveling Interplanetary Phenomena*, D. Reidel Publ. Co, Dordrecht, Holland, p. 227.

Steinolfson, R.S. and Tandberg-Hanssen, E. : 1977, *Solar Phys.* 55, p. 99.

Suess, S.T. : 1976, in D.J. Williams (ed.), *Physics of Solar Planetary Environment*, Proc. Internat. Symp. on Solar-Terrestrial Physics, Boulder, Colorado, Vol. 1, A.G.U., p. 443.

Suess, S.T. : 1979, *Space Sci. Rev.* 23, p. 159.

Svalgaard, L., Wilcox, J.M. and Duvall, T.L. : 1974, *Solar Phys.* 37, p. 157.

Svalgaard, L., Duvall, Jr., T.L. and Scherrer, P.H. : 1978, *Solar Phys.* 58, p. 225.

Timothy, A.F., Krieger, A.S., Vaiana, G.S. : 1975, *Solar Phys.* 42, p. 135.

Vaiana, G.S., Krieger, A.S. and Timothy, A.F. : 1973, *Solar Phys.* 32, p. 81.

Wolfe, J.H. : 1972, in NASA Spec. Publ. 308, *Solar Wind*, p. 170.

Winckler, C.N., and Bedijn, P.J. : 1976, *J. Geophys. Res.* 81, p. 3198.

Yoshimura, H. : 1977, *Solar Phys.* 54, p. 229.

Zirker, J.B. : 1977, *Rev. Geophys. Space Phys.* 15, p. 257.

DISCUSSION

Bird: You presented a slide showing a velocity profile across a coronal hole, demonstrating that the solar wind speed peaks near the center of the hole where the magnetic field divergence was less than near the boundaries. It is not true, however, that the degree of magnetic field divergence affects only the radial rate of increase in velocity and not its asymptotic value?

D'Uston: This picture intended to show two open diverging coronal magnetic structures, the way the divergence of the field lines is smaller in the center of larger coronal holes and that observed velocity increases are larger there. It does not presume theoretical meanings; it is a schematic example of what is observed.

Suess: (Comment) I would like to add to the speaker's reply to the question from M. Bird regarding slide #8. The slide shows a Figure I prepared to show the well known empirical relationship between coronal hole size and solar wind speed at 1 AU, and also that there are empirical and model results which suggest that the energy supplied to the solar wind flow in the coronal hole depends somehow on the magnetic field strength and direction. This is not the same as saying the field divergence effects the solar wind speed at 1 AU directly - it does not, as Bird just stated.

Schatten: Your slide showing the small coronal holes and their relation to interplanetary sector structure appears to show that there are no "solar sectors", but the interplanetary structures arise from coronal holes. Do you agree? Also, sometimes small holes give rise to large sectors. Why?

D'Uston: Interplanetary sectors are connected to Unipolar Magnetic Regions of the Sun's surface. Coronal holes are embedded in them, thus there is no difference between the polarity of the high speed streams and that of the UMR from which it comes. I mentioned open diverging magnetic structures of the corona seem much more fundamental in giving birth to high speed streams. There are observations of interplanetary sectors without fast streams. Such interplanetary magnetic fields are carried out from UMR of the Sun by solar wind and this shows that not all the solar wind comes from coronal holes. It can be seen that small coronal holes in large UMR and the related magnetic sector is large.

STELLAR MASS FLUX AND CORONAL HEATING BY SHOCK WAVES

P. Couturier and A. Mangeney
Observatoire de Meudon , FRANCE

P. Souffrin
Observatoire de Nice, FRANCE

1. INTRODUCTION

The heating of the solar corona and the solar wind phenomenon are basically related, however, the two parts are generally modelised independently : the models of the transition zone and corona are restricted to levels lower than the temperature maximum and the solar wind models begin above it. Here we study a self-consistent oversimplified model which maintains the global balance of energy sources and sinks from the chromospheric level to the interplanetary medium. The heating mechanism chosen is the shock wave dissipation ; it was shown by Gonczi et al (1977) that overlapping shock waves could carry a significant mechanical energy flux towards a static corona ; here we apply the same mechanism to an expanding corona. The model includes self-consistently the different coupling between convective energy flux, conductive flux, radiative losses in optically thin atmosphere, shock wave pressure and dissipation terms. The input parameters are the base pressure, the base temperature and the mechanical flux introduced at chromospheric level in form of shock waves. If these three parameters allow a solar wind expansion, the output results are the radial variations of the density, of the temperature, of the solar wind velocity and of the mechanical flux. Due to the presence of a boundary layer associated to the steep temperature gradient in the transition zone, the three input parameters cannot be arbitrarily fixed, in fact when we impose two of them, the third one cannot vary within a large interval (i.e. within a factor of two or less), this point has been qualitatively discussed in a previous paper : Couturier et al. (1979)

2. BASIC EQUATIONS AND METHOD.

An extended paper covering these topics is in preparation ; due to the restricted room given to contributed papers, we shall only give the general structure of the differential system which is solved in our model and we shall not define the notations currently used.

M. Dryer and E. Tandberg-Hanssen (eds.), Solar and Interplanetary Dynamics, 127-130.
Copyright © 1980 by the IAU.

$$F_m = \rho v r^2 \tag{1}$$

$$\frac{dT}{dr} = -\frac{F_c}{\varkappa T^{5/2}} \tag{2}$$

$$\rho v \frac{dv}{dr} = -\frac{d}{dr}(p + P_*) - \frac{\rho GM}{r^2} \tag{3}$$

$$\frac{1}{r^2}\frac{d}{dr}(r^2 \Phi_*) = -\rho v P_* \frac{d}{dr}\left(\frac{1}{\rho}\right) - \frac{\rho T \Delta s_*}{\tau_*} \tag{4}$$

$$\frac{1}{r^2}\frac{d}{dr}\left\{\left(\frac{v^2}{2} - \frac{GM}{r} + \frac{5p}{2\rho}\right) + \frac{r^2 F_c}{F_m} + \frac{r^2 \Phi_*}{F_m}\frac{2v+c}{v+c}\right\} = -\frac{\rho^2 \varphi(T)}{F_m} \tag{5}$$

$$\quad\quad\quad A \quad\quad B \quad\quad C \quad\quad\quad D \quad\quad\quad\quad E$$

We assume a stationary radial expansion of the flow, with a mass flux F_m (Eq 1). We include the gradient of the shock wave pressure in the momentum (Eq 3). The radial variations of the mechanical flux Φ_* (Eq 4) are given by the work done by the shock pressure and by the dissipative term : Δs_* is the entropy production for overlapping shock waves of period τ_* ; Φ_*, Δs_* and P_* are related to the thermodynamical variables and to the shock strength : explicit expressions for a static corona are given in Gonczi et al.(1977), in the present case the detailed expressions will be given in the extended version in preparation. The energy conservation equation(5) will be used as a differential equation for the conductive flux F_c. The quantities A, B, C, D, E are used for reference in Table 2, they are respectively related to the kinetic energy flux, the potential energy flux, the enthalpy flux, the conductive flux and the mechanical flux. The radiative losses $\rho^2 \varphi(T)$ has been evaluated by various authors (for instance McWhirter et al. 1975). The optically thin medium hypothesis limits the validity of the function $\varphi(T)$ to temperatures above 50000°K.

Equations 2 to 5 form a system of four first-order differential equations, the independent variables are v, Φ_*, T and F_c ; ρ is related to these variables through equation 1, so F_m is a free parameter. The boundary conditions are the following : three input parameters fix the base temperature T_0 , the mass density ρ_0 and the mechanical flux Φ_0 at the chromospheric level $r = R_0$. Two free parameters F_{c0} and F_m are adjusted in order to get a solution i) which crosses the critical point for supersonic expansion, ii) which gives asymptotic temperature T $(r \to \infty) \to 0$. The choice of the mass flux and of the initial conductive flux is very "sensitive" ; for computation we use an adaptation of the shooting-splitting method described by Couturier (1977).

3. RESULTS AND CONCLUSION.

For one set of input parameters, tables 1 and 2 give some characteristic quantities obtained at four levels : the chromospheric level (I) the temperature maximum (II), the critical point (III) and the earth orbit (IV). To get this solution we find F_{c0}=-2.10^3 erg cm^{-2} s^{-1}, F_m = 7.7 10^{10} g s^{-1} sterad^{-1}. Quantities in table 2 are normalized with the total energy flux per unit mass which remains constant in the solar wind: 4.10^{14} erg g^{-1}. The radiative losses above level I amount to 9.10^4 erg cm^{-2} s^{-1}.

TABLE 1	$r/R_\odot$	n (e cm^{-3})	T (°K)	Φ_*(erg cm^{-2}s^{-1})	v(km s^{-1})
I	1.	2.2 10^9	5.25 10^4	1.18 10^5	0.04
II	1.29	1.75 10^7	1.25 10^6	1.13 10^4	2.92
III	8.25	1.3 10^4	5.72 10^5	3. 10^{-2}	97.3
IV	214.	7.88	1.41 10^5	0.	239.

TABLE 2	$r/R_\odot$	A	B	C	D	E
I	1.	0.	4.76	0.054	205.	-0.001
II	1.29	0.0001	3.68	1.29	3.35	0.
III	8.25	0.118	0.577	0.591	0.0005	0.867
IV	214.	0.711	0.022	0.145	0.	0.166

Taking into account the fact we have only three degrees of freedom for the input of this oversimplified model, we consider that the results fit reasonably the observations.Even if the observational support for sufficient mechanical fluxes in form of shocks remains questionable,the heating mechanism through another process will give the same structure of differential system,and the computing method is at hand to solve the problem. The extended version will discuss the degree of flexibility in the choice of input parameters and will apply the model to stellar winds.

REFERENCES

Couturier,P. : 1977, Astron. Astrophys. 59, pp239-248.
Couturier,P., Mangeney,A,Souffrin,P.:1979, Astron. Astrophys.74, pp9-11
Gonczi,G., Mangeney,A., Souffrin,P. :1977,Astron. Astrophys. 54, pp689-7C
McWhirter,R.W.P., Thonemann,P.C., Wilson,R. : 1975, Astron. Astrophys.
                              40, pp63-71

DISCUSSION

Kuperus: There seems to be increasing evidence that the propagation and dissipation of shock waves in a plane parallel or radially symmetric atmosphere, not taking into account the magnetic field, cannot satisfactorily explain the hot corona with its multitude of structures.

Couturier: I agree with your remark; the purpose of our work is to show that the energy balance through the whole corona and transition zone imposes severe constraints on the chromospheric parameters. That point represents some progress for the study of stratified stellar atmospheres. The next step will be to introduce just MHD shock waves as soon as we have performed the treatment of the evolution of such waves in a stratified atmosphere. The description of expanding flux tube of open magnetic fields will also be possible with some crude assumptions. I do not think, however, that we could get in that way a self consistent description of the inhomogeneous structures of the transition zone, but, for other atmosphere and solar mass loss studies, it is more important to reduce the number of input parameters in a self-consistent model than to develop a complex model with more degrees of freedom which could be fitted to solar observations. Smoothing the inhomogeneous structures of the corona would not affect the global energy balance in open field regions as long as non-resistive dissipation of magnetic fields is not a predominant mechanism.

Lemaire: At which altitude in your model is the mean free path of a thermal proton becoming larger than the density scale height?

Couturier: Above the critical point!

REFLEXION AND TRANSMISSION OF ALFVEN WAVES IN AN ATMOSPHERE.

Bel, N. and Leroy,B.
Observatoire de Paris
Département d'Astrophysique Fondamentale
92190   Meudon
France

If we want to study the transfer of energy by waves through a stellar atmosphere, we first need to know its reflectivity with respect to each of the three MHD modes. Here, we will consider linear Alfvén waves only. The problem is set as follows. An energy flux being given at the bottom (z=0) of the atmosphere, we ask for how much of it is recorded at an altitude z.

The atmosphere is assumed to be isothermal and permeated by a uniform, vertical magnetic field B. (The vertical is taken as the z-axis of a Cartesian system of reference.) The equations of motion can be put into a matrix form which exhibits the linear couplings between ascending and descending "modes" (Leroy,1979). It is the so-called coupled modes description (Budden,1961). They then write:

$$W' = \begin{vmatrix} ik(z) & -1/4H \\ -1/4H & -ik(z) \end{vmatrix} W, \qquad (1)$$

where the prime (') denotes differentiation with respect to z, $k(z) := \omega \exp(-z/2H)/v_A(0)$ ($v_A(0)$ is the Alfvén velocity at z=0 and ω is the frequency), and H is the scale height. A time dependence $\exp(i\,\omega\,t)$ is assumed throughout. In the limit $H \to \infty$, W_2 turns out to be an ascending wave and W_1 a descending one. In the case of a finite scale height it has been shown (Leroy,1979) that W_2 (W_1) can still be regarded as an ascending (a descending) wave.

From Eq.(1) it can be shown that the components W_1 and W_2 each satisfy a linear second-order differential equation; it follows that W_1 and W_2 may each be expressed as a linear combination of two particular solutions, say F,f and G,g, respectively. These particular solutions cannot be arbitrary; they must be coupled by Eq.(1). A convenient choice of initial conditions is

$$F(0) = g(0) = 1 \text{ and } f(0) = G(0) = 0.$$

131

M. Dryer and E. Tandberg-Hanssen (eds.), Solar and Interplanetary Dynamics, 131-133.
Copyright © 1980 by the IAU.

Then, the solutions such that

$$W_1(0) = W_{10}, \qquad W_2(0) = W_{20}$$

may be expressed in matrix notation in the form:

$$\begin{vmatrix} W_1 \\ W_2 \end{vmatrix} = \begin{vmatrix} F & f \\ G & g \end{vmatrix} \begin{vmatrix} W_{10} \\ W_{20} \end{vmatrix},$$

or, inverting this relation:

$$\begin{vmatrix} W_{10} \\ W_{20} \end{vmatrix} = \begin{vmatrix} g & -f \\ -G & F \end{vmatrix} \begin{vmatrix} W_1 \\ W_2 \end{vmatrix} =: M(z) \begin{vmatrix} W_1 \\ W_2 \end{vmatrix}. \qquad (2)$$

Such a formulation is well known in the Optics of stratified media (Abelès, 1950), where the matrix $M(z)$ is called the <u>characteristic matrix</u> of the stratified medium.

Now, we calculate the reflectivity and transmissivity of a slab of atmosphere extending from $z=0$ to an arbitrary height z. To do this, it must be prescribed that no wave is descending into this slab from above. If we denote the amplitudes of the incident (ascending) and reflected waves at the bottom of the slab by A and R, respectively, and the amplitude of the transmitted wave by T, Eq.(2) yields:

$$\begin{vmatrix} R \\ A \end{vmatrix} = M(z) \begin{vmatrix} 0 \\ T \end{vmatrix}. \qquad (3)$$

By definition, the reflexion and transmission coefficients are $r = R/A$ and $t = T/A$, respectively. It is easy to see that, in our case:

$$r = -f/F \qquad \text{and} \qquad t = 1/F.$$

We have shown that the energy flux density associated to a wave of amplitude W_i $(i=1,2)$ is

$$B^2 \omega^2/8\pi \; v_A^2(0)) \; |W_i|^2 ;$$

the reflectivity and transmissivity are therefore given by $|r|^2$ and $|t|^2$, respectively.

We apply this to the particular case of the solar atmosphere. This is described as a sequence of two isothermal (lossless) layers; the first one represents the photosphere and chromosphere, the second one the corona. The discontinuity in temperature simulates the transition region. In order that the dynamical equilibrium be preserved the continuity of pressure is written at the interface between the two layers ($z=h$). As in Optics, each layer is described by its characteristic matrix, and the two-layer sequence by the product of these matrices (Abelès, 1950). In Eq.(3) the matrix $M(z)$ is then:

$$M(z) = M_1(h) M_2(z-h), \qquad\qquad (4)$$

where indices 1 and 2 refer to the first and second layer, respectively.

The numerical results for the solar case are presented in Table 1. We adopted the following figures:

$$H_1 = 200 \text{ km}, \quad T_1 = 5000 \text{ °K},$$
$$H_2 = 1000 \text{ km}, \quad T_2 = 10^6 \text{ °K},$$
$$h = 2000 \text{ km}, \quad \rho_1(0) = 10^{-7} \text{ g.cm}^{-3}.$$

The results are presented for two values of the magnetic field intensity, B = 1 G and B = 10 G, as well as for two oscillation periods, $2\pi/\omega = 300$ and $2\pi/\omega = 20$ h.

Whatever the strength of the magnetic field the Alfvén waves of shorter frequency (possibly generated by supergranulation motions) are almost totally reflected downwards at an altitude z= 5000 km. As for the waves of higher frequency, the weaker the magnetic field, the lesser reflected they are.

Table 1. Reflectivity of a slab of solar atmosphere
extending from the bottom of the photosphere
to an altitude z= 5000 km.

	$2\pi/\omega = 300$ s	$2\pi/\omega = 20$ h
B = 1 G	.03	.98
B = 10 G	.45	.99

References:

Abelès, F.: 1950, Annales de Phys. 5, pp. 598-640.
Budden, K.G.: 1961, Radio waves in the Ionosphere, University Press, Cambridge.
Leroy, B.: 1979, Astron. Astrophys., in press.

SOLAR RADAR OBSERVATIONS

A.O. Benz
Radio Astronomy Group, Microwave Laboratory ETH Zürich,
Switzerland

Abstract

Radar observations of the sun have been made extensively at decameter and low meter wavelengths (Eshleman et al., 1960, and James,
1966). Their interpretation by specular reflection on high density
structures with "corner reflector" shape is unlikely from the echo
spectral broadening and range depth. Gordon's (1973) interpretation of
the scattering by a 4 wave interaction between radar and coronal Langmuir waves requires a level of 10^{-2}nKT (thermal energy density) of the
Langmuir waves. A radar experiment in microwaves with the 300 m dish
in Arecibo*) is described, which was able to test this hypothesis. It
was based on the idea of scattering radar waves on Langmuir waves by
the much more efficient 3 wave interaction. The echo at the beat
frequency of the radar (2380 MHz) and the Langmuir wave (170 -270 MHz)
is then to be expected at 2600 MHz. The results, however, show the
absence of echos, from which an upper limit of 6.10^{-4}nKT for the level
of Langmuir waves is derived. First results will soon be published
(Benz and Fitze, 1979).

Here I report from an other microwave radar experiment in Arecibo
which was receiving at the transmitted frequency, similar as the decametric radar observations. We have probed a coronal streamer, a coronal
hole, and an active region (emitting noise storm radio bursts).

No echo has yet been detected. Previously, radar observations have
been suggested to be possible only at low frequencies, since the optical
depth of the plasma layer increases rapidly with frequency, absorbing
any echo mirrored at this plasma. However, James' result clearly shows
that the reflection occurs well above the plasma layer.

I propose that decametric and low metric radar echos are produced
by interactions of the radar wave with ion acoustic waves. Such low
frequency electrostatic waves have been suggested by Benz and Wentzel
(1979) to be present in type I radio burst sources due to a current-

135

M. Dryer and E. Tandberg-Hanssen (eds.), Solar and Interplanetary Dynamics, 135-138.

driven instability, which dissipates free magnetic energy. The requirement of this model on magnetic flux into the corona, 10^{15} gauss cm^2s^{-1}, fits very well the observations. The absence of an echo from Langmuir waves is then due to their low level, and the difficulty to detect echos at high frequency due to the high collisional damping of ion acoustic (and radar) waves at lower heights.

References

Benz, A.O., and Fitze, H.R.: 1979, Astron.Astrophys., in press.

Benz, A.O., and Wentzel, D.G.: 1979, Bull.Am.Astron.Soc.11,pp.141, submitted to Astron.Astrophys.

Eshleman, V.R., Barthle, R.C., and Gallagher, P.B.: 1960, Science 131, pp. 329-330.

Gordon, I.M.: 1973, Space Sci.Rev. 15, pp. 157-204.

James, J.C.: 1966, Astrophys.J. 146, pp. 356-372.

*) The Arecibo Observatory is part of the NAIC which is operated by Cornell University under contract with the US National Science Foundation.

DISCUSSION

Degaonker: How accurately can the height or level of reflection of radar echoes be determined? I would like to know the height from the photosphere from which these decametric echoes are received? I am interested in knowing the height of the transition region from closed to open field lines and its variation.

Benz: The accuracy of the height of reflection is determined by the change in group velocity of the radar waves near the plasma level and the unknown position of the scattering center on the sun. Its distance from the plane through the center of the sun is typically 1.4 solar radii with a maximum range from 0.6 to 2.0. The mean value (1.4 solar radii) may well correspond to the transition you refer to.

Dryer: Is it possible to indicate any possible relevance of the radar observations of wave activity above active regions to the observations of the (so-called) solar flare precursors?

Benz: In principle, this is possible. One would not do this by an experiment like mine with high spatial resolution (and therefore low probability to hit the target), which is more appropriate to a stable source. J. C. James has observed extremely strong echoes during the passage of a coronal shock (type II solar radio burst).

Kuperus: Concluding that a large level of low frequency turbulence exists in a more steady situation could well be interpreted as good evidence for anomalous conductivity in large parts of the active corona.

Benz: Note that the very high level of ion acoustic waves in our noise stress level is very localized in space and time. The low frequency turbulence responsible for the decametric radar echo may be

much weaker. Nevertheless, it would enhance the resistivity above
Spitzer's in a large region.

Bratenahl: Low level, low frequency could still represent
anomalous resistivity if current path is very narrow with high current
density.

Benz: Indeed, I have assumed a source volume V of 10^{28} cm ^{3} for the
radar scatterer and the value for the energy density of Langmuir

waves scales like $10^{-4} \left(\dfrac{10^{28}}{V} \right)$. Furthermore, I have assumed that in

the average, one source was present during the integration time of
15 min, and that the electron temperature in these sources is 5×10^6 K.

Moore: You seem to favor ion acoustic waves as the basic phenomenon
for generating the echo. How do you select this non-magnetic mode over
others in which the magnetic field is directly involved?

Benz: I do not. The Benz and Wentzel noise storm model suggests
ion acoustic waves as the cause of the bursts. For the radar
reflection process lower hybrid or ion cyclotron waves may well be
more important.

Gergely: I believe that reconnection related to noise storms cannot
be a primary dissipation mechanism. There are active region complexes
which are seen on the disk for seven or eight rotations. They sometimes produce a storm during one rotation, no storm during the next,
and a storm again during the third. So there must be something in the
field geometry which determines if there is or is not a storm. Some
of the strongest storms are related to old, decaying regions, not
young regions.

Benz: The energy release in the Benz and Wentzel type I model is
less than 10% of the energy input into an active region. The density
in noise storm sources is between 10^7 and 10^9 cm^{-3}, which makes it
difficult to detect them in X-rays. The fact that noise storms also
occur in decaying regions may simply be explained by coronal field
rearrangements, which are also necessary in the decay phase of a spot.

Bhonsle: In your presentation you have mentioned two extreme
frequencies for solar radar echoes, namely, at 38 MHz and 2380 MHz.
In the former case one obtains strong echoes and in the latter case
no echo is observed. I would like to know whether there have been
attempts to obtain echoes at other intermediate frequencies. I think
it should be of considerable theoretical interest to determine
"transition" frequency at which a change over from "echo" to "no echo"
condition occurs.

Benz: Note that the microwave radar experiment probes Langmuir
waves between 170 and 270 MHz; the decametric radar, however, waves at
places with plasma frequencies above 38 MHz. These are the numbers
you have to compare. The attempt to find echoes from the 408 MHz level
has been made in Arecibo. Since it was done during a sunspot
minimum, its negative result is no surprise.

Stewart: Why did you choose a radar frequency about 10 times the plasma frequency?

Benz: The coupling between radar and Langmuir waves is proportional to the square of the wave number, k_L, of the Langmuir wave. The largest possible k_L is about one third of the inverse Debye length. Using the resonance condition ($k_L \simeq 2k_r$), one finds $\omega_r \simeq 10\omega_L$ as optimal condition.

Stewart: Your radar experiment only puts an upper limit on plasma turbulence because you did <u>not</u> receive an echo. So any talk of a high level of turbulence is speculative.

Benz: The microwave radar experiment has practically ruled out the old interpretation of decametric radar echos in terms of Langmuir turbulence. A new theory for decametric echos is necessary. The presence of low frequency turbulence is more evidently suggested by the good association of strong radar echos with noise storms.

MODE-COUPLED MHD WAVES IN THE CORONA AND SOLAR WIND

M. Heinemann
Department of Physics, Boston College, Chestnut Hill, MA
02167
S. Olbert
Department of Physics and Center for Space Research, Massachusetts Institute of Technology, Cambridge, MA 02139

The purpose of this paper is to outline a model of mode-coupled MHD compressional waves in the corona and solar wind. The eventual aim of this work is to be able to compute how MHD waves propagate through the corona and into the solar wind beginning with a source of Alfven or fast mode waves at the base of the corona. The necessity for consideration of mode coupling arises because of typical scalelengths in the corona. For wave sources, such as supergranulation, with wave periods of about a day, the different modes do no propagate independently, as in the WKB approximation, but are coupled because the ratio of wavelength to scalelength is of the order of one or greater.

The mathematical basis for this work is a model of small amplitude, poloidal, axisymmetric waves in an axisymmetric background flow in which solar rotation is neglected. The main effort has been to relax the WKB assumption: that is, we treat the wave equations directly. The formulation is based on the poloidal components of the ideal MHD equations: the mass conservation equation, Euler's equation (with a scalar pressure and including gravity), the ideal Ohm's law, and a polytropic pressure relation. We have written the equations in terms of the Fourier amplitudes of the fluctuations of the total pressure, the components of the flow and Alfven velocities parallel and perpendicular to the background magnetic field, and the density. In terms of these amplitudes, the equations in hyperbolic regions are

$$\partial_{\pm}\left\{ \rho\sqrt{(V_A^2-V^2)\left[V^2(V_A^2+c_s^2)-V_A^2c_s^2\right]}\,\delta V_{\perp} \pm \sqrt{(V^2-c_s^2)}\,\delta P_T \right\} = \cdots \qquad (1-2)$$

$$\partial_{\parallel}(V\delta V_{A\perp} - V_A\delta V_{\perp}) = \cdots \qquad (3)$$

$$\partial_{\parallel}\left[\rho(V\delta V_{\parallel} - V_A\delta V_{A\parallel}) - \tfrac{1}{2}V_A^2\delta\rho + \delta P_T\right] = \cdots \qquad (4)$$

$$\partial_{\parallel}\left[\rho(V_A\delta V_{\parallel} - V\delta V_{A\parallel}) - \tfrac{1}{2}VV_A\delta\rho\right] = \cdots \qquad (5)$$

$$\partial_{\parallel}\left[(V_A^2+c_s^2)V\delta\rho + \rho V_A^2\delta V_{\parallel} - V\delta P_T\right] = \cdots \qquad (6)$$

M. Dryer and E. Tandberg-Hanssen (eds.), Solar and Interplanetary Dynamics, 139-141.
Copyright © 1980 by the IAU.

The right hand sides of the equations, which contain terms proportional to the amplitudes and determine the coupling, are omitted for
brevity. Even though these equations are complicated, they have several simple features and are transparent enough to outline a method of
solution. Of the six equations, the first two are written as partial
derivatives along two families of characteristics, shown as the curved
lines in the hyperbolic regions in Figure 1. They are drawn for radial
field lines only because they are much more complicated for nonradial
geometries. The equations contain explicit singularities which occur

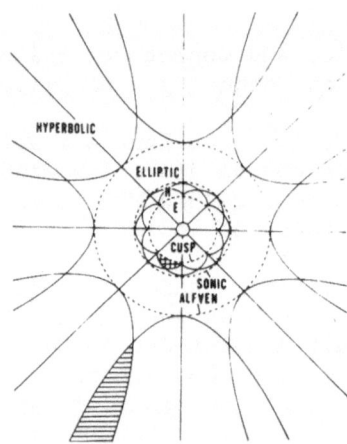

Figure 1. The characteristics and singular lines.

where the flow speed is equal to one of the MHD group velocities along
the field direction: the Alfven speed, the sound speed, and the speed
of the cusp at the rear of the slow mode wave front. These singularities lie along lines defined by the background flow; the singular lines
are shown as dotted lines in Figure 1. The remaining four equations
are written as partial derivatives along the field lines. In the
hyperbolic regions, the equations can be integrated as ordinary differential equations along the characteristics and the field line. In
elliptic regions, the characteristics are not defined; the first two
equations take a different form there.

These equations describe the coupling of the MHD fast and slow
mode waves and two entropy modes. This identification can be made by
replacing $\underline{\nabla}$ by $i\underline{k}$, retaining terms involving $i\omega$, and setting the gradients on the RHS to zero. The result is that there are two fast and two
slow wave modes which occur in forward and backward propagating pairs
for a given $\underline{k}$. The physical significance of the characteristics is
that they represent the envelopes of the fast and slow mode wave fronts:
they are the Mach "cones" of the magnetized plasma (shown as cross-
hatched areas in Figure 1). By a standard construction, shown in
Figure 2, one can show that when the flow velocity lies inside the
slow mode wave front or outside the fast mode wave front there are

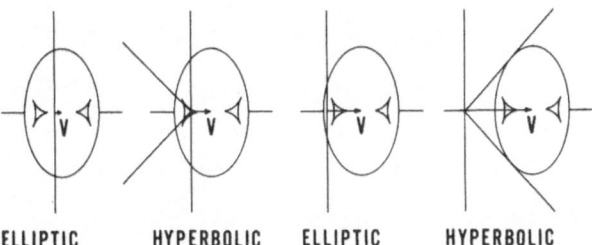

ELLIPTIC   HYPERBOLIC   ELLIPTIC   HYPERBOLIC

Figure 2. Relation of the wave modes to the characteristics.

real characteristics, but not otherwise. It is clear from Figure 2
that the inner hyperbolic region represents the envelope of slow mode
waves propagating toward the Sun while the outer hyperbolic region
represents the envelope of fast mode waves being convected away from
the Sun.

The terms on the right hand side describing the coupling are quite
long and tedious. However, there are only a few essentially different
quantities which appear. They are: (1) the curvature of the field
lines, (2) derivatives of the density, and (3) derivatives of field
line constants characterizing the background flow.

In attempting to solve the equations, the singular lines are of
particular importance. We have obtained regularity conditions, in the
form of the vanishing at the singular lines of linear combinations of
quantities related to the amplitudes, which are required for smooth
flow through the lines. Physical interpretations are: the cusp line is
singular for slow mode waves propagating across (i.e., $\underline{k}$ perpendicular
to) the field, the sonic line is singular for slow mode waves propagating backward along the field, and the Alfven surface is singular for
fast mode waves propagating backward along the field. The regularity
conditions, which are similar to the Parker critical point condition,
are equivalent to boundary conditions: each one reduces by one the number of free boundary conditions that may be stated at the base of the
corona.

Based on this analysis, it is possible to outline a method of numerical solution. Methods in each type of region are standard. For
given boundary conditions, solutions can be obtained in the elliptic
regions by the method of relaxation. In the hyperbolic regions they
can be obtained by integrating along the characteristics. Because the
boundary conditions and the regularity conditions are not stated on the
same lines, however, the solutions cannot be obtained all at once. The
novel feature of this problem is that it will be necessary to iterate
among solutions in the different regions in order to satisfy simultaneously the regularity conditions and the boundary conditions at the base
of the corona.

PROPERTIES OF MAGNETOHYDRODYNAMIC TURBULENCE IN THE SOLAR WIND

M.Dobrowolny (Laboratorio Plasma Spazio, CNR, Frascati, Italy)
A.Mangeney (Observatoire de Meudon, Paris, France)
P.L.Veltri (Dipartimento di Fisica, Università della Calabria, Italy)

The observations of MHD turbulence in the solar wind indicate that this is in a state characterized, to a good degree by the absence of non linear interactions. It is argued that this is a general property of incompressible MHD turbulence in a magnetized plasma.

A well known result of the measurements of hydromagnetic turbulence in the solar wind (Belcher and Davis, 1971; Burlaga and Turner, 1976) is the correlation between velocity ($\delta \underline{v}$) and magnetic field ($\delta \underline{B}$) fluctuations, which satisfy to a good degree (especially in the regions of the streams' trailing edges) the relation

$$\delta \underline{v} = \pm \delta \underline{B}/\sqrt{4\pi\rho} \tag{1}$$

with a sign implying that, of the two possible modes, only those propagating away from the Sun are present.

This property can be shown (Dobrowolny et al., 1979) to imply that the solar wind MHD turbulence is in a very peculiar state characterized by the absence of non linear interactions. This follows from the equations of incompressible magnetohydrodynamics written in the form

$$\frac{\partial}{\partial t} \delta \underline{Z}^{\pm} \mp (\underline{C}_A \cdot \nabla)\delta \underline{Z}^{\pm} + (\delta \underline{Z}^{\mp} \cdot \nabla)\delta \underline{Z}^{\pm} = -\nabla(p + B^2/8\pi) \tag{2}$$

$$\Delta^2(p + B^2/8\pi) = -\nabla \cdot \left[(\delta \underline{Z}^+ \cdot \nabla)\delta \underline{Z}^- \right] \tag{3}$$

where

$$\delta \underline{Z}^{\pm} = \delta \underline{v} \pm \delta \underline{B}/\sqrt{4\pi\rho} \tag{4}$$

and $\underline{C}_A = <\underline{B}>/\sqrt{4\pi\rho}$ is the Alfvénic speed in the average field $<\underline{B}>$. Note that, in the case of infinitesimal fluctuations, $\delta \underline{Z}^{\pm}$ represent the two possible Alfvénic waves propagating away and toward the Sun. The property (1) is then equivalent to having, either $\delta \underline{Z}^+ = 0$ and

143

M. Dryer and E. Tandberg-Hanssen (eds.), Solar and Interplanetary Dynamics, 143-146.
Copyright © 1980 by the IAU.

$\delta \underline{Z}^- \neq 0$ or viceversa. As the non linear terms in eq. (2) are of the type $(\delta \underline{Z}^{\mp} \cdot \nabla) \, \delta \underline{Z}^{\pm}$, it follows that, in fact, the magnetohydrodynamic turbulence in the solar wind is, to a good approximation, in a state without non linear interactions.

We argue now that this peculiar property is not a particular one of the turbulence in the solar wind, but must be rather considered a general property of developed incompressible MHD turbulence in the presence of a background magnetic field. The point is the following: waves generated at the Sun must be necessarily outwardly propagating, if observed at 1 AU (Belcher and Davis, 1971). However the maximum power in Alfvén waves generated from motions in the convection zone and getting out in the solar wind, is found to be mainly in periods of the order of one to few hours, which correspond also to typical scales of supergranulation (Hollwegg, 1978). Much less power can get out at higher frequencies which, on the other hand, may suffer severe photospheric damping (Osterbrock, 1961). In contrast to this, the power spectra measurements (extending to periods of $\sim$ 1 s), indicate that there is appreciable power in the shorter periods also (from 1 s to 1 h). Adding to this the fact that there are mechanisms of local wave generation, at the higher frequencies, for example from velocity shear or microscopic instabilities, we conclude it is unlikely that the higher frequencies observed in the wind are of solar origin. The argument of solar origin, as an explanation for having only outwardly propagating waves does not therefore apply to such higher frequencies which, however, consist still of outwardly propagating waves.

Thus the property of being in a state characterized by the absence of non linear interactions, seems to be a general outcome of the development of incompressible anisotropic MHD turbulence.
We will now show that physical arguments of the same dimensional type than those used to derive the $k^{-3/2}$ law for the spectrum of isotropic turbulence (Kraichnan, 1965), indicate in fact an evolution of developed anisotropic MHD turbulence toward a state where one of the possible modes has disappeared.

Consider the interaction between vortices of the same scale ℓ. The elementary interaction time, determined by the velocity C_A in the background magnetic field is $\tau_{int} \sim \ell/C_A$. From eq. (2) we can derive, in order of magnitude, the variation $dZ^{\pm}$ in amplitude of a given vortex ($\delta Z^{\pm}$), in one interaction time, due to its interaction with a vortex of the opposite type ($\delta Z^{\mp}$). This will be

$$dZ^{\pm} \sim \tau_{int} (\delta Z^{\pm} \delta Z^{\mp})/\ell$$

In N such stochastic interactions the amplitude variation will be $\Delta Z^{\pm} = \sqrt{N}\ dZ^{\pm}$. Hence, the number of interactions $N^{\pm}$ it takes to obtain a variation, for a given vortex $\delta Z^{\pm}$, equal to its initial amplitude (i.e. $\Delta Z^{\pm} \sim \delta Z^{\pm}$) will be given by

$$N^{\pm} \sim \ell^2 / (\delta Z^{\mp})^2 \tau_{int}^2$$

The corresponding time $T^{\pm}$ needed to obtain such a variation is

$$T^{\pm} \sim c_A \ell / (\delta Z^{\mp})^2 \tag{5}$$

and should be considered as a typical time for a significant local energy transfer across the spectrum for a given type of mode.

Using (5), we could easily reproduce the power law $-3/2$ for the spectrum of the symmetric case ($\delta Z^{+} \sim \delta Z^{-}$). As this is a known result (Kraichnan, 1965), we refer rather to the asymmetric case $\delta Z^{+} \neq \delta Z^{-}$, i.e. a case where there is, at some time, an unbalance between the energy in the $\pm$ modes. For example, suppose $\delta Z^{+} > \delta Z^{-}$. Then, from (5)

$$T^{+}/T^{-} \sim (\delta Z^{+}/\delta Z^{-})^2 \qquad > 1 \tag{6}$$

which means that the mode δZ^{-} transfers energy across the spectrum faster than the mode δZ^{+} (and then dissipate at the short wavelengths). As the energy per unit time pumped at the source is constant, the energy in the modes δZ^{-} will further decrease. We thus see that, starting from an initial unbalance $|\delta Z^{+}| > |\delta Z^{-}|$, we have necessarily a cascade towards a situation where all the energy available remains in the mode δZ^{+}. This is precisely the state indicated by the observations of MHD turbulence in the interplanetary plasma and the above physical reasoning derives it quite generally as a consequence of the nature of the non linear terms in the equations of incompressible MHD.

REFERENCES

Belcher, J.W., and Davis, L.: 1971, J. Geophys. Res. 76, pp. 3534-3563.
Burlaga, L.F. and Turner, J.M.: 1976, J. Geophys. Res. 81, pp. 73-77.
Dobrowolny, M., Mangeney, A. and Veltri, P.L.: 1979, to be published in Astron. and Astrophys.
Hollweg, J.V.: 1978, Solar Phys. 56, pp. 305-333.
Kraichnan, R.H.: 1965, Phys. of Fluids 8, pp. 1385-1387.
Osterbrock, D.E.: 1961, Astrophys. J. 134, pp. 347-388.

DISCUSSION

Heinemann: Belcher and Davis' original argument depended, essentially, on the solar wind flow speed; that is, that the flow speed eventually became greater than the Alfven speed. I believe you neglected the flow speed in your equations: all that appeared was the Alfven speed. Would your results change in a qualitative way if you retained the flow velocity?

Dobrowolny: My arguments refer to the case of an average uniform velocity of the fluid as you can locally suppose in the interplanetary medium on the scale length of Alfven waves. On the other hand, the argument of solar origin, to explain the presence of only outwardly propagation waves does not probably apply at least to the highest frequencies observed (which may be locally generated) and which are also found to be outwardly propagating.

AN EMPIRICAL RELATION BETWEEN DENSITY, FLOW VELOCITY
AND HELIOCENTRIC DISTANCE IN THE SOLAR WIND

M. Eyni and R. Steinitz
Department of Physics, Ben-Gurion University
Beer-Sheva, Israel

1. INTRODUCTION

Solar wind flow quantities such as matter flux, momentum flux and energy flux, may be closely related to the mechanism responsible for the evolution of the solar wind. In the highly supersonic flow regime their study is facilitated by the fact that the contribution of thermal motions to the momentum and energy fluxes is negligibly small, and thus all three quantities are expressible in terms of proton density n and flow velocity u. If a relation between n, u and heliocentric distance r can be established, the study of these quantities is further simplified. In the following we point out that such a relation does in fact exist, and comment on its implications.

2. THE (n,u,r) RELATION

In looking for a relation between n, u and r we have the suspicion that such a relation may be lost with heliocentric distance due to stream-stream interactions. For our analysis we therefore used data from Helios 1 (Rosenbauer et al., 1977) which is the most recent data available to us, and includes data measured as close as 0.3 AU to the sun. The published data were averaged over time intervals ranging typically over half a day to five days, restricting the variance in velocity in any given interval to less than about 100 km/s. The details will appear elsewhere.

We fitted by least squares the data from Helios 1 to the relation

$$\log n = a \log u + b \log r + C . \tag{1}$$

Here n is in protons/cm^3, u is in km/s and r is in AU. From the fit we obtain a = -2.0; b = -2.0 and C = 6.1, yielding the result

$$n = 1.3 \times 10^6 \, r^{-2.0} \, u^{-2.0} \tag{2}$$

Relation (2) gives the sought after dependence of n on u.

147

M. Dryer and E. Tandberg-Hanssen (eds.), Solar and Interplanetary Dynamics, 147-150.
Copyright © 1980 by the IAU.

The average flow velocity u measured by Helios closer to the sun was smaller than at larger distances and correspondingly n attained higher values closer to the sun. This velocity bias simulates a stronger decrease of n with r than the correct one, giving $n \sim r^{-2.35}$ instead of $n \sim r^{-2}$. However, we can see from relation (2) that nu^2 does not depend on velocity and thus we obtain $nu^2 \sim r^{-2.0}$. In figure 1 we plot $\log(nu^2)$ against log r. The slope of the fitted line is exactly -2.0. The figure also shows that the spread in the distribution of $\log(nu^2)$ increases with heliocentric distance.

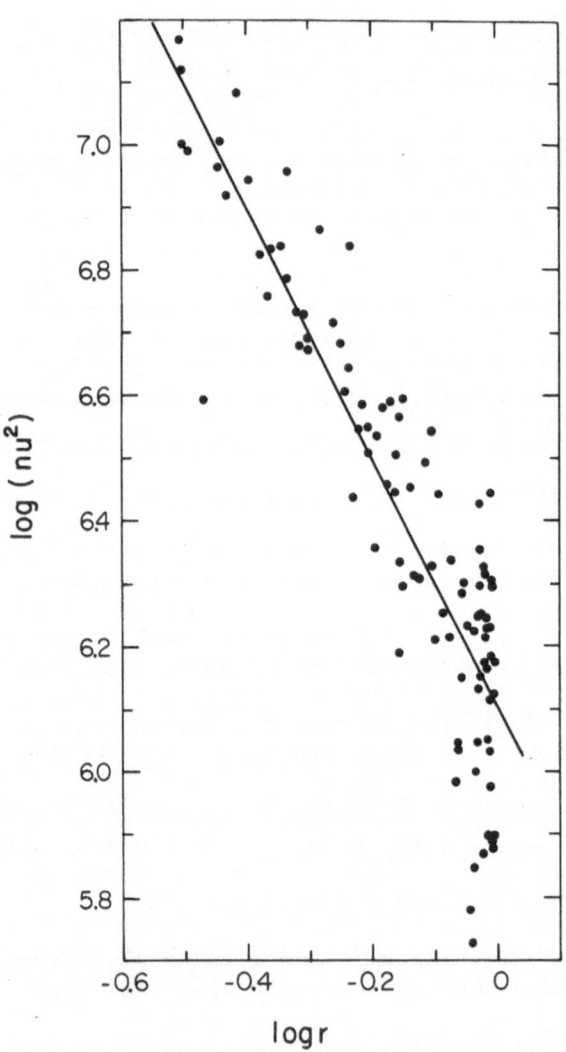

Figure 1. $\log(nu^2)$ vs. log r from Helios 1 data.

3. DISCUSSION

The relation given by equation 2 can be interpreted as an invariance of the momentum flux density $F_\mu = mnu^2$, relative to stream structure at a given heliocentric distance. The relation also implies that the matter flux density $F_m = mnu$ decreases with increasing flow velocity.

Coronal holes are characterized by the smaller radiative losses in the U.V. compared to other parts of the corona. Thus our finding that the kinetic energy flux density ($\frac{1}{2} mnu^3$) increases with flow velocity, supports the notion that fast streams originate in coronal holes.

From the increase of the dispersion of $\log(nu^2)$ with r, we conclude that our initial suspicion is correct, and that the relation between n and u may indeed be lost due to stream-stream interaction. In passing we note that also the dispersion of the logarithm of proton temperature T increases with r.

The dispersion in $\log(nu^2)$ or $\log T$ may be smaller for streams originating in higher solar latitudes, presumably because the stream-stream interaction due to solar rotation becomes less effective with higher solar latitude of the wind origin. This idea can be verified in the future, when measurements out of the ecliptic will become available.

REFERENCE

Rosenbauer, H., Schwenn, R., Marsch, E., Meyer, B., Miggenrieder, H.,
    Montgomery, M.D., Mühlhäuser, K.H., Pilipp, W., Voges, W., and
    Zink, S.M.: 1977, *J. Geophys.*, 42, pp. 561-580.

DISCUSSION

LaBonte: (1) Is the increasing fluctuation of the quantity nu^2 with radial distance real or instrumental? (2) What is its cause? (3) Do data from spacecraft at radial distances >1 AU show a continuation of this effect?

Eyni: (1) The increasing fluctuations of the quantity $\underline{\log nu^2}$ with radial distance is real. The proton temperature shows a similar behavior.

(2) The cause could be stream-stream interaction.

(3) We did not make a similar analysis for radial distances >1 AU. The effect has to diminish at large heliocentric distances, because the interaction between streams makes the stream pattern disappear at large distances from the Sun. Indeed, Intriligator has shown that the variance of velocity decreases with the distance.

Ahluwalia: What is the range of velocities (u) used by you in your correlation analysis? I ask this question because if u varies over a very <u>narrow range</u> then the way you have set-up your equations, the correlation analysis becomes independent of u. We must then end up with

the relation: $n \propto \dfrac{1}{r^2}$, which is not a very surprising result!

Eyni: The velocity range of the present Helios 1 analysis is 300–720 km/s. This range (see, also, the accompanying paper by *Steinitz and Eyni*) cannot be considered to be a narrow range for u.

ARE SOLAR WIND MEASUREMENTS OF DIFFERENT SPACECRAFT CONSISTENT?

R. Steinitz and M. Eyni
Department of Physics, Ben-Gurion University
Beer-Sheva, Israel

1. INTRODUCTION

Results of solar wind measurements by different spacecraft are not always in full accord. Such measurements are in general not from one and the same distance r from the sun, nor are they taken at the same phase of the solar activity cycle. One would like to be able to discriminate between spacecraft calibration effects on the one hand, and solar wind variations which reflect true spatial gradients or changing boundary conditions at the sun on the other hand. Accordingly, we examine in this paper the possibility of reconciling the apparent discrepancies.

In the following, we first compare density measurements and relate their differences to the velocities with which they have been sampled. Furtheron we compare proton temperature gradients obtained from Helios 1 and Mariner 2 and suggest that the different results are due to stream-stream interactions in the solar wind. We conclude with a brief summary of the more obvious as well as the subtle effects velocity sampling has on the evaluation of some solar wind parameters and their gradients.

2. COMPARISON OF DENSITIES

Table 1 lists average proton number densities $<n>$ from Mariner 2 (Neugebauer and Snyder, 1966), Vela 3 (Bame et al., 1971), Helios 1 (Rosenbauer et al., 1977), Heos (Formisano and Moreno, 1974) and Imp 6, 7,8 (Feldman et al., 1978). Making use of the result that $n \sim r^{-2}$ (Eyni and Steinitz, 1979), the densities given in the table are normalized to 1 AU. Also average velocities $<u>$ and average momentum flux densities per unit mass are given. For Mariner 2, Vela 3 and Helios 1 it is imme-diately apparent that the differences in average densities can simply be attributed to velocity biasing: for Mariner 2 higher densities closer to the sun are accompanied by lower average velocities, an effect which is further enhanced for Vela 3. In full accord with momentum flux density invariance (Eyni and Steinitz, 1979), the last column in the table indi-cates the high degree of consistency between Mariner 2 and Vela 3, as

M. Dryer and E. Tandberg-Hanssen (eds.), Solar and Interplanetary Dynamics, 151-154.

Table 1.   Proton Number Densities

	No. of points	Range (AU)	$\langle n \rangle$ (cm^{-3})	$\langle u \rangle$ (km s^{-1})	$\langle nu^2 \rangle / 10^{16}$ (cm^{-1} s^{-2})
Mariner 2	35	0.70-0.83	6.75	461	1.18
	50	0.84-1.00	4.45	529	1.14
Vela 3	122	1.00	7.01	410	1.14
Helios 1	37	0.30-0.50	7.22	492	1.35
	61	0.50-1.00	5.88	514	1.33
Heos	2513	1.00	4.17	410	0.70
Imp 6,7,8	56	1.00	9.13	482	-

well as the internal consistency of the Mariner 2 data. Thus, the momentum flux density enables a meaningful comparison to be made between different density measurements. Similarly, the Helios 1 data show good internal consistency, but a slight difference with Mariner 2 and Vela 3 (a calibration effect?). The Heos results appear to be inconsistent with the other measurements.

The mean velocity from Imp is similar to Helios 1 in the 0.3-0.5 AU range, yet the densities are significantly higher. The data we used for the Helios 1 analysis have velocities well represented in the range 300-720 km/s, but the Imp data are restricted to the range 365-560 km/s. The lower velocity range for Imp may be the result of (a) the fact that the published Imp data are averages over complete solar rotations, and (b) high velocities not persisting over complete rotations. The variance of velocity over a whole solar rotation together with the averaging process will thus result in higher mean densities corresponding to a given mean velocity, than would be obtained without averaging over a large velocity variance. Another effect of this averaging procedure is the weakening of the anticorrelation of density and flow velocity. A least-squares fit to the Imp data (56 points) yields

$$\log n = -1.12 \log u + 3.95 ,$$

which is considerably weaker than the $n \sim u^{-2}$ dependence.

3.   COMPARISON OF TEMPERATURE GRADIENTS

In a previous analysis (Eyni and Steinitz, 1978) we demonstrated the presence of proton cooling in the Mariner 2 data for velocities below 500 km/s, but for higher velocities cooling is not evident. However, the Helios 1 data show cooling in all velocity ranges.

Representing the temperature T for a given velocity as

$$T = T_0 r^{-\alpha} ,$$

we can regard the cooling index α as a measure of the temperature gradient. For u ~ 600 km/s, the Mariner 2 data yield α = 0.15, while the Helios 1 data yield α = 0.45 in the same range (0.7-1.0 AU). We suggest that the weak temperature gradient from Mariner 2, may be due to masking of the true cooling present, by stream-stream interactions. Presumably these interactions have been more effective in Mariner 2 than in Helios 1: in Helios 1 velocities above 550 km/s persisted unintermittently for typically 5-9 days, while in Mariner 2 they persisted only for 2-4 days.

4. CONCLUDING REMARKS

The effect velocity sampling has on the interpretation of solar wind densities and proton temperatures, can be summarized as follows:
(a) simple velocity biasing (i.e. Table 1) can result in erroneous values of densities and may mask the presence of temperature gradients;
(b) averaging over samples with a large velocity variance yields excessive densities and weakens the density-velocity anticorrelation;
(c) the presence of a large velocity variance is also accompanied by real physical effects - in the form of stream-stream interactions and a possible masking of the true temperature gradients by local heating.

We propose that the momentum flux density is a useful quantity for comparison purposes, since its variance is small. The suggestion that temperature gradients are masked by local heating through stream-stream interactions in the ecliptic plane, can be tested by probing the solar wind outside the ecliptic.

REFERENCES

Bame, S.J., Asbridge, J.R., Felthauser, H.E., Gilbert, H.E., Hundhausen, A.J., Smith, D.M., Strong, I.B., and Sydoriak, S.J.: 1971, *A compilation of Vela 3 solar wind observations, 1965 to 1967.* Los Alamos Scientific Report LA-4536.

Eyni, M., and Steinitz, R.: 1978, *J. Geophys. Res.,* 83, 215-216.

Eyni, M., and Steinitz, R.: 1979, this conference.

Feldman, W.C., Asbridge, J.R., Bame, S.J., and Gosling, J.T.: 1978, *J. Geophys. Res.,* 83, 2177-2189.

Formisano, V., and Moreno, G.: 1974, *J. Geophys. Res.,* 79, 5109-5117.

Neugebauer, M., and Snyder, C.W.: 1966, *J. Geophys. Res.,* 71, 4469-4483.

Rosenbauer, H., Schwenn, R., Marsch, E., Meyer, B., Miggenrieder, H., Montgomery, M.D., Mühlhäuser, K.H., Pilipp, W., Voges, W., and Zink, S.M.: 1977, *J. Geophys.,* 42, 561-580.

DISCUSSION

Stix: Couldn't it be that the discrepancy between the Mariner and Vela results on the one side and the Helios results on the other side is due to some deviation from the r^{-2} law which you assumed for your calibration to 1 AU?

Steinitz: As shown in the previous paper (Eyni and Steinitz) the r^{-2} is not an assumption, but a result. Therefore, I doubt that the difference between Helios 1 and Mariner 2 and Vela 3 is due to the r^{-2} dependence.

OBSERVATION OF DUST GENERATED HYDROGEN IN THE SOLAR VICINITY

H.J. Fahr, H.W. Ripken, and G. Lay
Institut für Astrophysik und Extraterrestrische Forschung
Universität Bonn
Auf dem Hügel 71, 5300 Bonn 1, Fed. Rep. Germany

Abstract. Solar wind protons impinging on interplanetary dust grains are
trapped, deionized, and subsequently desorbed. The steady state distribution of desorbed neutral hydrogen inside of 0.4 AU can be deduced by
observation of resonantly scattered solar 121.6 nm radiation. Calculated
integral intensities and spectral profiles are given, showing the clear
spectral separation of the different radiation components. Given a specific solar wind proton flux, the interplanetary dust distribution can
be determined. Conversely, dust density profiles from zodiacal light
measurements can be used to deduce solar wind proton fluxes at heliocentric distances of 0.4 to 0.15 AU. Observations of latitudinal and short-
term temporal proton flux variations seem feasible.

Solar wind protons impinge on interplanetary dust grains orbiting
the sun. Depending mainly on grain constitution and proton energy, they
penetrate about 10 to 30 nm into the molecular lattice of the grains.
With a high efficiency ε the protons are trapped ($\varepsilon > 0.9$) and subsequently deionized. The newly formed hydrogen atoms are retained, possi-
bly expelling other, previously trapped hydrogen atoms from the lattice.
Saturation of the dust grain surface layer is reached, depending on the
grain temperature, at irradiation levels of less than $5 \cdot 10^{17}$ protons
cm^{-2}. For a mean solar wind proton flux of $4.8 \cdot 10^9$ cm^{-2} s^{-1} at 0.2 AU
this occurs after 1200 d; thus it is safe to assume that all existing
dust grains within the inner solar system are saturated by hydrogen.
This condition of the dust grain surfaces implies that virtually no
additional hydrogen can be retained, and that, on the average, for each
impinging solar wind proton one hydrogen atom is immediately released
from the surface. Only a minor fraction of the impinging protons is expected to be "reflected" as charged particles by quasi-elastic collision
processes in the top surface layer (Lord, 1968; Bühler et al., 1966).

Assuming a continuous production of hydrogen by "dust deionization",
the production rate P_h and the loss terms L can be written as (see Ripken and Fahr, 1979):

$$P_h = N_p \, v_{rel} \, \varepsilon \, \Gamma(r) = L_{ph}(N_h) + L_{ex}(N_h) + L_{eli}(N_h) \qquad (1)$$

(N_p: solar wind proton density; v_{rel}: relative velocity between dust
grains and impinging protons; $\Gamma(r)$: effective geometrical cross section

of dust grains; L_{ph} : photoionization rate of dust generated hydrogen; L_{ex}: charge exchange rate; L_{eli}: electron impact ionization rate; N_h: density of dust generated hydrogen. Eq. (1) describes the steady state condition of the dust generated hydrogen atoms. For a known interplanetary dust distribution including $\Gamma(r)$ (e.g., optically determined by zodiacal light measurements), the solar wind proton flux $N_p\,v_{rel}$ can be deduced from hydrogen density observations. It is thus possible to determine solar wind proton fluxes at heliocentric distances of r>0.15 AU. Conversely, in assuming a specific solar wind flux, the dust distribution in interplanetary space can be determined without resorting to optical observations.

The feasibility of hydrogen density determinations inside the orbit of the Earth is examined next. Analogous to the backscatter of solar EUV radiation by interstellar neutral gas (intensities I_i), solar 121.6 nm radiation is scattered by dust generated hydrogen (intensities I_d). Model calculations locate the main radiation sources along the line of sight: for the dust component they lie close to the sun, whereas for the interstellar component, depending on the viewing position and orientation of the instrument, they lie either within 0.4 AU of the observer or about 2 AU to 20 AU away from him. Integrated intensities along the line of sight are shown in Fig. 1 as functions of solar offset angle γ and time of observation from Earth. Clearly, summer observations reveal dominant intensities I_d for $|\gamma|<15°$, while spring observations yield $I_i>I_d$ for reasonable solar offset angles ($\gamma>7.5°$). In order to determine N_h accurately by means of backscatter observations, it will thus be necessary to separate the two signals I_d and I_i spectroscopically. For both observing positions and an offset angle $\gamma=10°$ the calculated spectra are given in Fig. 2, exhibiting strongly doppler-shifted components. Employing suitable instrumentation (resonance absorption cells or EUV spectrometers; Blamont et al., 1975; Artzner, 1978), the geocoronal, interstellar, and dust generated components can readily be separated and analyzed.

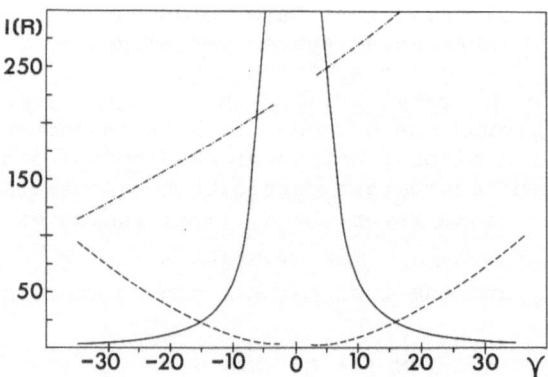

Figure 1. Calculated integrated 121.6 nm radiation intensities as functions of solar offset angle γ. Full lines: I_d; dashed lines: I_i, observation time June 21; dashed-dotted line: I_i, observation time March 21.

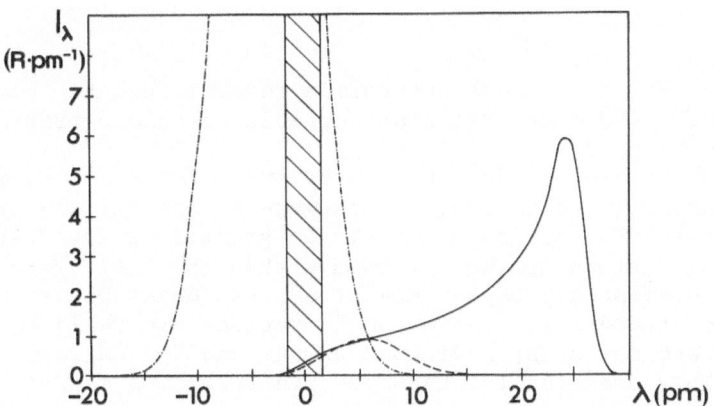

Figure 2. Calculated spectral profiles of resonantly scattered solar 121.6 nm radiation. Solar offset angle of line of sight is 10°. Full line: dust generated component, strongly doppler-shifted due to orbital velocity of dust grains; dashed line: interstellar component, observation time June 21; dashed-dotted lines: interstellar component, observation time March 21, with maximum at 29.25 R·pm^{-1}; hatched area: geocoronal component, temperature 10^3 K, optically thick.

The aforementioned desorption of hydrogen from saturated dust grain surfaces indicates that solar wind flux changes are instantaneously reflected in equilibrium hydrogen densities. This seems to indicate that determinations of temporal and spatial variations of the proton flux close to the sun are feasible. The line-of-sight source function of dust generated hydrogen backscatter radiation is peaking strongly and the corresponding spectrum exhibits maximal doppler shifts at the closest approach to the sun. Thus, for small solar offset angles ($\gamma < 10°$) the linear extent of the source region producing the spectral intensity maximum of I_d is less than 0.1 AU, and flux variations in solar wind structures of limited spatial extent are expected to be observable already at solar distances of about 0.15 AU.

Analogous calculations for interstellar and dust generated neutral helium and solar wind helium ions are currently being carried out, as well as extensions of the models out of the dust plane of symmetry.

References

Artzner, G.: 1978, Astron. Astrophys. 70, pp. L11-L14.

Blamont, J.E., Cazes, S., and Emerich, C.: 1975, J. Geophys. Res. 80, pp. 2247-2265.

Bühler, F., Geiss, J., Meister, J., Eberhardt, P., Huneke, J.C., and Signer, P.: 1966, Earth Planet. Sci. Lett. 1, pp. 249-255.

Lord, H.C.: 1968, J. Geophys. Res. 73, pp. 5271-5280.

Ripken, H.W., and Fahr, H.J.: Solar Wind Interactions with Neutral Hydrogen inside the Orbit of the Earth, Proc. of 4th Solar Wind Conf., Burghausen, 1978, in: Lecture Notes in Physics, Springer-Verlag, Heidelberg, 1979.

DISCUSSION

Callebaut: Why is there a saturation for the dust to absorb the
impinging ions? (Some materials are capable to absorb tremendous
amounts of hydrogen.)

Ripken: Saturation of dust grains by impinging and subsequently
trapped and neutralized protons occurs when an approximate 1-to-1
correlation with the lattice atoms of the grain is reached (Lord, 1968).
No more hydrogen atoms can be retained within the dust grain crystal
lattice, and desorption takes place at a rate comparable to the
trapping rate of protons. These considerations are valid for freshly
exposed silicate surfaces; radiation damage and "space weathering" will
reduce, possibly drastically, the hydrogen retention capabilities of
the grains.

Kuperus: Could the presence of the neutral particles influence the
dissipation of MHD waves, e.g., by ambipolar diffusion?

Ripken: The model calculations we have performed are for a steady
state problem. Possible time dependent phenomena have not been
examined yet. However, I do not believe that neutral particles can
appreciably influence MHD waves in the solar wind, or in general
solar wind dynamics, since the maximum density of dust generated
hydrogen, derived from our standard model at about 0.4 AU, is only
$2.5 \cdot 10^{-4}$ cm^{-3}. This compares with a solar wind proton density of
about 35 cm^{-3}, yielding a ratio of $N_h/N_p = 7.1 \cdot 10^{-6}$.

MODEL CALCULATIONS OF SOLAR WIND EXPANSION INCLUDING AN ENHANCED FRACTION OF IONIZING ELECTRONS

E. F. Petelski, H. J. Fahr, and H. W. Ripken
Institut für Astrophysik und Extraterrestrische
Forschung, Universität Bonn, Auf dem Hügel 71
D-53 Bonn, W. Germany

Abstract. Collective interactions of the solar wind and newly ionized interstellar gas cause turbulent electron heating to ionizing energies analogous to laboratory experiments on the critical ionization velocity effect. Implications for solar wind and interstellar gas dynamics are calculated by simultaneously solving continuity equations for solar wind protons, interstellar hydrogen atoms, and energetic electrons. Electron impact ionization is shown to be practically as important as photoionization, giving rise to a stronger deceleration and heating of the distant solar wind, a weaker terminating shock, a smaller stand-off distance of the heliopause, and implying higher densities of the outer solar wind and the interstellar neutral gas.

It is known from laboratory experiments (Danielsson, 1973; Himmel et al., 1976) that counterstreaming of a magnetized plasma and a marginally ionized neutral gas results in considerably enhanced ionization of the neutrals and strong braking of the relative motion if a critical velocity $v_c = (2W/m_i)^{1/2}$ is exceeded (W=ionization energy of the gas, m_i=ion mass of the moving species). The physical mechanism envisaged to explain this "critical ionization velocity effect" is a transfer of ion free kinetic energy to ambient electrons via turbulent heating, e.g. through a modified two-stream instability, subsequent ionization of the neutral gas by the energetic electrons, assimilation of the newly formed ions into comotion with the plasma after a short time determined by the instability growth rate, and deceleration of the plasma. By comparing typical parameters of laboratory experiments, the Apollo 13 lunar impact event (Lindeman et al., 1974), and the heliosphere (Tab. 1), it can be shown that this mechanism must also operate in the interaction region of the solar wind and the interstellar neutral gas (Petelski et al., 1979). - Consequences for the heliospheric

M. Dryer and E. Tandberg-Hanssen (eds.), Solar and Interplanetary Dynamics, 159-162.
Copyright © 1980 by the IAU.

Table 1. Comparison of parameters (l_{ch}=characteristic length, v_{rel}=plasma-neutral gas relative velocity, n_n/n_p=neutral gas /plasma density, B=magnetic field, T_e=background electron temperature, $E_{e,f}$=fast electron kinetic energy (estimated for the heliosphere, to be at the lower edge of the laboratory energies), $\sigma_{e,f}$=fast electron impact ionization cross section).

Parameter	Lab.	Apollo 13 Lunar Impact	Heliosphere		
			1 AU	10 AU	100 AU
Gas	He	plastics ($\approx$80a.m.u.)	H		
l_{ch} (m)	0.1	2×10^{5}	1.5×10^{11}	1.5×10^{12}	1.5×10^{13}
v_{rel} (ms^{-1})	4×10^{5}	3.3×10^{5}	4×10^{5}	4×10^{5}	3×10^{5}
n_n (m^{-3})	10^{20}	10^{12}	2×10^{3}	5×10^{4}	1×10^{5}
n_p (m^{-3})	10^{18}	3×10^{6}	5×10^{6}	5×10^{4}	7.5×10^{2}
B (T)	0.5	3.6×10^{-8}	5×10^{-9}	5×10^{-10}	1.0×10^{-10}
T_e (K)	5.8×10^{4}	1.2×10^{5}	1.5×10^{5}	4×10^{4}	$>10^{4}$
$E_{e,f}$ (eV)	100	50	27		
$\sigma_{e,f}$ (m^{2})	3.5×10^{-21}	2×10^{-20}	5×10^{-21}		

parameters are assessed by simultaneously solving the coupled continuity equations for the mass, the momentum, and
the energy of solar wind protons, the mass of interstellar
hydrogen atoms, and the number of ionizing electrons created
by critical velocity effects, the proton source terms being
calculated from photoionization, charge exchange, and fast
electron impact ionization. To simplify the algorithm, a con
stant energy transfer factor of 0.35 as well as a constant
fast electron energy of $\approx$30 eV are assumed in compliance with
laboratory findings. Furthermore, critical velocity phenomena
in the heliosphere are presupposed to involve the modified
two-stream instability, the growth rate of which is set to
the lower hybrid frequency. It is ascertained that the electrons of the outer solar wind are effectively prevented from
adiabatically cooling to non-ionizing energies. Accordingly,
a significantly stronger deceleration of the solar wind, a
weaker radial decrease of its density, and a steeper radial
increase of the solar wind proton temperature and the interstellar hydrogen density are obtained than calculated from
photoionization and charge exchange alone. Also, the solar
wind terminating shock is weakened, or even eliminated if the
asymptotic interstellar hydrogen density is raised to 0.285,
and the radius of the heliopause is reduced (cf. Fig. 1 and

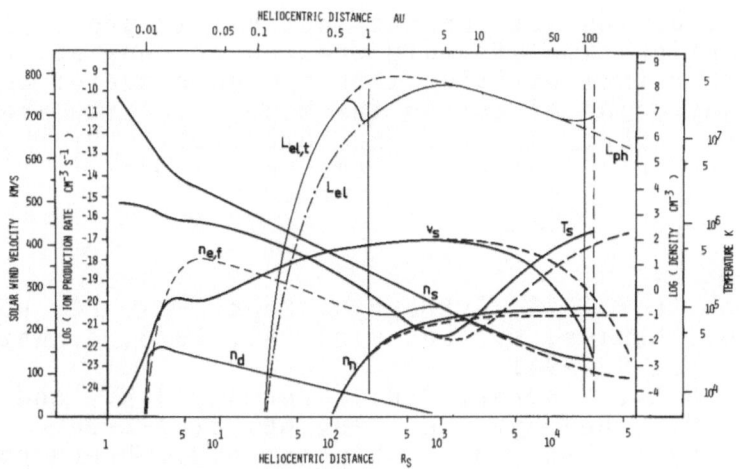

Figure 1. Heliospheric parameters along the stagnation line;
----: standard results exclusively based on photoionization
and charge exchange, ———: results including v_c effects.
($n_s/v_s/T_s$=solar wind density/velocity/proton temperature,
$L_{ph}/L_{el}/L_{el,t}$=photoionization rate/electron impact ionization rate due to v_c effects/total electron impact
ionization rate including ionizations by naturally hot electrons inside 0.8 AU, $n_{e,f}$=density of fast electrons energised
by v_c effects, n_d= density of dust-generated hydrogen)

Table 2. Comparison of standard data based on photoionization
and charge exchange alone with critical velocity data.

Parameter	Standard	Critical velocity	Change
$v_{s,sh}$ (m s^{-1})	2.8×10^5	1.9×10^5	- 32%
$T_{s,sh}$ (K)	5×10^5	6.5×10^5	+ 30%
$M_{MA,sh}$	3.8	2.3	- 39%
r_{sh} (AU)	118	98	- 17%
$n_{s,sh}$ (m^{-3})	5.7×10^2	1.4×10^3	+146%
$n_{n,\infty}$ (m^{-3})	1.0×10^5	1.7×10^5	+ 70%

($v_{s,sh}/T_{s,sh}/n_{s,sh}$=solar
wind parameters at the
shock, $M_{MA,sh}$=solar wind
magnetoacoustic Mach num-
ber at the shock, r_{sh}=
shock distance, $n_{n,\infty}$=as-
ymptotic interstellar hy-
drogen density)

Tab. 2). Coupling between the instability growth time and the thermalization time of newly created protons allows one to choose other instabilities with a wide range of growth times without any major effect on the model. In situ measurements of electron energy spectra and interstellar hydrogen densities in the outer solar wind are recommended to test these results.

References.
Danielsson, L.: 1973, Astrophys. Space Sci. 24, pp. 459-485.
Himmel, G., Möbius, E., and Piel, A.: 1976, Z. Naturforsch. 31a, pp. 934-941.
Lindeman, R.A., Vondrak, R.R., Freeman, J.W., and Snyder, C. W.: 1974, Geophys. Res. 79, pp. 2287-2296.
Petelski, E.F., Fahr, H.J., Ripken, H.W., Brenning, N., and Axnäs, I.: 1979, Astron. Astrophys. (submitted).

DISCUSSION

Ahluwalia: What are the relative contributions of charge exchange, photoionization, and ionization by fast electrons? If I remember correctly, charge-exchange is the dominant process. I have a second question: I note that according to your calculations the heliospheric boundary is at ~100 A.U. How much would the boundary shift, if only charge exchange is considered?

Petelski: On average, the charge exchange rate is 2.5 to 3 times larger than the photoionization rate which in turn is comparable to the electron impact ionization rate if the critical ionization velocity process is assumed to operate in the solar wind.

If only charge exchange is considered, the shock radial distance is increased by 20 AU (and the shock Mach number is doubled) over the standard case.

Tandon: In your model proton temperature near the sun comes out to be higher than a few times a million, °K. Comment, please.

Petelski: In our model, a solar wind temperature of 1.5×10^6K is adopted close to the sun.

CORRELATED VARIATIONS OF PLANETARY ALBEDOS AND SOLAR-INTERPLANETARY PARAMETERS

G. W. Lockwood
Lowell Observatory, Flagstaff, Arizona

S. T. Suess
NOAA/ERL, Boulder, Colorado

D. T. Thompson
Lowell Observatory, Flagstaff, Arizona

The brightnesses of Titan and Neptune have been monitored photo-electrically at 472 and 551 nm since 1972 at the Lowell Observatory, yielding annual mean magnitudes accurate to 0.3 percent (0.003 mag). Both objects increased steadily in brightness until 1976 and declined thereafter (Lockwood 1977, Lockwood and Thompson 1979). The range of variation was about 0.08 mag for Titan and 0.03 mag for Neptune.

This period of observation coincides with the decline of solar activity in cycle 20 and the subsequent rise in cycle 21. Because there is no evidence to suggest that the solar flux in the visible region varies by more than a small fraction of one percent (White 1977), we hypothesize that some variable component of the solar output influences the albedos of Titan and Neptune. Correlation studies (Suess and Lockwood, in preparation) of the observed planetary brightness variations and various solar and solar-interplanetary parameters have allowed us to rule out some of these as candidates for a cause-effect relationship but do not lead yet to a conclusive identification of the controlling parameter and the mechanism by which it operates upon planetary atmospheres.

The observed seasonal mean planetary variations are shown in Figure 2 along with the monthly mean Zurich sunspot number and an index of the monthly number of flares of importance ≥ 1. In Figure 1, monthly mean planetary magnitudes have been scaled and merged into a single unified data set and are shown along with a number of solar and solar-interplanetary parameters: sunspot number, Ca plage area index, 10.7-cm radio flux, 21-cm radio flux, geomagnetic aa index, solar wind speed, and geomagnetic recurrence index. The aa index is an indirect measure of solar wind speed and the recurrence index is a measure of the tendency of the solar wind to be dominated by recurrent streams.

From Figure 1 it can be seen that the rises in sunspot number, 10.7- and 21-cm radio flux and Ca plage index following sunspot minimum

163

M. Dryer and E. Tandberg-Hanssen (eds.), Solar and Interplanetary Dynamics, 163-166.
Copyright © 1980 by the IAU.

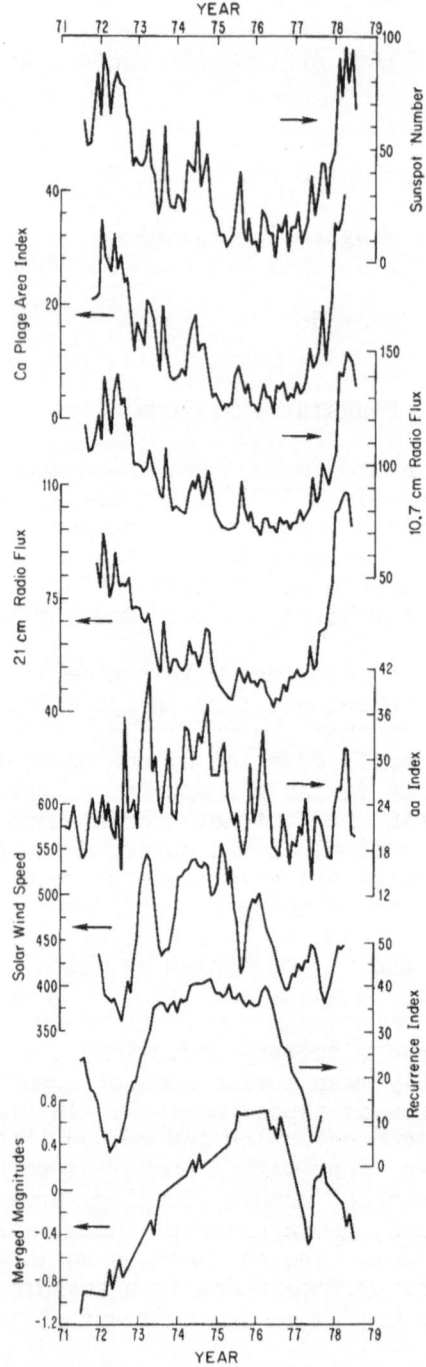

Figure 1. The variation of scaled and merged planetary magnitudes and various solar and solar-interplanetary parameters.

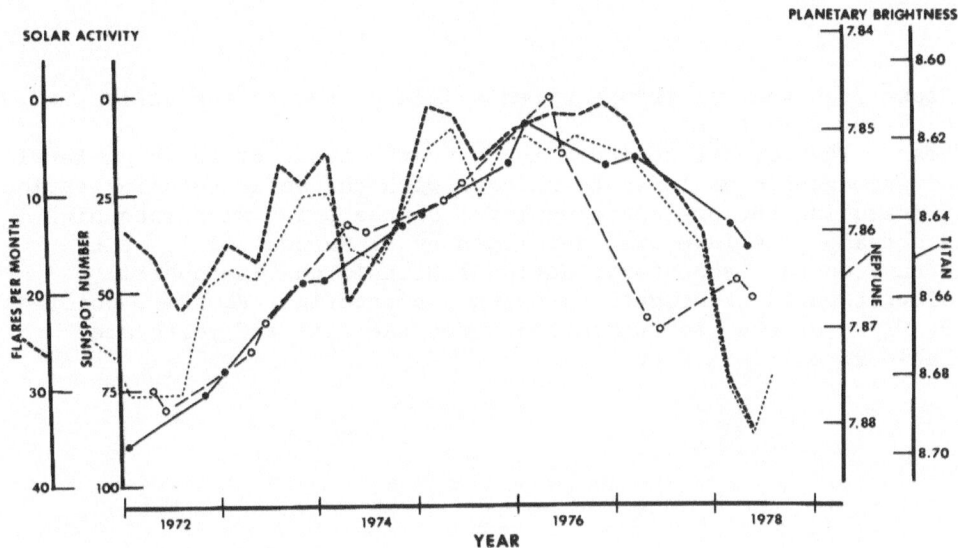

Figure 2. *Pre- and post-opposition mean b and y magnitudes of Titan and Neptune, monthly mean Zurich sunspot number and number of flares per month of importance 1 or greater.*

in 1976 are preceded by the planetary variations and hence are unlikely sources of a direct solar-planetary effect. However, the solar wind speed, aa index, and recurrence index correlate well with the merged magnitudes and lead them in phase by as much as 1.5 years. The available solar EUV data suggest that its most recent minimum may have occurred about a year prior to the 1976 maxima of Titan and Neptune (Hinteregger 1979).

Hence, our conclusion is that, among the parameters studied, the solar EUV output and the solar wind are the most viable candidates for the cause of the observed planetary variations.

This research is supported by the Atmospheric Sciences Program of the National Science Foundation.

References

Hinteregger, H. E.: 1979, *J. Geophys. Res.* 84, 1933.
Lockwood, G. W.: 1977, *Icarus* 32, 413.
Lockwood, G. W. and Thompson, D. T.: 1979, *Nature* 280, 43.
White, O. R. (ed.): 1977, *The Solar Output and Its Variations,* Colorado Associated University Press, Boulder.

DISCUSSION

Ahluwalia: What is the recurrence index? How do you define it?
Is this your invention?

Suess: The recurrence index measures the tendency for high levels
of the geomagnetic aa index to reoccur with the solar rotation period-
hence measuring the organization level of the solar wind into high
speed streams. The index was developed by T. Sargent, III, at the
Space Environment Laboratory, National Oceanic and Atmospheric
Administration/Environmental Research Laboratories, Boulder, Colorado
80303. Plots of the recurrence index for the last 100 years are
available from Mr. Sargent.

LARGE-SCALE MAGNETIC FIELD STRUCTURE AT THE EARTH'S ORBIT, ITS CORRELATION WITH SOLAR ACTIVITY AND ORIENTATION AND MOTION OF THE SOLAR SYSTEM IN THE GALAXY

G. J. Vassilyeva, M. A. Kuznetsova and L. M. Kotlyar
Main Astronomical Observatory, USSR Academy of Sciences
Pulkovo, Leningrad USSR

ABSTRACT

Interplanetary magnetic field data from the different satellites obtained during the period 1963-1973 at 1 A.U. and compiled by J. King have been analysed in heliocentric ecliptic coordinates. The peculiarities of the background interplanetary magnetic field (BIMF) are discussed in relation to the orientation of the solar system in the Galaxy and the variable helioefficiency of the planets. The results of the direct cosmic experiments are evidence of the solar activity being a complex phenomenon of the solar system as a whole.

The main objective of this investigation is an attempt to reconcile interplanetary magnetic field spatial structure with the spatial structure of the variable helioefficiency of the planets. The spatial structure of the variable helioefficiency in the solar system seems to be associated with two important directions: the line of nodes of the galactic equator and ecliptic ($\lambda = 87 \div 267$) and the ecliptic projection of the galactic magnetic field direction ($b'' = 0$, $1'' = 70$; $\lambda = 135 \div 315$) (Vassilyeva et al., 1979). This second direction is specifically manifested itself in the variable helioefficiency of Jupiter and the Earth (for solar maximum and solar minimum) (Figure 1).

In spite of many publications devoted to analysis of IMF data obtained during the period 1963-1973 at 1 A.U., a new treatment of these data have been undertaken. Analysis of the hourly values of B_z^+, B_z^-, B_z, B_R^+, B_R^-, B_R, B_T^+, B_T^-, B_T – component and B, averaged over 30- degrees ecliptic longitude intervals for the 1967-1973 period confirmed the existence of the large scale background interplanetary magnetic field (BIMF). This BIMF reveals itself as a weak signal against the background of the uncorrelated noise, created by the well-known sector structure of IMF. The existence of BIMF with the following features is discussed.

M. Dryer and E. Tandberg-Hanssen (eds.), Solar and Interplanetary Dynamics, 167-172.
Copyright © 1980 by the IAU.

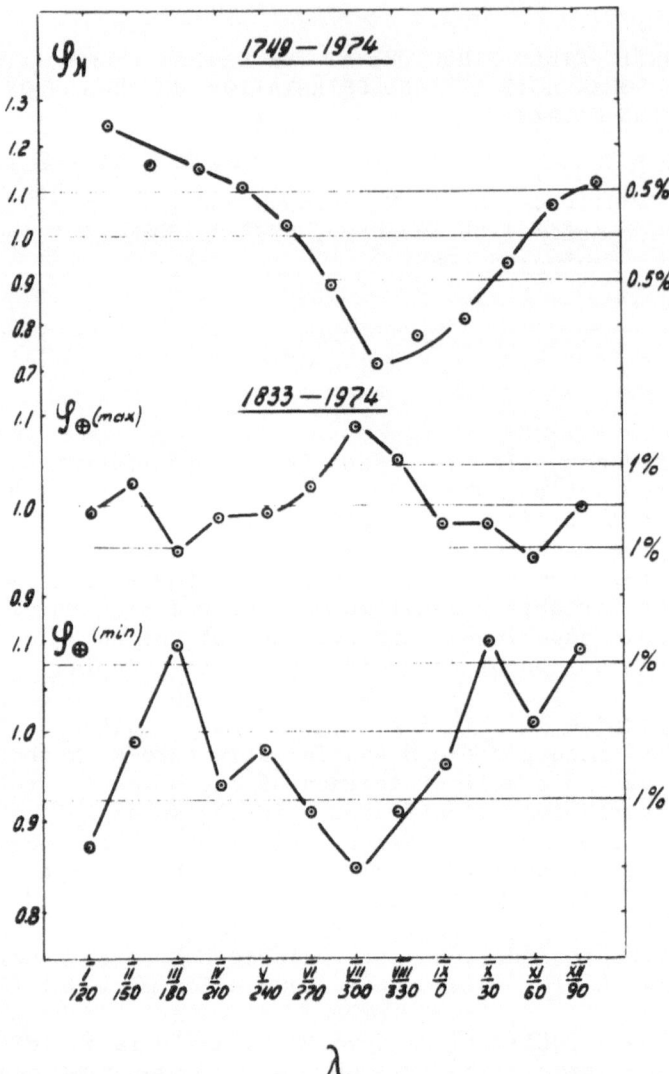

Figure 1.   Helioefficiency of Earth and Jupiter.

1.   BIMF does not co-rotate with the Sun.
2.   The B_R and B_T dominant magnetic polarity are found to reproduce
their values and signs at the same longitudes during some years.
3.   The vortex structure of BIMF seems to occur, i.e., the circulation
of B_T-component over the Earth's orbit is found not to be equal to
zero and to be equivalent to the current of $J_z \sim \oint B_T dl \sim 10^9$ amperes
within the Earth's orbit.

4. The reversal of the Sun's dipolar field is associated not only with the well-known reversal of the Rosenberg-Coleman dominant polarity effect in B_R but it is also associated with the reversal B_T-circulation over the Earth's orbit and its breakdown in the 1970-1971 period (Figure 2). The nature of B_T-circulation and its reversal is closely correlated with the relative N-S solar corona rotation (Stepanov and Tyagun, 1976).

5. The distributions of the hourly values of B_R and B_T components over the specific ecliptic longitude intervals, summarized over the period including the reversal of the Sun's dipolar field, shows identical maxima of dominant polarity near two nodes of the galactic equator and ecliptic (Figure 3).

6. Magnetic field strength values of all three components

$$|B_R| = \frac{1}{2}(|B_R^-| + |B_R^+|); \quad |B_T| = \frac{1}{2}(|B_T^+| + |B_T^-|); \quad |B_z| = \frac{1}{2}(|B_z^+| + |B_z^-|),$$

derived separately from "+" and "-" hourly sets averaged over 30 degree longitude intervals during seven years (1967-1973) have the same very accurate character of variation along the Earth's orbit. The amplitude of variations relative to the middle level as high as 4-5% (Figure 4). Maxima and minima of $|B_R|$, $|B_T|$, $|B_z|$ ecliptic variations indicate the intersection of the Earth's orbit with the ecliptic projection of two coincident directions: solar motion to apex relative to the stars of 14-15 magnitude and galactic magnetic field direction (b" = $0°$, 1" = $70°$). The $|B_R|$, $|B_T|$, $|B_z|$ variations along the Earth's orbit are in very good agreement with the variations of the Earth's velocity of rotation for the same period (Yagudin, 1978).

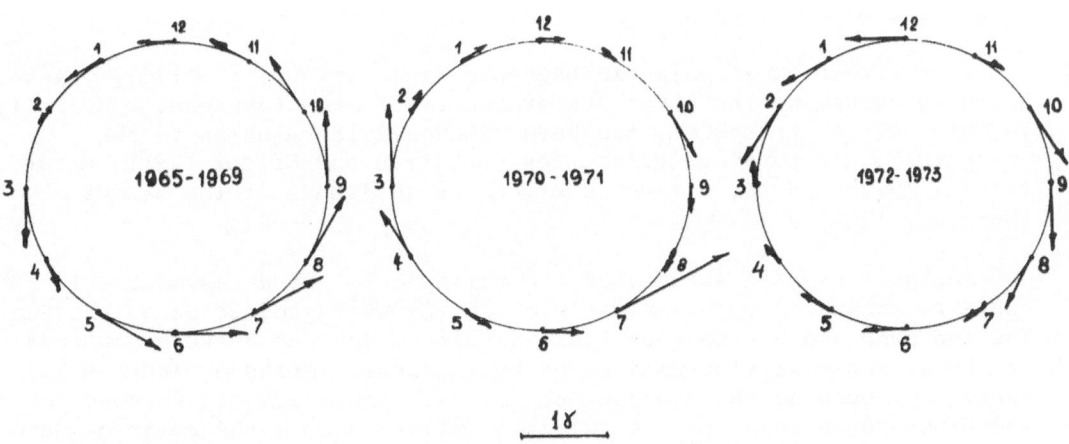

Figure 2.   Circulation of B_T along orbit of Earth.

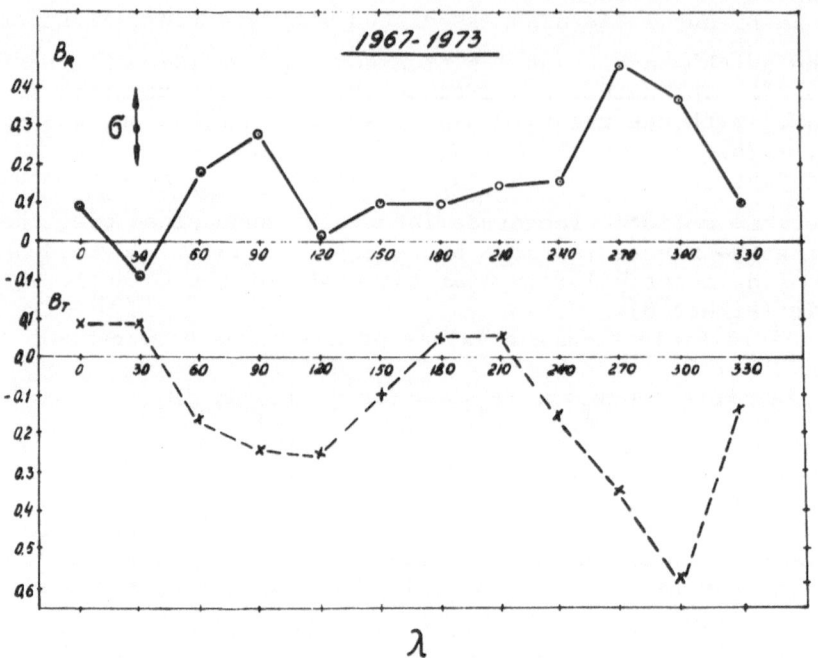

Figure 3. Hourly values of B_R and B_T.

So, side by side with the well-known sector structure of the IMF, which co-rotates with the Sun and is associated with the large scale background solar magnetic field, the large scale structure of BIMF exists also. This structure is correlated with the magnetic field of sunspots, solar dipolar field and the orientation and motion of the solar system in the Galaxy.

This direction of galactic magnetic field (b" = 0, l" = +70), which seems to determine the structure of the solar magnetosphere, especially in the outer solar system, has been discussed in relation to the galactic cosmic rays' peculiarities (Schatten and Wilcox, 1969; Marsden et al., 1976) and the direct geoeffective influence of the Galaxy (Morozov, 1944).

Analysis of BIMF shows that the ecliptic longitude asymmetry in IMF could be exhausted by the effects of the Earth's projection on the Sun. The independence of the mean field magnitude and the average amplitude of the directional fluctuations of heliographic latitude within ± 7.3° range have been noted by Hedgecock (1975). The discussed features of the BIMF are evidence of the galactic influence upon the solar magneto-sphere. Moreover, reversal of the Sun's dipolar field seems to be controlled by the Galaxy. Correlation of the parameters of BIMF with

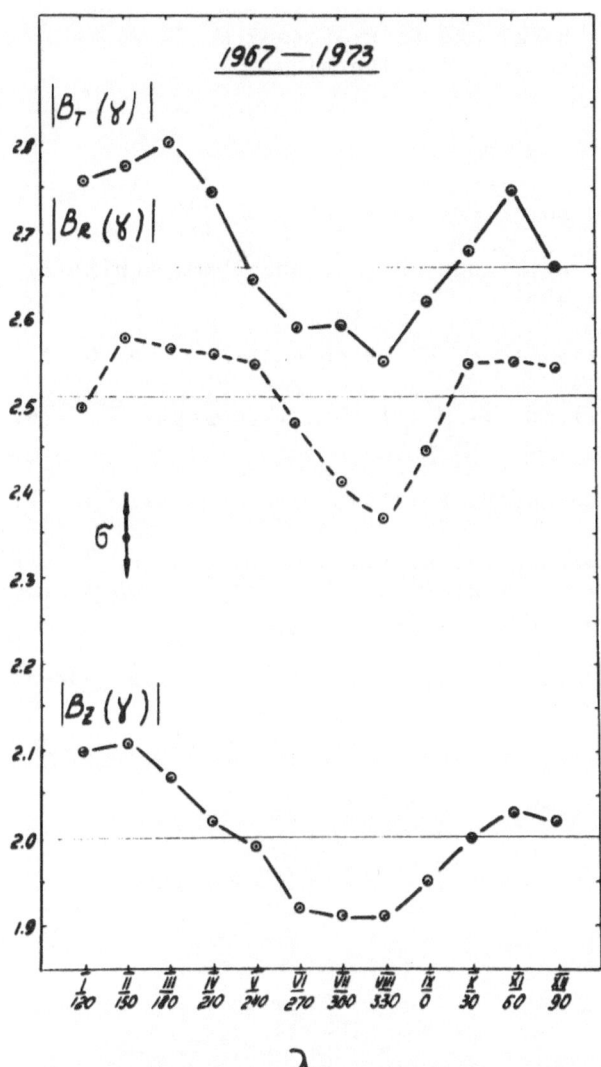

Figure 4. Variation of mean IMF components (see text).

the variable helioefficiency of planets opens an interesting chance to employ planets as probe particles in cosmic experiments.

ACKNOWLEDGMENTS

This paper contains results of new treatments of interplanetary magnetic field data compiled by Dr. J. King. Magnetic type recording of these data has been kindly placed at our disposal by the World Data Center A for Rockets and Satellites (Greenbelt, Maryland, U.S.A.). We are indebted to Dr. P. C. Hedgecock for the stimulating advice and the

HEOS I, II magnetometer data, and to Professor M. I. Pudovkin for fruitful discussions at Leningrad University.

REFERENCES

Hedgecock, P. C.: 1975, *Solar Phys. 44*, pp. 208-224.

Marsden, R. G., Elliot, H., Hynds, R. G., and Thambyahpillai, T.: 1976, *Nature 260*, pp. 491-495.

Morozov, N. A.: 1944, *Izv. Acad. of Science USSR 8*, pp. 63-71.

Schatten, K. H., and Wilcox, J. M.: 1969, *J. Geophys. Res. 74*, pp. 4157-4161.

Stepanov, V. E., and Tyagun, N. F.: 1976, *IAU Symp. 71*, pp. 71-73.

Vassilyeva, G. J., Schpitalnaya, A. A., and Petrova, N. S.: 1979, *Int. Solar-Terrestrial Predictions Proc. and Workshop Program*, Preprint 98, Boulder, CO, U.S.A.

Yagudin, L. I.: 1978, *Astron. Zournal (Letters), 4*, pp. 332-336.

ENERGY AND MASS INJECTED BY FLARES AND ERUPTIVE PROMINENCES

O. Engvold
Institute of Theoretical Astrophysics
University of Oslo, Norway.

1. INTRODUCTION

The dynamic nature of the outer corona has been revealed in
recent years by improved techniques of studying the corona
with ground based instruments and from space. Rapid changes
in the coronal brightness pattern takes the form of mass
ejections in which the outward moving material generally
appears to escape the Sun. These events are called coronal
transients. Observations have shown that the excess mass
and energy in coronal transients are supplied by material
expelled from the chromosphere and lower corona in
conjunction with Hα surface activity. The transients are the
coronal response to release of mass and energy which are
manifested by flares and eruptive prominences.

Many forms of mass motion are accessible for direct
observations, such as: (1) surges, (2) flare sprays, (3)
ascending prominences or suddenly disappearing filaments,
and, possibly, (4) flare-waves (Moreton waves). Such events
are commonly referred to as high-speed ejections. The
equally spectacular, although less vigorous, post flare
loops or loop prominence systems (Bruzek, 1964) have been
classified as slow ascent events.

Munro et al. (1979) investigated 77 mass ejection
Skylab transients and how they related to Hα surface
activity. Thirty-four events were definitely or probably
associated with Hα surface activity. Their results may be
summarized as follows: (1) Coronal transients are more
closely associated with erupting prominences than they are
with flares. (2) About 70% of the transients were associated
with erupting filaments or disappearing filaments. (3)
Three-quarters of the number of transients were rooted in or
near active regions. Hence, coronal transient production
seems more probable when the surface magnetic fields are
strong and complex (Hildner et al., 1976).

M. Dryer and E. Tandberg-Hanssen (eds.), Solar and Interplanetary Dynamics, 173-188.

We shall describe and discuss various types of dynamical events at chromospheric levels and in the lower corona, and also attempt to evaluate their significance with regard to coronal disturbances. The study by Rust et al. (1979) did not detect any close relation between surges and coronal transients. There are, on the other hand, strong evidences that ascending prominences and flare sprays are esential in the process which brings about observable coronal transients. Thus, it will be of particular interest to consider these two types of events when discussing the dynamics of the corona and the interplanetary medium.

2. SLOW ASCENT EVENTS

It has been suggested (Kopp and Pneuman, 1976) that loop prominence systems (LPS)might manifest a slow reconnection of the magnetic lines of force torn by the flare blast. As such the loop prominences will be the result of another dynamic event rather than its cause. Some LPS appear without being in observable connection with a flare. The gradual growth and rise of a loop system appears to be the result of cooling of the matter contained in the magnetic arches rather than a physical motion of individual loop structures.

An upward expansion and subsequent contraction ("rise and fall") which is observed in some limb flares does not seem to involve true mass motions (Švestka, 1976).

3. HIGH SPEED EJECTIONS

3.1 Radio bursts.

The presence of flare-associated streams of particles is established from observations of frequency drift rates in dynamic spectra of solar radio bursts. Groups of Type III bursts, which are thought to arise from high speed electrons propagating along coronal streamers, tend to occur during the flash phase of a flare (Sheridan, 1979).

Radio-interferometric observations of "moving Type IV mA" sources of emission are evidently generated by clouds of plasma that are ejected from the flare region (Dulk and Altschuler, 1973). In the case of two events recorded on October 3-5, 1977, the moving radio sources were very close to the leading edges of the associated eruptive prominences (Stewart et al., 1978). They found that the magnetic energy density of the two moving Type IV sources was larger than the kinetic energy density of the associated eruptive prominences. The temporal and spatial correlation between moving Type IV mA and identifiable features in a flare spray is clearly illustrated by Riddle et al. (1974).

The presence of Type II bursts suggests that flares produce shock waves which presumably propagate ahead of mass ejection coronal transients (Hildner, 1977)

3.2 Flare waves.

A fast wave front-like disturbance is frequently seen to come out from the flare site. It is manifested as a line-of-sight motion in the Hα line wing, and it propagates over distances as large as 1 $R_\odot$ at the average speed of 600-800 km s^{-1} (Smith and Harvey, 1971). It is also detected as triggers of sympathetic flares (Becker, 1958), and occasionally as sequential brightenings of small points in the chromosphere (Rust et al., 1979). It has been suggested that this type of flare wave may be due to mass ejections along low loops, which lit up the chromosphere at the foot-points of the loops. Martin (1978) argues that this mechanism cannot account for the observed properties of some flare waves. Uchida et al. (1973) propose that flare waves are the skirts of coronal shock waves sweeping over the chromosphere. In that case, the mass and energy associated with flare waves will be that of coronal transients, i.e., respectively, of the order of 10^{16}g and 10^{31} erg (cf. Webb et al., 1979)

The existence of a slow wave disturbance which propagates at speeds of 60-200 km s^{-1} has been inferred from activation and disruption of filaments by distant flares (Bruzek, 1969; Rust and Švestka, 1978).

3.3 Surges.

Surges are ejections of chromospheric material from regions close to the flare proper (Westin, 1969). No clear evidence has been presented for mass motions associated with the flare core itself (Cheng, 1978; Webb et al., 1979). The average upward speed of surges ranges from 50 to 300 km s^{-1} and they may reach up to heights of 20 000 - 200 000 km before they fall back to the chromosphere along their original trajectories. Their initial acceleration is found to be 1-5 times $g_\odot$ (Roy, 1973; Platov, 1973). According to Schmahl (1978) the pressure gradient along the surge trajectory may be sufficient to drive a surge upwards.

In Figure 1 matter is seen to be injected upwards (Hα-0.9Å) near the flare, and at the same time a down stream (Hα + 0.9Å) is detected close to the spot. The observations indicate that the mass flow takes place within a closed magnetic loop.

Tandberg-Hanssen and Malville (1974) measured magnetic

field strengths between 30 and 100 gauss in surges. Assuming a particle number density 10^{11} cm^{-3} in the "cool" ejecta implies that only the most violent surges (200 km s^{-1}) will be capable of disrupting the weakest fields. Mass motion in surges evidently takes place without causing noticeable perturbations to the local magnetic field.

Estimates of mass contained in surges are based on very uncertain values of their volumes as well as of mass density. In the case of the Sept. 5, 1973 event, Webb et al. derived $M_{surge} = 10^{15} - 10^{16}$ g. Somewhat lower values were obtained by Rust et al. (1979), and Smith et al. (1977) concluded from their X-ray and Hα data that about 10^{14} g was injected by a surge of Aug. 21, 1972.

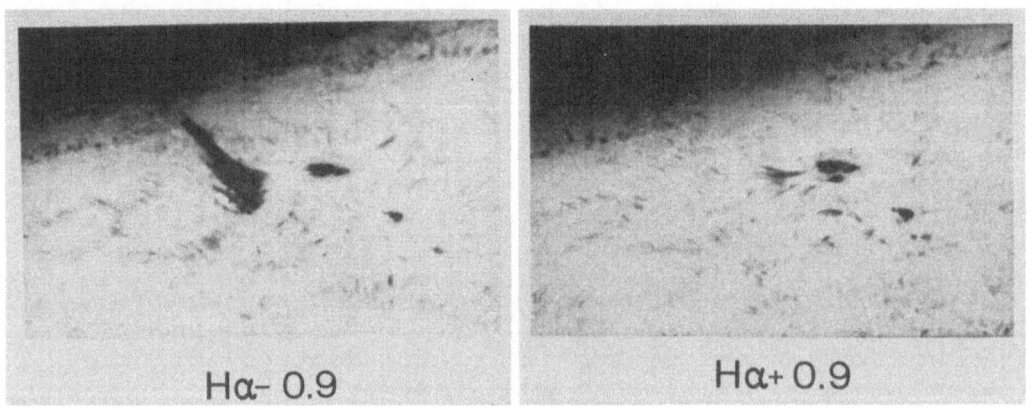

Figure 1. Hα filtergrams of a small surge of January 13, 1974, recorded by the author with the UBF of the vacuum tower telescope of Sacramento Peak Observatory.

In only 6 of 54 cases was there a noticeable change in brightness or size of X-ray emitting regions with surge activity, which indicates that very little heating of the corona takes place. A maximum value for the sum of kinetic and potential energy was found to be $5 \cdot 10^{29}$ erg, or possibly more, in a surge associated with the Sept. 5 flare (Webb et al., 1979).

Surges are neither causing nor seen to be associated with coronal transients (Rust et al., 1979).

3.4 Eruptive prominences.

Ascending prominences and flare sprays.

Very high speed ejecta occurring in conjunction with flares were coined flare sprays by Warwick (1957). The sprays

seemed to emanate from the flare itself, and the material
showed a characteristic clumpiness as it was ejected
(Valniček, 1964; Smith, 1968). The velocity as a rule
exceeded the velocity of escape. The slowly accelerating,
ascending prominences seen at the limb were found to be
identical to suddenly disappearing filaments on the disk.
Hence, they consist of "cool" material situated in the low
corona prior to the eruption. Valniček (1964) concluded
that ascending prominences and sprays were distinctly
different types of ejecta.   However, Tandberg-Hanssen et al.
(1980) found that sprays are extant active region filaments.
We will also show in the following that the speed-height
curves for ascending prominences and sprays do not group
into separate types (Engvold, 1980).

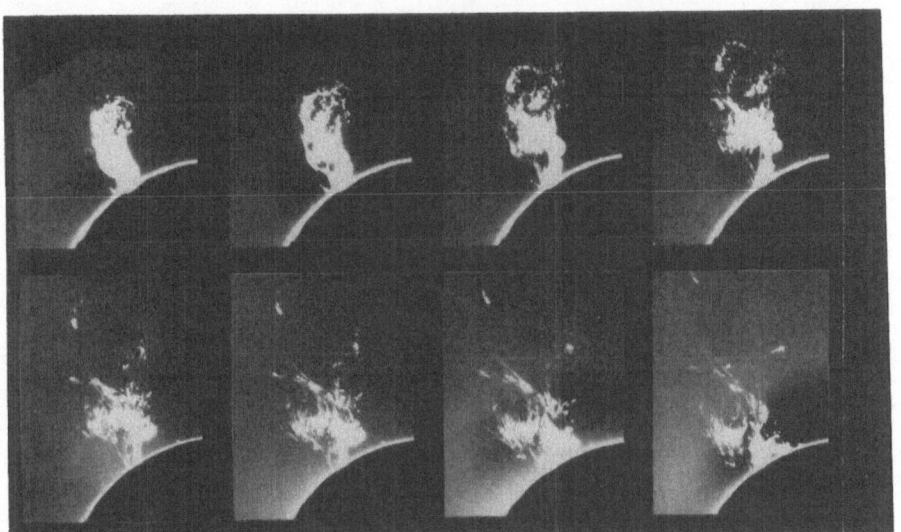

Figure 2.   Hα photographs of the flare spray of October 28,
1972 (R. T. Hansen, HAO, Mauna Loa Observatory).

Observations of eruptive prominences.

     Excellent observations of eruptive prominences have been
obtained and studied over the past 100 years. The nature of
the phenomena has made them rather difficult observational
objects on which to compile a systematic surveillance. The
first systematic morphological investigation was carried out
by Pettit (1925, 1940). His study was based on about 60
events observed from 1885 to 1939. A spectacular event of
May 29, 1919, from Pettit's data is shown in Figure 3.

     We have studied Pettit's data, as well as more recent
compilations of observations by various authors, and a
number of events which have been reported and described in

the literature (Engvold, 1980). It amounts to a total of
about 300 individual eruptives. Speed-height relations could
be derived for 118 of the events.

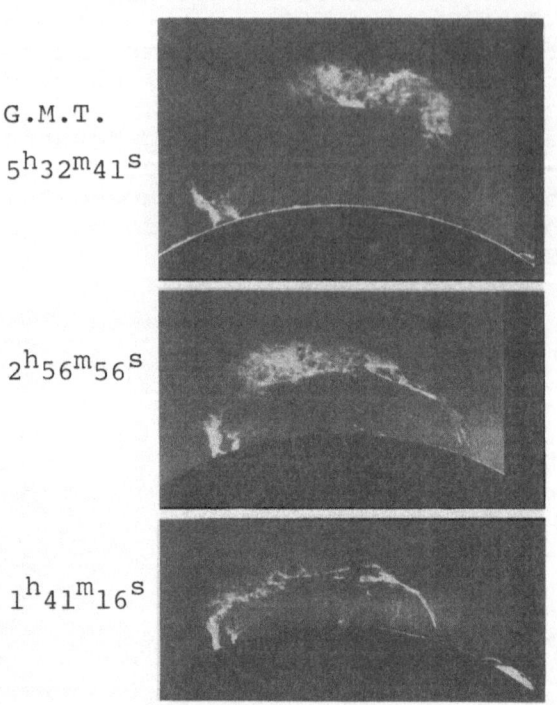

G.M.T.

$5^h32^m41^s$

$2^h56^m56^s$

$1^h41^m16^s$

Figure 3.   A large eruptive prominence of May 29, 1919
(Pettit, 1925).   (Yerkes Observatory)

The d'Azambujas (1948) comprehensive study of prominences
and filaments covers the period 1919-1937. They introduced
the term "disparition brusques" (DB) for sudden disappearance
of a filament. Extensive and detailed studies of DB's have
been made by Bruzek (1951), Smith and Booton (1962), Smith
and Ramsey (1964), and Westin and Liszka (1970). We have
examined, with special regard to DB's, the synoptic charts
and catalogues of filaments and active regions published by
Observatoire de Paris, Meudon (cf. Martres and Zlicaric,
1977). Our study covers the Carrington rotations 1389-1662,
i.e. from July 1957 through December 1977.

The speed-height relations.

The variation of ascending velocity with height is
intimately connected to the propelling force acting on the
matter. As such it is of interest to study the shape of
acceleration curves of various eruptive prominences.

Measurements of ascending velocities often refer to the upper and leading edge of a prominences where also the velocity tends to be highest (McCabe, 1971; Engvold and Rustad, 1974). Various parts and fragments of a prominence may move differently (Waldmeier, 1958; McCabe and Fisher, 1970; Sakurai, 1976). Curved trajectories of ejected matter are evidence of strong influence by magnetic fields. The initial rise of erupting prominences takes place without substantial distortion of their shapes. Schmahl and Hildner (1977) concluded that the prominences and the corona around it are permeated by a magnetic field which becomes unstable everywhere nearly simultaneously.

The flare spray shown in Figure 2 is typical of a high speed case. A rather unique high speed event was observed by Valniček (1962) on Sept. 16, 1961. The ejected matter, which showed little tendency to fragmentation, was accelerated up to 1400 km s^{-1} at a height of 250 000 km before it faded. Another peculiar high speed event was recorded with the OSO-7 coronagraph on Dec. 14, 1971. Three plasma clouds were seen moving outwards in the corona at speeds of about 1000 km s^{-1} (Brueckner, 1972; Kosugi, 1976). The event was preceeded by an eruptive prominence and the clouds were evidently parts of a peculiar coronal transient.

We have compared the speed-height relations of 118 events (Engvold, 1980). Some examples are shown in Figure 4. Smooth curves are derived on the basis from 2 to 5 points obtained from published tables and graphs.

There are no particular regions in the speed-height diagram that seem to be void of observed cases. This is in conflict with the concept of two particular curves of motion (Valnicek, 1964). The many different speed-height relations that have been measured can hardly be attributed to the effect of motion at an angle to the plane of the sky.

Eruptive prominences appear with widely different upward acceleration, ranging from 5 m s^{-2} to nearly 3 km s^{-2} (ten times $g_\odot$). The acceleration varies with time.

Some prominences decelerate in their upward motion (Pittini, 1979; Waldmeier, 1979). An initial acceleration of a flare spray of March 1, 1969, was followed by a persistent deceleration of all fragments of the ejecta (McCabe and Fisher, 1970). At least 15% of all recorded cases exhibit a deceleration in the later phase of the eruption.

We divide the speed-height diagram into three equally large sectors and count the number of recorded events in

each. The result of such a rather arbitrary division becomes
53% low-speed, 33% medium-speed, and 14% high-speed cases.

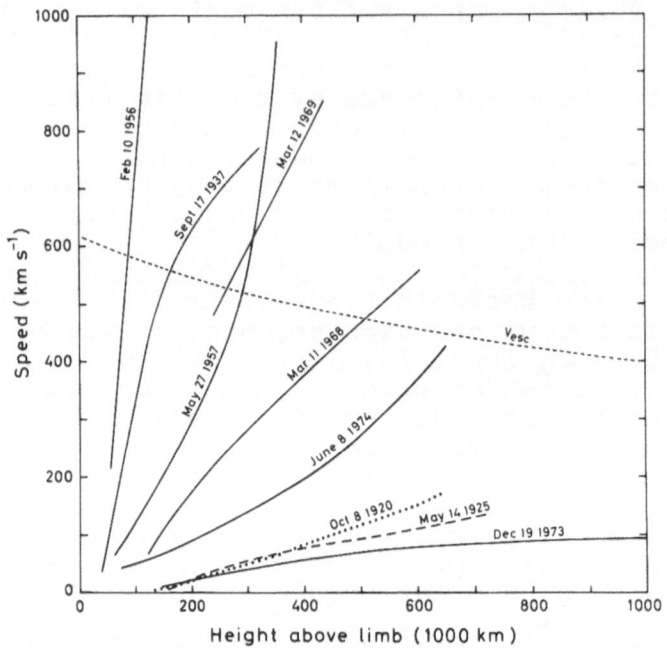

Figure 4. Some speed-height relations for eruptive
prominences.

Active region relation.

The association of rapidly ascending prominences with
flares (flare sprays) is well established (Waldmeier, 1939;
Warwick, 1957; Smith and Booton, 1962). Tandberg-Hanssen et
al. (1980) concluded from their study of Hα disk observations
that sprays are eruptive active region filaments.

We have sorted DB filaments (Paris Observatory catalogues) into three groups according to their location
relative to the nearest active region; I: polar filaments
and filaments located more than 8-10° away from active
regions, II: filaments lying next to and within about 8-10°
from an active region, and III: filaments located within an
active region. We use the definition of active regions as
outlined by the Ca$^+$K emission which is adopted in the
catalogues of Paris Observatory. The distribution of
suddenly disappearing filaments is: 49.7% (I), 37.6% (II),
and 12.8% (III).

The fraction of active region filaments among DB's is

the same as previously found for high-speed eruptive
prominences. The equality of the two numbers (12.8% and
14%) is fortuitous, but it suggests that between 10% and
20% of all eruptive prominences are likely to be high-speed
events (sprays).

Frequency of occurrence.

The frequency of occurrence of eruptive prominences is
essential in an evaluation of the mass and energy they
contribute to the corona and the interplanetary medium.
Determination of an absolute rate of eruption for a given
period of time calls for a systematically recorded sample
and good time coverage. Such data is not easily obtained.

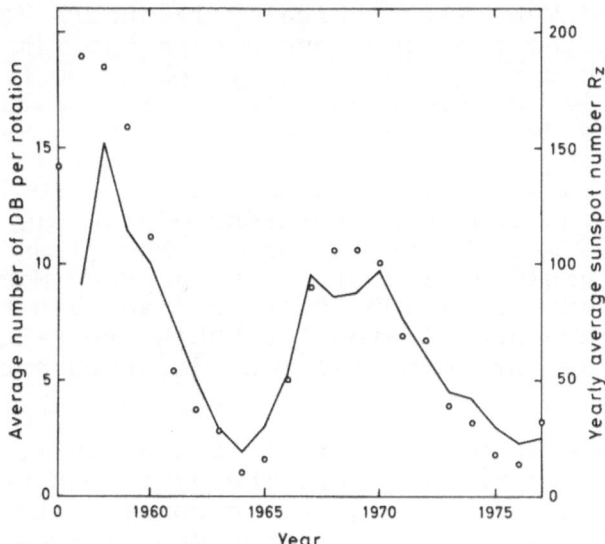

Figure 5. The variation of number of "disparition brusques"
during two solar cycles.

Many more eruptives occur about maximum compared to
minimum sunspot number (Engvold, 1980). About 40% of the
events appear within three yearly intervals centred about
the times of sunspot maxima as compared to only 6% about
sunspot minima. Figure 5 shows the yearly average number of
of DB's per solar rotation. It may be seen from the same
Figure that the average numbers of DB is proportional to the
relative Sunspot Number (Waldmeier, 1978). A similar good
correspondnece is obtained for the DB's recorded in the
period 1919-1931 and which are tabulated by d'Azambuja and
d'Azambuja (1948). The results suggest that the number of
DB's increases by factors 10-15 from sunspot minimum to
maximum.

An average of about 14% of all filaments observed during one solar rotation were reported to exhibit a sudden disappearance. The percentage is evidently larger, and it varies little with phase of the solar cycle.

In possibly as many as half of all cases a new filament will reform and later erupt from the same region. The average time intervals between two such eruptions is found to be about 5 days (Engvold, 1980).

The chances of detecting an eruptive prominence at the limb (EPL) is critically dependent on the time coverage. From tables of EPL observations during 1957 by Koeckelenbergh and Peeters (1957) we see that the number of recorded events per month is proportional to the duration of observations. Since and EPL lasts typically 1-3 hours (Pettit, 1940), we may define the time coverage as the time actually spent at the telescope during one day, plus two hours. A total of 143 EPL's was recorded with a time coverage of nearly 1000 hours during 1957 at the Royal Belgium Observatory (Koeckelenbergh and Peeters, 1957). The rate thus becomes approx. 1300 yr^{-1}. Since the observed EPL's are associated with the limb near portions of the sun the true rate of eruptions must be still larger. The filament catalogues for 1957 yield a rate of 460 DB's yr^{-1}. Similarly, we get for the year 1971; 700 EPL's yr^{-1} and 200 DB's yr^{-1}. The fact that the deduced rates for EPL's are larger than for DB's indicates that many filament disappearances pass unobserved.

The available data does not yield a definite number for the rate of erupting prominences. The derived number must be regarded as lower limit values. We conclude that there may be more than 3-4 eruptive prominences/filaments every day during years of maximum solar activity.

About 350 coronal transients occurred during the Skylab period (Hildner et al., 1976). A corresponding number of 100 DB's was recorded. According to the results above one might expect the number of erupting prominences and the number of coronal transients to be compatible.

The mass ejected by eruptive prominences.

The mass supplied to the corona by one eruptive prominence may be estimated as the pre-eruption mass minus the downfalling part. The Hα coronagraph observations of a flare spray of March 1, 1969 (McCabe and Fisher, 1970) shows clearly that some fragments move outwards whereas the trajectories for others turn back towards the solar surface.

Schmahl and Hildner (1977) inferred a column density of ground-state hydrogen in the pre-eruptive state of the Dec. 19, 1973, prominence, based on an estimate on LyC opacity at 625Å. By integrating over the areal extent of the prominence they derived a total mass $M_{prom} = 2 \cdot 10^{15}$ g. Allowing for the filamentary appearance we estimate about $5 \cdot 10^{27}$ cm^3 for the volume occupied by a typical quiescent prominence. We assume 10^{11} cm^{-3} for the number density of hydrogen (Tandberg-Hanssen, 1974; Hirayama, 1978) and get $M_{prom} = 10^{15}$ g. The results suggest that $2 \cdot 10^{15}$ g may be adopted as a typical value for the pre-eruption mass of a prominence (see also Rust et al., 1979).

Possibly less than 10% of the prominence mass was actually observed to fall back into the chromosphere in a case of June 8, 1974 (Engvold et al., 1976). The many spectral lines recorded in the Dec. 19, 1973 case (Schmahl and Hildner, 1977; Rust et al., 1979), allowed one to study how temperature and density changed as the eruption progressed. Up to a height of about 2 $R_\odot$ the prominence material experienced only a slight warming and rarefaction. Above approx. 3 $R_\odot$ the matter was subject to rapid dispersion and heating. The downfalling material along fine vertical threads which appear in EUV lines was estimated to be only about 1% of the pre-eruptive mass. The comprehensive Skylab observations illustrate clearly the short-coming of data consisting of one "cool" line only, if one wishes to determine what becomes of the initial pre-eruptive matter.

The two-ribbon flare-like brightenings which are frequently seen to accompany prominences are further evidence for downfalling material (Hyder, 1967).

Thus, there are evidences that some of the matter contained in pre-eruptive prominences returns to the solar surface. We cannot specify very accurately what this fraction typically may be. The available data suggests that close to $2 \cdot 10^{15}$ g of matter may be dispersed in the corona and interplanetary medium by one single eruptive prominence. This is a rather insignificant fraction, i.e. about 1%, of the total mass contained in the corona (Tandberg-Hanssen, 1974). However, the proton flux at the base of solar wind streams (10^{14} cm^{-2} s^{-1}; Zirker, 1976) taken over the surface area associated with a prominence (approx. $3 \cdot 10^{19}$ cm^2) corresponds to the mass of one eruptive prominence every three days.

The energy of eruptive prominences.

The study of mass motion made by the Skylab Workshop on Solar Flares (Rust et al., 1979) concludes that the magnetic

field must be the ultimate source of energy which drives
the observed phenomena. Stewart et al. (1978) estimated the
magnetic energy density of two moving Type IV sources of
radio emission and concluded that both the eruptive
prominence and the source motions are controlled by magnetic
fields.

From an observed speed-height relation one may measure
the speed and height of a prominence immediately before it
fades and disappears in Hα or Ca$^+$ K (Figure 4). From the
observations compiled by Engvold (1980) we adopt typical
values for the speed (U) and the corresponding height (R) in
the corona to estimate the kinetic and potential energies,
respectively,

$$E_k = \tfrac{1}{2}M_p U^2 \qquad \text{and} \qquad E_p = M_p\, g_\odot\, R_\odot (1-R_\odot/R)$$

of the "cool" matter of eruptive prominences.

Table I

Kinetic and potential energies of eruptive prominences

Type of event	U (km s^{-1})	R	E_k	E_p	E_k+E_p
			(	10^{30} erg	)
High-speed (Spray)	800	1.35 $R_\odot$	6.4	1.0	7
Medium-speed	450	1.65 $R_\odot$	2.0	1.5	4
Low-speed (Ascend. prom.)	200	1.85 $R_\odot$	0.4	1.8	2

The results for, respectively, high-speed, medium-speed,
and low-speed eruptive prominences are presented in Table I.
(We have neglected the negative sign in the gravitational
potential energy since we are interested in only the change
of the E_p.) The mass $2\cdot10^{15}$ g is assumed for all eruptive
prominences.

The relative importance of kinetic and potential energy
appears to change from high-speed to low-speed events.
However, within the accuracy of our estimates they are
compatible. Particular cases, such as of February 10, 1956
(Menzel et al., 1956; Warwick, 1957), may have exceeded the
tabulated values substantially.

The tabulated energies exceed by several orders of magnitude the amount of energy liberated as radiation during the eruptive phase.

The persistent acceleration of many eruptive prominences implies a steady transformation of electromagnetic energy into motion and gravitational potential energy. This is in agreement with the fact that $E_k + E_p$ of coronal transients appears to be higher than of the associated eruptive prominences (Rust et al., 1979).

Webb et al. (1979) estimated that $>10^{31}$ erg of magnetic energy was delivered into the interplanetay space by transport of ambient magnetic field by a coronal transient.

4. CONCLUDING REMARKS

A study of energy and mass injected by flares and eruptive prominences requires data with good spectral, spatial, and time coverage. In the cases of a few individual events these requirements have been met partly. A large number of cases, which are recorded in spectral lines of typically chromospheric temperatures, may be used to map the initial phases of mass ejection events. We have not attempted to estimate the mass and energy carried in particle streams at the flare site as evidenced by radio bursts.

Flare sprays and ascending prominences all consist of matter situated in the low corona prior to the disruption. Sprays and ascending prominences may be united in the term eruptive prominences. The leading edge of eruptive prominences shows upward acceleration in the range 0.01-10 times $g_\odot$. Fragments of eruptive prominences occasionally slow down. Eruptive prominences are driven by and move with the magnetic field. They preceed coronal transients. Surges follow preexisting magnetic lines of force. Surges are neither causing nor associated with coronal transients. The mass of eruptive prominences is typically about $2 \cdot 10^{15}$ g. The sum of their kinetic and potential energies amounts to $10^{30} - 10^{31}$ erg. The mass and energy of surges are evidently less than, but not very different from, eruptive prominences. The number of eruptive prominences varies proportionally with the sunspot number. Possibly more than 3-4 prominences may erupt every day during years of maximum solar activity.

REFERENCES

Becker,U.:1958,Z.f.Astrophys. 44, pp. 243-248.
Brueckner,G,E.:1972,"Conference on Flare-Produced Shock Waves...", HAO/NCAR, Boulder, Sept. 11-14.
Bruzek,A.:1951,Z.f.Astrophys. 28, pp. 277-295.

Bruzek,A.:1964, Astrophys. J. 140, pp. 746.
Bruzek,A.:1969,"Solar Flares and Space Research",ed.C.de
                Jager and Z. Svestka, pp. 61.
Cheng,C.:1978, Solar Phys. 56, pp.205.
d'Azambuja,L. and M.:1948, Ann.Obs.Paris, Meudon,6,fasc.7,1.
Dulk,G.A., and Altschuler,M.:1971, Solar Phys. 20, pp. 438.
Engvold,O.:1980, Inst.Theoret.Astrophys.Report (in prep.)
Engvold,O. and Rustad,B.M.:1974, Solar Phys. 35, pp. 409.
Engvold,O., Malville, J.M., and Rustad,B.M.:1976, Solar
                Phys. 48, pp. 137.
Gosling,J.T., Hildner,E., MacQueen,R.M., Munro,R.H., Poland,
        A.I., and Ross,C.L.:1976, Solar Phys. 48, pp. 389.
Hildner,E.:1977, Astrophys. and Space Sci. Library 71.
Hildner,E., Gosling,J.T., MacQueen,R.M., Munro.R.H., Poland,
        A.I., and Ross,C.L.:1976, Solar Phys. 48, pp. 127.
Hirayama,T.:1978, IAU Coll. No. 44, pp. 4.
Hyder,C.:1967, Solar Phys. 2, pp. 49.
Koeckelenbergh,A. and Peeters,J.:1957, Com.1'Obs.Roy.Belg.147.
Kopp,R.A., and Pneuman,G.W.:1976, Solar Phys. 50, pp. 85.
Kosugi,T.:1976, Solar Phys. 48, pp. 339.
Martin,S.F.:1978, IAU Coll. No. 44, pp.268.
Martres,M.-J., and Zlicaric,G.:1977, Cartes Synoptiques de
        la Chromosphere et Catalogues des Filaments et des
        Centres d'Activité, Obs. de Paris, VI, fasc.2.
McCabe,M.K.:1971, Solar Phys. 19, pp. 451.
McCabe,M.K., and Fisher,R.R.:1970, Solar Phys. 14, pp. 212.
Menzel,D.H., Smith,E.v.P., deMastus,H., Ramsey,H., Schnable,
        G., and Lawrence,R.:1956, Astron. J. 61, pp. 186.
Munro,R.H., Gosling,J.T., Hildner,E., MacQueen,R.M., Poland,
        A.I., and Ross,C.L.: 1979, Solar Phys, 61, 201.
Pettit,E.:1925, Publ. Yerkes Obs. 3, pp. 205.
Pettit,E.:1940, Publ. Astr. Soc. Pacific 52, pp. 172.
Pittini,A.:1979, Astr.Mitt.Eidgen.Zürich No. 373.
Platov,Yu.V.:1973, Solar Phys. 28, pp. 477.
Riddle,A., Tandberg-Hanssen,E., and Hansen,R.T.:1974, Solar
                Phys. 35, pp. 171.
Roy,J.-R.:1973, Solar Phys. 28, pp. 95.
Rust,D.M., and Hildner,E.:1976, Solar Phys. 48, pp. 381.
Rust,D.M., and Švestka,Z.:1978, IAU Coll. No. 44, pp. 276.
Rust,D.M., Hildner,E., Dryer,M., Hanssen,R.T., McClymont,A.
        N., McKenna Lawlor,S.M.P., McLean,D.J., Schmahl,E.J.,
        Steinolfson,R.S., Tandberg-Hanssen,E., Tousey,R., Webb,D.
        F., and Wu,S.T.:1979, Proceedings of the Skylab Workshop
        on Solar Flares, Chapt. 7, ed. P.A.Sturrock.
Sakurai,T.:1976, Publ.Astr.Soc. Japan 28, pp. 177.
Schmahl,E.J.:1978, Solar Phys. (submitted)
Schmahl,E.J., and Hildner,E.:1977, Solar Phys. 55, pp. 473.
Scheridan,K.V.:1979, Paper presented at present Symposium.
Smith,E.v.P.:1968, Nobel Symposium No. 9, ed. Y.Öhman, pp.137
Smith,H.J., and Booton,W.D.:1962, GRD,Res.Note No. 58, AFCRL
Smith,J.B.,Jr., Speich,D.M., Wilson,R.M, Tandberg-Hanssen,E.
        and Wu,S.T.:1977, Solar Phys. 52, pp. 379.

Smith,S.F., and Harvey,K.L.:1971, Physics of the Solar
     Corona, ed. J.Macris, pp. 156.
Smith,S.F., and Ramsey,H.E.:1964, Z.f.Astrophys. 60, pp. 1.
Stewart,R.T., Hansen,R.T., and Scheridan,K.V.:1978, IAU
     Coll. No. 44, pp. 315.
Švestka,Z.:1976, Solar Flares, D.Reidel.
Tandberg-Hanssen,E.:1974, Solar Prominences, D.Reidel.
Tandberg-Hanssen,E., and Malville,J.M.:1974, Solar Phys.
     39, pp. 107.
Tandberg-Hanssen,E., Martin,S.F., and Hansen,R.T.: 1980,
     Solar Phys. 65, 357.
Valniček,B.:1962, Bull.Astr.Inst. Czech. 13, pp. 91.
Valniček,B.:1964,   ibid   15, pp. 207.
Waldmeier,M.:1939, Z.f.Astrophys. 18, pp. 241.
Waldmeier,M.:1958,   ibid   44, pp. 213.
Waldmeier,M.:1978, Astr.Mitt.Eidgen.Zürich No. 358.
Waldmeier,M.:1979,   ibid   No. 371.
Warwick,J.W.:1957, Astrophys. J. 125, pp. 811.
Webb,D.F., Krieger,A.S., and Rust,D.M.:1976, Solar Phys.
     48, pp. 159.
Webb,D.F., Cheng,C.C., Dulk,G.A., Edberg,S.J., Martin,S.F.,
     McKenna-Lawlor,S., and McLean,D.J.:1979, Proceedings of
     the Skylab Workshop on Solar Flares, Appendix B, ed.
     P.A.Sturrock.
Westin,H.:1969, Solar Phys. 7, pp. 393.
Westin,H., and Lizska,L.:1970, Solar Phys. 11, pp. 409.
Uchida,Y., Altschuler,M.D., and Newkirk Jr.,G.:1973, Solar
     Phys. 28, pp. 495.
Zirker,J.B.:1976, IAU Coll. no. 36, pp. 421.

DISCUSSION

Moore: What is your opinion, as an observer, on whether filament
eruptions and flares are essentially the same thing; i.e., whether both
are driven by the same basic mechanism?

Engvold: Two-ribbon flares, which seem to be a very common type
morphologically, are concurrent with large eruptions, filaments and
with formation of loop prominence systems. The evidences suggest to me
that the infall-impact mechanism, which was proposed by Hyder, is valid.
Hence, two-ribbon flares as well as eruptive prominences (and so-called
'post-flare' loops) both seem to be the result of the action of some
other agent.

McIntosh: The statistics of increasing filament disappearance with
increasing sunspot number are misleading for two reasons: (1) the
frequency of disappearing filaments per solar rotation is not correlated
with sunspot number, flare occurrence or rate of emerging magnetic flux;
(2) 50% of filament eruptions occur away from sunspots. It is clear from
observations used to compile the 1964-1974 atlas of Hα synoptic charts
that large-scale merger among long-lived magnetic features is related to

filament eruption between merging features. Also, large-scale shear may be another factor of importance to erupting filaments. Would you please comment?

Engvold: I have not checked the correlation of number of suddenly disappearing filaments with the sunspot number per solar rotation, in the data used here. I agree, the fact that the correlation is strong in the yearly averages does not necessarily imply that sunspots or active regions are responsible for the eruptions. It may be important to note that the number of filaments (or rather the total filament area projected into the disk) is nearly proportional to the sunspot number (when properly averaged). The good correlation mentioned above may, therefore, simply mean that the probability that a given filament will erupt is always nearly the same and independent on solar activity.

Tandon: You gave the estimates of kinetic and gravitation energy to calculate the total energy of eruptive prominences. What is the order of magnetic energy associated with them, and will it not upset the total energy computations?

Engvold: The observations suggest that erupting prominences are magnetically controlled and, surely, we would expect the associated magnetic energy to be greater than the sums of kinetic and gravitational energy listed in my Table I.

Webb: In your discussion of DBs in active regions, you noted that 13% of all DBs in your sample occur in active regions. This % seems remarkably small since filaments in the ARs come and go all the time. Do you believe there could be an observability effect here?

Engvold: We cannot exclude the possibility that there may be a selection in the data in the sense that suddenly disappearing active region filaments more easily escape detection than those occurring outside active regions. Suddenly disappearing filaments may not be recorded as such in the data used here if they re-form within one day.

Martres: The filaments are listed as "plage filaments" when their lifetime is 2 days or longer.

X-RAY EVIDENCE OF CORONAL PREFLARE EMISSION

D. F. Webb
American Science and Engineering, Inc.
Cambridge, Massachusetts 02139 USA

Flares are believed to derive their energy from localized nonpotential magnetic fields; this energy is somehow stored and initially released in the low corona. Therefore, knowledge of coronal conditions prior to flares is essential to our understanding of the flare process. The coronal plasma is constrained by the magnetic field and its characteristic emission is easily observed at soft X-ray wavelengths. Therefore, X-ray imagery is well suited for observing changes in emission influenced by nonpotential magnetic fields in the low corona. This study used soft X-ray images from the Skylab AS&E telescope to search for evidence of coronal preflare emission.

In previous studies Kahler and Buratti (1976) and Kahler (1979) found that there were no systematic preflare X-ray brightenings at the locations of subsequent small flares, and therefore no requirement for coronal preflare heating of the flare loops. However, there exist in the literature many specific examples of preflare activity in the form of discrete brightness changes and filament activation, which are interpreted as magnetic field changes. For example, Martin and Ramsey (1972) found that about half of the Hα flares they studied exhibited preflare filament activity. And during Skylab several examples were found of EUV and X-ray preflare brightenings in the active region (AR) where a flare occurred. The emerging flux model of flares (Heyvaerts, Priest and Rust, 1977) predicts preflare soft X-ray emission outside the flare site tens of minutes before flare onset. These observations and the existence of a specific model predicting preflare activity motivated this study.

This study addresses the following questions: Do systematic preflare X-ray brightenings exist either at the flare site or in adjacent parts of the AR; what are the characteristics of such brightenings; and what do they tell us about preflare coronal conditions? The study had two parts: (1) a statistical study of preflare X-ray events and (2) an analysis of the best observed events including comparison with high time resolution Hα data.

M. Dryer and E. Tandberg-Hanssen (eds.), Solar and Interplanetary Dynamics, 189-193.
Copyright © 1980 by the IAU.

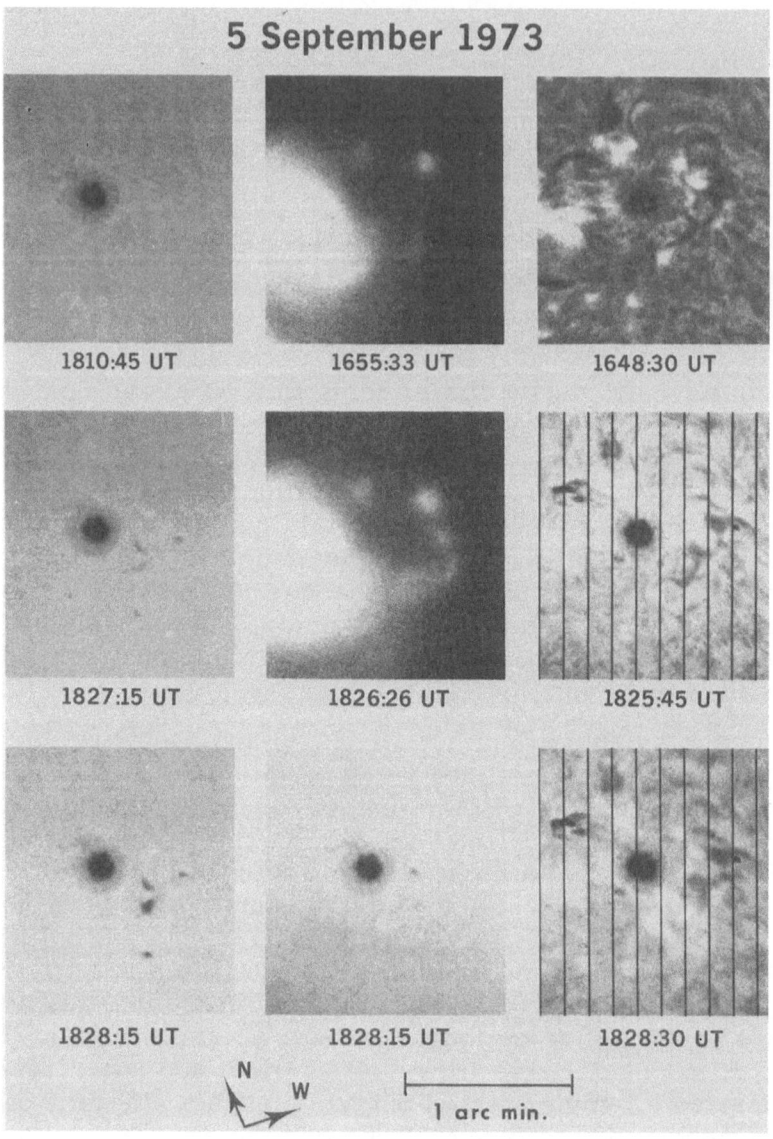

Figure 1. The central X-ray image at 1826 UT shows preflare emission associated with the activation of the filament to the lower right of the sunspot. Hα blue-wing filtergrams in the left column show the ascending filament mass before flare onset and displaced to the west of the flare site (bottom images). X-ray images are with the Skylab AS&E telescope with a passband of 2 - 17 Å. Hα images are courtesy of Sacramento Peak and Lockheed observatories.

Our criteria for selection of events for this study were:

1. Observing period 28 May – 27 November 1973,

2. The X-ray flare profile and onset time (T_o) could be determined from Solrad and/or the Skylab XREA whole-sun detector,

3. At least two X-ray images available, one during flare rise phase and one during preflare period T_o -30 min, and

4. Flare not within 20° of solar limb.

My study differed from the Kahler and Buratti, and Kahler studies primarily in that they used a preflare period of 20 min. I chose a 30 min period because the best observed evidence of preflare activity, filament motion, begins an average of 30 min before flare onset (Martin and Ramsey), and because this interval provided a reasonable number of events with several X-ray sequences.

The Table summarizes the results. A total of 25 events satisfied the criteria. Seventeen of the 25 events (2/3) had probable or possible preflare features (defined as observable transient brightenings) within the AR where the flare occurred. Eleven of the 17 events contained a single preflare structure (5 had 2 features and one had 3 features). Of the 17 events with preflare features, in 9 cases the feature was not at the flare site (in 5 cases it was at the flare site and in 3 it was mixed). Again, of the 17 events, in 11 at least one of the preflare features was not observed to reach flare intensity (in 7 cases at least one feature did flare and 1 was uncertain). In nearly all of the events, the preflare structure was either loop-like or a kernel. X-ray kernels have been defined as compact $(5 - 7^m)$, transient (5 – 10 min) knots of flare brightness occurring in ARs typically during the rise phase of flares (Kahler, Petrasso and Kane, 1976).

Evidence from my preliminary study comparing these events with Hα and from other studies suggests that the X-ray kernels are associated with bright Hα knots. Also at least some of the elongated preflare X-ray features are associated with the activation of filaments. Figure 1 shows an example of coronal preflare emission associated with the activation of a filament.

In conclusion I find that a majority, but not all flares studied, may have had X-ray preflare features. The preflare feature typically was not at the flare site, but was adjacent to the AR neutral line. A single feature was evident in most of the preflare events.

These observations suggest that preflare X-ray brightenings are associated with changing magnetic structures, and are evidence of magnetic energy release and coronal heating prior to the flare. One type of feature,

X–RAY PREFLARE EMISSION: STATISTICAL STUDY

Total Number of Events	25	

	Yes or Maybe	No
Preflare Feature in Flare AR ?	17	8
Preflare Feature at Flare Site ?	8	9
Preflare Feature is a Flare ?	8	11

	1	2	3
Number of Features per Event	11	5	1

	Loop	Kernel	Sinuous
Morphology of Preflare Features	9	8	3

filament activation, is the signature of a major disruption of the magnetic field prior to the flare. Another preflare feature, an X-ray kernel,
has been interpreted as a small emerging flux loop overlying an Hα
kernel (Kahler, Petrasso and Kane). In a model of a simple loop flare
an emerging flux loop might contact a pre–existing larger loop which
bridges the neutral line. A neutral sheet might be formed which heats
up due to a plasma instability. The initial (preflare) heating is first
observed as the kernel because of its small volume. If the large loop
is replaced by a filament lying along the neutral line, we have the situation of an activated filament preceding a two–ribbon flare. Therefore,
the results of this study appear to be compatible with the emerging flux
flare model for the majority of flares studied.

I thank Stephen Kahler and Sara Martin for their assistance providing data
for this study and for many helpful discussions. I thank Jess Smith for
providing S–056 images. This work was supported by NASA under contract NAS8–27758.

REFERENCES

Heyvaerts, J., Priest, E.R., and Rust, D.M.: 1977, Astrophys. J.
    216, 123.
Kahler, S.W.: 1979, Solar Phys. 62, 347.
Kahler, S.W. and Buratti, B.J.: 1976, Solar Phys. 47, 157.
Kahler, S.W., Petrasso, R.D., and Kane, S.R.: 1976, Solar Phys.
    50, 179.
Martin, S.F. and Ramsey, H.E.: 1972, in P.S. McIntosh and M. Dryer
    (eds.) Solar Activity Observations and Predictions, MIT Press,
    Cambridge, MA.

DISCUSSION

Jackson: How do you know that the pre-flare brightenings are associated with the subsequent flare and not just random brightenings of adjacent features?

Webb: My study was limited by a lack of high time resolution data during Skylab. We do know that the X-ray brightenings are associated with filaments which show preflare motion and sometimes erupt at flare onset. Other brightenings, such as X-ray kernels, are associated wtih bright Hα bursts lying along the neutral line in the active region where the flare occurs. We need to perform a study to statistically check if such brightenings are random or not. We also will examine good quality Hα data to determine where the X-ray features are with respect to the flare site.

Levine: (Comment) In support of one of your interpretations, I would like to point out that I was able to find unambiguous pre-flare brightenings in EUV emission within pre-existing small loops prior to a small flare on November 28, 1973. Further, examination of solar magnetograms showed a direct association with emerging flux. (See Levine, *Solar Phys.*, *56*, 185, 1978).

SPICULES AND MACROSPICULES

W. van Tend
The Astronomical Institute at Utrecht, Sterrewacht,
Zonnenburg 2, 3512 NL UTRECHT, The Netherlands.

ABSTRACT

The transition zone overheating model and the melon seed model are compared with observations of spicules, macrospicules, surges and sprays.

1. Introduction

For about ten years two essentially different models for spicules have existed (Beckers, 1972): Firstly the overheating model of the lower transition zone, and secondly the melon seed model. In this paper we consider these two models in the light of recent observational and theoretical developments.

2. The Overheating Model

In the overheating model a Rayleigh-Taylor like instability occurs, when the lower transition zone receives more energy by heat conduction from the corona than can be radiated away locally. The instability manifests itself as spicules (Kuperus and Athay, 1967).

During spicule generation by overheating of the lower transition zone, the coronal energy content diminishes. The corona thus describes an orbit in the energy balance diagram depicted in Figure 1. The time needed for completing one cycle in this diagram is given by the heating time τ of the corona: $\tau = pH/F$, where p is the coronal base pressure, H is the scale height and F is the energy flux. For the solar corona this time is 1000 sec, which is indeed the observed spicule life time.

In flares a similar time interval is observed between the filament activation and the Hα flash. If the Hα flash is the response of the chromosphere/transition region to excess heating of the corona, this is what is expected, since it takes 1000 sec to heat the corona. The Hα ribbons might then be the base of surges, collimated along field lines as large spicules.

M. Dryer and E. Tandberg-Hanssen (eds.), Solar and Interplanetary Dynamics, 195-197.

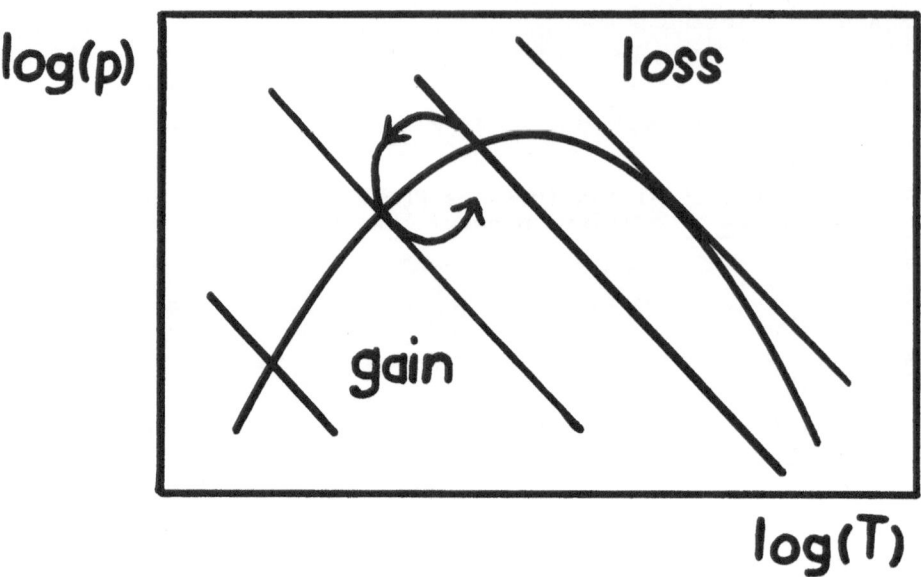

Figure 1
A plot of coronal base pressure against temperature, showing the
energy balance curve. Straight lines are contours of constant coronal
energy content. A possible orbit of a corona perturbed from energy
balance is indicated.

More details on the overheating model are found in Van Tend (1979).

3. The Melon Seed Model

    The melon seed model describes spicules as a result of the
stretching of field lines after reconnection at the photospheric level.
Then a field reversal is required at every spicule. That many field
reversals are not observed. The model may apply however to the more
recently discovered macrospicules, which are much rarer. Observational
evidence for this is the association of macrospicules with small flares
(Moore et al., 1977). Flares are always associated with neutral lines,
and thus macrospicules would be associated with field reversals, as
required.

    Also theoretical work (Van Tend, 1980) indicates a possible
application of the melon seed model to macrospicules. Magnetic flux
leaves flux tubes (like sunspots) during their decay. Only by reconnection at neutral lines does photospheric magnetic flux disappear.
The fastest reconnection can be shown to occur at the photospheric
level with fields (re)concentrated in flux tubes. Then a configuration
as shown in Figure 2 occurs. This field line pattern is stable up to a

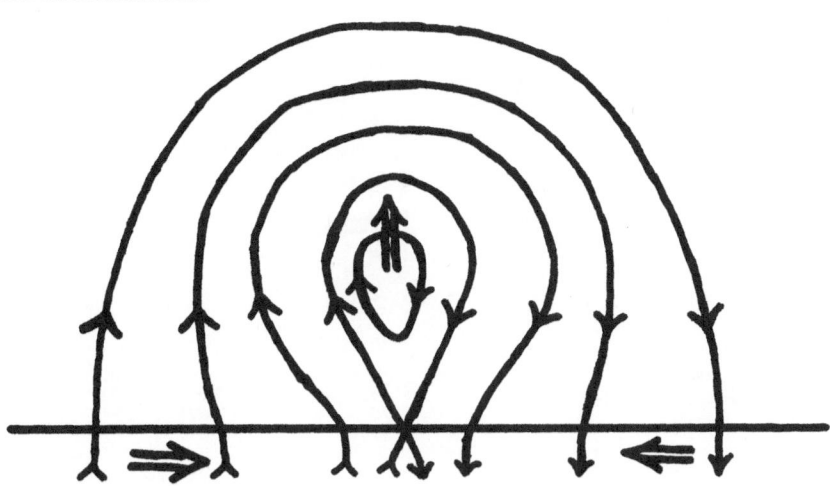

Figure 2
The field line configuration of the melon seed model.

certain threshold height of the coronal O-type neutral point, as shown
by the Van Tend and Kuperus (1978) flare build-up model. If the
threshold is at a low height, a macrospicule associated with a small
flare will result by the reconnection of one pair of flux tubes. If the
threshold is high, a larger flare (associated with a spray and prominence eruption) results after several flux tube pairs have reconnected.

Acknowledgement

      This work was supported by the Netherlands Organization for the
Advancement of Pure Research (Z.W.O.).

References

Beckers, J.M.: 1972, *Annual Review of Astron. and Astrophys.* 10, 73.
Kuperus, M. and Athay, R.G.: 1967, *Solar Phys.* 1, 361.
Moore, R.L., Tang, F., Bohlin, J.D. and Golub, L.: 1977, *Astrophys. J.*
    218, 286.
Van Tend, W.: 1979, *Solar Phys.*, 61, 89.
Van Tend, W.: 1980, Ph. D. Thesis, University of Utrecht, to be
    published.
Van Tend, W. and Kuperus, M.: 1978, *Solar Phys.* 59, 115.

UV EMITTING SPICULES

Giannina Poletto
Osservatorio Astrofisico di Arcetri, Firenze, Italia

Abstract

Extreme ultraviolet observations of the chromospheric network in a coronal hole obtained in 1973 by the Harvard College Observatory experiment aboard Skylab are analyzed. Upper and lower limits to the actual emission measure in UV spicules have been obtained, and the consistency of the derived values with the hypothesis that UV spicules are Hα spicules falling back after being heated is discussed.

Introduction

EUV spectroheliograms from Skylab show evidence of a concentration of emission at the boundaries of the supergranular network, where it is well known that spicules originate. As their average cross-sections are well below the instrument spatial resolution, spicules cannot be individually resolved. However, evidence for the identification of the brightest network elements with spicules comes from the close spatial correspondence between dark Hα mottles and bright UV regions, as well as from their comparable time evolution (Feldman et al. 1976, Reeves et al. 1976). Significant UV emission from spicules is also required to match observed and theoretical limb-brightening curves, which result in a good agreement if the variation with height of the number of UV spicules is the same as for Hα spicules (Withbroe and Mariska, 1975; Kanno, 1978).

Even if their presence is well ascertained, the physical parameters of UV spicules are largely unknown. The purpose of this short note is to show how, from low-resolution data, upper and lower limits to the actual emission measures in UV spicules, could be derived, and how they compare with current ideas about their connection with Hα spicules.

The Method.

The data which are here analyzed were acquired by the EUV Harvard spectrometer on Skylab. Eight spectroheliograms obtained at different times on May 31, 1973 give the brightness distribution on a 5x5 arcmin coronal hole area located near the center of the Sun. Each spectro-

199

M. Dryer and E. Tandberg-Hanssen (eds.), Solar and Interplanetary Dynamics, 199-201.

heliogram has been produced at the seven wavelengths of the normal
polychromatic setting, and consists of a number of picture elements,
partially overlapping, with individual resolution of 5 x 5".

As spicules are concentrated in network areas, the network should
be preliminarily identified. The network area percentage at increasing heights can be evaluated from the frequency vs. intensity distribution of chromospheric and transition region lines originating from
ions formed in different temperature ranges. The histogram data have
been interpreted as reproducing a symmetric distribution, representative of the cell plus an asymmetric distribution, obtained by subtraction from the total of the Gaussian-like distribution, representative
of the network and spicule emission. The asymmetric distribution is
found to occupy an area percentage between 35% (OIV) and 48% (CIII).
In the following a constant value of 40% will be assumed for all of
the ions, due to the order-of-magnitude character of the present
considerations.

It is now necessary to derive the number of spicules a typical
spectroheliogram is likely to contain. If, following Beckers (1968),
the number of spicules per unit area, is assumed to be about 3×10^{-8}
km^{-2}, a representative spectroheliogram will contain $\sim 1.5 \times 10^3$
spicules. Their distribution along the network can only be guessed,
due to their unobservability, and two different hypotheses will presently be made: 1) the network itself consists of spicules evenly
distributed along the supergranular boundaries, 2) spicules are
clumped together and arranged so as to completely fill a fraction of
the network picture elements.

In the first case all of the network elements will include a
spicule (due to the already mentioned partial overlapping of picture
elements), and the average intensities derived from the frequency vs.
intensity distribution and usually ascribed to the network emission
would rather be representative of the spicular emission. The spicule
contribution to the global solar UV emission should therefore be identified with the "network" contribution.

In the latter case, average intensities in spicules could be derived, on the assumption that spicules are responsible for the brightest network elements, isolating a number of points, in the tail of the
frequency vs. intensity distribution, equal to the ratio of the total
spicule area to the network area. This allows us also to evaluate the
spicule contribution to the global UV emission.

Differential emission measures $Q(T)$ for UV spicules can be easily
derived once the spicule intensity is known, through the relationship
(e.g. Withbroe, 1977):

$$I \ (erg/cm^2/sec/ster) \ = \ 1.7 \ 10^{-16} \ Af \int Q(T) gG(T) dT$$

where symbols have their usual meaning, and $Q(T)$ is defined by

$$\int Q(T) dT \ = \ \int Ne^2 dv$$

Taking into account that in case 1) the spicule intensity I is spread,
due to the low instrument resolution, over a larger area, and therefore
correcting it for this effect, it follows that the actual differential

emission measures for spicules appear to be higher by a factor of $\sim$ 2 with respect to those usually ascribed to the network.

Discussion.

Assumption 1) and 2) should be considered as representative of extreme situations. Actual spicule behavior will somehow be an average between 1) and 2), where 1) is likely to give an upper limit to the actual $Q(T)$ values, as spicules are known to frequently clump together, while 2) gives a lower limit, as no account has been taken of the intensity spreading caused by the instrument. As an order-of-magnitude estimate, it will be assumed that actual $Q(T)$ spicular values are about 10 times greater than the usually reported network values.

Without further assumption, it is not possible to establish whether the $Q(T)$ increase is due to a density increase or a temperature gradient decrease. However, since we know that in the brightest regions of the UV network strong redshifts are observed (Bonnet, 1978), we can admit that UV spicules are $H\alpha$ spicules which fall back again after being heated to high temperatures. Moreover, since we know that only 1% of spicular material escapes in the solar wind, it follows, from mass conservation, that UV spicules should have about the same density as the $H\alpha$ spicules (unless spicular material is able to traverse magnetic fieldlines).

To an increase of about a factor of 10 in the $Q(T)$, it corresponds, if the temperature gradient does not change, an increase of about a factor of 3 in the densities. This will bring the average density in the $4.6 \lesssim \log T \lesssim 5.6$ interval close to a value of 2.10^{10} cm^{-3}, which favorably compares with the density in the upper part of cool spicules.

Therefore if UV spicules are to be interpreted as $H\alpha$ spicules falling back after being heated, we are led to an average density of some 2×10^{10} cm^{-3}, and to a gradient comparable with that usually quoted as "network" gradient. Future high-resolution observations, allowing one to check these tentative values, will give a definite answer to the problem of the UV spicule nature and origin.

Acknowledgment.

It is a pleasure to acknowledge helpful and stimulating discussions with G. L. Withbroe.

References.

Beckers, J.M.: 1968, Solar Phys. 3, 367.
Bonnet, R.M.: 1978, Space Sci. Rev. 21, 379.
Feldman, U., Doschek, G.A., and Patterson, N.P.: 1976, Astrophys. J. 209 270.
Kanno, M.: 1978, Publ. Astron. Soc. Japan 30, 581.
Reeves, E.M., Vernazza, J.E., and Withbroe, G.L.: 1976 Phil. Trans. Roy. Soc. London A281, 319.
Withbroe, G.L.: 1977, Proc. 199 Nov. 7-10, OSO-8 Workshop, pp. 1-27.
Withbroe, G.L. and Mariska, J.T.: 1976, Solar Phys. 48, 21.

ON A PECULIAR TYPE OF FILAMENT ACTIVATION

ANTON BRUZEK
Kiepenheuer Institut, Freiburg FRG

Abstract. Hα observations of a peculiar type of filament activation are presented and discussed. Ejected filament material was
stopped at a distant position and formed a new stable filament for
1/2- 1 hour until it returned to its source or faded in situ. A
similar event has been observed at the solar limb.

Various types of active prominences – such as surges, sprays
and eruptive filaments – show fast, large scale displacements of
material. Common to all of them is that the material does not survive the motion: it either fades in the corona or it falls back
to the chromosphere where is disappears immediately at impact; no
material is seen coming to rest and to survive. Here I want to
talk about observations of filaments which, on the contrary,
settled down after displacement for some time as more or less
stable filaments.

A multiple activation occurred with the complex flare event
of 15 June 1972 (Bruzek 1975). The small filament shown in Figure 1
had two phases of special activity: 1) After slight changes in the
early morning hours, material started moving South along a curvilinear trajectory at about 0800 UT. It partly darkened, partly
brightened and showed dopplershifts indicating upward-downward motion along a large arc. The velocity of displacement was about
50 km/s. The material came to rest at position P forming a nice,
rather stable filament which had its largest size between 0845 –
0900 UT. After 0900 UT the material returned along the former trajectory to its place of origin at position F – as indicated by
dopplershifts and displacements – and remained there in a rather
active state. 2) A second ejection of material started about 0940
UT before the onset of a class 2 flare in that region, and material
moved again South along the same trajectory as before. After 1010
UT it settled down as a strong filament again at position P although with considerable internal motions. This time the material
did not return to its source but faded slowly in situ during the

M. Dryer and E. Tandberg-Hanssen (eds.), Solar and Interplanetary Dynamics, 203-206.
Copyright © 1980 by the IAU.

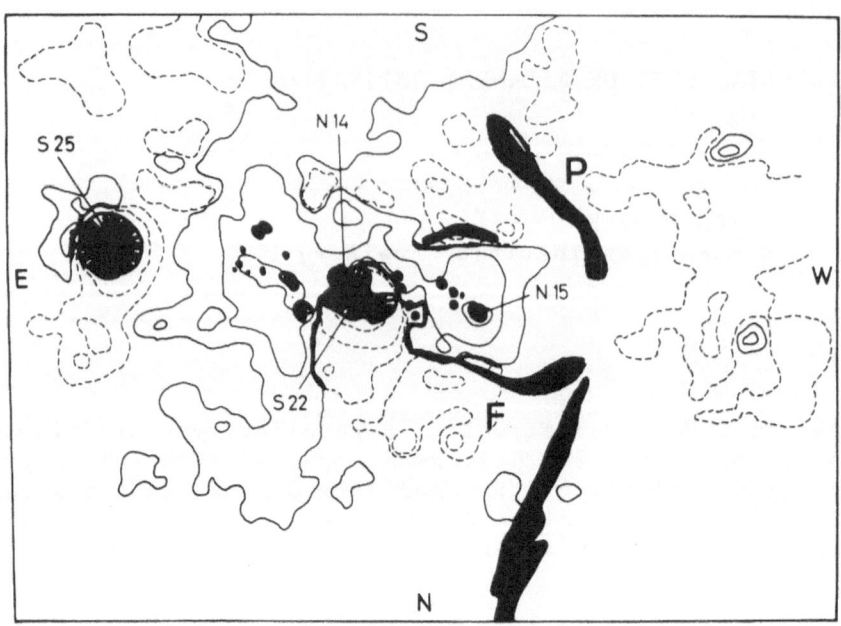

Figure 1. Spots with polarities and magnetic field strengths, and main filaments in McMath region 11926, 15 June 1972 superposed by isogausslines taken from a magnetogram of Sacramento Peak Observatory of 14 June, 1754 UT. Filament F moves to position P.

next hour. - Another disk observation of a temporary stabilization of ejected material has been reported by Zirin (1976).

From disk observations, however, it is not quite clear where the ejected material came to rest: in higher levels in the corona or just above the chromosphere, as most filaments do. The dopplershifts observed in our case indicate,however, descending motion at the end of the trajectory and therefore are strong evidence for the lower position. Direct evidence is provided by a limb observation made 12 September 1966 (Fig. 2). In that case material from an active prominence moved along a flat arc to a distant place where it settled down slowly and faded (or merged with the chromosphere) within half an hour. Thus it appears that ejected prominence material in some cases can stay at low levels for at least half an hour until it fades or returns to its source.

Let us turn now to the configuration of the magnetic field associated with our filament activations which is fundamental for the support and motion of prominence material. Figure 1 shows the filament in its original (F) and its displaced position (P) in a magnetogram taken at Sacramento Peak Observatory on 14 June which, however, very likely is also representative for the 15 June situa-

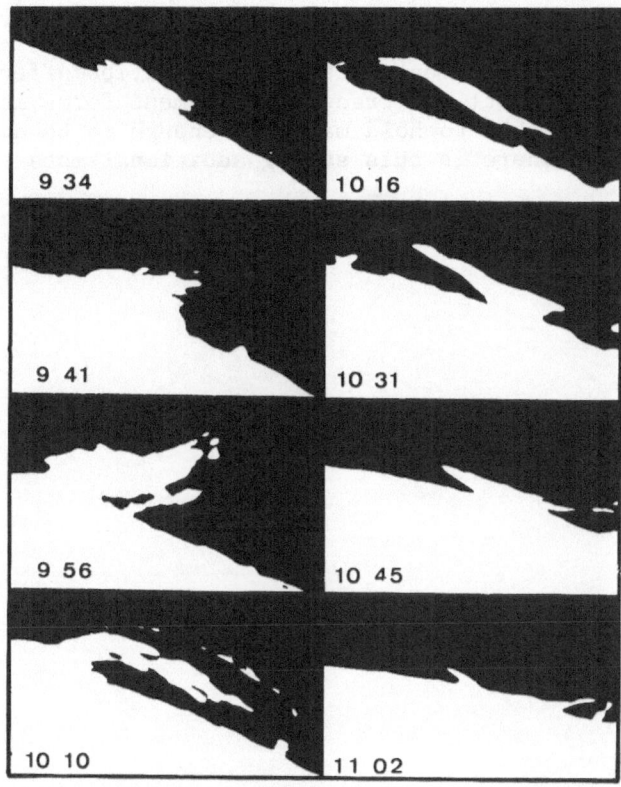

Figure 2 Active Prominence of 12 Sep. 1966; prominence material is ejected to the right and slowly descends and fades.

tion. The original filament F is rooted in a neutral line between strong opposite magnetic fields. The moving filament material runs first between weaker opposite fields but settles down in a purely southpolar region. From that, it is difficult to arrive to a consistent interpretation. It is clear that the material did not follow the direction of the magnetic field; it rather seems to move across field lines which is hard to conceive. The temporary stabilization of the filament material above a unipolar field is not understood either. Zirin's filament was also formed in an apparent unipolar field, but he found some arguments that there might have been a neutral line. In our case the field seems unambiguously unipolar. At best, one could assume that a rather oblique and curved neutral sheet existed above the U-shaped neutral line around spot N15 where the filament could float high in the corona; but this contradicts the above conclusion about the low position of the filament and could not explain why the motion is stopped there.

References:
Bruzek, A.: 1975 in: Shea, M.A. and Smart, D.F. (eds) Results Obtained During the Campaign for Integrated Observations of Solar Flares, AFCRL-TR-75-0437, Special Reports No 193, p.43
Zirin, H.: 1976, Solar Phys. 50, p.399

DISCUSSION

Moore: (Comment) I think this event fits the explanation offered by Zirin for a quite similar event: the transient filament forms in a filament channel which is not able to hold material enough to be a visible filament except when there is this strong additional mass supply.

THE FILAMENT ERUPTION IN THE 3B FLARE OF JULY 29, 1973: ONSET AND MAGNETIC FIELD CONFIGURATION

R.L. Moore
Caltech, Big Bear Solar Observatory, Hale Observatories

B.J. LaBonte
Hale Observatories

We present direct observational evidence for the preflare magnetic field configuration, the nature of the filament destabilization and triggering of the flare, and the magnetic field configuration after the filament eruption in the large, well-ordered, expanding two-ribbon flare of July 29, 1973.

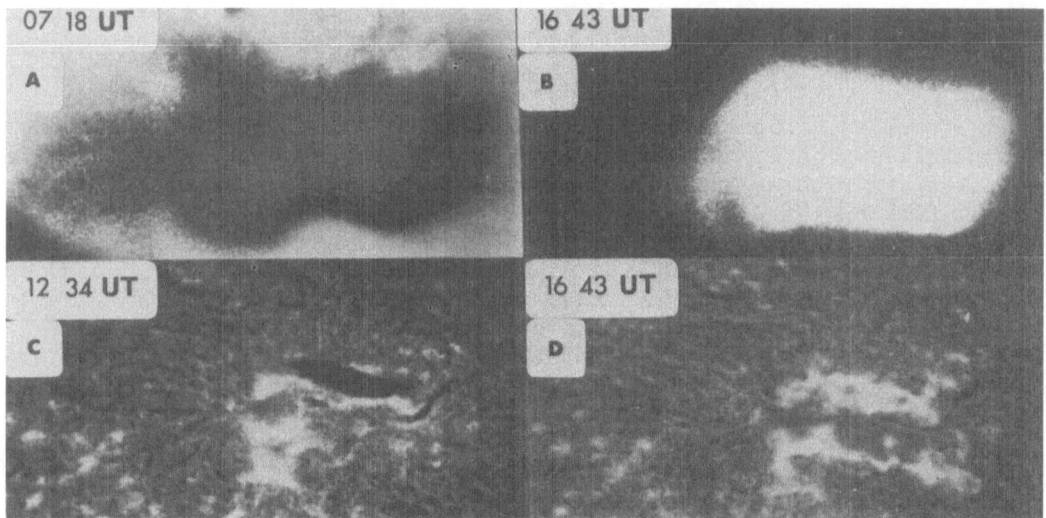

Fig. 1. Preflare (left) and late phase (right) coronal X-ray structure (top) and chromospheric Hα structure (bottom). A and C show that before eruption the large filament was enclosed in an arcade of coronal loops; B and D show that a similar, but brighter, arcade of hot flare loops, rooted in the separating Hα ribbons, was again present in the late phase. A,B,C,D are all of the same scale, orientation and area. A and B are from AS&E Skylab X-ray filtergrams of bandpass 2-32 Å plus 44-45 Å and of 16s exposure; A is a positive-negative photographic subtraction which enhances features of low contrast, such as individual loops in the arcade. C and D are Hα filtergrams from the NOAA observatory in Boulder, CO. In this and all subsequent Figures, east is up, north is to the right.

M. Dryer and E. Tandberg-Hanssen (eds.), Solar and Interplanetary Dynamics, 207-211.
Copyright © 1980 by the IAU.

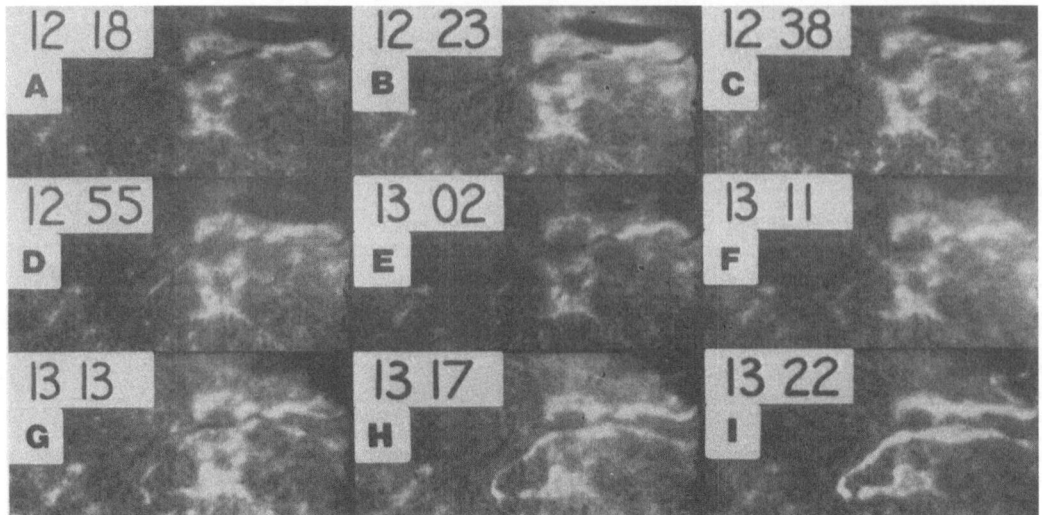

Fig. 2.   Precursor activity, filament eruption, and onset and early
development of the Hα flare ribbons.   A, at about 45 min before flare
onset, shows a narrow filament which ran along the neutral line and well
below the large filament.   B,C,D show the precursor activity: break-up
of the low filament accompanied by brightenings along the neutral line
below the large filament and a bright mass ejection to the left along the
neutral line.   In E, the large filament is erupting and the first traces
of the flare ribbons are visible; these embryo ribbons are below the
rising filament, and their distance from the neutral line is much less
than the pre-eruption height of the bottom of the large filament.   In F,
the rising filament is higher and only faintly visible at the top edge
of the frame, and the ribbons have increased in brightness and extent.
G,H,I show the premaximum development of the flare ribbons; note the
reverse curls on opposite ends of the two ribbons. Filtergrams from Big
Bear Solar Observatory (BBSO) field station in Tel Aviv, Israel.

     The observations show the following (Figures 1, 2 and 3). (1) Prior
to the eruption, the filament was under an arcade of closed magnetic-
field lines.   (2) The magnetic field in the chromosphere and in the
filament was strongly sheared across the neutral line.   (3) The eruption
of the filament and the onset of the two-ribbon Hα flare were preceded
by precursor activity in the form of small Hα brightenings and mass
motion along the neutral line and well below the bottom edge of the
filament.   (4) The onset of the flare ribbons occurred simultaneously
with the filament eruption.   (5) The initial distance of the Hα ribbons
from the neutral line was much less than the height of the filament
above the chromosphere.   (6) The precursor Hα brightenings and the first
brightenings in the flare ribbons were in the vicinity of the steepest
magnetic field gradient in the flare region.   (7) There was no evidence
for emerging magnetic flux in the flare region.

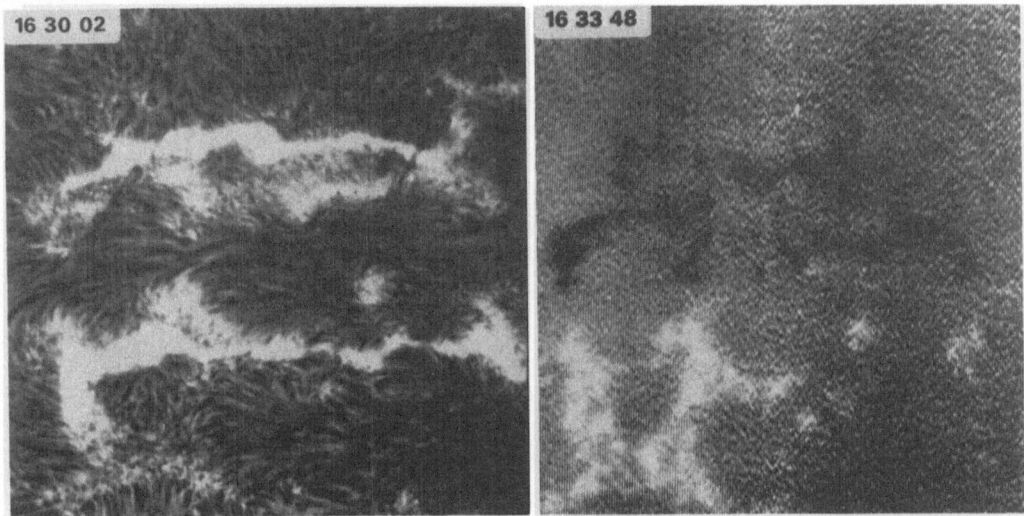

Fig. 3. High-resolution Hα filtergram (left) and videomagnetogram
(right) taken in late phase at BBSO. Comparison with Figure 2 shows
that the flare started in the vicinity of the steepest gradient in the
magnetic field across the neutral line. The Hα fibril structure here
and in Figure 2 shows that the magnetic field in the chromosphere and
in the large filament was strongly sheared across the neutral line
before the flare and was still highly sheared in the chromosphere after
the filament erupted. Neither the Hα filtergram nor the magnetogram
show any sign of emerging magnetic flux.

    We interpret the above empirical results as follows with regard to
the magnetic field configuration and how it changed in the flare
(Figure 4). (1) The preflare field configuration was similar to that
proposed by Heyvaerts et al. (1977), except that there was no emerging
flux. The essential aspect is that the field near the neutral line and
supporting the filament is strongly sheared, and the degree of shear
decreases with distance from the neutral line, so that the strongly
sheared field is enclosed in an envelope of loops which are much more
nearly perpendicular to neutral line. (2) Both the destabilization of
the filament and the initial flare ribbons resulted from magnetic field
reconnection below the filament; this initial reconnection triggered the
flare. (3) The reconnection began above the neutral line in the region
of greatest shear in the magnetic field; a gradual increase in the shear
to an untenable degree was the immediate cause of this flare, not the
emergence of new magnetic flux as proposed by Heyvaerts et al. (1977).
(4) Following the initial reconnection which started the filament
eruption, the eruption set up the "inverted Y" configuration for the
decay phase. Figure 4 shows how this transformation could occur in
three dimensions and in accord with the observed chromospheric and
coronal structure prior to and during the flare; cf Figure 1 of
Hirayama (1974). The initial reconnection causes the filament to start

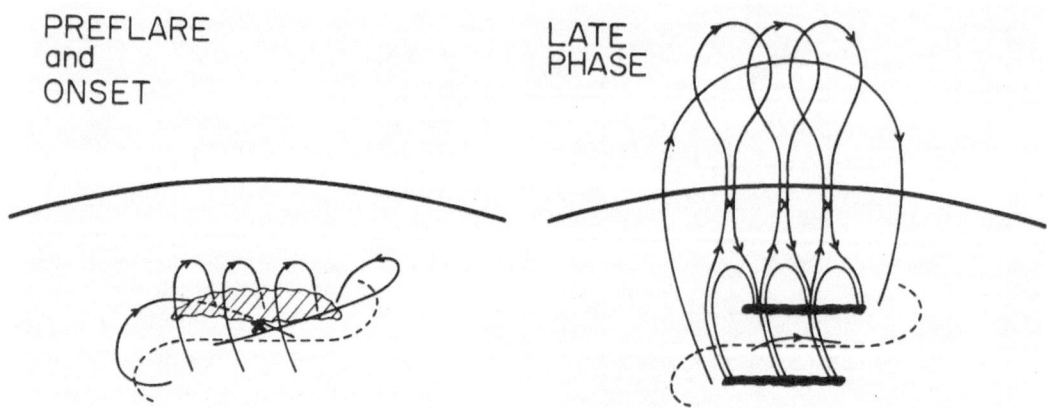

Fig. 4. Sketch of the inferred magnetic field configuration before and
after the filament eruption. The heavy arc is the limb; the dashed line
is the photospheric neutral line. Lines with arrowheads are magnetic
field lines, and X's mark places where reconnection occurs. At flare
onset and in the early phase, reconnection occurs under the filament
between "closed" field lines having their remote ends in the curls (not
shown here) of the flare ribbons. In the late phase, the reconnecting
"open" field lines are rooted at the leading edges of the spreading
ribbons. The upper reaches of the expelled field lines may be much
higher in the decay phase than shown here.

to erupt, and from the start the change in the field configuration
facilitates more reconnection below the rising filament, which leads to
further expulsion of the filament, and so forth; the overall configuration is in this way unstable and thus produces the flare. In the late
phase, the field lines which were initially closed over the filament
and were "opened" by the eruption reclose by reconnection in the wake of
the expelled filament.

      We thank S. Kahler and D. Webb of AS&E for supplying the filter-
grams for Figure 1. This research was supported by the National Aeronautics and Space Administration under Contract No. NAS8-33215 and by
the Air Force Geophysics Laboratory under Contract No. F1962-77-C-0106.

REFERENCES

Heyvaerts, J., Priest, E.R., and Rust, D.M.: 1977, Astrophys J 216,
        pp. 123-137.
Hirayama, T.: 1974, Solar Phys 34, pp. 323-338.

DISCUSSION

Pallavicini: I agree with your interpretation for this particular flare. However, this flare is a typical example of a class of flares, characterized by long-duration and long-decay time and associated with prominence eruptions and white-light transients. I do not think it is safe to extend the same interpretation to all flares.

Moore: Perhaps you are right. But, I think that this flare is more similar to most flares than you think. Many flares, perhaps most, begin with a filament eruption, or an outward eruption in the sheared field along the neutral line with no filament, at the onset of the two Hα ribbons.

Pneuman: I think you touched on a very fundamental point here, which is: what comes first? Does the prominence lift, perhaps due to some internal instability allowing field lines to collapse and reconnect underneath - or, does the reconnection begin first and push the filament upward?

Moore: The evidence for this flare is that reconnection and the onset of the filament eruption are practically simultaneous. What comes first is the build-up of shear which leads to the onset of reconnection which is the same as the onset of the filament eruption.

DYNAMICS OF A QUIESCENT FILAMENT

B. Schmieder, M.-J. Martres, P. Mein, I. Soru-Escaut
Observatoire de Paris, 92190 Meudon, France

The topic of this paper is to present some results concerning the dynamics of quiescent filaments. Observations with the Meudon Heliograph in three wavelengths in the Hα line have already shown two kinds of perturbations : (1) a fast one lasting about 10 minutes and (2) a slow one which lasts a few hours. These qualitative observations have been completed recently with the Multichannel Subtractive Double Pass Spectrograph (MSDP) which operates on the Solar Tower in Meudon. This instrument allows a good spatial and temporal resolution which gives quantitative results.

We have got observations in the Hα line of a quiescent filament on October the 11th 1977 (set of observations during 4 minutes with a 20 seconds sample) and on October the 13th 1977. From the measures of chromospheric radial velocities we obtain the following conclusions :

A. Fast Perturbation
The fast perturbation lasts about 10 minutes and occurs only in a part of the filament. In that perturbed region we observe cells of upward and downward radial velocities V_r (Figure 1). The horizontal gradient of V_r is strong. The velocity cells have 10 to 50 arc second each $(0.7 - 3.6 \times 10^{+4}$ km). There are associated to brightness without any correspondance with photospheric magnetic field (the magnetic field is lower than 20 gauss). The amplitude maximum values of the velocity are $\pm$ 7 km s^{-1}. The presence of high velocities : red and blue shifts simultaneously, and the presence of brightness between the cells suggest, by their short existence, that this perturbation is similar to the "Arch filament systems" or to "the preflare phenomena".

B. Slow Perturbation
The Heliograph patrol and the MSDP pictures taken two days later show no more high velocity cells but only faint upward velocities in the whole absorbing feature. There is no associated brightness. The lifetime (a few hours) and the velocity amplitude of the slow perturbation (3 km s^{-1}) seem in contradiction with the apparent stability of the filament. That suggests that some matter travels through the filament.

M. Dryer and E. Tandberg-Hanssen (eds.), Solar and Interplanetary Dynamics, 213-215.

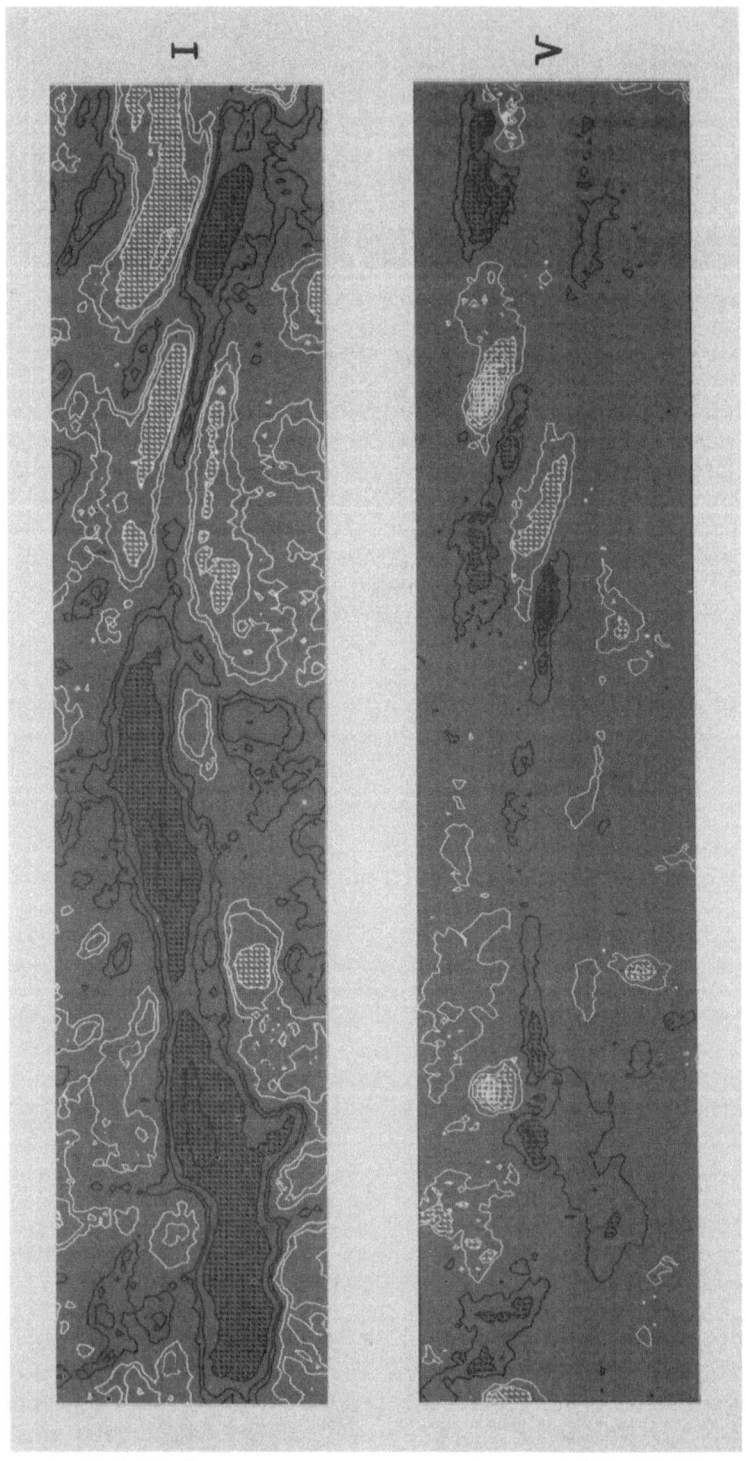

Figure 1. October 11, 1977 : maps of Hα intensity fluctuations I (black lines – absorbing regions, white lines – bright regions) and Doppler shifts V at $\Delta\lambda = \pm 0.27$ Å (black lines – upward velocities, white lines – downward velocities). The size of the maps corresponds to 1 by 5 arc minutes.

Conclusion
Quantitative results are given in Table I.

Table 1

Perturbation	Fast	Slow
Lifetime	10 mn	several hours
Size of velocity features	10" x 50"	whole filament
Radial velocity	± 7 km s^{-1}	+ 3 km s^{-1}

PARTICLE ACCELERATION IN THE PROCESS OF ERUPTIVE OPENING AND RECONNECTION OF MAGNETIC FIELDS

Z. Švestka
SRL Utrecht, Holland, and UCSD, Calif., U.S.A.

S. F. Martin
San Fernando Observatory, CSUN, Calif., U.S.A.

R. A. Kopp
Los Alamos Scientific Lab., New Mexico, U.S.A.

In a series of papers on the flare of 29 July 1973 (Nolte et al., 1979; Martin, 1979; Švestka et al., 1979) it has been shown that Hα "post-flare" loops are the cooled aftermath of previously hot coronal loops which were visible in x-rays in the same position earlier in the flare. Kopp and Pneuman (1976) have proposed that these post-flare loops are formed by a process of successive magnetic field reconnections of previously distended magnetic field lines as illustrated in Figure 1. Each successive reconnection of the magnetic field yields a closed magnetic loop that forms above and concentric with previously formed loops. A shock wave created during each sudden reconnection travels down both legs of each loop and provides energy for ionizing chromospheric mass at the footpoints of the loop. Subsequent condensation of the ionized mass at the tops of the loops renders them visible as this mass falls to the chromosphere.

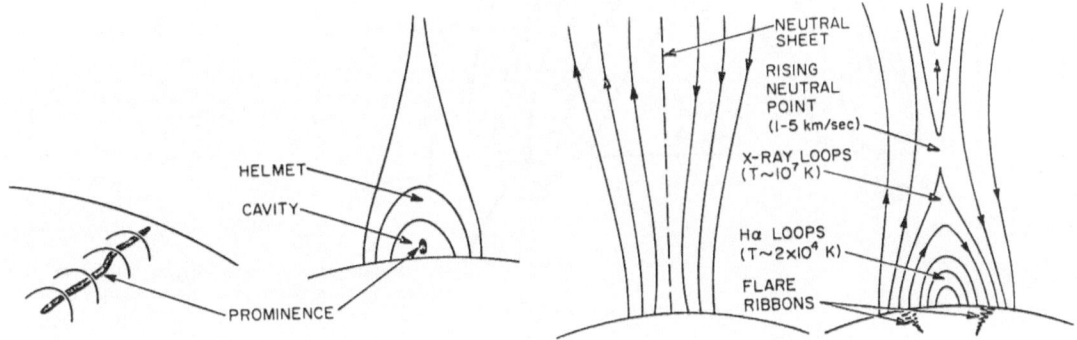

Figure 1. Kopp and Pneuman model for post-flare loops.

M. Dryer and E. Tandberg-Hanssen (eds.), Solar and Interplanetary Dynamics, 217-221.

      The 29 July 1973 flare observations show that the Kopp and
Pneuman model is entirely consistent with the observations of both
the cool Hα flare loops and the hotter x-ray loops since both the cool
and hot loop events are actually the successive formation of individual
loops at increasingly greater heights. Therefore, we suggest, as
already discussed in Švestka (1979), Martin (1979), Pneuman (1979) and
Švestka et al. (1979), that the Kopp and Pneuman model is applicable
to the early phases of flare loop development as well as to the later
phase .

      The Kopp and Pneuman model has several key features in common
with some other reconnection models which make it an attractive model
for two-ribbon flares in general. First, it yields loops. As illustrated in Figure 2, Skylab observations have shown that loops are the
fundamental form of the coronal part of EUV and x-ray flares.
Secondly, when reconnection occurs, high energy particles and/or
thermal waves would be accelerated along the lower half of the reconnected magnetic field. The impact of such particles and/or heat waves
with the chromosphere can produce the typical two-ribbon flare.

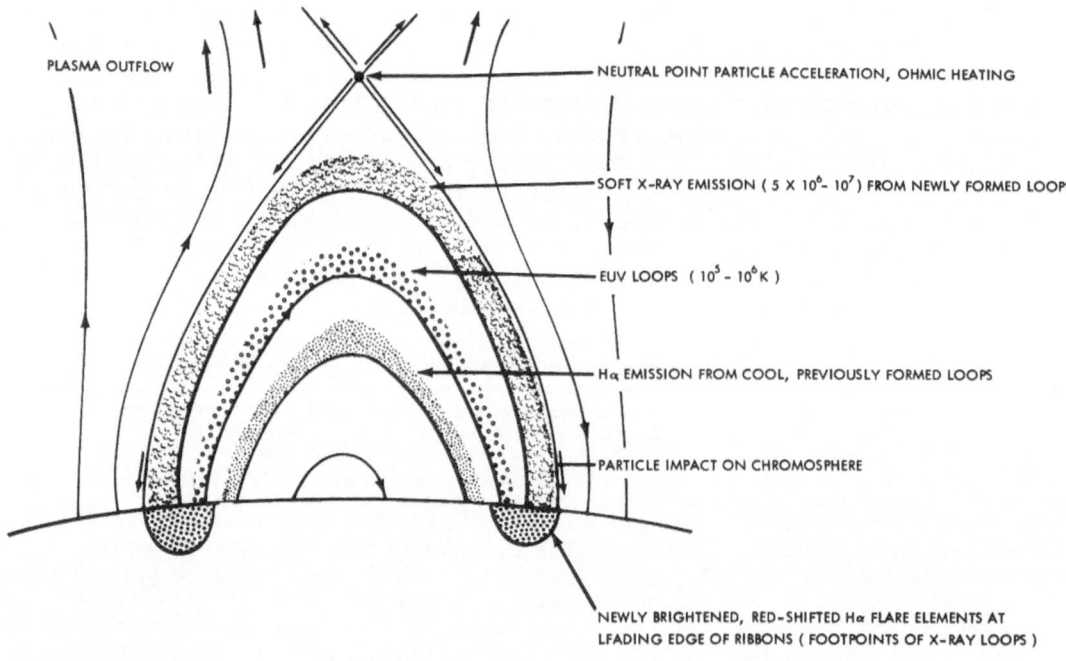

Figure 2.   Interrelationship of various coronal flare loops and
            the chromospheric flare.

In support of this aspect of the model, we show in Figure 3a

a two-ribbon flare photographed on 11 April 1973 from the slit-jaw
images of the Hα multi-slit spectrograph at the Rye Canyon site of
the Lockheed Solar Observatory. Figure 3b shows the corresponding
Hα spectral lines below each slit in Figure 3a. Along only the outer
border of such two-ribbon flares, the multi-slit spectra show very
short-lived red-shifts in the flare profile whenever a slit is located
on a newly formed flare element (3rd slit from the left).

Figure 3a,3b.   Slit-jaw image (upper half) allows the identification
                of a tiny, red-shifted flare element seen at the
                outer border of the flare in the 3rd slit from the
                left in the array of Hα spectra (lower half).

These very tiny red-shifted flare elements are consistent with the
idea that each successive magnetic reconnection provides sufficient
particle or thermal-wave energy to temporarily depress the chromosphere
at points along the outer, newly developing border of two-ribbon flares.
Furthermore, the Hα multi-slit spectra also show that this red-shift
at points along the outer border is a continuous characteristic
throughout most of the lifetime of two-ribbon flares.

Two-ribbon flares have been shown to be powerful sources of
particle acceleration (Švestka, 1976): Essentially all cosmic-ray and
proton flares are of this type; out of 50 flares which were clearly
identified as sources of protons in space from 1956 through 1969
(Švestka and Simon, 1976), 31 were definitely, and 14 most probably,
two-ribbon flares. Thus one can suppose that the two-ribbon flare
process is characteristic for particle acceleration on the Sun.
Because several observations indicate that the acceleration is
accomplished in loops (white-light flare patches, impulsive kernels,
also see Hudson, 1979), it is logical to associate particle acceleration
with magnetic reconnection when new loops are formed (Figure 4). Some
particles are trapped in the loops, whereas others escape upwards and
can be accelerated to still higher energies as passing through the shock
wave that precedes the coronal transient.

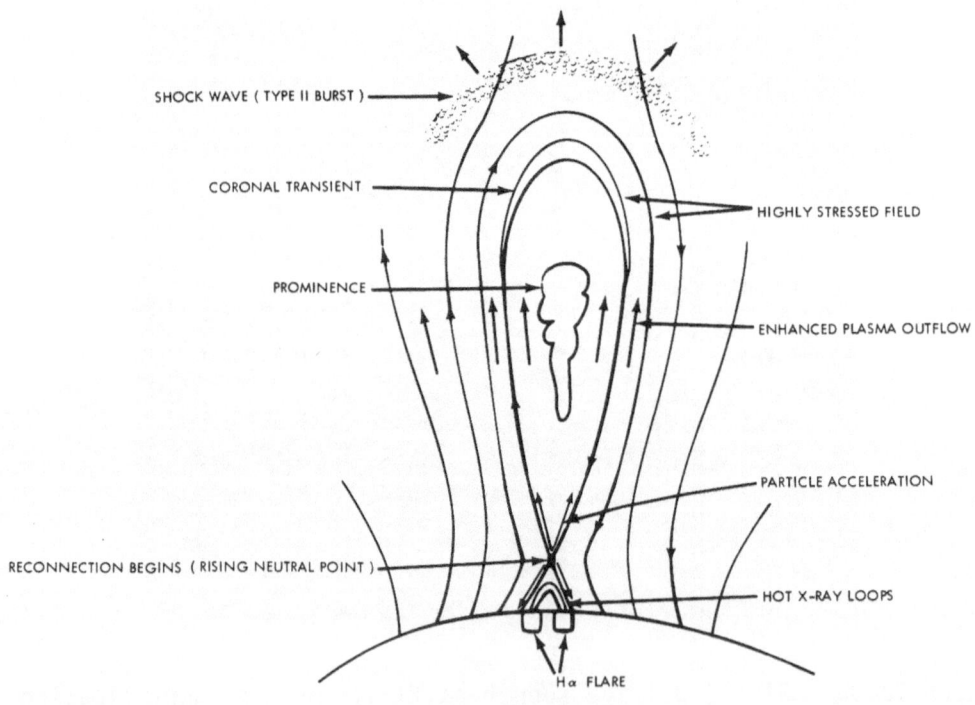

Figure 4. Flare onset.

An important feature of this model is that particle acceleration would be a long-enduring process with the most rapid acceleration occurring during the flash phase of the flares but with continued acceleration and injection occurring throughout the whole flare life. After the flash phase, the decreasing intensity and rate of growth of flare loops and chromospheric flare ribbons suggests that the reconnection rate gradually decreases, but theprocess continues for a very long period of time: The continuous formation of loops at the limb are often seen long after the Hα flare has faded, and in the flare of 29 July 1973, observed on Skylab, new loops were formed still 12 hours after the flare onset (Švestka et al., 1979).

A closely related result is that all disparition brusques, also without any obvious chromospheric flare, have been found by Rust and Webb (1977) to be followed by flare-like systems of X-ray loops. This suggests the possibility that non-optical X-ray loop flares may follow all disparition brusques. Hence, we are led to think that disparition brusques might be associated with the frequent occurrence of long-lived low-energy particle events in space.

ACKNOWLEDGMENTS

The contribution of S.F.M. was supported by AFOSR Contract F49620-78-C0025 and that of Z.Š. by NASA Contract NAS8-32984.

REFERENCES

Hudson, H.: 1979, Paper presented at the Workshop on Particle Acceleration, La Jolla, California.
Kopp, R. A. and Pneuman, G. W.: 1976, *Solar Phys.* 50, 85.
Martin, S. F.: 1979, *Solar Phys.*, *64*, 165.
Nolte, J. T., Gerrassimenko, M., Krieger, A. S., Petrasso, R., Švestka, Z.: 1979, *Solar Phys.* 62, 123.
Pneuman, G. W.: 1979, in E. Priest (ed.), *Solar Flare Magnetohydro- dynamics,* Gordon and Breach Publ. Co., London, in press.
Rust, D. M. and Webb, D. F.: 1977, *Solar Phys.* 54, 403.
Švestka, Z.: 1976, *Solar Flares,* D. Reidel Publ. Co., Dordrecht, Holland.
Švestka, Z.: 1979, Proceedings of a discussion at Royal Society of London, *On the Sun and Heliosphere,* April 1979, in press.
Švestka, Z. and Simon, P. (eds.): 1976, *Catalog of Solar Particle Events, 1956-1969,* D. Reidel Publ. Co., Dordrecht, Holland.
Švestka, Z., Dodson-Prince, H. W., Martin, S. F., Mohler, O. C., Moore, R. L., Nolte, J. T., Petrasso, R.: 1979, *Solar Phys.*, in prep.

ON THE THERMALISATION OF FLARE-TIME ENERGETIC ELECTRONS OBSERVED AT RADIO AND X-RAY WAVELENGTHS

S. S. DEGAONKAR, H. S. SAWANT and R. V. BHONSLE
Physical Research Laboratory
Ahmedabad-380 009, India

ABSTRACT

An interesting microwave event at 2800 MHz was recorded at
Ahmedabad on September 19, 1977 at 1026 UT at the same time as the
H-Alpha solar flare of importance 3B. The microwave burst was of
impulsive nature, with as many as twenty impulses in seventy minutes
with a quasi-periodicity of 1 to 5 minutes. An X-ray burst recorded by
GOES Satellite in 1-8A band showed at the same time a smooth soft X-ray
profile with apparently no sign of hard X-ray bursts. This indicates
that the acceleration of discrete electron streams which produced
impulsive microwave bursts was not sufficient to produce the hard
X-ray component but got thermalised to produce soft X-ray emission,
with a gradual rise and a slow decay covering a long duration of more
than $2\frac{1}{2}$ hours.

INTRODUCTION

A Dicke type microwave radiometer operating at 2800 MHz is in
operation at Ahmedabad to record the daily solar radio flux with an
accuracy of ± 3 per cent. Region 889 (McMath 14943, N08, L = 197, Class/
area E/730 on 13 September 1977) on the solar disk became active on
16 September 1977. A major flare of optical importance 3B took place
at 1026 UT on 19 September 1977 from Region 889 which was recorded at
Ahmedabad as a complex microwave event in all its details.

Figure 1 shows the complex microwave event which started exactly
at 1026 UT as that of the optical flare reported in the Solar Geophysical Data Report, Boulder, Colorado, USA. We can distinguish in
Figure 1 as many as 20 microwave impulsive bursts, within a span of
70 minutes, beginning 1026 UT and each having a duration of about 1 to
5 minutes. The largest burst occurred at 1105 UT with a peak flux of
~1252 sfu which is comparable with 1100 sfu at 2700 MHz observed at
Sagamore Hill in USA (1 sfu = 10^{-22} Wm^{-2} Hz^{-1}). The X-ray flux in 1-8A
band observed by GOES-2 satellite started rising at 0950 UT and again

223

M. Dryer and E. Tandberg-Hanssen (eds.), Solar and Interplanetary Dynamics, 223-226.
Copyright © 1980 by the IAU.

at 1016 UT, reaching maximum at 1052 UT with a peak flux of 2×10^{-4} ergs/cm^2/sec as shown in Figure 1.

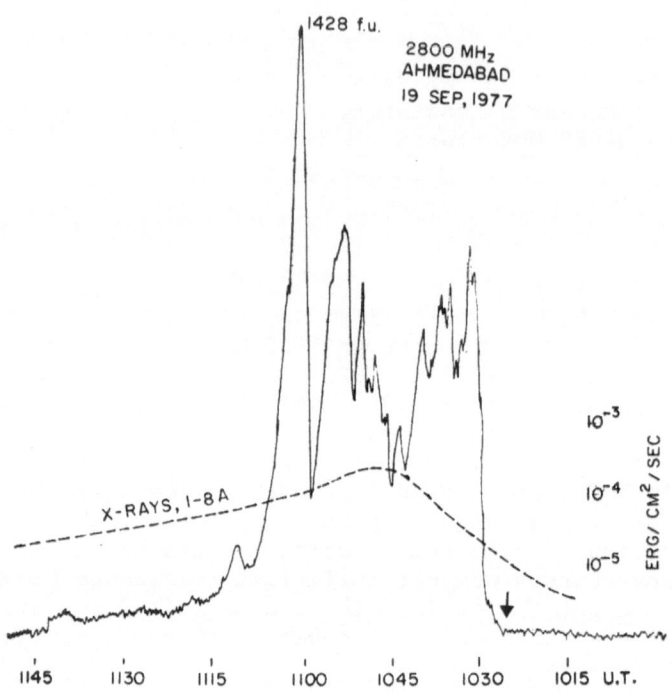

Figure 1. Microwave bursts at 2800 MHz showing quasi-periodicity. Associated X-ray flare is shown by broken line.

INTERPRETATION

The interesting feature regarding the flare on 19 September 1977 is that it displayed strong impulsive phenomenon in the microwave radiation. It is known that hard X-ray bursts are almost associated with microwave impulsive bursts and are of very short duration in flares (~10–50 sec.). The hard X-rays occur mostly after the onset of the soft X-ray emission (Kane, 1969). It is found that the soft X-ray emission starts earlier than the hard X-ray and microwave bursts (Švestka, 1975). It appears that since only soft X-rays were recorded during the event at 1026 UT on 19 September 1977, the non-thermal electrons got quickly thermalised due to collisions with the ambient plasma as they plunged along magnetic flux tube to lower levels.

Another fact which strengthens our conclusion regarding the X-ray emission on 19 September as being due to a single thermal enhancement is that the soft X-ray flux decayed very slowly (more than 2½ hours) as compared to the microwave event (~70 min). The impulsive nature of microwave bursts is indicative of acceleration of discrete electron streams whereas the smooth profile of soft X-rays indicates thermalization of electrons. The emission measure required for soft X-ray burst is ~10^{50} cm^{-3} and the electron temperature ~10^7 K (DeFeiter, 1975; Švestka, 1975). For impulsive microwave bursts and hard X-rays, these values will be comparatively higher depending on the intensity of the flaring region.

CONCLUSION

It is difficult to say whether the observed periodicities are truly periodic phenomena representing acceleration of electron streams successively or are due to random flaring of several bright points within the disturbed region. This microwave event which showed strong impulsive behaviour had no counterpart in hard X-rays. Such a situation could be understood if the non-thermal electrons responsible for microwave emission got thermalized by collisions with the ambient plasma as is borne out by the absence of hard X-rays.

ACKNOWLEDGMENT

We thank Dr. S. R. Kane of University of California, Berkeley, California, U.S.A. for helpful discussions. Our thanks are due to Professor K. R. Ramanathan for his keen interest and Professor D. Lal, Director of PRL, for encouragement. Operational facilities for Solar Radio Astronomy provided by the Space Applications Centre, ISRO, Ahmedabad are gratefully acknowledged. This work is supported by the Department of Space, Government of India.

REFERENCES

DeFeiter, L. D.: 1975, Solar Gamma, X- and EUV Radiation, S. R. Kane (Ed.), D. Reidel Publ. Co., Dordrecht-Holland, pp. 283-294.

Kane, S. R.: 1975, Solar Gamma, X- and EUV Radiation, S. R. Kane (Ed.), Dordrecht-Holland, pp. 385-409.

Švestka, Z.: 1975, Solar Gamma, X- and EUV Radiation, S. R. Kane (Ed.), D. Reidel Publ. Co., Holland, pp. 427-439.

DISCUSSION

Haug: You assume that the nonthermal electrons responsible for the impulsive microwave emission have been thermalized quickly by collisions in the ambient plasma. Why is there no hard X-ray production in these collisions?

Degaonkar: Maybe part of the non-thermal electron flux escaped from the apex of the flux tube and part got thermalized. Since the flux of soft X-rays was not very large, and hard X-rays were not produced, I feel that the thermalization might have taken place by electron collisions with ambient electrons and not with dense matter, which is necessary for production of hard X-rays.

Tandon: Would you comment on a wide time lag between the soft X-ray peak and the maximum peak of microwaves at 2800 MHz?

Degaonkar: The time lag could be due to the different production locations of microwaves and X-rays or there might be fresh generation of microwaves without further enhancements in X-rays. We are looking into this problem further.

Hoyng: Have you actually observed that hard X-rays were absent?

Degaonkar: I have not seen anybody reporting to have observed hard X-rays on that day. I don't know if there was any difficulty in measuring it, which is unlikely.

Kahler: The 1-8A X-ray flux is only a little more than 10^{-4} erg cm^{-2} sec^{-1} at the event peak. This seems to be an unusually small event considering the size of the microwave burst.

Degaonkar: I agree. That is why this event appears to be uncommon. The peak X-ray flux is 2×10^{-4} erg cm^{-2} s^{-1}; but the impulsive microwave burst is quite strong.

RECENT OBSERVATIONS OF ENERGETIC ELECTRONS IN SOLAR FLARES

S.R. Kane
Space Sciences Laboratory, University of California,
Berkeley, California 94720

SUMMARY

It has been apparent for the last few years that a large fraction of the total energy released during a solar flare appears initially in the form of energetic electrons accelerated during the impulsive phase. An estimate of the energy of these electrons is based on the observed hard x-ray spectra as well as the assumed form (thermal or non-thermal) of the electron distribution. Even after the basic form of the electron distribution is assumed, additional assumptions, such as the low energy cut-off in the case of the power law energy spectrum or existence of a multi-thermal source in the case of the thermal spectrum, are usually required. In order to test these assumptions, measurements of the hard x-ray spectrum with spatial resolution and covering a wide range of x-ray energy are essential. In absence of good spatial resolution, as is the case with most of the presently available hard x-ray observations, the impulsive x-ray emission at energies $h\nu \lesssim 10$ keV is often unobservable because of the presence of a large background of relatively intense gradual emission associated with most flares. Observations made in the past suffered either because of the lack of a clearly identifiable impulsive x-ray emission at low energies (Peterson et al, 1973) or an adequate spectral resolution (Kahler, 1973). Thus so far it has not been possible to measure unambiguously the spectrum of impulsive x-rays $\lesssim 10$ keV and hence to deduce a possible low energy cut-off in the energetic electron spectrum. Here we report briefly such an observation made with the ISEE-3 x-ray spectrometer experiment and its implications with regard to the characteristics of energetic electrons in solar flares.

The x-ray spectrometer experiment aboard the International Sun-Earth Explorer-3 (ISEE-3) spacecraft has been described in detail elsewhere (Anderson et al, 1978; Kane et al, 1979). It consists of two detectors: a xenon-filled proportional counter covering the energy range 4.8-14 keV and a NaI (Tl) scintillator covering the energy range 12-1264 keV.

M. Dryer and E. Tandberg-Hanssen (eds.), Solar and Interplanetary Dynamics, 227-230.
Copyright © 1980 by the IAU.

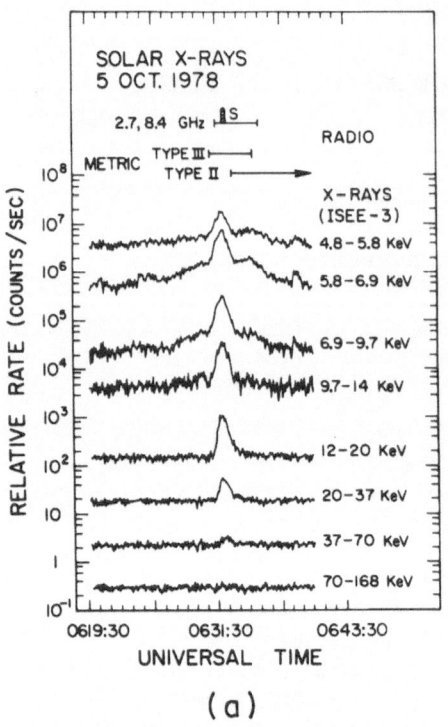 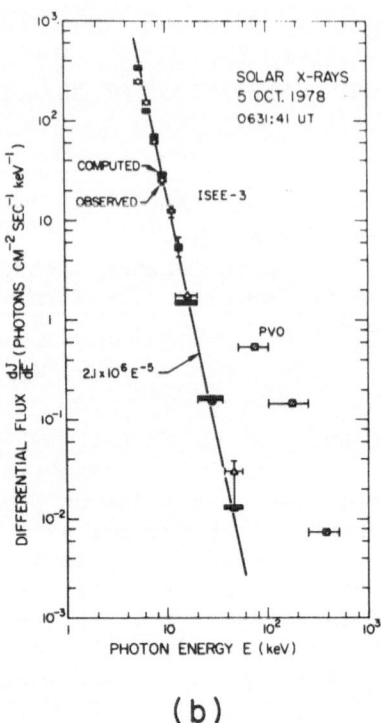

(a)   **(b)**

Fig. 1. Observations of an impulsive x-ray burst on 5 October 1978 attributed to a relatively large solar flare located $\sim 15°$ behind the east limb of the Sun. (a) Time intensity profile: the impulsive emission can be clearly i-
dentified down to the lowest x-ray energy (~ 5 keV) observable with the ISEE-3 spectrometer. (b) Spectral plot at
the time of maximum: note that the x-ray flux observed by
ISEE-3 is much smaller than that observed by the PVO detectors. Also note that the impulsive x-ray spectrum observed
by ISEE-3 is consistent with a power law down to ~ 5 keV
energy (Kane et al, 1979).

Fig. 1 shows an impulsive solar x-ray burst observed by the ISEE-
3 experiment on 5 Oct. 1978. This x-ray burst was also observed by a
detector aboard the Pioneer Venus Orbiter (PVO) and it has been estimated that the associated solar flare was located $\sim 15°$ behind the
east limb of the Sun (Kane et al, 1979). Thus only the part of the
x-ray source located at a height $\gtrsim 25,000$ km above the photosphere was
visible to the ISEE-3 detector, the lower part of the source being occulted by the photosphere from the ISEE-3 field of view. From Fig.
1(a) it can be seen that the impulsive emission from the coronal source
can be identified down to x-ray energies ~ 5 keV. This has been possible because most of the gradual emission, presumably emitted at much
lower altitudes, was occulted, making the impulsive emission dominant

even at x-ray energies ∿ 5 keV. The x-ray spectrum, shown in Fig. 1(b), is consistent with a power law electron spectrum with no apparent low energy cut-off up to energies ∿ 5 keV. Although an explanation of the observed x-ray spectrum in terms of the emission from a multi-thermal electron spectrum cannot be ruled out, we believe that the present observation lends new support to the existence of non-thermal electron spectra during the impulsive phase of solar flares.

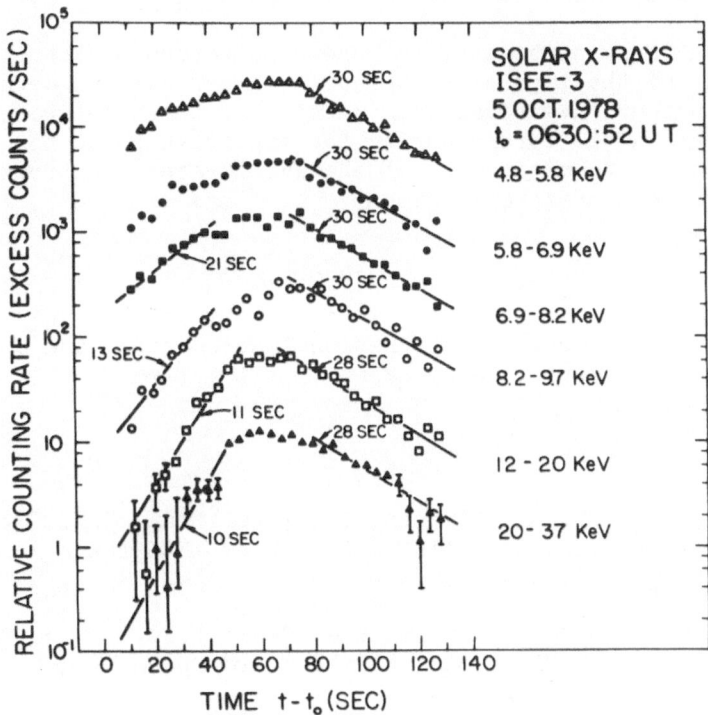

Fig. 2. Rise and decay characteristics of the impulsive x-ray burst shown in Fig. 1(a). Note that the decay time is essentially independent of energy for 5-35 keV x-rays.

Details of the rise and decay of the x-ray burst are shown in Fig. 2. Both the rise and decay times are larger than those in the case of on-the-disk flares. Further, whereas the rise time does decrease with increase in x-ray energy, the decay time is relatively constant for x-rays in 5-35 keV range. This suggests that the coronal part of the impulsive x-ray source probably consists of a relatively large region in which energetic electrons are injected more or less continuously during the impulsive phase. Because of the relatively low ambient density at coronal altitudes, the collisional losses are expected to be negligible for these electrons. If electrons are injected uniformly into the impulsive x-ray source, which extends from the upper chronosphere/transition region to the corona, the low-altitude part will be an intense thick-target x-ray source and the coronal part will be a relatively weak thin-target x-ray source. If the injected electron spectrum is a power law in energy, comparison of the ISEE-3 measurements of the coronal source with the PVO measurements of the total source shows the following: (1) $n_i \tau = 2 \times 10^8$ sec cm^{-3} where n_i is the average ion density inside the coronal source and τ is the lifetime of energetic electrons in that source; (2) the lifetime τ is not determined by coulomb collisions but by escape of the electrons from the coronal source into outer corona (Kane <u>et al</u>, 1979).

Thus there is evidence that the energy spectrum of the electrons accelerated during the impulsive phase of a flare extends down to ~ 5 keV energy. Further, a substantial fraction of the accelerated electrons is present in the corona during the impulsive phase thus indicating only a partial precipitation of the accelerated electrons in the upper chronosphere/transition region.

ACKNOWLEDGEMENTS

This research was supported by the National Aeronautics and Space Administration under Contract NAS 5-22307.

REFERENCES

Anderson, K.A., Kane, S.R., Primbsch, J.H., Weitzman, R.H., Evans, W.D., Klebasadel, R.W., and Aiello, W.P., 1978, IEEE Trans. Geosc. Electronics, <u>GE-16</u>, 157.

Kahler, S.W., 1973, in R. Ramaty and R.G. Stone (eds.), <u>High Energy Phenomena on the Sun</u>, NASA SP-324, NASA Goddard Space Flight Center, Greenbelt, Maryland 20771, p. 124.

Kane, S.R., K.A. Anderson, W.D. Evans, R.W. Klebasadel, and J. Laros, 1979, Astrophys. J. Letters, (in press).

Peterson, L.E., Datlowe, D.W., and McKenzie, D.L., 1973, in R. Ramaty and R.G. Stone (eds.), <u>High Energy Phenomena on the Sun</u>, NASA SP-342, p. 132.

AN ENERGY STORAGE PROCESS AND ENERGY BUDGET OF SOLAR FLARES

K. Tanaka, Z. Smith, and M. Dryer
Tokyo Astronomical Observatory, University of Tokyo and
Space Environment Laboratory, National Oceanic and Atmospheric
Administration, Boulder, respectively

The flare energy is generally considered to be stored in stressed (twisted or sheared) magnetic fields. Origin of the stress may be either intrinsic or due to horizontal shear motion (Tanaka and Nakagawa 1973) or due to propagation of twist from below (Piddington 1974). Characteristic magnetic configurations in the great activities (inverted, twisted δ-configuration; Zirin and Tanaka 1973) suggest an inherent shape of fluxtube for these regions: a twisted magnetic knot. Further, evolutionary characteristics such as rapid growths of spots and growth of twist in parallel with apparent shear motion of spot, together with the fact that the shear motion is associated with upward velocity (Tanaka and LaBonte 1979), suggest a continuous emergence of such a twisted knot from below throughout the activity (Tanaka 1979). In this model (Fig. 1) the flare energy may be supplied directly into the corona as the twisted portion of the fluxtube emerges out. The amount of energy supplied between t_0 and t may be equated to the energy contained in the twist(ϕ) between z_1 and z_2,

$$M(t) = 1/4\pi \int_0^r r\phi B_\phi B_z 2\pi r dr. \tag{1}$$

Observationally ϕ may be evaluated from the growth of the penumbral twist, which is related empirically to the apparent horizontal shear velocity v by $r\phi=1.5v(t-t_0)$ (Tanaka 1979). Assuming the force-free field we have $<B_\phi B_z>\approx 0.22B_p^2$ with B_p equal to the peak field strength of the moving spot. Then,

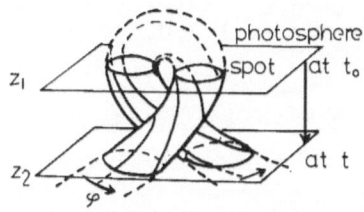

Figure 1. An emerging twisted magnetic knot.

M. Dryer and E. Tandberg-Hanssen (eds.), Solar and Interplanetary Dynamics, 231-234.
Copyright © 1980 by the IAU.

$$M(t) \simeq \int_{t_0}^{t} 0.0173 \ B_p^2 Av \ dt, \tag{2}$$

where A is area of the moving spot. Eq.(2) is equal to the evaluation of accumulated energy due to the horizontal shear motion except numerical factor (cf. Tanaka and Nakagawa 1973).

We evaluated the energy supply for a very flare-rich and fast-evolved active region McMath 13043 (1974 July) which showed three successive sunspot motions in good spatial and temporal correlations with the activities, and compared it with released flare energies (thermal and kinetic). Time-integrated thermal energy was evaluated from the total radiated energy:

$$E_T(t) \simeq \int_{t_0}^{t} F_{obs}(1-8A) \cdot C(1-8A,T) \cdot F(total,T)/F(1-8A,T) \ dt \tag{3}$$

$$\simeq \int_{t_0}^{t} 12.4 \ F_{obs}(1-8A) \ dt, \tag{4}$$

where $F_{obs}(1-8A)$ is the observed flux in the 1-8A band (GOES), C is a correction factor to obtain true flux (Dere et al. 1974), the last term is a theoretical ratio of the total to the 1-8A fluxes (Raymond et al. 1976). We adopted eq.(4), an empirical result from eq.(3) for a well-studied 2b flare of Sep.7 1973 (Withbroe 1978). The kinetic energy was evaluated from the interplanetary shock wave data (IMP 7 and 8). From the potential and kinetic energy fluxes shown in Fig.2 the integrated kinetic energy was obtained assuming a constant area S equal to a solid angle $\pi/2$ at 1 A.U. and correcting for transit times of the shock waves:

$$E_K(t) = S \int_{t_0}^{t} \rho v^3/2 \ dt \ + \ S \int_{t_0}^{t} dt \int_{R_\bullet}^{1AU} \rho v GM/r^2 dr. \tag{5}$$

For three periods corresponding to the three motions, M(t), $E_T(t)$, and $E_K(t)$ are shown in Fig.3. M(t) proves to show quite consistent time profiles with $E_T(t)$ and $E_K(t)$. In particular remarkable is the similarity between M(t) and $E_T(t)$ shown in Fig.4. $E_T(t)$ scales half of M(t) precisely in the whole period. For 8 major flares net increase of M(t) between the flares are compared with total thermal (E_T) and kinetic (E_K) released energies in Table 1.

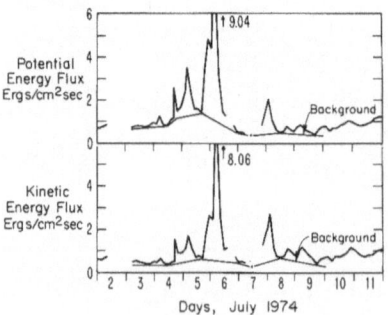

Figure 2. Potential(upper) and kinetic(lower) energy fluxes of shock waves at 1 A.U.

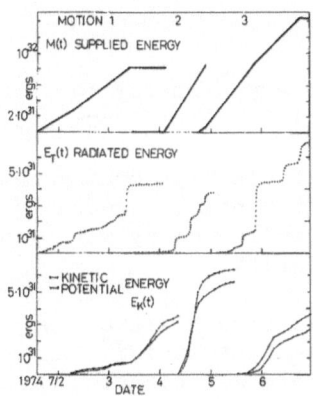

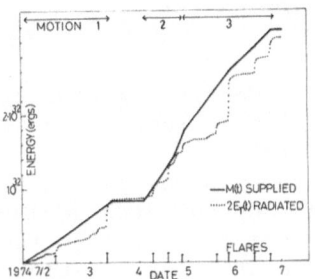

Figure 4. M(t):thick line and
E_T(t):dotted line in the whole
period: July 2-July 7 1974.

Figure 3. M(t),E_T(t) and E_K(t),
respectively.

Table 1. Energy Budget of 8 Major Flares

flare	July 2(1n)	3(2b)	4(1b)	4(1b)	4(1n)	5(2b)	6(1b)	6(1b)
$M(10^{30}$ergs)	26	56	22	28	34	98	32	22
E_T	13	30	9	11	15	43	11	14
E_K	9	43		115		32	7	14
E_T+E_K	22	73		150		75	18	28

One finds E_T/M = 0.46±0.09, E_K/M = 0.61±0.42, (E_T+E_K)/M = 1.09±0.45.
Note the small scatter in E_T/M. Similar ratio (E_T/M = 0.57±0.19) is
obtained for the 1972 August flares (Tanaka 1979). Large scatter in E_K/M
may be due to the assumption of constant area for different shock waves.
The solid angle of $\pi/2$ may be suggested from this result. We note that
agreement between energy input and output was obtained using qualitativel
different sources of data. Present result may support the direct energy
supply due to the emergence of the twisted fluxtube which is produced
in the convection zone by some flow patterns.

REFERENCES

Dere,K., Horan,D. and Krepline,R.: 1974, J.Atmosph.Terr.Phys. 36,989.
Piddington,J.: 1974, Solar Phys. 38, 465.
Tanaka,K.: 1979, Publ.Astron.Soc.Japan, in press.
Tanaka,K. and LaBonte,B.: 1979, in preparation.
Tanaka,K. and Nakagawa,Y.: 1973, Solar Phys. 33, 187.
Withbroe,G.: 1978, Astrophys.J.225, 641.
Zirin,H. and Tanaka,K.: 1973, Solar Phys. 32, 173.

DISCUSSION

Stix: Concerning your last figure, is there any symmetry (or anti-symmetry) with respect to the equator of the tilt of newly formed sun-spot pairs?

Tanaka: In the past cycles most of the great flare-producing regions appeared in the north hemisphere, so no statistics exist concerning your question. But in a few examples there is antisymmetry with respect to the relative orientation of the P-and f- polarity.

Pneuman: The picture you showed was of a two-ribbon flare. To explain such a flare <u>totally</u> in terms of emerging untwisting flux would require enormous changes in the photospheric field distribution. I don't believe such large scale changes are observed in the magnetograms.

Tanaka: The changes of magnetic fields which would occur when the once-emerged twisted fluxtube relaxes are mainly changes of magnetic field orientations, and so to detect them we need high time and spatial resolution observation by the vector magnetograph. (Large scale changes of magnetic field orientations have been reported in the magnetograph observation of a large flare near the limb, by Tanaka, <u>Solar Phys.</u>, <u>58</u>.)

FLARE ASSOCIATED ERUPTIVE PROMINENCE ACTIVITY OF FEBRUARY 1, 1979

A. BHATNAGAR, R. M. JAIN, D. B. JADHAV and R. N. SHELKE
Vedhshala Udaipur Solar Observatory
Udaipur

R. V. BHONSLE
Physical Research Laboratory
Ahmedabad

ABSTRACT

Observations and analysis of solar flare activated ascending "Fountain type" prominence of 1 February 1979 are presented. This "Fountain" prominence rose to 180,000 km above the solar surface and gave rise to a number of ascending loops and helical structure. These "helicals" are clear manifestation of magnetic field configuration. From these observations it is shown that, as the "Fountain" prominence rises, it carries along with it the complex magnetic field which unfolds as the prominence material expands into distinct magnetic field lines. Several type III radio bursts were also seen associated with this event. No type II or IV radio emission was reported.

1. INTRODUCTION

The observations of solar limb activities, such as active loop, eruptive prominences, and surges provide a good aid for studying the coronal transients and dynamics of plasma motion and associated radio emission. Recently, considerable interest has been shown for detailed study of these phenomena. Tandberg-Hanssen and Hansen (1973) have carried out an observing programme at Mauna Lao Observatory of HAO, to specially study in detail the ascending prominences through a wideband Hα filter (10A half width) so that large Doppler shifted (~250 km/s) prominence features do not escape detection.

2. CINEMATOGRAPHIC Hα OBSERVATIONS

Time-lapse Hα solar observations are made through a 15 cm aperture telescope and a Halle 0.5A passband filter, from the island Solar Observatory at Udaipur. The normal rate for Hα observations is generally kept to 1 to 2 per minute, but for fast moving energetic events the rate

M. Dryer and E. Tandberg-Hanssen (eds.), Solar and Interplanetary Dynamics, 235-240.
Copyright © 1980 by the IAU.

of observations is increased to 10-12 frames per minute.   16 mm time-
lapse movies are made from the original 35 mm pictures for studying the
dynamics of solar phenomena.

2.1 1 February 1979 Ascending 'Fountain' Prominence

In McMath plage region 15808, which was behind the southeastern
limb on 1 February 1979, an interesting ascending prominence or
following Tandberg-Hanssen's et al., nomenclature - a "Fountain"
prominence appeared between 0856-1036 UT.   Initially at 0855 UT a bright
spray appeared moving outwards at a small angle to the limb with a
velocity of 88 km/s, and expanded into a loop structure A, shown in
Figure 1. This closed loop A expanded at a rate of about 20 km/s and
finally disintegrated into small bits-and-pieces within 10 minutes of
its first appearance. During its journey through the corona, it retained
the loop configuration.

Following this event a major flare behind the limb occurred at
0905 UT in the same region (Solar Geophysical Data, No. 415). Although
this active region was almost 14°-15° behind the limb on 1 February,
an intense bright blob of material was seen appearing just above the
limb around 0904 UT; this indicates that the flare might have been a
major flare behind the limb as is further confirmed by enormous
erupting prominence activity.   The Solrad 11 data also indicate a
strong enhancement of X-ray flux around 0905 UT.

Soon after the appearance of a bright 'blob' on the limb, another
closed loop feature B (Figure 1, 0908 UT) started ascending with a
velocity of about 100 km/s.   Within 3 minutes, loop B developed into a
complex 'helical' structure (Figure 1 , 0913 UT). At first two
distinct knots 1 and 2 forming a 'spiral' feature could be identified.
As loop B rose to greater height, the structure opened up and two more
knots on the 'helical' structure could be easily seen.   A line tracing
showing helical structure on the frame at 09.16 is shown in Figure 2.
During the ascending motion of feature B, the 'spiral' configuration
of the loops remained intact.   Finally, as the 'spirals' expanded the
knots became diffuse and were not seen (Figure 1 , 0921 UT).

The feature C (Figure 1, 0913 UT) near the base of loop B
ascended in a curvilinear path and soon (Figure 1, 0915 UT) formed into
a well-defined helical structure moving with a velocity of nearly
75 km/s. Around 0921 UT (Figure 1) a vertically directed spray was
noticed, eminating from the active region.   This spray apparently
lifted the prominence material to greater height (100,000 km) with much
higher velocities on the order of 240 km/s.

Feature D (Figure 1, 0921 UT) appeared first as a bright knot on
the tip of the "Fountain" which expanded into a distinct closed-loop
structure at a rate of about 60 km/s.   It appears that the vertically
moving spray (Figure 1, 0921 UT) pushed the loop D to greater height and
speed of nearly 150 km/s.   Both features A and D expanded into a loop

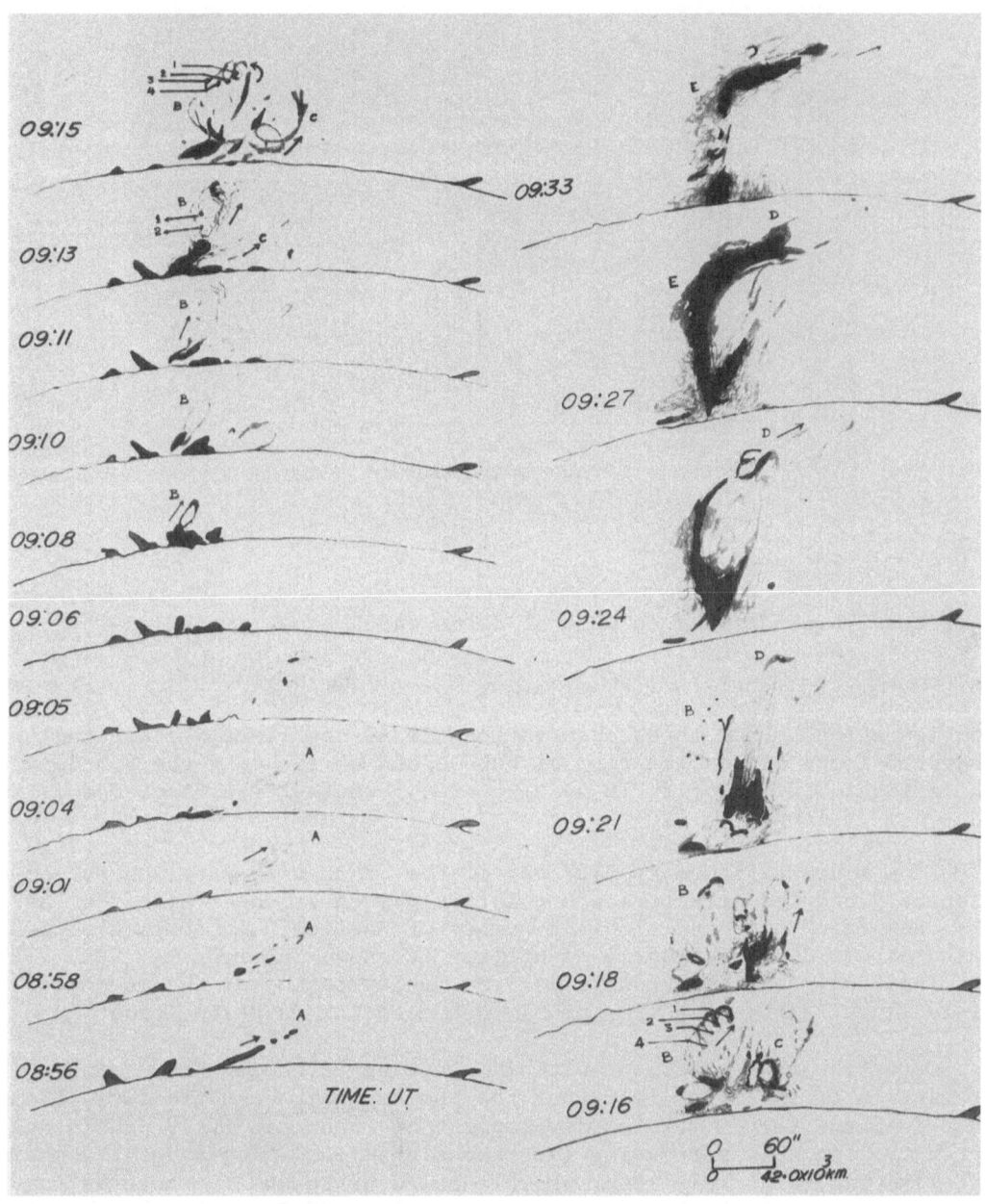

Figure 1. Line drawings of the 'Fountain' prominence of 1 February 1979. Frame times are in UT.

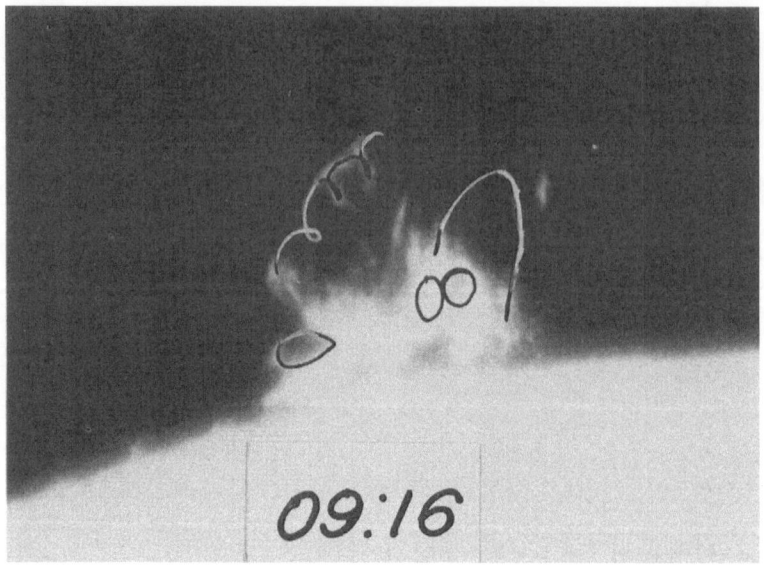

Figure 2.   Line drawing on 0916 UT frame which shows the helical
            structure.

configuration. From these observations it is not clear whether the
observed loops are in the form of bubble, and we see only the boundary
where sight-line velocity is nearly zero or whether the flux rope is in
the form of loop.

     The side E (Figure 1, 0927 UT) of the "Fountain" prominence, shows
extremely complex structure and relatively much stronger emission. As
the "Fountain" ascended slowly, a number of 'helical' or 'coiled' like
features could be distinctly identified as shown in Figure 1 (0933 UT).
The 'helical' configuration is a clear manifestation of the magnetic
field lines, along which the prominence material is delineating.

     From these observations of this erupting ascending "Fountain"
prominence it is amply clear that the photospheric magnetic field is the
most dominant factor for the development of "Fountain type" prominence
activity. In this particular case the sight-line velocity of the moving
material was relatively less, which enabled us to observe interesting
motions even through the narrow band Hα filter (0.5A). In this event
we see clearly a number of cases wherein the prominence material is
trapped and contained by magnetic field; this would perhaps indicate
a near balance of kinetic and magnetic energy density, i.e.,

$$\frac{1}{2}\,\rho\,v^2 = \frac{B^2}{8\pi}\,.$$ (1)

From this equation the observed velocity of 100–200 km/s of various prominence features, and assuming the number density of 10^{10} cm^{-3} (which may not be correct) will be controlled by magnetic fields of 2–3 G.

From these observations, we notice that the plasma rising from the flare active region starts as bright and dense 'blobs' of plasma, permeated with complex magnetic field. Due to the flare-shock wave the dense plasma ascends rapidly carrying with it the magnetic field. As the spray plasma ascends and expands, the observed helical and loops open-up to manifest the magnetic field configuration.

ACKNOWLEDGMENT

This research work has been supported by the financial grant received from the Department of Science and Technology, Government of India, under the Science and Engineering Research Council.

REFERENCES

Tandberg-Hanssen, E., and Hansen, R. T.: 1973, 13th Int. Cosmic Ray Conf., Denver 2, pp. 1178.

Tandberg-Hanssen, E.: 1974, Solar Prominences, Reidel Publ. Co., Dordrecht, Holland.

Tandberg-Hanssen, E., Hansen, R. T., and Riddle, A. C.: 1975, Solar Physics 44, pp. 417.

DISCUSSION

Stewart: Do you see unwinding of the helical structure in the eruptive prominence just described?
Bhatnagar: Yes. As the spray prominence material rises up and expands, helical structure appears to unwind or, so to say, simplifies and opens-up.

Moore: What was the local time of day when these observations were made, and what is the approximate spatial resolution of these pictures?
Bhatnagar: Local time was 14.00 hrs and solar seeing would have been 2" to 4" of arc.

Steinitz: Could the interpretation also be that subsequent magnetic loops are radiating and there is only an illusion of mass motion?
Bhatnagar: Of course, line of sight velocities have not been measured, but results from similar observations definitely show that these are actual mass motions in eruptive prominences.

Tandberg-Hanssen: (Reply to Steinitz) In similar prominences we observe Doppler shifts; hence, Bhatnagar et al.'s interpretation in terms of mass motions seems reasonable also in this case.

THE DISRUPTION OF EUV CORONAL LOOPS FOLLOWING A MASS EJECTION TRANSIENT

E. J. Schmahl
Astronomy Program, University of Maryland
  College Park, Maryland

Abstract. A classic filament disruption/coronal transient event ocurred on 10 January 1974. After the prominence liftoff, "gradual" x-rays were recorded by Solrad 9. A white light coronal ejection, interpreted as a loop seen edge-on, followed. During the mass outflow, Hα loops formed at the original site of the prominence. The loops appeared also in EUV spectroheliograms, and rose rapidly before vanishing abruptly. During the disintegration of the loops the apices showed great enhancements and vertical spike structures. The overall behavior of this loop prominence system is compatible with reconnection models.

1. INTRODUCTION

It has been shown that eruptive prominences are the most common precursor of coronal transients (cf. Munro et al, 1979) and that loop prominence systems (LPS), or their X-ray and microwave analogues (cf. MacCombie and Rust, 1979), are the frequent followers of mass ejections. These near-surface phenomena are tracers of coronal magnetic activity and restructuring in the wake of transients.

The 10 January 1974 event, observed in white light, Hα, X-rays, and EUV illustrates the well-known sequence: prominence eruption, flare, transient, LPS.

2. OBSERVATIONS

The first relevant Hα phenomenon was the liftoff of a small prominence at S09E90 near McMath 12702, starting at $\sim$ 08:50 and fading out at $\sim$ 09:15. At $\sim$ 09:40 a 1N flare occurred $\sim$ 8° S of the prominence. Hα surging was reported and EUV infall was seen later. At $\sim$ 10:10 Hα LPS became visible at the site of liftoff.

Presumably at this time the white light transient was moving outwards, since at 11:38 the HAO coronagraph photographed coronal material moving at a speed of $\sim$ 400 km s^{-1} radially above the prominence site (Munro et al 1979, Hildner, 1977). The narrowness and polarization

M. Dryer and E. Tandberg-Hanssen (eds.), Solar and Interplanetary Dynamics, 241-244.

structure of the transient led Munro (1978) to interpret the shape as a loop seen edge-on. Assuming this, the overall geometry of the coronal magnetic field could have considerable shear, since the LPS may lie close ($\stackrel{<}{\sim}$ 45°) to the plane of the sky.

EUV spectroheliograms were first acquired at 09:53 during the 1 N Hα flare. The EUV flare appeared similar in size ($\sim$ 20") and shape (mound) to Hα. The EUV pointing shifted to the site of the prominence where continuous 5 min spectroheliograms showed loops forming from 10:13 on, their size increasing at an initial rate of $\sim$ 10 km s^{-1} and later at $\sim$ 40 km s^{-1}. (Such acceleration is unusual for LPS).

While the LPS formed, Solrad-9 recorded soft X-rays which peaked at $\sim$ 09:50 and declined smoothly to background at $\sim$ 10:50. (No imaging X-ray telescopes were in operation.) Since the LPS emitted in the EUV coronal MgX line, the X-rays presumably arose in a simmilar volume.

In Hα and EUV "transition zone" lines the loops were co-spatial (to $\stackrel{<}{\sim}$ 5"). The Hα loops vanished between 10:44 and 10:46. In EUV the loops were disrupting at 10:48. Large spikes appeared at their apices, and enhancements at the tops increased remarkably. Subsequent EUV images in the next orbit show that the low corona returned to the pre-loop conditions.

3. ANALYSIS AND DISCUSSION

MacCombie and Rust (1979) have shown the intimate relation between X-ray and Hα LPS. We therefore assume that the Solrad X-rays are emitted from the same or similar volume as the MgX emission, and determine temperatures, emission measures, and densities (see Table I). The temperature falls from $\sim$ 6 x 10^6 to $\sim$ 4 x 10^6 K between 09:50 and 10:50, and the emission measure falls from $\sim$ 2 to 1 x 10^{48} cm^{-3} during the same interval. These values are similar to that of the 13/14 August 1973 event, whose parameters may be compared in Table I.

The abrupt termination of this loop system after an increase of its apparent velocity seems more consistent with reconnection (e.g. Kopp and Pneuman 1976) than with loop expansion. Reconnection and filling of loops progresses from low heights upward, until the height is reached where the pre-transient magnetic field was open, and then reconnection stops. Free-fall times of order $\stackrel{<}{\sim}$ 10^2 km s^{-1} are sufficient to deplete the highest loops in the observed time of $\sim$ 2 min (cf. Foukal 1978). The source of mass for the loops is, however, not understood.

The large vertical spike-like enhancements at the apex of the LPS (Fig. 1) during the disruption suggest a relationship to vertical neutral sheets above the highest closed loop in a Kopp and Pneumann helmet streamer. We conjecture that these spikes trace magnetic field lines which open into the solar wind above the helmet.

TABLE I

Coronal Transient and Near-Surface Phenomena

	10 January 1974	13/14 August 1973*
Flare: Hα/X-ray	1F/< C0	1N/C1
Eruptive prominence height:	$\gtrsim$ .05 R$_\odot$	0.18 R$_\odot$
Loop prominence system:	Hα + EUV + X-rays	EUV + X-rays
Max volume (cm^3)**	6 (27)	5 (28)
Height (cm)	4 (9)	1 (10)
X-ray em. meas. (cm^{-3})	1-2 (48)	1.6 (48)
Max temperature (K)	6 (6)	5 (6)
Density x-ray (cm^{-3})	1 (10)	5.5 (9)
Mass source (g)	1 (14)	4.9 (14)
Apparent velocity (km s^{-1})	10→40	1→0.5
Mass upflow required (km s^{-1})	35-150	$\sim$ 3
K. E. of upflow (erg)	1 (28)	> 2 (25
Radiative energy (erg)	2 (29)	$\sim$ 1 (30)
Lifetime (hr)	$\sim$ 1	> 33
Radio events	none	Type II
Coronal Transient*: speed (km s^{-1})	400	$\sim$ 175
mass (g)	1 (15)	2 (15)
P.E.+ K.E.(erg)	1.6 (30)	2.3 (30)

*Data from MacCombie and Rust (1979), Hildner (1978).
**Assumes x-ray volume equals that of Mg X emission.

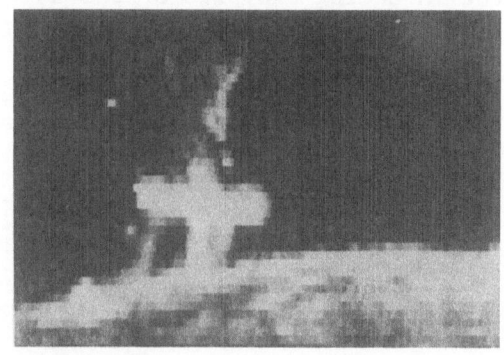

Figure 1.

OIV λ554 spectroheliogram
at 10:48 U.T. The image
size is 3 x 5 arc min.

REFERENCES

Foukal, P.: 1978, *Astrophys. J.* 223, pp. 1046-1057.
Hildner, E.: 1977, in *"Study of Travelling Interplanetary Phenomena"*
    pp.3-21 (eds. M. Shea et al.) Reidel, Dordrecht-Holland.
Kopp, R. A. and Pneuman, G. W.: 1976, *Solar Phys.* 50, p. 85.
MacCombie, W. J. and Rust, D. M.: 1979, *Solar Phys.* 61, 69.
Munro, R. H.: 1978, *Bull. Am. Astron. Soc.*, 9 (abstract).
Munro, R. H., Gosling, J. T., Hildner, E., MacQueen, R. M., Poland, A.I.,
    and Ross, C. L.: 1979, *Solar Phys.* 61, pp. 201-215.

DISCUSSION

Moore: Why did you entitle this paper "The <u>Disruption</u> of EUV Coronal Loops..."?

Schmahl: The "disruption" refers to the disappearance of the EUV loops, which are formed (in Kopp and Pneuman's model) by the reconnection of earlier loops disrupted at the time of the mass ejection.

Webb: Do you see any evidence for a temperature gradient in this event? (By this I mean a difference in the height of the loops in different transition zone lines?)

Schmahl: The "transition zone" ($5 \times 10^4 \leq T \leq 5 \times 10^5$K) line emission is apparently co-spatial with the Hα loop. The coronal emission in the MgX line ($T \sim 1.5 \times 10^6$K) is complicated by the presence of stationary, intervening coronal structures, but the MgX loop emission is certainly compatible with the distribution in X-rays, which appear slightly above the cooler loops.

Gaizauskas: You drew attention to large spiky features in an EUV spectroheliogram for this event which had no apparent counterpart in Hα. This absence may be due to lack of resolution in the filtergrams. When observed at high spatial resolution, large active regions on the limb often show fine spiky features extending more or less radially to coronal heights.

Schmahl: It is my impression that most Hα loop prominences are smooth at the top, although I have seen small "nubbins" at the apex in one case. I would be very interested in seeing examples of the Hα spiky structures that you mention.

DECAMETER RADIO AND WHITE LIGHT OBSERVATIONS OF THE 21 AUGUST 1973
CORONAL TRANSIENT.

T. E. Gergely and M. R. Kundu
Astronomy Program, U. of Md., College Park, MD

Observations from SKYLAB have shown that coronal transients, which
involve mass ejections occur quite frequently, possibly up to three
times a day at solar maximum (Hildner et al., 1976). An estimated mass
of $\sim 10^{15} - 10^{16}$ g (Stewart et al., 1974) and a total mechanical energy
in excess of 2×10^{31} ergs (Webb et al., 1978) is expelled from the Sun
during each event. The transients therefore play a major role in the
dynamics of the outer corona and of the interplanetary medium. Joint
radio and white light observations provide the best opportunity to
derive the physical parameters, such as the electron density and magnetic
field in different parts of the transients, and consequently to estimate
the forces driving the ejecta.

The coronal transient which occurred on 21 August has been well
observed and extensively analyzed (e.g. Poland and Munro, 1976; and
references therein). Radio observations of the event were obtained
with the two-dimensional, swept-frequency array (called the Teepee Tee)
of the University of Maryland. White-light observations consisting
of a series of photographs were taken by the High Altitude Observatory's
coronagraph aboard SKYLAB. The radio emission associated with the
transient was continuum in nature, and lasted for almost 5 hours.
Simultaneous radio and white-light measurements show that the radio
source was cospatial with one of the secondary white-light loops. To
establish the association of the radio source with a particular feature
of the transient, we compared the white light and the radio "pictures"
of the event. The earliest available radio position corresponds to
1555 UT, while the white-light picture corresponding most closely in
time was taken at 1511 UT. An overlay shows the radio source to coincide
in position with the lower part of a secondary white-light loop, and with
the northern edge of the primary loop system. The correspondence in time
between the radio and the white-light pictures is better than 1 minute
for the three other times when white-light pictures were taken. The
position of the radio source corresponds closely with the densest portion
of the secondary white-light loop on all of these measurements. Since
the pictures cover a period of more than two hours, we conclude that the
radio source was associated with the secondary white-light loop.

M. Dryer and E. Tandberg-Hanssen (eds.), Solar and Interplanetary Dynamics, 245-249.
Copyright © 1980 by the IAU.

The columnar electron density at the source of the radio emission, combined with information about the frequency spectrum of the source, can be used to determine a lower limit of the thickness of the loop as follows. For radio emission to escape from the coronal plasma, the condition $f > f_p$ must be satisfied, where f is the observed frequency of emission and f_p is the local plasma frequency, given by:

$$f_p = 9 \times 10^{-3} \, N_e = 9 \times 10^{-3} \, (\sigma/L)^{\frac{1}{2}}; \tag{1}$$

where N_e (cm^{-3}) is the local electron density, σ(cm^{-2}) is the columnar electron density, L(cm^{-1}) is the extent of the transient along the line of sight and f_p is in MHz. Since f and σ are observed quantities, a lower limit for L may be obtained from the above relationship. As a first approximation, we neglect the contribution of the background corona to the electron density. The lowest frequency at which radio emission was observed was $f = 32$ MHz. The radio emission originated entirely within the region enclosed by the columnar electron-density contour $\sigma > 10^{17}$ cm^{-2}. The columnar electron density at the site of the radio source ranges from 10^{17} cm^{-2} to 10×10^{17} cm^{-2}. Assuming an average value of 5×10^{17} cm^{-2}, we derive a lower limit for the depth (extension along the line of sight) of the transient $L > 4 \times 10^{10}$ cm $\stackrel{\sim}{\sim} 0.6$ R$_\odot$ ($N_e < 1.2 \times 10^7$ cm^{-3}). It appears, therefore, that the extension of the coronal transient along the line of sight was comparable to its characteristic extension on the plane of the sky.

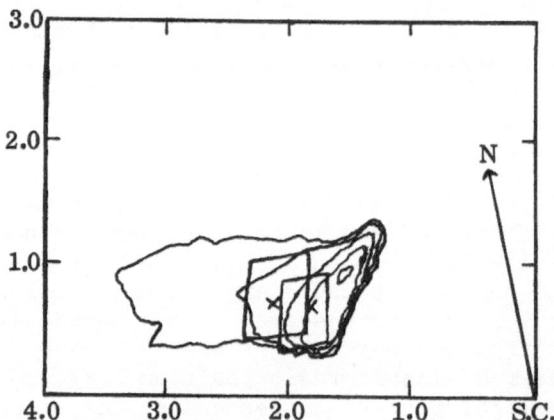

Contours of transient mass (electrons cm^{-2}) as a function of position in solar radii at 1835 UT, 21 August 1973. Contours are at 1×10^{16}, 2×10^{17}, 5×10^{17}, 1×10^{18} and 2×10^{18} electrons cm^{-2}. The two crosses and the quadrilaterals indicate the centroid and extension to a level of approximately 0.05 peak intensity of the low ($f < 50$ MHz) and high frequency ($f > 50$ MHz) radio source.

The depth has remained unknown until now, since coronal transients in white light can be observed only near the limb and not in projection on the disk. The radio source showed no dispersion of height with frequency, and therefore we attribute the emission to gyrosynchrotron radiation. Another argument can be made in support of the gyrosynchrotron origin of the emission. Fundamental plasma radiation would have been heavily attenuated as the burst occurred close to, and possibly behind, the limb. Second harmonic radiation, on the other hand, would have required the depth of the source L to be greater than 1.2 $R_\odot$, and consequently greater than its extent on the plane of the sky. This seem rather unlikely. Gyrosynchrotron emission from mildly relativistic electrons occurs at fairly low harmonics ($\sim$ 4-10) of the gyrofrequency. Since the peak intensity of the emission occurred at about 50 MHz we estimate from the relation $f_{peak} \approx$ 4-10 f_H = 11.2-28.0 B, the magnetic field strength in the secondary loop to be in the 2.0-4.5 gauss range at 2.0 $R_\odot$.

Radio measurements can clearly provide an estimate of the magnitude of the magnetic field in coronal transients when the mechanism giving rise to the emission can be established. Owing to the lack of accurate information about the field geometry, however, it is not possible to use these results to distinguish directly between the various models proposed to account for the observed dynamics of coronal transients (i.e. Mouschovias and Poland, 1978; Anzer, 1979; Dryer et al., 1979). If we assume a coronal temperature of 1.5 x 10^6 K, our results yield approximate equality of the gas and magnetic pressures at 2.0 $R_\odot$ (i.e. $\beta \sim 1.0$). For this transient, therefore, the gas is strongly influenced by changes in the magnetic field.

ACKNOWLEDGEMENT

This work was partially supported by NASA grants 21-002-199, and NAS-8-33105 and NSF grant AST 77-12282. Computations were supported by the Computer Science Center of the University of Maryland.

REFERENCES

Dryer, M., Wu, S.T., Steinolfson, R.S. and Wilson, R.M.: 1979, *Astrophys.J.* 227, 1050.
Hildner, E., Gosling, J.T., MacQueen, R.M., Munro, R.H., Poland, A.I. and Ross, C.L.: 1976, *Solar Phys.* 48, 127.
Mouschovias, T. Ch. and Poland, A.I.: 1978, *Astrophys.J.* 220, 675.
Poland, A.I. and Munro, R.M.: 1976, *Astrophys. J.*, 209, 927.
Stewart, R.T., Howard, R.A., Hansen, S.F., Gergely, T.E., and Kundu, M.R.: 1974, *Solar Phys.*, 36, 219.
Webb, D.F., Cheng, C.C., Dulk, G.A., Edberg, S.J., Martin, S.F., Mackenna-Lawlor, S. and McLean, D.J.: 1978, in 'Proceedings of the SKYLAB Workshop on Solar Flares', to be published.

DISCUSSION

Pneuman: Have you made a comparison between the spatial location and heights of the secondary loop you mentioned and the location of the X-ray loop system observed during the same event?

Gergely: We have not made a comparison between the location of the X-ray loop system and the secondary loop studied by us. The X-ray loop system was of course much lower in the corona than the features I described.

Uchida: What spectral type does the radio emission related to the secondary feature show? I presume that it is stationary type IV. Is there any trace of type II's related to the primary loop feature at all?

Gergely: There is no evidence of a type II burst related to the primary loop. The early part of the radio event was detected by G. Dulk with the Colorado University interferometer, and he doesn't see any type II burst. I guess you may classify the burst as a stationary type IV.

Benz: The fact that the radio positions coincide at different frequencies does not necessarily mean that you observed gyro-synchrotron emission. If the density enhancement in the transient is a factor of ten above the background, the plasma levels of your range of frequencies are very close to each other. Do you have other evidence for gyro-synchrotron emission?

Gergely: I agree with your comments. There are arguments against both fundamental and second harmonic plasma emission, details of which are given in our paper. We feel that gyro-synchrotron emission describes better the observations, but the evidence is not conclusive.

Anzer: From your data and $\beta = 1$, one would have $n \sim 10^{10}/cm^3$ at $2R_\odot$, which seems extremely high.

Gergely: From our data and considering $\beta = 1$, one gets $n_e \approx 4 \times 10^8$ to $10^9 cm^{-3}$. I agree that this is still quite high, β is probably in the range 0.1–1.0. Of course the temperature may be wrong too. In any case, the transient is magnetically controlled.

Degaonkar: What is the duration of continuum; i.e., whether it was of same duration as that of the transient? Did the continuum show high or low frequency cut-off?

Gergely: The continuum did not extend above 90 MHz on the high frequency side and approximately 30 MHz on the low frequency side. The emission lasted for about 5 hours, but the intensity varied during this time. Details are given in Gergely *et al.* (*Ap. J.*, 230, 575, 1979).

Sheeley: Are you sure that the secondary feature in the August 21, 1973 coronal transient is really a loop, or could it have some other structure?

Gergely: No, I use the term "loop" in a general sense. Perhaps it would be more correct to speak of a secondary feature.

Petelski: Concerning source heights, can one really determine the radial position or does one just obtain the projection onto a plane perpendicular to the line of sight?

Gergely: One determines the position on the plane of the sky. However, the white light polarization measurement shows the features to be at less than $10°$ from the plane of the sky; therefore, projection effects are negligible.

RADIO DATA AND COMPUTER SIMULATIONS FOR SHOCK WAVES GENERATED BY SOLAR FLARES

Alan Maxwell
Harvard College Observatory, Cambridge, MA 02138, USA
Murray Dryer
Environmental Research Laboratories, NOAA, Boulder, CO 80302, USA

Solar radio bursts of spectral type II provide a prime diagnostic for the passage of shock waves, generated by solar flares, through the solar corona. In this investigation we have compared radio data on the shocks with computer simulations for the propagation of fast-mode MHD shocks through the solar corona. The radio data were recorded at the Harvard Radio Astronomy Station, Fort Davis, Texas. The computer simulations were carried out at NOAA, Boulder, Colorado.

It is generally agreed that type II solar radio bursts result from the passage outward through the solar corona of fast-mode MHD shocks generated by relatively intense solar flares. The radio emission comes from plasma oscillations, is generally confined to a narrow band of radio frequencies, is often seen at both the fundamental and second harmonic, and is randomly polarized. In many cases, the fundamental emission is first observed at frequencies of approximately 150 MHz, about 5 minutes after the explosive phase of a flare (generally indicated by emission of impulsive bursts in the microwave band and associated hard X-ray bursts). The radio emission then drifts slowly downward in frequency, taking approximately 15 minutes to drift from 150 MHz to 25 MHz, say. With the assumption of an appropriate electron density model, it is possible to interpret the observed emission frequencies in terms of emission from corresponding plasma heights in the solar atmosphere: for the assumed density models, the frequency range 150 to 25 MHz then transforms to an equivalent height range of approximately 1.5 to 2.5 $R_\odot$. From the drift rate of the radio burst it is then possible to determine the outward radio velocity component of the shock generating the burst and this is usually of the order 1000 to 2000 km s^{-1}. Velocities of the outward-traveling shocks have also been determined from positional data, taken as a function of time, recorded by the radioheliograph at Culgoora, Australia, while operating at frequencies of 43, 80, and 160 MHz, and these data give shock velocities of the same order (Nelson 1977).

Data taken at Culgoora also show that the type II emission regions may be widely distributed, at times over as much as 180 arc deg around the solar disk (Smerd, 1970). Uchida (1974) has shown, by computer simula-

M. Dryer and E. Tandberg-Hanssen (eds.), Solar and Interplanetary Dynamics, 251-255.

tions, that fast-moving MHD waves tend to be refracted into regions of
low Alfven velocity in the corona, where they strengthen into shocks,
and in this manner give rise to type II emission sources of the sort
recorded at Culgoora. Uchida's model was developed for a wave generated
by a blast. In the case of large flares, however, radio evidence (in
the form of type IV emission falling closely behind the type II burst
both in time and position) suggests that the shock may often be driven
by a piston.

Computer models for the propagation of fast-mode MHD shocks through
the solar corona, and then out of the interplanetary plasma have been
developed at NOAA in Boulder and the University of Alabama in Huntsville
(cf. Wu et al., 1978; Steinolfson et al., 1978). The models simulate
the coronal mass motions and shock waves resulting from solar flares.
The model that was used for the present investigation was two-dimensional,
time-dependent, and was applied to the meridional plane. The assumed
magnetic topology was that of a hexapole embedded in the solar corona.
An input pulse was then applied at a region where the magnetic field
lines appeared to open into the interplanetary plasma. The pulse was
applied over 5 degrees in heliographic latitude, and it was applied at
the base of the corona where the ambient temperature was assumed to be
2×10^6 K, the magnetic field 2 Gauss, and the plasma beta was assumed
to be 1. The density in the solar atmosphere was assumed to decrease in
a quasi-exponential manner (the computer density model approximated the
density model used for the interpretation of the radio data at heights
of about 2 to 3 $R_\odot$). The computer simulation was terminated at 6 $R_\odot$.

The computer simulation provides information on the global response
of the corona, after the input of a large energy pulse, in terms of
density, temperature, particle velocity, and the redistribution of
magnetic field. Various forms of input pulses could be applied: for
example, a series of rapid pressure pulses, a square-wave pressure pulse,
a magnetic pulse, etc. In the present investigation, the best simulation for the observed radio data was given by an input pulse which had
the form of a square-wave pulse of duration 10 min, containing a temperature (or pressure) increase of 40 times the ambient value. It is also
interesting to note that this input pulse is similar to the pressure
changes observed in X-ray loops, at the time of a flare, where the pressure may increase by a factor of about 50 for a period of several
minutes (Pallavicini et al., 1977). The energy in the applied pulse
was of the order of 2×10^{32} erg.

A comparison of the shock velocities derived from the computer model
with shock velocities derived from radio data for a given flare is shown
in Figure 1. Density ratios and velocity magnitudes in the corona determined from the computer simulation for a time corresponding to 6 min
after the input pulse was first applied (that is, at about the time the
type II burst was first seen on the radio records) are shown in Figure 2.
The ejected mass, as indicated by the computer simulation, was 6.4×10^{16} g.
Full details of this work may be found in a paper by Dryer and Maxwell
(1979).

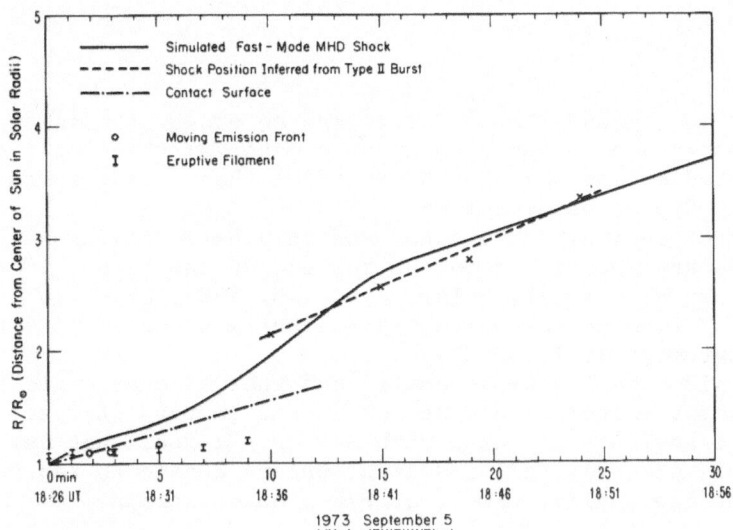

Figure 1. Comparison of radio data on a shock wave generated by a solar flare with the computer simulation for fast-mode MHD shock in the solar corona.

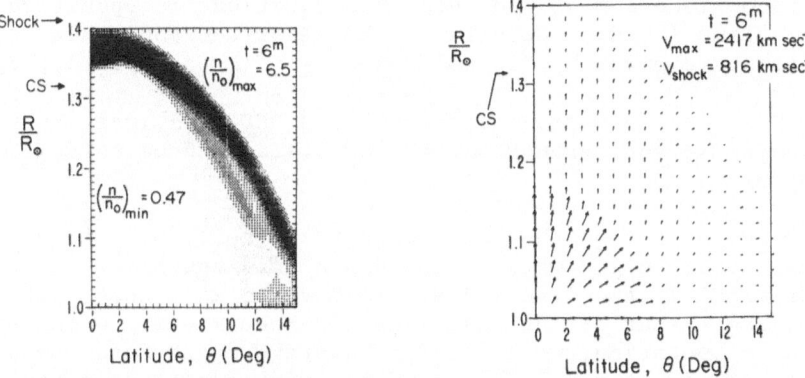

Figure 2. Density distribution and bulk plasma velocities in the solar corona inferred from the computer simulation, showing position of shock and contact surface (piston) 6 min after input pulse is first applied.

REFERENCES

Dryer, M. and Maxwell, A. 1979, Astrophys. J., 231, 945.
Nelson, G.J. 1977, Proc. Astron. Soc. Australia, 3, 159.
Pallavicini, R., Serio, S., and Vaiana, G. 1977, Astrophys. J., 216, 108.
Smerd, S.F. 1970, Proc. Astron. Soc. Australia, 1, 305.
Steinolfson, R.S., Wu, S.T., Dryer, M. and Tandberg-Hanssen, E. 1978, Astrophys. J., 225, 259.
Uchida, Y. 1974, Solar Phys., 39, 431.
Wu, S.T., Dryer, M., Nakagawa, Y., and Han, S.M. 1978, Astrophys. J., 219, 324.

DISCUSSION

Jackson: 6×10^{16} g and 2×10^{32} ergs are an order of magnitude more mass and more energy associated with ejected material than for the transients measured during Skylab. Do you feel that your results can be scaled to fit the Skylab observations?

Maxwell: There are many transients, and they have varying masses and energies. We are interested in the top end of the range. (For example, Gosling et al., (1975, Solar Phys., 40, 439), give 2.4×10^{16} g and 1.1×10^{32} erg for the white-light transient associated with the flare of 1973 September 7, 11:40 UT).

Dryer: Note also that magnetic energy and thermal energy are included in our energy calculations but not in the energy estimates for the white light transients measured with Skylab. I feel that our results can be scaled (via properly tailored input pulses chosen to represent the flare) to fit the Skylab, OSO-7, or P78-1 observations.

Petelski: Could you extend a bit on the algorithm employed in the calculations and, in particular, on the approximations used for solving the MHD equations?

Dryer: The complete set of nonlinear MHD equations is specified to the meridional plane (r, θ). Details of the algorithm and other numerical matters are given by Steinolfson et al. (Astrophys. J., 225, 259 (1978)).

Nakagawa: The basic equations are idealized MHD equations. They are totally nonlinear and, except for the numerical scheme used, involve no approximation.

Moore: How do you assign the total energy and total mass from your two-dimensional model; e.g., what is the depth into the plane?

Dryer: Our model is two-dimensional. We have to make some suitable assumption for the extent of the disturbance perpendicular to the plane. For the present computation, we assumed an average depth into the plane of $0.2\ R_\odot$ at $t = 30$ min ($6\ R_\odot$). This conforms with depth estimates made for white-light transients by Poland and Munro. Since the observational estimates are uncertain (cf., an estimate of $0.6\ R_\odot$ by Gergely and Kundu, this symposium) and considering further that our model is not 3-dimensional, our estimates of total mass and energy are also uncertain by, say, a factor of three or so. Progress toward a time-dependent MHD, 3-dimensional model is being made by Nakagawa, Wu and Han (this symposium). We believe, however, that the present 2-dimensional model is adequate for providing insight into some of the basic physical processes present in the temporal mass motion.

Kahler: A temperature increase of a factor of 40 and a duration of 10 min. suggests that you are assuming the pulse is the impulsive phase of the flare. Have you attempted to match the start time and duration of the pulse to the hard X-ray or microwave burst?

Maxwell: We take the starting time of hard X-ray bursts and impulsive microwave bursts as defining the start of the explosive phase

of the flare and as time zero for the initial application of the
input pulse for our computer simulation. Our pulse length (10 min)
may be a little longer than the observed durations of the hard X-ray
bursts and microwave bursts.

Nakagawa: Rising loops do not necessarily represent the cause of
the matter ejected outward. Rising loops could result from heating that
has already occurred as seen, say, in hard X-ray bursts.

Uchida: I fully understand the existing mathematical and computational restrictions, but I would like to point out that a type II
burst is a more complicated phenomenon of somewhat different nature.
For example, the sources of type II bursts show more "ragged" structure,
appearing sometimes even beyond the horizon, instead of your bowshock
type surface. This seems to favor my previous work, in which I suggested
that the type II burst may be caused by a weak blast-pulse, emitted in
the impulsive phase of a flare, which "illuminates" favorable regions
where the shock is strengthened due to the low Alfvén velocity. Your
calculation, however, may apply to the interplanetary type IIs, which
we expect in front of the mass ejecta in the form of a bowshock far
out in the interplanetary space.

Maxwell: I agree. I did not have time in my talk to discuss
shocks generated by short-lived blasts. Our work, of course, concerns
piston-driven shocks, generated by large flares giving large input
pulses.

Dryer: I would like to remind the audience that Dr. Uchida's
pioneering work is confined to the regime of linear (ray-tracing)
analysis. Hence one can only infer, as he has just noted, the
situation regarding non-linear steepening of the MHD waves into MHD
shock waves. Our calculations are fully non-linear; hence, a shock
wave can indeed develop anywhere within the low corona as well as interplanetary space. Thus, the model can be applied to coronal, as well as
interplanetary, type II spectral radio emission.

Somov: Your calculations of the dynamic response of corona on a
pulse show that very large mass can be ejected. Is this mass contained
in the corona before a pulse or ejected from below the boundary placed
on the coronal base? In the last case the calculation assumes
tacitly some unspecified reason for ejection of chromospheric plasma.
Then, your resulting solution is the continuation of the solution for
the corona into the chromosphere.

Dryer: This is correct. We have in mind the fact that the chromosphere (which is actually below our "coronal base" in the model) can
provide substantial mass flux. Thus we allow mass flow across our lower
boundary such that the time-dependent conservation requirements are
continually satisfied.

ESTIMATION OF SHOCK THICKNESS FROM DYNAMIC SPECTRA OF TYPE II BURSTS

H. S. SAWANT, S. S. DEGAONKAR, S. K. ALURKAR and R. V. BHONSLE
Physical Research Laboratory
Ahmedabad-380 009, India

ABSTRACT

Twenty type II solar radio bursts were observed during the period 1968 to 1972 by a solar radio spectroscope (240-40 MHz) at Ahmadebad. Intensity variations in type II bursts as a function of frequency and time are sometimes observed in their dynamic spectra. This fine structure enables determination of the shock thickness of the order of a few hundred to a few thousand kilometers. In a few cases, an interaction between streams of fast electrons and propagating shocks is clearly evidenced by simultaneous observations of short duration narrow band structures in type III bursts and type II bursts.

INTRODUCTION

Type II bursts are characterized by slow frequency drift with time as compared to fast frequency drifts observed in type III. They occur less frequently than type III bursts and are generally associated with large solar flares. The velocity of a type II disturbance as estimated from the frequency drift with time is of the order of 10^3 Km/sec. The drifting feature of type II is identified with a collisionless shock wave set up as a result of an explosion at the time of a flash phase in a solar flare (Kundu, 1965; Wild and Smerd, 1972).

With the help of a solar radio spectroscope (frequency range 240-40 MHz; bandwidth $\sim$300 KHz; time resolution $\sim$0.5s) at Ahmedabad, twenty type II events were recorded between 1968 and 1972. A typical example showing interaction between type II and III is shown in Figure 1.

RESULTS

Besides the usual macroscopic features some microscopic features in type II bursts are observed. These are: (i) short duration narrow band patchy structures, and (ii) simultaneous appearance of short

257

M. Dryer and E. Tandberg-Hanssen (eds.), Solar and Interplanetary Dynamics, 257-259.

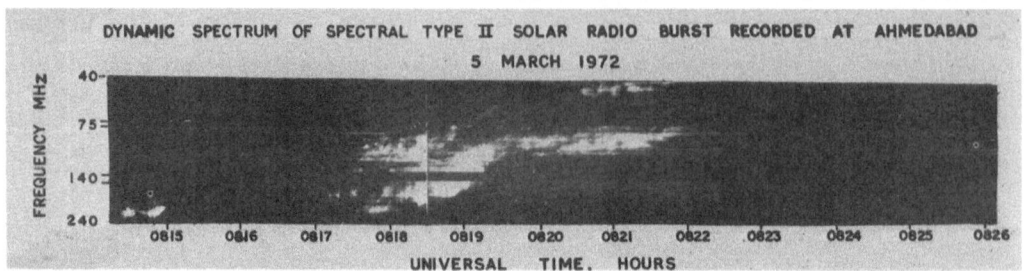

Figure 1. Dynamic spectrum of type II burst showing fine structure
and interaction with type III burst.

duration narrow band patches in type III and type II bursts. The
patchy structures have the following characteristics: (a) they occur
predominantly below 100 MHz; (b) their bandwidth was as low as 500 kHz,
to tens of megahertz; and (c) their duration varied from 1 to 5 seconds.

INTERPRETATION

The short-lived patchy structures within type II bursts can be
understood from the following mechanism. As the shock front encounters
electron density irregularities, the enhancement in the emission takes
place as a result of the interaction between the shock and the
irregularity. It is possible to estimate the scale size of the density
irregularities and obtain a lower limit to the shock thickness.
Radiation can be generated ahead of the shock front by the escaping
electrons from the shock (McLean, 1974). Knowing the velocity of the
shock from the frequency drift and the electron density model, and
from the time duration of the patches seen in type II, we have obtained
the linear dimensions of $10^2 - 10^3$ km of the irregularity. This puts an
upper limit to the shock thickness since the shock has to interact with
the density irregularity to produce enhanced intensity. Observation of
shock thicknesses by satellites have shown that the shock thickness
varies from 600 to 1600 km (Dryer, 1975).

CONCLUSION

From the intensity variations of narrow band short duration
patches, sometimes observed in type II bursts, the thickness of the
shock front from a hundred to a few thousand kilometers has been
derived. These values are a lower limit to the shock thickness and are
consistent with *in situ* satellite measurements.

ACKNOWLEDGMENT

We thank Professor D. Lal, Director of PRL, for his interest in
this work. Financial support for this work has come from the Department of Space, Government of India. Thanks are also due to our staff
who assisted in the maintenance of the equipment at SAC campus.

REFERENCES

Kundu, M. R.: 1965, *Solar Radio Astronomy*, Interscience Publ., N.Y.

McLean, D. J.: 1974, *IAU Symp. No. 57*, G. Newkirk, Jr. (Ed.) D. Reidel
    Co., pp. 301-327.

Wild, J. P., and Smerd, S. F.: 1972, *Ann. Rev. Astron. Astrophys., 10*,
    pp. 159-196.

Dryer, M.: 1975, *Space Sci. Rev., 17*, pp. 277-325.

Lacombe, C., and Moller-Pedersen, B.: 1971, *Astron. & Astrophys., 15*,
    pp. 404-418.

DISCUSSION

Gergely: Do you have any positional measurements of the type II
and type III's? If not, one may not know for certain if the shock
and the electron beams interact, since they may be in entirely
different parts of the corona.
Sawant: True, we didn't have positional information of type II and
associated type III bursts. But considering the fact that the type II
shock fronts generally have large heliolongitudinal extent, we have
assumed that the electrons causing type III bursts might have interacted
with the type II shock front. Further, the narrow frequency range over
which type III's have been observed just prior to the type II burst
may justify our above assumption.

Petelski: Is there any <u>direct</u> evidence for the shocks?
Sawant: Yes, the occurrence of type II burst is a sufficient
evidence for the shock.

Newkirk: Since one does not know the scale of the inhomogeneities,
your derivation of the thickness of the shock must be regarded as
upper limit.
Sawant: Yes, I agree.

Dryer: What shock velocities did you observe?
Sawant: Usually type II bursts imply shock velocities of 1000 to
2000 km/s but at the time of August 1972 events, shock velocities of
about 4000 km/s were inferred.

EVIDENCE FOR OPEN FIELD LINES FROM ACTIVE REGIONS: SHORT COMMUNICATION

K. V. SHERIDAN
Division of Radiophysics
CSIRO, Sydney, Australia

A paper that has considerable relevance of the subject matter of this symposium is the following: "Evidence for Extreme Divergence of Open Field Lines from Solar Active Regions," by G. A. Dulk (Division of Radiophysics, CSIRO, Sydney, Australia and Department of Astro-Geophysics, University of Colorado, Boulder, Colorado), D. B. Melrose (Department of Theoretical Physics, University of Sydney, Australia) and S. Suzuki (Division of Radiophysics, CSIRO, Sydney, Australia).

The paper will appear in the Proceedings of the Astronomical Society of Australia for 23 May 1979.

This paper includes a review of the evidence on the structure of the open magnetic field lines that emerge from solar *active regions* into interplanetary space. The evidence comes mainly from the measured sizes, positions and polarization of type III and type V bursts, and from electron streams observed from space. They find that the observations are best interpreted in terms of a strongly-diverging field topology, with the open field lines filling a cone of angle $\sim 60^{\circ}$.

These observational results are in agreement with the extrapolation of the field above the Sun's surface based on potential theory as discussed by R. Levine (see "Evolution of Coronal and Interplanetary Magnetic Fields," R. Levine, this issue).

M. Dryer and E. Tandberg-Hanssen (eds.), Solar and Interplanetary Dynamics, 261.

MHD ASPECTS OF CORONAL TRANSIENTS

U. Anzer
Max-Planck-Institut für Physik und Astrophysik, Munich

1. INTRODUCTION

If one defines coronal transients as events which occur in the solar
corona on rapid time scales ($\lesssim$ several hours) then one would have to
include a large variety of solar phenomena: flares, sprays, erupting
prominences, X-ray transients, white light transients, etc. Here we
shall focus our attention on the latter two phenomena; solar flares
have been discussed at great length in a recent Skylab workshop and IAU
Colloqium No. 44 was devoted to the study of prominences. Coronal
transients, in the narrower sense, were first seen with the instruments
on board of Skylab, both in the optical and the X-ray part of the
spectrum.

The X-ray observations in the range between 2 and 50 Å were described by Webb et al. (1976) and Rust and Webb (1977). They report a
total number of 156 observed X-ray enhancements. Their general behaviour
can be summarized as follows: most of them have loop-like structures
with lengths between 50 000 and 500 000 km and an average diameter of
15 000 km. They last between 3 and 40 hours. The loops expand initially
with velocities up to 50 km/s but slow down rapidly to 1-10 km/s. The
estimated temperatures lie in range from 2 to 5×10^6K, the densities
between 10^9 and 10^{10} cm^{-3}. The events which occur away from active
regions are very often associated with the disappearance of an H_α filament. Their total thermal energy content is of the order of 10^{29} erg.

The white light coronal transients were first reported by Gosling
et al. (1974). Detailed studies of many different aspects of these
transients were performed and a good summary can be found in the paper
by Hildner (1977). Due to the instrumental design of the coronagraph
the white light events could only be seen from ~ 1.6 R$_\odot$ to 6 R$_\odot$. This
of course makes it hard to establish correlations between transients
and phenomena which occur near the surface of the sun, where one assumes
that transients originate. Like the X-ray transients many of the white
light events (115) also show loop-like structures. The tops of these
loops move rapidly through the field of view whereas their legs remain

263

M. Dryer and E. Tandberg-Hanssen (eds.), Solar and Interplanetary Dynamics, 263-277.
Copyright © 1980 by the IAU.

visible for several days. The width of the loops is of the order of
several tenths of $R_\odot$. The densities decrease from $\sim 2 \times 10^7$ cm^{-3} to
$\sim 5 \times 10^5$ cm^{-3} as the transients move from 1.6 to 6 $R_\odot$. The temperatures
cannot be measured directly, but it has been concluded from polariza-
tion measurements that most of the material must be at temperatures
higher than 10 000 K. The velocities are in the range 100 to 800 km/s,
which is much higher than those of the X-ray transients; typical velocity curves are shown in fig. 1.

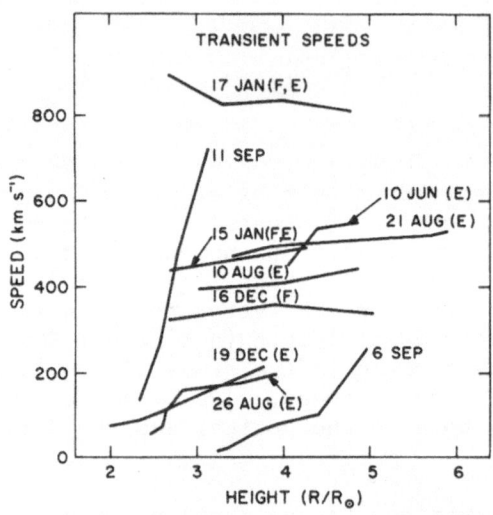

Fig. 1. Speeds of loop transients' leading edge versus height. Uncertainties in the velocity measurements range from ± 50 km s^{-1} to ± 100 km s^{-1}. All but the January 1974 events occurred in 1973.
The letter E or F indicates that the ejection was associated with an eruptive prominence or flare,
respectively. See Hildner (1977).

Most of the transients show little or no acceleration during their
passage through the field of view. The energies associated with these
motions are between 2×10^{30} erg and 7×10^{31} erg, and masses between
10^{15} g and 2×10^{16} g are ejected.

Rust and Hildner (1976) describe an event (13 Aug., 1973) for
which both X-ray and white light data were available. Fig. 2 shows the
relative positions of the loop structures, fig. 3 the temporal evolution. It should be noted that for this particular event accelerating
X-ray structures are observed, whereas in general X-ray loops show a
deceleration. Unfortunately there is a large data gap between 1.4 $R_\odot$
and 3.8 $R_\odot$. Therefore we cannot be absolutely sure that white light
and X-ray event are identical. The masses estimated for both are comparable which speaks in favour of the interpretation that they are
identical structures.

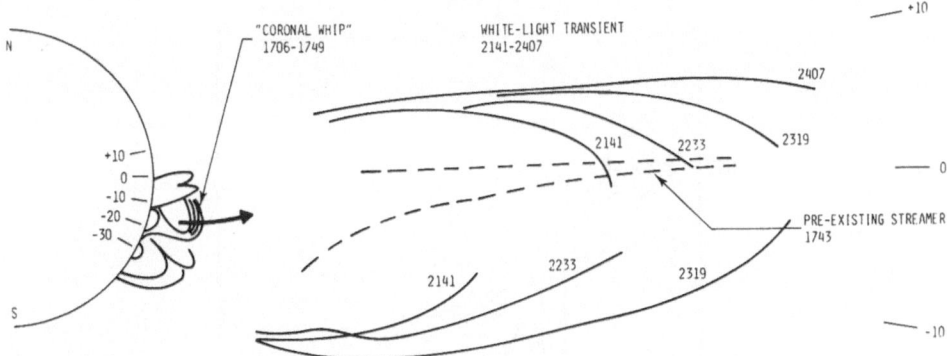

Fig. 2. Composite drawing of the mass ejection as deduced from the X-ray photographs and white light images. Heavy lines indicate the edges of the white light transient at the times indicated.

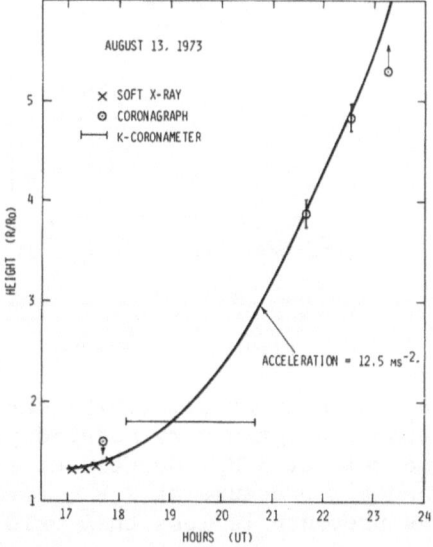

Fig. 3. Height vs time plot showing progress of the expanding X-ray arch in the inner corona and of the leading edge of the white light bubble in the outer corona. A curve for constant outward acceleration at 12.5 m s^{-2} appears to fit the points well; however, other curves with slightly different assumed start times could describe the event, too.

In order to understand the dynamics of coronal transients it is necessary to obtain information on the strength and configuration of the magnetic field in the corona. This information, however, is very indirect and the values derived are based on many assumptions. Dulk and McLean (1978) gave a review on the fields estimated in the corona, fig. 4. This diagram represents a composite of all kinds of different field estimates and shows a large scatter. For a radius of 1.1 $R_\odot$ e.g. one may deduce a field of 5 to 20 G.

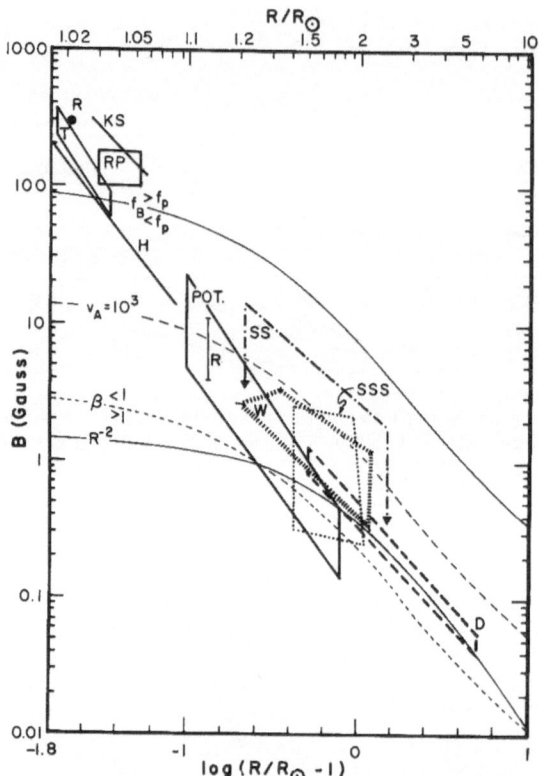

Fig. 4.   Magnetic field strength vs height above active regions. A coronal density model twice that given
by Newkirk (1967) or Saito (1970) for the equatorial corona at sunspot minimum has been assumed.
(This assumption affects only the positions of boxes 'SS', 'SSS' and 'W', and the curves for $f_B = f_p$,
$v_A = 10^3$ km s^{-1} and $\beta = 1$.) The various lines and boxes are identified in the text.

Dulk et al. (1976) and Gergely et al. (1979) derived estimates for individual transients. These are again very model-dependent. Dulk et al. found B = 3G at 1.8 $R_\odot$ and < 1G at 3 $R_\odot$, Gergely et al. give B ∼ 2-4.5G at 2 $R_\odot$. The resulting magnetic pressure at 2 $R_\odot$, is of the order of 1 dyn/cm^2, whereas the gas pressure is less than 5×10^{-3} dyn/cm^2. This shows clearly that magnetic forces have to be taken into account in theoretical models.

2.   THEORETICAL MODELS

The existing models can be devided into two groups. Models of the first group describe transients as single structures which move through the corona, the surrounding corona only providing the driving magnetic field, other interactions between corona and transient are not considered. The other approach is to assume that coronal transients are perturbations of a stationary corona caused by rapid changes at the lower boundary (i.e. the solar chromosphere). Models with gas pressure, temperature and magnetic pressure pulses are studied.

2.1 Models with single structures

Calculations based on the assumption that a transient is a large loop
were presented by Mouschovias and Poland (1978). They started from the
observation that loop-like transients show no (or only very small)
acceleration during their passage through the field of view. They took
a constant velocity which in their model implies that the magnetic
forces exactly balance gravity. They used a helical field inside the
transient. To avoid pinch instabilities of their configurations, they
assumed that the ratio of the azimuthal to the longitudinal field is
less than 1.4. On the other hand this ratio must be larger than unity
to produce an outward force. Under the assumption that this ratio is
constant in time they were able to deduce the evolution of the loop.
They found that both the width of the loop and it's radius of curvature
are proportional to the distance of the transient from the center of the
sun. These results are in agreement with the observations made for the
transient of 10 Aug., 1973, (see fig. 5). The results are not too conclusive when one takes into account the large error bars of the measure-

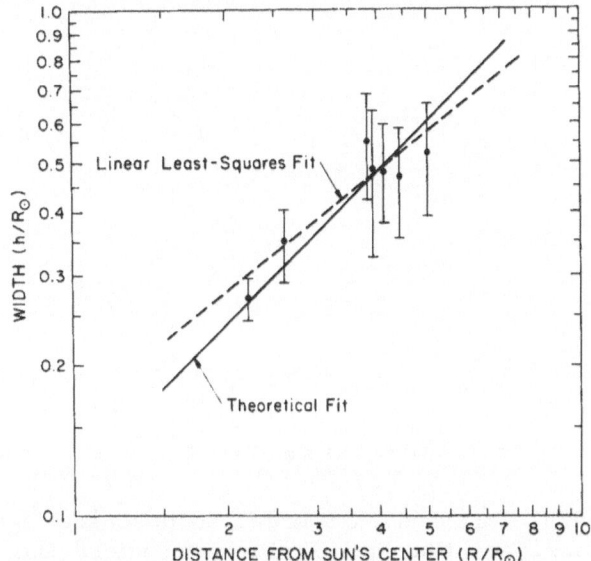

FIG 5.—The measured width (*h*) of the top portion of the
1973 August 10 loop transient as a function of distance (*R*)
from the Sun's center. The linear least-squares fit (*dashed line*)
has a slope of 0.8. The theoretical curve (*solid line*) has a slope
of 1.

ments. Another weakness of this paper is that it only considers the
magnetic field inside the transient, leaving the surrounding field
completely unspecified. But, of course, this field has eventually to
provide all the driving.

Anzer (1978) has addressed the question as to whether magnetic
fields of the magnitude observed in the corona can accelerate and propel
loop transients into interplanetary space with the observed velocities.

He used a very simple model where the transient is a current ring. Then
the magnetic force is just the one which this ring current excerts on
itself. Because the corona is a very good conductor the magnetic flux
through the ring will be conserved. This then allows one to calculate
the driving force acting on the loop and the resulting acceleration.
Velocity curves for this model are shown in fig. 6. The different
curves are for different initial radii and modifications of the circular geometry. The intention of this model was to demonstrate that
coronal loop transients can be driven magnetically and should not be
taken as an exact representation of the field configuration around
loop transients. The main question which is still unanswered is whether
initial field configurations of the type used here can be generated in
the solar atmosphere. One possibility which one could imagine is that
inside a loop with a longitudinal magnetic field currents are induced
by rapid rotation of the foot points. This would lead to a situation
similar to the one discussed above.

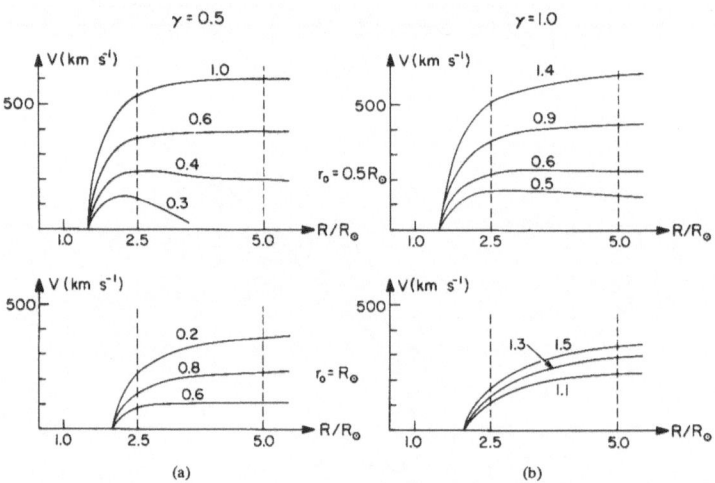

Fig. 6.   Calculated speed versus height curves. (a) gives the results for $\gamma = 0.5$ and two different values
of r_0. Each curve is labelled with its value of the parameter a. (b) same as (a) except $\gamma = 1.0$.

    Van Tend (1979) extended this model to describe the onset of
transients. He considers the equilibrium case where the force on the
ring current balances the gravity of the loop. Calculating the equilibria for different distances, R, of the loop from the sun's center he
finds a range $R_1 < R < R_2$ for which the equilibrium is unstable. This
instability then is assumed to initiate the transient. The results
should not be taken too literally because the original model for the
driving of transients is very crude, as mentioned earlier. Therefore,
it is not clear how realistic a stability analysis of such configurations is (e.g. this sort of equilibrium will only hold near the top of
the loop; the geometry of the loop could change as it evolves slowly,
etc.).

    Pneuman (1978, 1979) has developed a different model for tran-
sients. His idea is that transients originate from the closed upper

parts of a helmet streamer and that they are propelled outward by
forces associated with erupting prominences. The model is shown in
fig. 7. Pneuman considers both loop configurations and arcades. He
derived equations of motions for the radial distance of the transient,

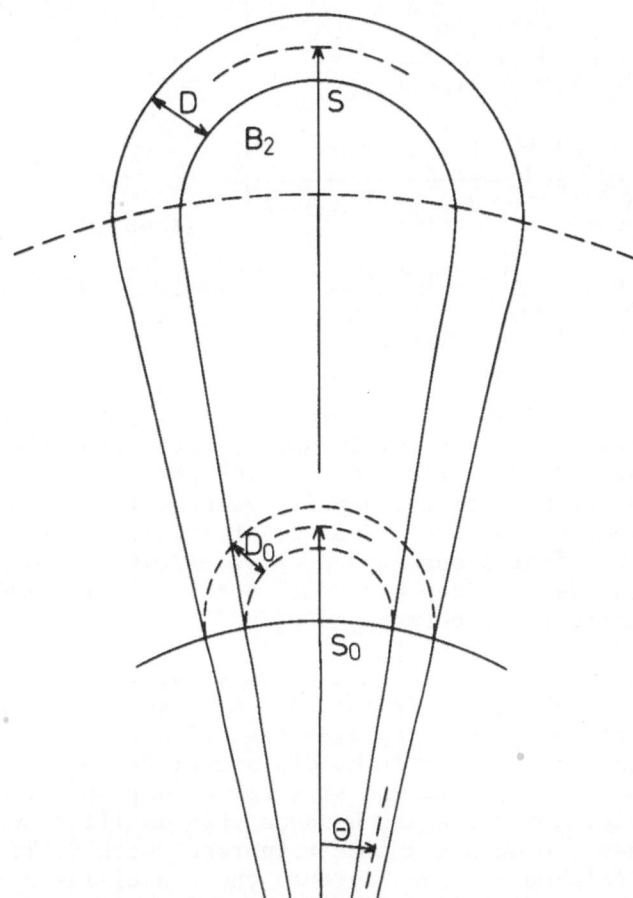

Fig. 7. Schematic of the idealized transient geometry.
The dashed curves denote the boundary of the fluxtube in
its equilibrium position with width D_0 and with its top at
a distance S_0 from the solar center. The solid curves show
the tube at some later time with width D and displacement S.
B_2 is the driving field behind the transient and θ is the
half-angle between the legs.

S, and for its thickness, D. His calculations are limited to the motion
of the top of the transient. He assumes that D is small compared to S
and that the gas pressure can be neglected everywhere. The magnetic
field above the transient is set to zero. Taking flux conservation

inside the transient and underneath it into account, one arrives at the following equations of motion:

$$\frac{d^2 S}{dt^2} = \frac{B_{2o}^2}{8\pi\rho_o D_o}\left(\frac{S_o}{S}\right)^2 - \frac{D_o B_o^2(1+\tan\theta)}{4\pi\rho_o DS \tan\theta} - \frac{GM_\odot}{S^2} \tag{1}$$
$$\text{(arcade)}$$

$$\frac{d^2 D}{dt^2} = -\frac{B_{2o}^2}{4\pi\rho_o D_o}\left(\frac{S_o}{S}\right)^2 + \frac{D_o B_o^2}{2\pi\rho_o D^2} \tag{2}$$

$$\frac{d^2 S}{dt^2} = \frac{B_{2o}^2 D}{8\pi\rho_o D_o^2}\left(\frac{S_o}{S}\right)^3 - \frac{D_o^2 B_o^2(1+\tan\theta)}{4\pi\rho_o S_o D^2\tan\theta} - \frac{GM_\odot}{S^2} \tag{3}$$
$$\text{(loop)}$$

$$\frac{d^2 D}{dt^2} = -\frac{B_{2o}^2 D}{4\pi\rho_o D_o^2}\left(\frac{S_o}{S}\right)^3 + \frac{D_o^2 B_o^2 S}{2\pi\rho_o S_o D^3} \tag{4}$$

Equilibrium solutions then can be found by setting the left hand sides of these equations equal to zero. Pneuman started from the following equilibrium values: $B_o = 5G$, $B_{2o} = 7G$, $n_o = 10^9$ cm^{-3}, $S_o = 1.2 R_\odot$ and $D_o = 0.24 R_\odot$. The assumed initial density seems rather high. At 2 $R_\odot$ this expanding loop would have a density of $\sim 2\times10^8$ cm^{-3}, whereas estimates of white light loops at 2 $R_\odot$ give $\sim 2\times10^7$ cm^{-3}. But if one lowers the initial density one can still obtain similar models by simply scaling the magnetic field correspondingly.

The eruption of a prominence then was simulated by a rapid increase of the supporting field (e.g. from 7 to 8 G in the model presented). The numerical calculations show a rapid acceleration of the transient and an almost constant velocity at large distances; between 2 and 6 $R_\odot$ the velocity increases from 500 to 750 km/s for a loop and from 300 to 400 km/s for an arcade. The thickness D shows some oscillations in the beginning but then approaches a linear increase with S. If one switches on the driving field more gradually then these oscillations are reduced.

The model is intended to describe a transient as being the response of the corona to an erupting prominence, but the assumption that initially the field underlying the loop is increased everywhere by a constant amount and that the resulting flux through this area is then conserved would better describe a situation where transients are driven by emerging photospheric flux.

An extension of this model by Pneuman and Anzer in which both dynamics of the prominence and driving by magnetic flux added through reconnection are considered is in progress. G. Pneuman will report on some aspects of this model during this conference.

2.2 Continuum models

Dryer, Han, Nakagawa, Steinolfson, Tandberg-Hanssen, Wilson and Wu have published a series of papers (e.g. Nakagawa et al. 1975, Steinolfson and Nakagawa 1976, Wu et al. 1978, Steinolfson et al. 1978, Dryer et al. 1979) in which they try to explain coronal transients as the response of the corona to a rapid pressure pulse at its base. Since their earlier models were purely gas dynamical and since there is strong observational evidence that transients are magnetically controlled, we can leave these models out of our discussion and concentrate here on their magnetohydro-dynamical models.

The papers by Wu et al. (1978) and Steinolfson et al. (1978) should be considered as preliminary studies. In both of them the numerical computations are carried only to $R < 2$ $R_\odot$ whereas reliable data for white light transients were obtained for $R \gtrsim 2$ $R_\odot$. Since these two papers are very similar we shall discuss them together. Wu et al. study the evolution of structures which lie in the solar equatorial plane (an assumption which does not apply to most white light transients). They start from initial configurations where the corona is in isothermal hydrostatic equilibrium with $T_0 = 1.5 \times 10^6 K$ and $n_0 = 2.7 \times 10^8$ cm^{-3} at the base. They take two types of potential magnetic field configurations: a) open fields and b) closed fields. The notion "open" here only means that no field line closes within the region where the numerical computations were performed, but the field lines do not go to infinity. It should be also mentioned that in this case no line of polarity reversal ("neutral line") occurs. Steinolfson et al. investigate structures which lie in meridional planes. Their initial corona is described by a polytrope with $\gamma = 1.2$; the values at the base are $T_0 = 1.5 \times 10^6 K$ and $n_0 = 3 \times 10^8$ cm^{-3}. Again both open and closed potential magnetic field structures are considered (fig. 8). The values of the magnetic field

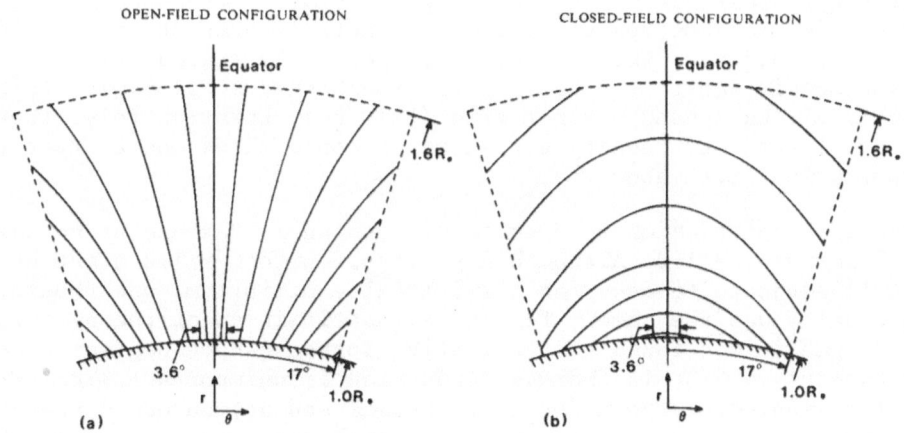

FIG. 8.—Schematic diagrams of open and closed magnetic field configurations in the meridional plane. The solar event is assumed to occur near the equator.

are chosen such that, at some reference point at the base either
$\beta = 1$ or $\beta = 0.1$ holds ($\beta = p_{gas}/p_{magn}$). All these models are very
special. The Lorentz force ($j \times B$) is set equal to zero and therefore
the corona must be in hydrostatic equilibrium. However, if one allows
for variations of the pressure (or density) at the base then one should
find deformations of the magnetic field such that they can balance the
pressure gradients.

In the models by Wu et al. the transient is initiated by a sudden
temperature increase from T_0 to $1.5 \times 10^7 K$ which lasts for 5s occuring
at the base of the corona in a region of 40 000 km x 170 000 km. This
produces a strong pressure pulse. Steinolfson et al. generate a similar
pressure pulse by increasing both temperature and density ($T = 4.2 \times T_0$
and $n = 1.2 \times n_0$), but they assume that the pulse lasts much longer
(5 min), and their area is only 40 000 km x 40 000 km. The perturbations of the solar atmosphere resulting from these pulses then are calculated numerically. An adiabatic index of $\gamma = 1.2$ is used for all
models. In the case of open magnetic fields Wu et al. find that the
contact surface between hot plasma and normal coronal plasma moves with
velocities ~ 150 km/s whereas the MHD shock ahead of this piston has a
velocity of 400 km/s. In the models by Steinolfson et al. the velocities of the leading edge (contact surface) are 290 km/s ($\beta = 1$) and
470 km/s ($\beta = 0.1$) for open configurations. No values for the shock
velocity are given.

The authors also find that, for closed configurations the hot
driving material cannot leave the solar atmosphere (a maximum height of
36 000 km is reached for $\beta = 1$), the closed magnetic field preventing
the plasma from moving out into interplanetary space. Therefore these
models with closed fields cannot produce transients. Wu et al. state
that their numerical calculations show the occurrance of shocks, but
the curves presented show no indication of discontinuities in density
or velocity. Gradients of the same magnitude as at the "shocks" occur
elsewhere in these curves as well. From their figures it is not clear
how well-defined the "shocks" actually are. It is stated that the
smearing out is due to the coarse grid (mesh size $\approx 2 \times 10^4 km$). Steinolfson et al. have used a finer grid ($7 \times 10^3 km$). Unfortunately, they
have not plotted any density curves which would allow one to test the
interpretation given above.

Another point which is open to discussion is the use of an adiabatic index of $\gamma = 1.2$. Although a polytrope with $\gamma = 1.2$ might be
perfectly adequate to describe the initial equilibrium, one should not
automatically use the same γ for the adiabatic index of the perturbations. Since it was found that radiative losses can be neglected one
would expect $\gamma = 5/3$, if thermal conduction is unimportant. The authors
argue that including the effects of thermal conduction would reduce the
value of γ. That this is not true in general can be seen from a consideration of the hot material which is ejected. Thermal conduction would
tend to cool this material. Since this gas expands, it will also exper-
ience adiabatic cooling which is larger for $\gamma = 5/3$ than for $\gamma = 1.2$,

Therefore, in this case taking $\gamma < 5/3$ is a step in the wrong direction, at least for the simulations of the hot ejected gas. Steinolfson et al. concluded from their test calculation ($\gamma = 5/3$) that changes in γ cause only small effects. This interpretation should be taken with some caution because the calculations only cover a period of 10 min, twice the duration of the heating pulse. These early phases of the coronal perturbations will be mainly determined by the driving pulse whereas, during later phases the effects due to adiabatic expansion will develop. How big these effects are should be determined by numerical calculations which cover a sufficiently long period.

In the paper by Dryer et al. (1979), an attempt was made to compare extended calculations (up to 10 $R_\odot$) with observations obtained for the transient of 21 Aug., 1973. This transient was associated with a flare for which temperature and density curves were obtained from X-ray observations (e.g. $T_{max} = 1.4 \times 10^7$K and $n_{max} = 10^9$ cm^{-3}). The equilibrium values for the model were chosen as $T_o = 2 \times 10^6$K and $n_o = 5 \times 10^7$ cm^{-3}, γ was set to 1.08. An open field configuration was taken, and the cases $\beta = 0.1$ and $\beta = 1$ were considered. The models were rotationally symmetric. Fig. 9 shows the position of the shock front and the location of the strongest density increases. Note that if one takes the size of the occulting disk (1.6 $R_\odot$) into account, then the similarity between this pricture and white light photographs becomes less pronounced.

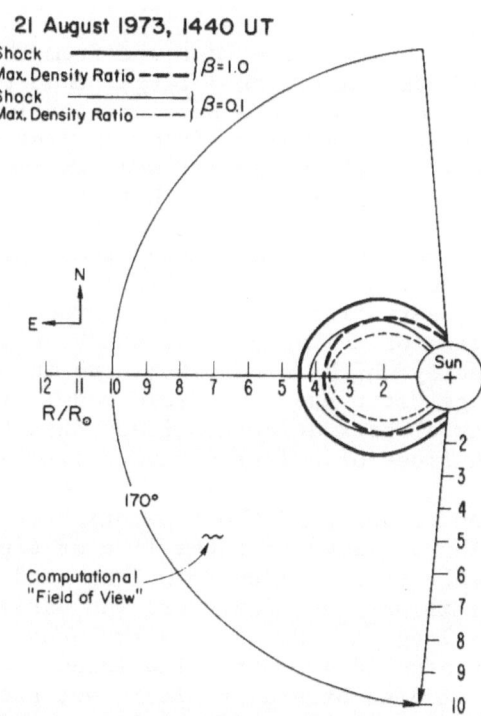

21 August 1973, 1440 UT

Shock ——— $\beta = 1.0$
Max. Density Ratio ———

Shock ——— $\beta = 0.1$
Max. Density Ratio ———

N

E

12 11 10 9 8 7 6 5 4 3 2   Sun +
R/R$_\odot$

2
3
170°   4
5
6
Computational   7
"Field of View"   8
9
10

FIG. 9.—Simulated shock and maximum-density ratio positions at $t = 1440$ UT for $\beta = 1.0$ and $\beta = 0.1$.

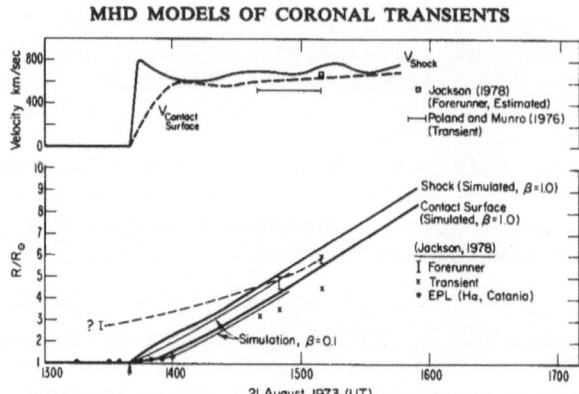

MHD MODELS OF CORONAL TRANSIENTS

Fig. 10.—Comparison (along the axis of symmetry) of the simulated shock and contact surface trajectories and velocities with the observed Hα eruptive prominence and coronal white-light transient and forerunner. The arrow at 1340 UT marks the moment of the simulated flare initiation.

Fig. 10 shows the motion of the transient. The calculated contours of excess density have a loop-like structure, but for β = 1, most of the mass remains concentrated in the legs of the loop. Only for β = 0.1 is a sizeable fraction of the total mass in the top of the loop moving outward. It seems, therefore, that low β models would describe the evolution of transients better.

The total mass of the transient derived from this model amounts to 2.3×10^{16}g which is at least a factor 3 larger than the observed mass. This large excess mass could cause a problem. The initial configuration considered here had very low density (5×10^7 cm^{-3} at the base). Taking more realistic densities would raise all mass estimates in the model and thus might lead to contradictions with the observations.

The main conclusion one must draw from these papers is that transients can only occur in regions of open fields. This assumption should be checked with the observations. If it is correct one should also include the motion of the solar wind which exists along open field lines and has a velocity which is comparable to that of transients. A point which speaks against these open field models is the lack of a "neutral line", because observations indicate that transients are associated with such lines of polarity reversal.

Another question is how good the assumption of azimuthal symmetry actually is. The authors assume that the line of sight depth of the transient is comparable to its width (~ 0.2 R$_\odot$ at 2 R$_\odot$). In this case the azimuthal extent is only $6°$. Therefore the variations in the ϕ-direction will be as large as the ones in the other directions - in contradiction to the assumed symmetry. The effects of different values of γ should also be studied carefully. It is not sufficient to compare the very early phases (t < 10 min) of models with different γ and then extrapolate to t $\approx$ several hours.

Finally, the interpretation of the observed fore-runners as being the calculated shocks seems to be a misinterpretation. These fore-runners rise very gradually out of the coronal background – it is even hard to define the outer edge of the fore-runners. On the other hand, shock fronts are associated with sharp density discontinuities which should also be observable.

3. CONCLUSION

Our understanding of coronal transients is still very fragmentary. The most serious limitation at present seems to me to be the fact that white light observations are confined to regions which are more than $1.6\,R_\odot$ from the sun's center. As a consequence of this, the cause of these transients is still an open question. The two candidates which have been discussed so far are flares and erupting prominences. One difficulty for the interpretation is that transients can be observed only if they lie close to the plane of the sky whereas many surface events are predominantly studied inside the solar disk. This could lead to a bias if one tried to find statistical correlations. It would favour the association of transients with limb phenomena. As far as the investigations of structures go one has to realize that the transients are optically thin and therefore that the observed intensities are always integrated over the whole line of sight. One can obtain only indirect estimates of the thickness of transients in the line of sight. Another important requirement is a knowledge of the coronal magnetic fields both prior to transients and during their occurance. Most theoretical models are based on magnetic fields, but little is known observationally.

The main difficulty which the theoretical models face is that the transients are 3-dimensional structures. 2-dimensional MHD-models may provide some insight into the basic mechanisms which drive transients, but it is not clear how close to reality these models actually are. On the other hand 3-dimensional MHD calculations are beyond the scope of any present investigation. For this reason modelling attempts based on discrete structures seem more promising. But since such models will be very idealized, it will be necessary to discuss thoroughly all omissions and their possible consequences.

ACKNOWLEDGEMENT

It is a pleasure to thank G. Pneuman for many clarifying discussions.

REFERENCES

Anzer,U.: 1978, Solar Phys. 57, pp 111-118
Dryer, M., Wu, S.T., Steinolfson, R.S., Wilson, R.M.: 1979, Astrophys. J. 227, pp. 10059-1071

Dulk, G.A., McLean, D.J.: 1978, Solar Phys. 57, pp. 279-295
Dulk, G.A., Smerd, S.F., MacQueen, R.M., Gosling, J.T., Magun A.,
    Stewart, R.T., Sheridan, K.V., Robinson, R.D., Jaques, S.: 1976,
    Solar Phys. 49, pp. 369-394
Gergely, T., Kundu, M., Poland, A.I., Munro, R.H.: 1979, Astrophys. J.
    (submitted)
Gosling, J.T., Hildner, E., MacQueen, R.M., Munro, R.H., Poland, A.I.,
    Ross, C.L.: 1974, J. Geophys. Res. 79, pp. 4581-4587
Hildner, E.: 1977, L.D. de Feiter Memorial STIP Syposium (Tel Aviv)
Mouschovias, T.Ch., Poland, A.I.: 1978, Astrophys. J. 220, pp. 675-682
Nakagawa, Y., Wu, S.T., Tandberg-Hanssen, E.: 1975, Solar Phys. 41,
    pp. 387-396
Pneuman, G.W.: 1978, IAU Colloquium No. 44, Physics of Solar Prominences
    (E. Jensen, P. Maltby, F.Q. Orrall, Eds.), pp. 281-301
Pneuman, G.W.: 1979, Preprint MPI-PAE/Astro 170, to appear in Solar Phys.
Rust, D.M., Hildner, E.: 1976, Solar Phys. 48, pp. 381-387
Rust, D.M., Webb, D.F.: 1977, Solar Phys. 54, pp. 403-417
Steinolfson, R.S., Nakagawa, Y.: 1976, Astrophys. J., 207, pp. 300-307
Steinolfson, R.S., Wu, S.T., Dryer, M., Tandberg-Hanssen, E.: 1978,
    Astrophys. J. 225, pp. 259-274
Van Tend, W.: 1979, Solar Phys. 61, pp. 89-93
Webb, D.F., Krieger, A.S., Rust, D.M.: 1976, Solar Phys. 48, pp. 159-186
Wu, S. T., Dryer, M., Nakagawa, Y., and Han, S. M.: 1978, Astrophys. J.
    219, pp. 324-335.

DISCUSSION

Petelski: Regarding the acceleration of particles by the magnetic field of a ring current, is that to be understood in analogy to the mirror effect in an inhomogeneous magnetic field whereby transverse kinetic energy is transformed into longitudinal energy?
Anzer: The model is purely MHD, with no calculation of individual particle motion.

Low: I concur with your criticism of models in which the flux tube has no external fields. The point is best illustrated by theoretical models of prominence where one finds that the Lorentz force acting on the prominence supporting its weight is due to the interaction between the internal field and the external (usually uniform) field. In the absence of external fields, there can be no net Lorentz force on the coronal transient flux tube.
Anzer: I agree.

Dryer: Your constructive comments regarding our model are appreciated. I would like to reply to several of them at this time: (i) Concerning the ejected mass estimate of Dryer et al. (for the 21 August 1973 event) of 2.3 x 10^{16} g... as compared to an observed mass ejecta which was less than a factor of 1/3 this number... please keep in mind that the simulated figure is based on a guess regarding the transient's thickness perpendicular to the planar plane of computation. Also, the observed esimate is made in a field of view above the $2R_\odot$ occulting disk. Hence variations by factors of 3 or so are not surprising. (ii) We have long recognized the eventual need for 3-D studies and are progressing toward that goal. (iii) Concerning our suggestion for identifying the leading edge of the white light forerunner with the shock, it is important to recall that the small density change indicated by the former is consistent with density jumps across weak shocks.

Anzer: (i) Your mass estimates were based on the assumption that the perpendicular thickness is the same as the width of the loop structure. I would be very surprised if these structures were much narrower in the perpendicular direction than in the radial direction. (iii) These shocks, then, must be very weak.

McLean: I would like to suggest that it is not possible on the basis of white-light observations to exclude the possibility that forerunners are shock fronts. If we envisage a spherical shock surface with a gradual density rise behind the discontinuity, then the two-dimensional density distribution, integrated along the line of sight, would be similar to that observed.

Anzer: If your interpretation is correct then forerunners are basically gradual density increases in front of a transient which may or may not have a shock in front of them.

Newkirk: It appears that neither of the theoretical interpretations of the mechanism driving coronal transients is satisfactory. With no recognizable distinction in morphology or other property, these events range in velocity over an order of magnitude and in kinetic energy over two orders of magnitude. This would suggest that a single mechanism provides the driving energy for all these events. Yet the impulsive mechanism proposed by Nakagawa *et al.*, has difficulty in accounting for the slow events, and the MHD models require that magnetic energy be continuously added as long as the transient continues to rise.

I have no specific model to propose; but it would appear that we are dealing with a response of the corona to a readjustment of the magnetic field in which photospheric motions have stored energy in the coronal field which then relaxes.

Anzer: The MHD models I discussed do not require a continuous energy input. They all are based on an initial increase in magnetic flux and evolutions which then conserve the flux. All these models also use some kind of instability for driving either flares or erupting prominences.

FLARE MODEL WITH FORCE-FREE FIELDS AND HELICAL SYMMETRY

Dirk K. CALLEBAUT
Physics Department, U.I.A., Universiteit Antwerpen.
Universiteitsplein 1. B-2610 Wilrijk-Antwerpen. Belgium.

ABSTRACT.

Physical arguments are given indicating that solar flare magnetic energy storage may happen through force-free fields with helical symmetry $(\partial_z + \lambda(r)\partial_\varphi = 0)$. The mathematical results turn out simple for helical fields whether general, in equilibrium or force-free. A preliminary stability analysis points to appropriate properties.

1. PHYSICAL ARGUMENTS.

1.1 Force-free magnetic fields.

The gas density and pressure are fairly low above the solar surface. Hence the great energy storage which becomes apparent in solar flares points strongly to force-free magnetic fields (curl $\bar{H}=\alpha\bar{H}, \alpha(\bar{r},t)$ being the current-to-field ratio and $\bar{H}.\nabla\alpha=0$) or at least to fields that are very nearly force-free. (Callebaut, 1976) The author has often advocated that the difference between force-free and nearly force-free may be relevant since a minor difference may have a strong influence on the stability and the energy release and even on the structure and the evolution. Nevertheless, here the restriction is made to pure force-free fields. However α is not restricted to be constant and a nice example will even turn up in which the energy storage corresponds to non-constant α.

1.2 Helical symmetry.

If one considers a cylindrical tube, with axis parallel to the solar surface (considered as flat) and partially or wholly above it, one is at first inclined to look for cylindrical symmetry $(\partial_z=0)$ or even symmetry of revolution $\partial_\Theta=0$. Much less restrictive is a combination of both $(\lambda\partial_\varphi+\partial_z=0$, λ will be taken to be a constant or a function of r (the distance to the axis) only), which means that no quantity varies

M. Dryer and E. Tandberg-Hanssen (eds.), Solar and Interplanetary Dynamics, 279-282.
Copyright © 1980 by the IAU.

along certain helices lying on circular cylinders. It seems plausible
indeed to have some kind of symmetry on general expectation grounds.
Moreover when the field emerges from the solar surface (Cfr. the
Kuperus and the Kuperus-van Tend models, 1978) it has to adapt itself
to become a force-free field. This adaptation process may impress the
same pitch on the field. Obviously there are some restrictions on this
helical symmetry: (a) the situation during the emerging phase changes
a bit; (b) the small pressure above the solar surface decays with
height; (c) the tube is not infinitely long but finite and curved into
the solar surface.

The idea is that a field may emerge from the solar surface and
with the axis parallel to it and lifting up the potential or force-
free field of the solar atmosphere. How much of the cylinder is above
the surface is not specified here, but a probable choice is to consider
half a cylinder. Furthermore of this half only a shell with thickness
say 1/3 of the radius is pervaded by the force-free field with helical
symmetry. The inner part may again contain a potential field. This
arcade configuration may be related to the two ribbon flare, the axis
and the "feet" being parallel to the ribbons.

2. MATHEMATICAL RESULTS FOR FIELDS WITH HELICAL SYMMETRY.

The helical symmetry reduces the situation to a 2-dimensional one
in which the variables are r and $\xi = \Theta - \lambda(r)z$.

2.1 General magnetic fields (Callebaut and Raadu, 1976)

2.1.1. λ= constant. Then there is a streamfunction so that:
$$rH_r = \partial_\xi \Psi \qquad \text{and} \qquad H_\Theta = \lambda r H_z - \partial_r \Psi$$

2.1.2. $\lambda(r) \neq$ constant. Then the solution is
$$H_r = C/r \qquad \text{and} \qquad H_\Theta = \lambda r H_z + A(r)$$

with C a constant (usually zero, see 3.1), H_r an arbitrary function of
r and ξ and A(r) an arbitrary function of r.
Pressure balanced fields were studied by Callebaut (1979)

2.2 Force-free fields.

2.2.1. λ = constant. A. $\alpha \neq$ cst, $\Psi \neq$ cst. Then one can show that
$r\lambda H_\Theta + H_z = f(\Psi)$ with $\Psi = \Psi(\alpha)$ and $d(r\lambda H_\Theta + H_z)/d\Psi = \alpha$. Spicer (1976) also
studied this case. In the very probable case (see below) that $rH_r =$
$\partial_\xi \Psi = 0$ one obtains in fact a one dimensional problem: $\alpha(r), H_\Theta(r), H_z(r)$.
B. $\Psi =$ cst. Then one obtains the force-free field of constant pitch
studied by Piet van der Laan (1968) for the pressureless region surrounding the plasma in the screw pinch in Jutphaas (Utrecht, the
Netherlands):
$$H_r = 0, H_z = C(\lambda^2 r^2 + 1)^{-1} , \quad H_\Theta = \lambda r C(\lambda^2 r^2 + 1)^{-1} , \quad \alpha = 2\lambda(\lambda^2 r^2 + 1)^{-1}.$$
C. α =cst. This case is well-known.

2.2.2. $\lambda \neq$ constant. For $H_r = 0$ one obtains again that α, H_Θ and H_z are functions of r only. E.g. if α is constant then only the Lundqvist field (or Bessel function field) $\bar{H} = H_0(0, J_1(\alpha r), J_0(\alpha r))$ is possible.

3. FURTHER CONSIDERATIONS.

3.1 Boundary conditions.

$H_r = 0$ can not always be inferred from the singularity at the axis, because the axis may be excluded, e.g. when only a cylindrical shell (arch) is considered or when the axis is still under the solar surface. However, if the external region is pervaded by a potential field it can be shown that H_r has to vanish at the boundary. Then H_r vanishes everywhere if $\lambda(r) \neq$ cst.

3.2 Stability.

The linear stability of these fields is not yet fully analyzed. The van der Laan field showed theoretically and experimentally a fair stability without being stable under all circumstances. However in the plasma experiment (torus) there was the stabilizing influence from the wall and the finite length. In the flare storage one may expect some stabilization from the anchoring in the solar surface. It may be expected that the field is quite a time stable, and then, by having evolved further to a less stable configuration or by some strong trigger, becomes unstable, releases energy and becomes after flaring a potential field or a force-free field. In this connection it has to be stressed that, for constant α, the lowest α compatible with the geometry is stable for fixed boundaries and also sometimes for non-fixed boundaries.

REFERENCES.

Callebaut, D.K.: 1976, CECAM Report of Workshop on Plasma Physics Applied to Active Phenomena on the Sun (I) Aug. 1-Sept. 30, pp. 8-11.
Callebaut, D.K. and Raadu, M.A.: 1976, CECAM Report, ibid., pp. 12-18.
Callebaut, D.K.: 1979, CECAM Report, ibid. (II), June 1-30, (to appear).
Spicer, D.S.: NRL Report 8036 (1976).
van der Laan, P.C.T.: 1968, Proc. II ème Coll. Intern. Interactions Champs Oscillants et les Plasmas, Vol. 4, p. 1095, Saclay, France.
van Tend, W. and Kuperus, M.: 1978, Solar Physics., 59, pp. 115-127.

DISCUSSION

Low: Gene Parker has pointed out from physical considerations that a magnetic field in static equilibrium must possess "suitable" invariance in its field pattern. I have recently derived an expansion to express these required invariances in terms of the Euler potentials defining the magnetic field. (Paper to appear in *Solar Phys.*) In principle, then, one can classify equilibrium fields according to their types of invariances. The problem is nonlinear and very difficult. Your interesting example of vertical invariance seems like an ideal class of equilibrium field to start classifying the field.

Callebaut: One of the aims of presenting this paper was precisely to bring to the attention: (a) some intuitive feeling for some underlying symmetry or invariance; (b) the possibility of extensively using helical symmetry with variable pitch: it is a wide class of fields and yet the solutions can be handled with care.

It is very pleasing to hear that you have already worked on the first feature and that you may be able to use the second one. I am very interested in this work.

Kuperus: Soloviev proved that a non-constant α force-free-field relaxes to a constant α force-free-field thus releasing the excess energy. The condition is that the Alfvén crossing time is much smaller than any photospheric perturbation time. How does this relate to your analysis?

Callebaut: That non-constant α force-free fields are unstable even when confined in rigid boundaries was already published in a book of Belgium Academy of Sciences by one of my students (T. Krüger) in 1967 or 1968. See also Kruger in *Journal of Plasma Physics* (1976).

This (weak) instability is precisely a desirable feature in my view. E.g., the van der Laan field is "fairly" stable, i.e., it has some weak instability for some perturbations which is useful to make the flare. This is again related to "nearly force-free" fields. I suppose that the original field has some small pressure and is stable in the beginning. Then, it evolves slowly to the van der Laan field (e.g.) by losing some matter, by cooling, by raising, by resistive evolution, etc., and thus becomes unstable. The MHD instability has (at least in the beginning) to be fairly weak, because the low pressure field and the force-free field (or a very low pressure field) may not be far from each other in these considerations. I am presently elaborating this process in detail.

THE FALSE EQUILIBRIUM OF A FORCE-FREE MAGNETIC FIELD[*]

B. C. LOW
Lau Kuei Huat (S) Pte. Ltd.,
55 Shipyard Road, Singapore 22
Republic of Singapore

ABSTRACT

It has been a customary assumption that any force-free magnetic field represents an equilibrium field in the solar atmosphere under the extreme condition $8\pi p/B^2 \ll 1$. An example of a force-free magnetic field is presented for which this assumption fails in the sense that no equilibrium is possible for the magnetic field if imposed with an arbitrary ambient pressure, however weak the pressure is. A simple mechanism is proposed for the onset of eruption in the course of otherwise quasi-static evolution of magnetic fields in the solar atmosphere.

1. INTRODUCTION

I show an interesting property of nonlinear force-free magnetic fields. I will first describe the property and then go on to point out its physical implication for the onset of eruptive processes in the solar atmosphere. The mathematical details are reported elsewhere (Low, 1979) and only the results are quoted here.

Consider the force-free equation for a magnetic field $\underline{B}$,

$$(\underline{\nabla} \times \underline{B}) \times \underline{B} = 0. \tag{1}$$

It has been customary to assume that any solution of this equation represents a magnetic field in static equilibrium under the extreme condition of low plasma beta. Let us construct an example below for which this seemingly reasonable assumption fails and find out what it physically means.

[*]This work was supported in part by the National Aeronautics and Space Administration under grant NGL 14-001-001 at the University of Chicago.

M. Dryer and E. Tandberg-Hanssen (eds.), Solar and Interplanetary Dynamics, 283-289.

2.   SYSTEM INVARIANT IN A GIVEN DIRECTION

For a magnetic field depending on only two Cartesian coordinates y and z, we may express it in the form,

$$\underline{B} = B_o \left(H, \frac{\partial A}{\partial z}, - \frac{\partial A}{\partial y} \right). \tag{2}$$

where B_o is a constant field strength and H and A are scalar functions. If $\underline{B}$ is force-free, equation (1) requires that H be a strict function of A,

$$H(y,z) = H\left[A(y,z)\right], \tag{3}$$

while A satisfies,

$$\nabla^2 A + H(A) \frac{d}{dA} H(A) = 0. \tag{4}$$

In the presence of a gas pressure p, the equation for magnetostatic equilibrium is,

$$\frac{1}{4\pi} (\underline{\nabla} \times \underline{B}) \times \underline{B} - \underline{\nabla}p = 0. \tag{5}$$

With dependence on only y and z, this equation requires that p is also a strict function of A,

$$p(y,z) = p_o P\left[A(y,z)\right], \tag{6}$$

where p_o is a constant pressure, while A satisfies,

$$\nabla^2 A + H(A) \frac{d}{dA} H(A) + \beta \frac{d}{dA} P(A) = 0, \tag{7}$$

where the plasma beta is defined to be,

$$\beta = 4\pi p_o / B_o^2 . \tag{8}$$

The question we are addressing here is the following. In the limit of $\beta \to 0$, a solution of equation (7) reduces to a solution of equation (4). In this sense, a force-free field is an approximation of the magnetostatic field in the limit of $\beta \ll 1$. The magnetostatic field differs from the corresponding force-free field by first order amounts needed to accommodate the weak pressure. However, there are force-free fields which have no neighboring magnetostatic equilibria to enable the field to accommodate arbitrarily imposed pressures, no matter how small

these pressures are. In other words, any one of these force-free
fields is not the limit of any solution of equation (7) as β goes
to zero. Since realistic pressures are never zero, these types of
pathological force-free fields do not represent static fields. There
is no equilibrium and force imbalance must develop dynamical motion
rapidly in consequence of the large Lorentz force the magnetic field
can produce in the low beta environment.

3.   THE EXAMPLE

To construct an example of the pathological force-free fields, let
us take the plane $z = 0$ to be the photosphere and confine our interest
to the region $z \geq 0$. Consider the force-free field,

$$\underline{B} = B_o \left[y^2 + (z + z_o)^2 + a^2 \right]^{-1} (a,\ z{+}z_o,\ -y) \tag{9}$$

where z_o and a are positive constants. This field extending over the
entire y–z plane was first considered by Gold and Hoyle (1960) and I
have also used it as a model for evolving bipolar solar magnetic
fields (Low, 1977a, b; 1978; 1979). It is a helical field invariant
in the x direction such that its field line projections on the y–z
plane are concentric circles centered at $y = 0$, $z = -z_o$. In the domain
$z \geq 0$, this field appears as bipolar arches with each pair of footpoints
rooted in the photosphere $z = 0$. It is easy to show that $\underline{B}$ is a
solution of equation (4) with $A = A_o$ and $H = H_o$ where,

$$A_o = \frac{1}{2} \log \left[y^2 + (z + z_o)^2 + a^2 \right], \tag{10}$$

$$H_o (A_o) = a \exp (-A_o). \tag{11}$$

For a fixed value of z_o, suppose $\underline{B}$ given in equation (9) is the
lowest order solution of equation (7) in the limit of $\beta \ll 1$. Then,
up to first order in β, we may expand,

$$A = A_o + \beta\, \delta A, \tag{12}$$

$$H = H_o + \beta\, \delta H. \tag{13}$$

Substitute the expansion into equation (7) to get the first order
equation,

$$\nabla^2 \delta A + \delta A \frac{d}{dA_o} \left[H_o (A_o) \frac{d}{dA_o} H_o (A_o) \right] + \frac{d}{dA_o} \left[H_o (A_o) \delta H(A_o) + P (A_o) \right] = 0. \tag{14}$$

If we change variables from y and z to A_o and s where s is the distance along a curve of constant A_o, the connectivity of the field lines requires that δH is related to δA by,

$$\delta H(A_o) = \int_{A_o = constant} \Lambda(A_o, S)\, \delta A(A_o, s)ds. \tag{15}$$

The curves of constant A_o are the force-free field lines projected on the y–z plane. In equation (15), the integration is to be performed along a field line projection from one footpoint to the other while Λ is a known function expressible in terms of A_o. Equation (14) then becomes,

$$\nabla^2 \delta A + \delta A \frac{d}{dA_o} \left[H_o(A_o) \frac{d}{dA_o} H_o(A_o) \right]$$

$$+ \int_{A_o = constant} \frac{\partial}{\partial A_o} \left[H_o(A_o) \Lambda(A_o, s) \right] \delta A(A_o, s)ds$$

$$= - \frac{d}{dA_o} P(A_o). \tag{16}$$

Field line connectivity also demands that,

$$\delta A = 0 \quad on \quad z = 0. \tag{17}$$

Finally, we expect the pressure not to affect regions far above the photosphere so that,

$$\lim_{(y^2 + z^2)^{\frac{1}{2}} \to \infty} \delta A = 0. \tag{18}$$

Equations (16)–(18) pose a boundary value problem for δA to calculate the first order correction to the force-free field to accommodate the weak pressure P appearing on the right side of equation (16) as a source term. Mathematical analysis of Low (1979) shows that for any choice of P, a unique solution δA always exists if $z_o > 0$. In this case the force-free field is physical in the sense that whatever ambient pressure which is present can be accounted for with a slight local departure of the field from being exactly force-free. This is intuitively what we would expect of most force-free fields. The slight departure from being exactly force-free goes to zero smoothly as the pressure is reduced to zero. On the other hand, the boundary value problem for δA does not have a solution for arbitrarily specified P when $z_o = 0$. The result is independent of the smallness of β. The force-free field in this case

does not represent true equilibrium since the pressure in a realistic
tenuous medium is never zero.

4. CONCLUSION

Consider the evolution of solar magnetic arches given in the
lowest order by the force-free field of equation (9) with z_o decreasing
with time from a finite value to zero. The evolution may be thought
of as the result of a horizontal helical field rising through the
photosphere, z=o or the motion of the photospheric footpoints in the
model of Low (1977a,b; 1978). At each point of the evolution, so long
as $z_o \neq 0$, the ambient pressure of the solar atmosphere has no dynamical
effect apart from contributing to first order correction to the force-
free field (9). When z_o attains zero, the presence of the ambient
pressure becomes crucial and in general, we no longer have equilibrium.
The state of non-equilibrium appears in the following form. There is no
way of deforming the field by first order amounts from the pathological
force-free configuration at $z_o = 0$ to balance pressure gradient everywhere. It was found that force imbalance is not reducible to zero in
the upper magnetic arches leading to either upward or downward uniform
motion of these arches. The onset of explosive flares or the so-called
coronal transient may be due to the appearance of a pathological force-
free configuration during the course of quasi-static evolution of
nearly force-free solar magnetic fields.

REFERENCES

Gold, T. and Hoyle, J., 1960, *M.N.R.A.S. 120,* 7.

Low, B. C., 1977a, *Astrophys. J., 212,* 234.

_____, 1977b, *Astrophys. J., 217,* 988.

_____, 1978, *Astrophys. J., 224,* 668.

_____, 1979, *Astrophys. J.* (submitted).

DISCUSSION

Van Hoven: (i) It is not a general result that one cannot find a
nearly force-free <u>equilibrium</u> with pressure gradients; a counter-
example is shown in Van Hoven et al., Ap. J. (1977). (ii) Ultimately,
one must test the stability of any equilibrium to evaluate its reality.
I would expect that such configurations as you describe are unstable.
    Low: (i) I shall have to read your paper to familiarize myself with
it. I would say in general, it is easy to construct nearly force-free

fields with pressure gradients, contrary to your remark. Take any
potential field in the form of arches rooted in an imaginary photospheric plane. It can be shown that putting any small pressure into
it merely requires the field to adjust by first order displacement to
achieve equilibrium. I believe the non-equilibrium I demonstrate is
the first example where such first order adjustment by a force-free
field is not possible. (ii) I emphasize that the configuration $z_o = 0$
is not even in equilibrium in the presence of pressure and it is "worse"
than an unstable equilibrium. There is just no equilibrium and that is
what is so interesting physically.

Tandon: Equation for steady state used by you is

$$-\text{grad } p + \frac{1}{4\pi} (\text{curl } \underline{B}) \times \underline{B} = 0.$$

For $\beta \ll 1$, the force-free solution is not explicitly a valid one.
Virtually, it envisages grad $p \sim 0$. The only valid solution is pressure
balance solution. In the steady state your analysis just approximates
the pressure balance solution with a near force-free solution which may
not always be possible. Failure to fit a near force-free solution even
for very low β with equation (1) merely points out that such a solution
is in no way an approximation to pressure balance solution in a steady
state. A force-free or near-force-free solution from equation (1)
implies that the steady state is not possible until or unless the
pressure term is balanced by another potential term (not included) in
the left-hand side of equation (1). We should not draw any further
conclusions from this analysis.

Low: You have misunderstood my paper. What I said is this. When
given a force-free field, it seems reasonable to assume, as we all
customarily do, that it approximates a magnetic field in static
equilibrium in the limit of $\beta \ll 1$. I have presented a first example
of a force-free field for which I showed rigorously that the above
assumption fails. This example has interesting physical implications
for the evolution of solar magnetic fields as discussed.

Callebaut: It may be that in your demonstration you have assumed too
many requirements. E.g., you have fixed the footpoints, if I understood
you correctly. It is *a priori* not excluded that a second-order motion
of the footpoints allows a first order solution for the problem to
turn up.

Low: Within the physical definition of the problem, the footpoints
must be fixed. I am thinking of a force-free field with the footpoints
rooted in an imaginary photosphere and asking whether the field can
adjust by first order above the photosphere to accommodate an imposed
ambient pressure.

Callebaut: (Comment) I have been a strong supporter of the use of
force-free fields for solar flares for more than a decade. Since several
years I advocate also the importance of nearly force-free fields (i.e.,
fields for which j is nearly parallel to $\overline{B}$) because it may be that some

of their properties (e.g., equilibrium, topology, stability, evolution, ...) are drastically different from those of the pure force-free field. E.g., one can think of a linearly stable force-free field and a pressure balanced field very close to it which is unstable. Your calculation, if correct, is very important because it shows that for some equilibria there is either a discontinuity (implying instability) or a need for a nonlinear treatment.

Kuperus: Dr. Low starts with a force-free field and then shows that adding a little pressure ($\beta \ll 1$) the force-free field can easily get out of *equilibrium*. This would indicate that the original ($\beta = 0$) force-free field is most likely already unstable from the very beginning and possibly not a good initial configuration to start with. The analysis could be summarized as an "anti-force free field" theorem.

Low: Let me mention the following result. If we subject each of the above force-free fields to infinitesimal perturbations which depend only on the same 2 Cartesian coordinates and calculate the energy increment ΔE, we find that ΔE has a non-negative minimum. Pressure is neglected here, so each of the force-free fields is stable within this framework. In fact, the critical force-free field with $z_o = 0$ is

marginally stable in that min $\Delta E = 0$.

ENERGY STORAGE AND INSTABILITY IN MAGNETIC FLUX TUBES

Takashi Sakurai
Department of Astronomy, University of Tokyo

1.   INTRODUCTION

Now it is known that the solar corona consists of many loops which are believed to represent the structure of the magnetic field. Since the plasma is very tenuous in the corona, the equilibrium of the magnetic field is approximated by the force-free field:

$$\text{rot } \underset{\sim}{B} \times \underset{\sim}{B} = 0 \quad . \tag{1}$$

In this paper we will propose a method of solution for equation (1), and will discuss on the energy build up and instability in the magnetic flux tubes.

2.   METHOD OF SOLUTION FOR THE FORCE-FREE FIELD

Our method (Sakurai 1979) is based on a variational principle for the force-free field. We represent the magnetic field $\underset{\sim}{B}$ in terms of Euler potentials u and v as

$$\underset{\sim}{B} = \underset{\sim}{\nabla}u \times \underset{\sim}{\nabla}v \quad . \tag{2}$$

Then the solution of the variational problem

$$W \equiv \int \frac{B^2}{8\pi} \, dV = \text{extremum} \quad , \tag{3a}$$

$$u \, , \, v = \text{given on the boundary} \tag{3b}$$

turns out to be the force-free field. Due to the boundary condition (3b), not only the distribution of magnetic flux but also the position of the footpoint of the field line is prescribed because every field line is labelled by a particular set of values of u and v.

This variational problem is solved directly by minimizing the

M. Dryer and E. Tandberg-Hanssen (eds.), Solar and Interplanetary Dynamics, 291-294.

magnetic energy W . The finite element method (Strang and Fix 1973) is
used to evaluate the integral (3a), and the minimization of the energy
is performed by the variable metric method (Kowalik and Osborne 1968).
Figure 1 shows the current free field of a pair of sunspots and the
force-free field when the footpoints are rotated 180°. As the tube is
twisted, the energy in the tube enhances as is shown in Figure 2.

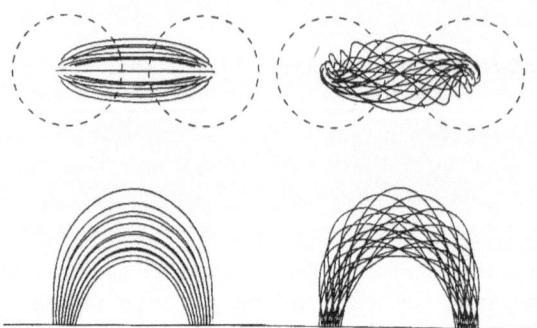

Figure 1 (left)
The current free field due to a
pair of sunspots (left) and the
force-free field when footpoints
are rotated 180° (right): top
view (above) and side view
(below).

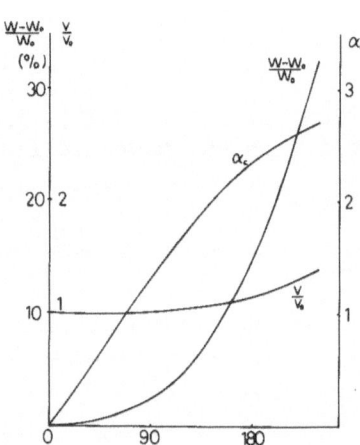

Figure 2 (right)
The change in energy W in the tube of
volume V , versus the angle of rotation
for the case shown in Figure 1. Subscripts 0 refer to the initial value.
α_c is the value of α on the field line
at the center of the tube. The unit of
α_c is the inverse of the radius of the
model sunspots.

A usual treatment for the force-free field is to write equation (1)
as

$$\text{rot } \underset{\sim}{B} = \alpha \underset{\sim}{B} \quad , \quad \underset{\sim}{B} \cdot \underset{\sim}{\nabla}\alpha = 0 \qquad (4)$$

and to assume that the quantity α does not vary in space (constant-α
force-free field, e.g. Nakagawa and Raadu 1972). On the contrary, α
is determined consistently in our method due to the boundary condition
(3b), and generally the quantity α is not a constant.

3.   INSTABILITY OF THE MAGNETIC FLUX TUBES

When the flux tube is twisted considerably, it may become unstable
to magnetohydrodynamical instabilities. Resulting motion of the tube

can be studied by a similar finite element technique applied to the
Lagrangian (Sakurai 1976). The time development of a kink-type instability is calculated in the case of straight flux tube for various pitch
of the helical field lines. Figure 3 shows the case when $P/R = 1/2$,
where R is the radius of the tube and $2\pi P$ is the pitch of the field
lines. It can be seen that the tube develops into a twisted loop just
like an over-twisted rubber string. For smaller values of P/R, the
initial deformation grows without much twisting, while for larger values
of P/R the instability does not grow markedly and the tube evolves into
a low arch. These kinds of motion are actually observed in eruptive
prominences.

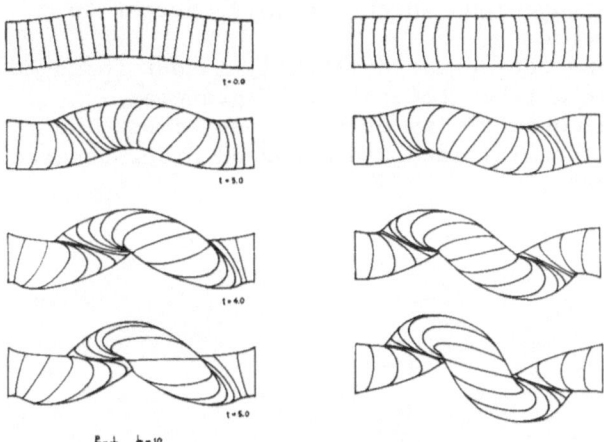

Figure 3
The motion of the twisted
flux tube due to a kink
instability: top-view
(right) and side-view
(left). The tube is divided
into small cylinders in
order to apply the finite
element method. The unit
of time is R/V_A, where R
is the radius of the tube
and V_A is the Alfvén speed
in the tube, respectively.

4.   SUMMARY

A new method to calculate the force-free field was proposed and
it was shown that the motion of the footpoint of the field line feeds
the energy into the flux tube. The instability of an over-twisted tube
was also studied, and the calculation could explain the motion of the
eruptive prominences. The energy build up and the instability in the
magnetic field discussed here can be an important process in solar flares.

REFERENCES

Kowalik, J., and Osborne, M. R. 1968, "Methods for Unconstrained
    Optimization Problems" (Americal Elsevier, New York), p.45.
Nakagawa, Y., and Raadu, M. A. 1972, Solar Phys. 25, 127.
Sakurai, T. 1976, Publ. Astron. Soc. Japan 28, 177.
Sakurai, T. 1979, Publ. Astron. Soc. Japan 31, 209.
Strang, G., and Fix, G. J. 1973, "An Analysis of the Finite Element
    Method" (Prentice-Hall, Englewood Cliffs, New Jersey).

DISCUSSION

Bratenahl: Were these calculations made in ideal MHD (i.e., infinite conductivity)?

Sakurai: Yes, and a variational formulation would be impossible for the system with dissipative processes.

Callebaut: (1) Does your variational analysis to determine a rotation correspond to an eigenvalue problem and does α correspond in some sense to an eigenvalue? (2) Is the theorem (well-known in quantum mechanics, etc.) that the eigenvalues are obtained with second order precision by using eigen functions which are only correct to first order, relevant to your work?

Sakurai: The quantity α does not appear explicitly in our formulation, but it can be said that α takes the role of Lagrangian multiplier in order to fix the connectivity of the field lines. The magnetic energy, a quantity to be minimized, has a second order accuracy but α will not have such property.

THE FILAMENT INSTABILITY IN A SHEARED FIELD

C. Chiuderi
Istituto di Astronomia, Università di Firenze (Italy)

G. Van Hoven
Department of Physics, University of California, Irvine (U.S.A.

Filaments and prominences are classical examples of condensations of nongravitational origin, their formation being due to the thermal instability. The physics of this instability in a uniform plasma at coronal temperatures, which exhibits increasing radiative output as it cools, is well known (Field, 1965). Filaments, however, form in regions of sheared magnetic fields, as evidenced by their location above photospheric polarity-inversion lines and their occurrence after a period of increasing fibril inclination (Tandberg-Hanssen, 1974). Physically, the importance of the field structure is readily understood if one recalls the capability of magnetic field of strongly collimating the local heat conduction. Thus, the preferred locations for the development of the thermal instability will be those where the field configuration inhibits the stabilizing effects of thermal conduction on the growing temperature perturbation.

In this brief report we summarize the results of a self-consistent calculation of this instability in a non-uniform field, showing how the dynamic response of density and temperature to the competing effects of optically-thin radiation and field-collimated thermal conduction leads to the formation of characteristic "knife-blade" filaments (Chiuderi and Van Hoven, 1979).

The equilibrium magnetic field in the region of filament formation is assumed to have the model force-free form:

$$\vec{B}_o(x) = B_o \left[\vec{e}_y \tanh(x/a) + \vec{e}_z \operatorname{sech}(x/a) \right] \quad . \tag{1}$$

This field exerts a constant magnetic pressure and is therefore consistent with a uniform equilibrium state, yet it exhibits a field reversal. The calculation is based on the use of the standard ideal MHD equations, to which the following energy equation is added:

$$\frac{dp}{dt} = -\gamma p (\nabla \cdot \vec{v}) + (\gamma - 1) \left[H - \rho^2 \phi(T) + \nabla \cdot (\kappa_o T^{5/2} \vec{e}_b \vec{e}_b \cdot \nabla T) \right] \quad , \tag{2}$$

M. Dryer and E. Tandberg-Hanssen (eds.), Solar and Interplanetary Dynamics, 295-298.
Copyright © 1980 by the IAU.

where H is a generalized heat function, $\phi(T)$ specializes the temperature dependence of the radiative loss, κ_o is the heat conduction coefficient and $\vec{e}_b = \vec{B}/B$ is a unit vector along $\vec{B}$. Every quantity appearing in the equations is now written as $A(\vec{r},t) = A_o(x)+A_1(x)\exp(\nu t-ky)$, $|A_1| \ll |A_o|$, and the whole set of equations is linearized. This leads to a system of seven equations, five being algebraic and the remaining two first-order coupled differential equations in the variables $\xi_x = v_{1x}/\nu$, the transverse displacement, and $q = p_1 + (\vec{B}_o \cdot \vec{B}_1)/\mu_o$, the total pressure perturbation. H is considered to operate only in equilibrium. The equations contain a number of characteristic frequencies:

$$\Omega_\rho = -(\gamma-1) \frac{\partial}{\partial p} (\rho^2 \phi(T)|_\rho \quad ; \quad \Omega_p = (1/\gamma)(1-2\frac{T_o}{\phi}\frac{d\phi}{dT})\Omega_\rho;$$

$$\sigma_\kappa = |(\gamma-1)\kappa_o T_o^{7/2}/p|k_\parallel^2 \ll \Omega_\rho .$$

In the range $10^{5.9} < T < 10^{6.8}$, $\phi \sim T^{-1}$ (Hildner, 1974) and we have

$$\Omega_\rho \simeq 10^{-3.6} n_9 T_6^{-2}, \ \Omega_p = 1.8 \ \Omega_\rho.$$

Finally, introducing the variable $\eta \equiv \tanh \zeta \equiv \tanh(x/a)$ and normalizing wavenumbers and frequencies as follows, $\alpha = ka$, $\tilde{\nu} = \nu a/c_a$ ($c_a^2 = B_o^2/\mu_o\rho_o$), we arrive at the second-order equation (' $\equiv d/d\zeta$):

$$q'' - 2\frac{\alpha^2\eta(1-\eta^2)}{\tilde{\nu}^2+\alpha^2\eta^2} q' - [\alpha^2+\alpha_\nu^2(\tilde{\nu},\alpha^2;\eta)]q = 0 , \qquad (3)$$

where α_ν^2 can be negative and contains all the physical parameters of the problem. Eq. (3) has the form of a non-standard eigenvalue equation and must be supplemented by suitable boundary conditions. It is assumed here that q is a localized, even-parity, monotonic function of ζ. It is then possible to show that these boundary conditions imply certain limitations on the possible frequencies and wavenumbers of the purely growing solutions. More precisely,

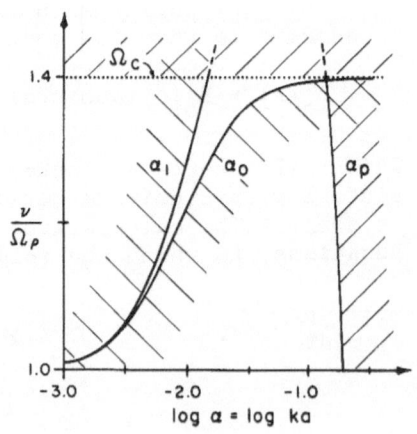

Figure 1. The allowed growth rate (ν) vs. wavenumber (α) parameter space, for monotonic filament profiles, is shown in the unshaded central window.

$$\Omega_\rho < \nu < \Omega_c \equiv (\Omega_\rho + \tilde{\beta}\Omega_p)/(1+\tilde{\beta}) \quad ; \quad \alpha_1^2(\nu) < \alpha^2 < \min\left[\alpha_o^2(\nu), \alpha_p^2(\nu)\right]$$

where $\tilde{\beta} = \gamma\beta/2 = \gamma\mu_o P/B_o^2$, and $\alpha_1^2(\nu)$, $\alpha_o^2(\nu)$, $\alpha_p^2(\nu)$ can be explicitly given in terms of β and the characteristic frequencies. The above conditions define a limited region in the (α, ν) plane where solutions may be found, as shown in Fig. 1.

A first set of numerical calculations have been performed, with the following choice of physical parameters, illustrative of the ambient coronal conditions before the start of filament condensation: $n_o = 10^{9.8}$ cm^{-3}, $T_o = 10^{6.2}$ K, $B_o = 7.5$ G and $a = 10^{8.5}$ cm. These give $\tilde{\beta} = 1$, $\Omega_\rho = 10^{-3.2}$ s^{-1}. The computed eigenvalues turn out to satisfy $\tilde{\nu}(\alpha) \sim \tilde{\nu}(\alpha_1)$ (Fig. 1) for $\nu < \Omega_c$. Concentrating on the fastest growing solutions, $\nu \simeq \Omega_c$, we can make the following comments.

(i) The computed eigenfunctions, $q(x)$, have the general aspect of decreasing exponentials (except for very small x) with a typical half-width of 10 a. This, however, is not the important empirical transverse scale, which is given by the variation of the physical observables ρ_1 and T_1. From the linearized MHD equation we find, in our regime,

$$\rho_1 = (\alpha_\nu/\nu a)^2 q \quad ; \quad T_1/T_o \simeq -(\gamma/\tilde{\beta} + 1)(\rho_1/\rho_o) \quad . \tag{4}$$

Since $\alpha_\nu^2(\eta)$ varies with x much faster than q, it actually determines the shape of ρ_1 and T_1 near the origin. We can now establish that the overall filament geometry corresponds to the empirical requirements that the horizontal width, δ (determined by the shape of α_ν^2) is much less than the vertical width, $\ell(\sim \pi/\kappa)$, which is much less than the horizontal length ($\sim \infty$ in this model). For our choice of parameters and $\Omega_c - \nu = 0.1 \Omega_\rho$, we obtain $\delta/\ell \sim 10^{-2}$.

(ii) The <u>initial</u> growth time of the fastest growing mode, $\tau_c \sim \Omega_c^{-1}$ is estimated to be $\tau \sim 10^{3.5} n_9^{-1} T_6^2$, which predicts a typical e-folding time somewhat shorter than is normally quoted. However, what is observed is the <u>nonlinear</u> stage of filament formation which usually has a much longer timescale.

(iii) Sheared magnetic fields are intimately connected with flares, both observationally and theoretically. The excess magnetic energy stored in the nonpotential field can be released due to finite resistivity effects, but the timescale of the process appears to be too long. The thermal instability timescale is much faster and it is therefore conceivable that the magnetic reconnection will take place in the nonlinear stage of filament development, at a considerably lower temperature and at a much faster rate, due to the strong temperature dependence of normal resistivity.

This research was supported in part by the Atmospheric Sciences Section of the National Science Foundation.

References

Chiuderi, C., and Van Hoven, G.: 1979, Ap. J. (Lett.) 232, L69.
Field, G.B.: 1965, Ap. J. 142, 531.
Hildner, E.: 1974, Solar Phys. 35, 123.
Tandberg-Hanssen, E.: 1974, "Solar Prominences," Reidel (Dordrecht).

DISCUSSION

Nakagawa: The effect of gravity is one of the crucial problems in the formation of condensations because the increased density required to induce thermal instability is immediately subject to gravitational acceleration downward. Your field configuration does not provide any support.

Van Hoven: I agree; this is only the initial step in a filament model, but it is the first dynamic calculation to reflect the sheared magnetic field and the finite geometry.

Foukal: The density assumed in calculations of thermal instability is critical in determining the growth rates, and the spatial wave numbers of the growing modes. Were you able to find reasonable growth rates for densities of the order of the background corona, from which such a filament would be expected to condense?

Van Hoven: The density in the example quoted here is $10^{9.8}$ cm^{-3}, and the growth rate is approximately proportional to this parameter.

A MODEL FOR IMPULSIVE ELECTRON ACCELERATION TO ENERGIES OF TENS OF kT_e

P. Hoyng, A. Duijveman and Th.F.J. van Grunsven
Space Research Laboratory
Astronomical Institute at Utrecht
         a n d
D.R. Nicholson
Department of Physics and Astronomy
University of Iowa, Iowa City, Iowa 52242.

1. OUTLINE

We describe a model for first stage electron acceleration based on quasi-linear interaction with Langmuir waves. The acceleration takes place in a MHD-unstable plasma region (length L, volume V) that contains many microscopically unstable current layers, which act as (quasi-stationary) sources of Langmuir waves, Figure 1. Our work does not depend on the details of the generation mechanism for Langmuir waves, nor on the details of the MHD instability. For the latter, one could think of multiple tearing instabilities.

An essential feature is that the Langmuir wave energy density w^ℓ is very inhomogeneously distributed in the acceleration region. As a result, during most of the time an electron does not interact with Langmuir waves. We take this into account in the following way. Instead of a Langmuir wave distribution w_κ which is large only in a small (filamentary) fraction εV of the volume V, we work with a diluted distribution εw_κ, spread homogeneously over V. The second feature that we include is escape of electrons from the source. Our model is an extension of earlier work by Benz (1976); it contains two phenomenological parameters, the filling factor ε, and ν (below). A full description is published elsewhere (Hoyng *et al.*, 1979).

2. EQUATION

The following equation for the tail of the electron velocity distribution is obtained:

$$\partial f/\partial t = x^{-2}\, \frac{\partial}{\partial x}\, d(x)\, \frac{\partial}{\partial x}\, f + b x\, (f_M - f) \tag{1}$$

$$d(x) = (\pi\omega_e/2x)\varepsilon \int_{x^{-1}}^{1} \kappa^{-3}\, w_\kappa\, d\kappa; \quad b = \omega_e\, (\nu L k_e)^{-1}$$

$$\int_0^1 w_\kappa\, d\kappa \equiv w^\ell \;\; ; \qquad x = v/v_e \;\; ; \quad \kappa = k/k_e$$

M. Dryer and E. Tandberg-Hanssen (eds.), Solar and Interplanetary Dynamics, 299-302.
Copyright © 1980 by the IAU.

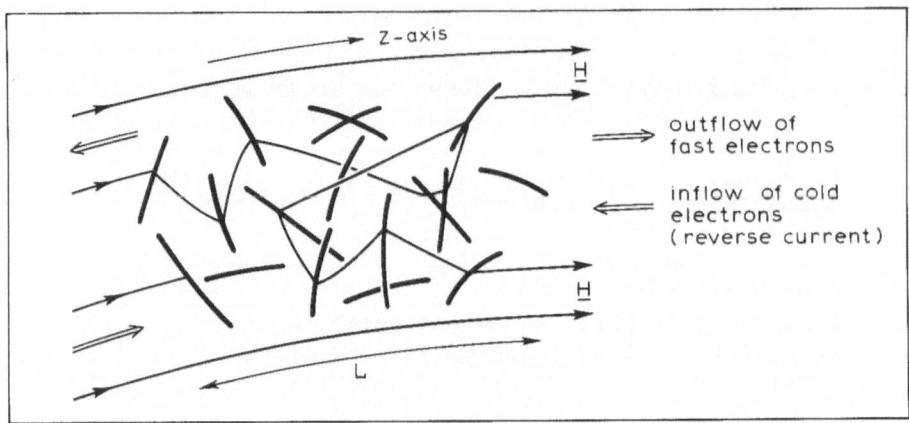

Figure 1. Outline of the acceleration region, containing many unstable
          current layers radiating Langmuir waves into the adjacent
          plasma. One 'stochastic' fieldline is sketched, but no attempt
          is made to draw a realistic field pattern. For application to
          type III bursts the region must be embedded in an open field.

ω_e, $v_e = (k_b T_e/m_e)^{\frac{1}{2}}$ are the electron plasma frequency and thermal
speed; $k_e = \omega_e/v_e$. w^ℓ is the (undiluted) energy density in Langmuir waves
relative to the electron thermal energy density. The electron temperature
is constant everywhere (corresponding maxwellian f_M). Escape is simulated
with $(f_M - f)/\tau$, where $\tau = \nu L/v$ is the lifetime of an electron in the
source ($\nu > 1$). This term expresses the fact that different values of the
<u>tail</u> distribution inside and outside the acceleration region equalize
by free streaming on a timescale τ. Reverse current effects and Coulomb
interactions are ignored. The (stationary) Langmuir wave distribution
w_K in (1) is supposed to be known. Strong or weak turbulence shows up
only through a possibly different w_K.

3. RESULTS
     Based on numerical simulations (Van Grunsven *et al.*, 1979) we choose
for w_K a singly peaked distribution (peak position $\bar{\kappa}$). We find that such
w_K allow a model $d(x)$ given by $d(x) \propto (\varepsilon w^\ell/\bar{\kappa}^2)(x\bar{\kappa})^\lambda$, $|\lambda| \leqslant 1$, $x\bar{\kappa} \geqslant 1$. For
this model $d(x)$ we determine the stationary solution of (1) for $x \geqslant x_0$,
requiring $f(x_0) = f_M(x_0)$. x_0 is selected such that the power generated
in Langmuir waves equals the power lost to escape of fast electrons
(usually $x_0 \sim 1/\bar{\kappa}$). There is little difference between solutions for
different λ, cf. Figure 2; they are characterized by a maximum velocity
x_i and a time t_i required to reach stationarity:

$$x_i \simeq \{ \frac{25\pi}{16} \frac{\varepsilon w^\ell}{\bar{\kappa}^2} \nu L k_e \}^{1/5} \ ; \ t_i \simeq (\nu L k_e)^{4/5} (\bar{\kappa}^2/\varepsilon w^\ell)^{1/5} \omega_e^{-1} \tag{2}$$

For $x > x_i$ all solutions of (1) decay exponentially. Note that only the
combination εw^ℓ matters and that our results depend weakly on ε and ν.

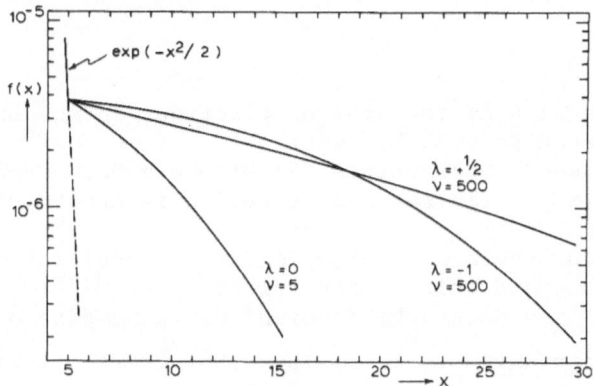

Figure 2. Stationary solutions of Eq. (1) for the model parameters (3). The discontinuity in $\partial f/\partial x$ at the boundary $x_0 = 5$ is not realistic.

As an example, we introduce the following model:

$$w^{\ell} = 5 \times 10^{-3} \quad ; \quad \bar{\kappa} = 0.2 \quad ; \quad \varepsilon = 10^{-4} \quad ; \quad \nu = 5 \text{ or } 500$$
$$n = 10^{10} \text{ cm}^{-3} \quad ; \quad L = 10^{8} \text{ cm} \quad ; \quad T_e = 10^{7} \text{ K}$$

(3)

It has a turbulent volume $\varepsilon V \sim \varepsilon L^3 = 10^{20}$ cm^3 and from (2) we find for $\nu = 5$: $x_i = 10.7$ ($\rightarrow E_{max} \simeq 60 k_b T_e \sim 50$ keV) and $t_i \simeq 0.05$ s. The flux of escaping electrons is $\simeq 10^{32}$ s^{-1} (scaling: x_i^4/ν). The very short switch-on times ($\sim 10^{-6}$ s) usually quoted for turbulent acceleration are inapplicable now.

Our model could be relevant for acceleration of type III burst stream electrons. The quasi-periodic nature of this phenomenon, in our view, would be determined by the time evolution of the MHD instability. We have also applied our model to a single, large area shock as is involved in a type II burst (Hoyng *et al.*, 1979). The model does not include heating of the acceleration region and therefore we do not apply it to solar X-ray bursts.

REFERENCES

Benz, A.O.: 1976, *Astrophys. J.* 211, 270.
van Grunsven, Th.F.J., Hoyng, P. and Nicholson, D.R.: 1979, *Astron. Astrophys.*, submitted.
Hoyng, P., Duijveman, A., van Grunsven, Th.F.J. and Nicholson, D.R.: 1979, *Astron. Astrophys.* submitted.

DISCUSSION

Moore: What fraction of the ambient electrons in the Langmuir wave region are accelerated to ~50 KeV energies?

Hoyng: For the source parameters (3) with $\nu = 5$, the relative density of fast electrons $(x > x_o)$ in the source region is about 10^{-3}.

Benz: Why do you have a maximum velocity of acceleration? What determines its value, and what is the value?

Hoyng: Using $\kappa^{-1} \leq x$ under the integral defining $d(x)$ one finds $d(x) \leq const.\ x^2$, <u>for any</u> w_κ. It follows that the acceleration term in equation (1) is at most of order x^{-2}, while the escape term is of order x. Apparently, the nature of the wave particle interaction is such that at low velocities acceleration dominates, while beyond a well-defined velocity given by x_i, (2), escape always dominates.

A MODEL FLARE AND THE CONTINUED POST-FLARE MASS RELEASE FROM THE FLARE
REGION

Yutaka Uchida
Tokyo Astronomical Observatory, University of Tokyo
Mitaka,Tokyo 181, Japan

ABSTRACT

     A mechanism for slow and enduring mass release well after the flare
onset, which is inferred from the enduring enhancement in the interplanetary mass flux after flare activity, is discussed in the modified scheme
of a "neutral sheet" model of flares. According to a possible new way of
looking at the neutral sheet model as proposed by Uchida and Sakurai, it
is argued that a slow and enduring mass release in the later phase of a
flare may come from the reconnection in the region of interleaved opposite polarity fields produced in the interchange collapse which is proposed to be responsible for the highly dynamical phase at the start of a
flare. Heated plasma, which is magnetically disconnected form the anchoring field in the photosphere through reconnection in this interleaved
region, is continuously produced during the thermal phase, and is expelled
from the flare region by the "buoyancy" in the gradient of the surrounding
high magnetic pressure. The trajectory of this mass leakage will be along
the steepest gradient of magnetic pressure, rather than along the field
lines as expected in other picture, and this process can supply mass to
the solar wind in open field regions.

1. A MODEL FLARE

     Uchida and Sakurai (1976, 1979a) have proposed a new way of looking
at the neutral sheet model of solar flares. In brief, the proposal is to
consider a general curved neutral sheet which itself is unstable to interchange instability but stabilized by the additional presence of a field
component lying along the sheet which prevents the instability. The remov-
al of this stabilizing field component, which is identified with the rise
of a dark filament before the flare start, brings the configuration into
an unstable one, and an interchange collapse takes place in a dynamical
time scale which is $\sim 10^1 \sim 2$sec, rather than in a magnetic diffusion
time scale which is intrinsically huge and many proposals have been made
for mechanisms to make it shorter. The dynamical or transient behavior
in the process of settling down of the configuration into a lower energy

303

M. Dryer and E. Tandberg-Hanssen (eds.), Solar and Interplanetary Dynamics, 303-306.

state as introduced in this model seems to explain the drastic and singular behavior around the flare start in a most natural way. Claimed presence of extraordinary high temperature region ($\sim 10^8$K) as observed in
thermal hard X-ray component at around the flare start (Ohki and Frost
1979) may be interpreted as due to the shocked high temperature region
produced in the plasma ahead of the interleaving collapse. The hot small
loop of 2 x 10^7K observed in soft X-ray and in FeXXIV line in EUV range
may be interpreted as being the wishbone type structure below the "neutral" sheet, which is filled with the chromospheric plasma heated and expanded by the heat input from the extraordinary high temperature source
thus produced above. The dynamical collapse also explains the emission
of MHD shock wave from the flare region in the form of Moreton waves or
type II burst shocks.

The interleaving property of the collapse is due to the characteristics of the instability that the growth rate is larger for smaller wavelength perturbation in the linear stage (Bernstein et al. 1958, Uchida
and Sakurai 1979b). This trend, though suppressed in the shortest wavelength range, holds even in the non-linear stage (Frieman 1954) in which
the collapse is driven by the release of the stress in the global distortion of the field (Uchida and Sakurai 1979b), the stress having been
piled up due to the footpoint motion in the presence of a stable "neutral
sheet" which hinders the relaxation.

Now in our picture, the theoretical counterpart of the later thermal
phase is as follows : As soon as the configuration settles down dynamically into the lower energy state, and we are left with an interleaved
structure, the magnetic annihilation occurs all over the extended interface of the interleaved opposite polarity regions. The total area is
$S_e \simeq$ 2hDN $\simeq$(2D/d)S_o, where $S_o \simeq$ hL is the area of the unperturbed sheet
with the height h and the length L, N $\simeq$ L/d is the number of leaves with
the thickness d, and D is the distance of excursion in the non-linear
stage of the interchange collapse, traversed by the outermost part of
each polarity region. d and D are estimated to be $\simeq 10^5$cm and $\sim$ 3 x 10^8cm,
respectively (Uchida and Sakurai 1979b), and therefore $S_e/S_o \sim 10^3 \sim$ 4.
The time scale of the annihilation, $\tau* = 4\pi\sigma*(d/2)^2/c^2$, is already as
short as $10^5 \sim 6$sec with $\sigma* = \sigma_n$, the classical conductivity, and can
easily be shortened to 10^4sec with assisting mechanisms which, eg., produce a rather weak anomalous resistivity effect, $\sigma* \sim 10^{-1} \sim -2$ σ_n. The
energy of the magnetic field which is involved in the annihilation in
this later phase is U $\simeq$ ($B^2/8\pi$)S_ed/2 $\sim$ ($B^2/8\pi$)DS$_o$ and amounts to $10^{31} \sim 32$
ergs for large flares. In small flares, L, and correspondingly U, is
smaller, and the 2 x 10^7K region may have the appearance of a single loop.

2. MASS LIBERATION IN THE LATER PHASE

One point to be noticed is that the reconnection in three-dimension
in our interleaved region of opposite polarity fields can produce magneto-
plasma whose magnetic field is disconnected from their photospheric roots
and is closed on itself (Fig. 1). Only two leaves with the field of one
polarity and the opposite polarity leaf sandwiched between them are shown
in the figure, and only lines of force reconnected to those on the middle

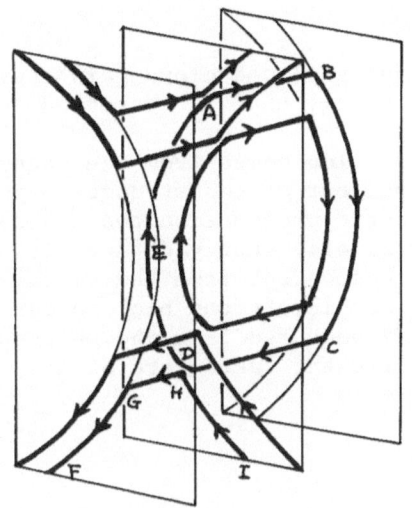

Fig. 1. Schematic picture for three-dimensional reconnection on interleaved structure

leaf are shown for the sake of clearness. The configuration will be composed of the repetition of such basic pattern, and closed loops of field lines disconnected from their photospheric roots are produced without entanglement. The reconnected field lines can not form such an isolated loops in conventional two-dimensional reconnection.

The plasma with isolated field lines like ABCDE in Fig. 1 can be squeezed out of this region by melon seed effect. The sum of the volume of the thus liberated plasma is V $\sim 2hDNd/2 \sim DS_0 \sim 10^{27.5} cm^3$, and the potential energy with which the plasma "rolls down" the hill of the magnetic pressure gradient (which reduces essentially to $(B^2/8\pi)V$ according to Parker (1958)) is estimated to be of the order of 10^{32}ergs. It may be noted that this energy is due to the buoyancy of the diamagnetic blob created in the gradient of strong magnetic field of active region, rather than the kinetic energy given to the plasma directly by the flare effects.

Reconnected field lines like FGHI in Fig. 1, on the other hand, will be able to account for the formation of loop-prominence-system appearing after flares. In our model, those lines of force with distant footpoints are folded in the inner side of the "leaves" in the interchange collapse, and come into contact with the opposite polarity field later than those with closer footpoints, corresponding to the observation.

References

Bernstein, I. B., Frieman, E. A., Kruskal, M. D. and Kulsrud, R. M., 1958, Proc. Roy. Soc., A 244, p 17.
Frieman, E. A., 1954, Astrophys. J., 120, p 18.
Ohki, K. and Frost, K., 1979, to be published in Solar Phys.
Parker, E. N., 1958, Astrophys. J. Suppl. 3, p 51.
Uchida, Y. and Sakurai, T., 1976, Abstr. US-Japan Cooperative Seminar, High Energy Phenomena in Solar Flare, p VI-4.
Uchida, Y. and Sakurai, T., 1979a, in Proc. Skylab Workshop on Solar Flares, Chapter III (Kahler, S., Spicer, D., Uchida, Y. and Zirin, H.) ed. P. S. Sturrock (University of Colorado Press).
Uchida, Y. and Sakurai, T., 1979b, sumbitted to Solar Phys.

DISCUSSION

Kuperus: How does the outward moving plasma cloud gain energy? Is
it magnetic energy or gravitational energy that is converted into
kinetic energy?

Uchida: The kinetic energy of the blob comes from the magnetic
buoyancy (a diamagnetic blob is pushed out of the stronger magnetic
region due to the difference of magnetic pressure on two sides). It
may be stressed that, although this kinetic energy derives itself from
the energy of active region, is is independent from the so-called
"flare energy." The flare does <u>not</u> accelerate the blob in our picture,
but simply produces a blob which is disanchored from the photosphere,
and the existing magnetic pressure gradient takes care of its
acceleration.

A MODEL OF SURGE

Giancarlo Noci
Oservatorio Astrofisico di Arcetri

ABSTRACT

This paper reports on the application of a siphon flow model to the late stage of a surge observed in the UV radiation by the S055 experiment on board Skylab. The MgX $\lambda625$ and OVI $\lambda1032$ emissions from the density distributions occurring in flows which become supersonic at the top of the loop agree with the observations, indicating a pressure drop, along the loop, of a factor of 4 for the plasma emitting the OVI line, and of a factor of 2 for the plasma emitting the MgX line.

INTRODUCTION

The observation of large velocity fields in regions of high magnetic field in the solar transition region chromosphere-corona (Brueckner, 1976; Bruner et al., 1976; Doscheck et al., 1976; Athay, 1979; Brueckner, 1979; Lite, 1979) has drawn attention to the study of fast plasma flows in that region and in the corona above.

Steady coronal flows have been studied by Cargill and Priest (1979), and by Noci (1979), and a model based on such flows is here used to interpret the MgX and OVI emission of a surge. Similar flows for much lower temperature regions (T < 6000 K) had been investigated by Meyer and Schmidt (1968) as a model of the Evershed motions; Pikel'ner has also used a model based on flows, though limited to subsonic flows, to try to explain prominence formation (1971).

CHARACTERISTICS OF THE FLOW

In the region considered, the magnetic field is supposed to be constant in time and large enough to give β (plasma energy/magnetic energy) << 1; hence the flows are confined inside given magnetic tubes. No wave force is assumed to act upon the plasma; furthermore the run of density

307

M. Dryer and E. Tandberg-Hanssen (eds.), Solar and Interplanetary Dynamics, 307-311.
Copyright © 1980 by the IAU.

and temperature is assumed to follow a polytropic law and the cross-
section of a magnetic tube to be constant.

The equations of the problem are clearly the same that apply to the
problem of the solar wind. If s is a coordinate along the loop axis and
s_m its value at the top of the loop, the topology of the solutions for
the velocity is characterized by a critical point at s_m. Therefore, beyond solutions always subsonic, characterized by the same pressure at the
two footpoints of the loop, there exists a subsonic-supersonic solution,
characterized by a pressure decrease, which adjusts itself to the boundary condition at the second footpoint of the loop by a stationary shock.
Large density decreases occur along the loop axis for these solutions,
large density increases across the shock. These density variations should
have a signature in large brightness variations along the loop.

THE SURGE OF OCTOBER 28, 1973 (Mc Math 584)

This model has been applied to the surge which followed the flare of
October 28, 1973 at 1758 U.T. (UV max. time) in active region Mc Math 584.
Observations were made by the Harvard spectrometer on board Skylab; I
thank E. Schmahl for making the data available to me through a preprint
(1979).

The application of the siphon model to a surge is prompted by the
consideration that pressure unbalances should be produced by flares (that
surges are driven by a pressure unbalance is suggested, e.g., by Schmahl,
1979). However, the model being a steady one, use has been made of observations sufficiently delayed from the flare maximum that subsequent observations did not show any evidence of travelling disturbances. Hence the
neglect of the time dependent term $\partial v/\partial t$ in the momentum equation, compared
with the stationary term $v\,\partial v/\partial s$, appears to be feasible.

The model assumes a shell structure of the loop, with temperature
increasing from the interior towards the exterior. The polytropic parameter α has been taken equal to 1.1; ionization equilibrium has been
assumed. The calculations have been therefore limited to the higher temperature ions for which the temperature decrease along the loop, in the
siphon model, is small, namely MgX and OVI. (The temperature decrease
is 7% in the shell producing the MgX emission, 14% in the cooler shell
producing the OVI emission.)

The comparison of the model with the observations is shown in Figure
1. The data used are brightness values from just beyond the flare position (i.e., from the point P_0) to the footpoint of the loop. It is seen
that the temperature is defined by the brightness gradient along the loop
axis, and the density by the brightness itself. Two model curves are
given for each ion to show the uncertainty of the parameter determination.
It appears that the temperature, in the case of MgX, is strongly dependent
on the weight one gives to the brightness distribution close to the flare
position. Consequently the density is also affected, since the temperature
value adopted determines the abundance of the ion: for MgX the increase

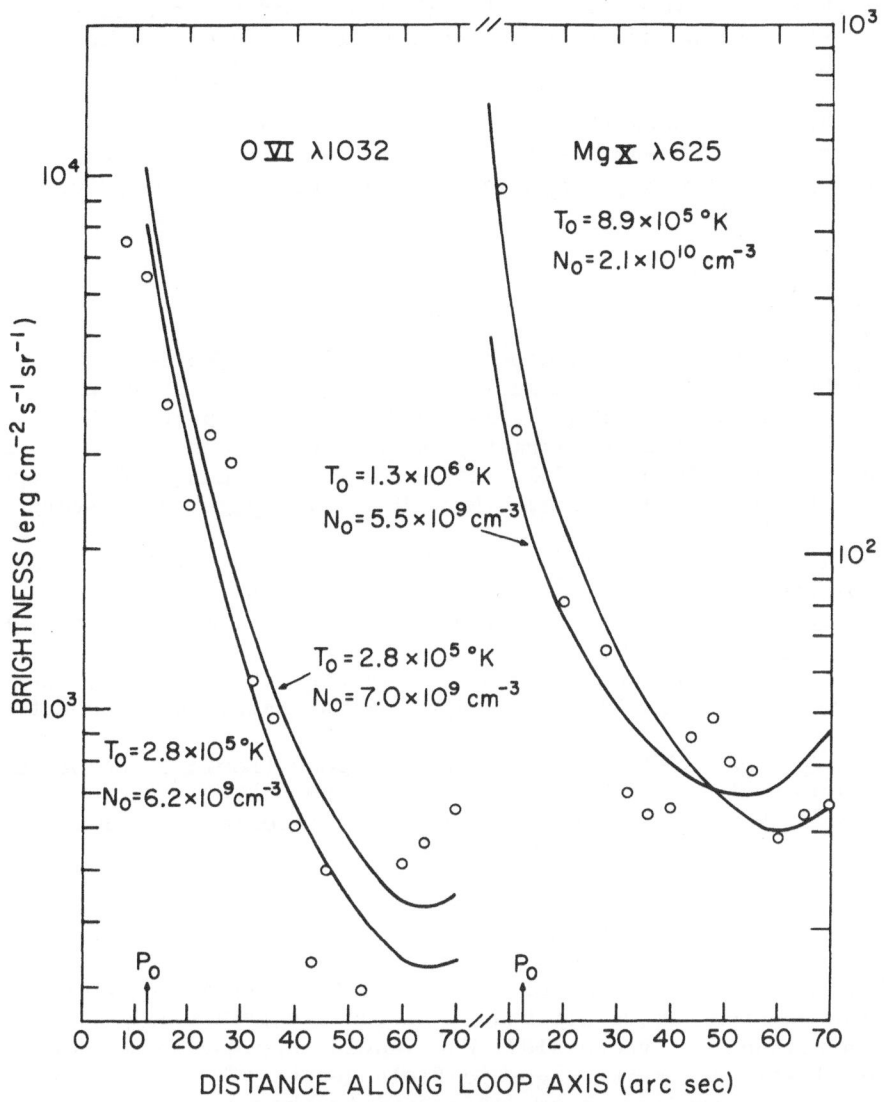

Figure 1. The surge of 28 October 1973, 18:09 (Mc Math 584). Observed
(open circles) and calculated (curves) brightness distributions. T_o, N_o
refer to point P_o (marked by an arrow, 13" from first footpoint). The
thickness of the emitting shells has been assumed δ=1" (The diameter of
the surge loop was ≤ 5").

in the density from the lower to the upper curve of Figure 1 compensates
for the decrease in ion abundance due to the temperature decrease.

About the density values, it must also be noted that they depend on
the assumed thickness (δ) of the emitting regions ($N_o^2 \sim 1/\delta$). It is worth

remarking, however, that if the thickness of the two emitting shells is the same, the pressure increases from the core of the loop towards the surrounding corona.

Figure 1 shows that the agreement between theory and observations is good, hence these results support the view that surges are driven by a pressure increase, connected with a flare, at some point inside a magnetic tube.

REFERENCES

Athay, R.G.: 1979, Workshop on Active Regions (Boulder, Colorado).
Brueckner, G.E.: 1976, unpublished; also quoted in Bruner et al.: 1976 and in White, O.R.: 1977, IAU Coll. 36, pp. 75-103.
Brueckner, G.E.: 1979, Workshop on Active Regions (Boulder, Colorado).
Bruner, E.C., Chipman, E.G., Lites, B.W., Rottman, G.J., Shine, R.A., Athay, R.G., and White, D.R.: 1976, Astrophys. J. Letters 210, pp. L97-L101.
Cargill, H., and Priest, E.: 1979, Workshop on Active Regions (Boulder, Colorado).
Doscheck, G.A., Feldman, U., and Bohlin, J.D.: 1976, Astrophys. J. Letters 205, pp. L177-L180.
Lites, B.W.: 1979, Workshop on Active Regions (Boulder, Colorado).
Meyer, F., and Schmidt, H.U.: 1968, Mitt. Astr. Gesellschaft 25, pp. 194-197.
Noci, G.: 1979, Workshop on Active Regions (Boulder, Colorado).
Pikel'ner, S.B.: 1971, Solar Phys. 17, pp. 44-49.
Schmahl, E.: 1979, preprint, Harvard-Smithsonian Center for Astrophysics.

DISCUSSION

Lemaire: In your model you assume that the flux tube has a constant cross-section versus altitude! Is there experimental evidence that this is a reasonable hypothesis? What would happen if you would assume a more likely diverging magnetic field geometry?

Noci: According to the observations the cross-section of a coronal loop is approximately constant. In a situation of changing cross-section the topology of the solutions will be influenced. For example, for loops diverging with height, but still symmetric (with respect to the top), if the cross-section variation is not very large, the critical point remains at the top and the velocity gradient decreases.

Heinemann: In ordinary hydrodynamic flow without gravitational field the flow goes supersonic only at a minimum of the cross-sectional area of the flow tube, but the flow tube here has constant cross-section. What changes this requirement? I wouldn't expect the gravitational field to do it because the top of the loop is well below the Parker critical point.

Noci: Let us think of the topology which applies to the case of the Parker solar wind, where the heliocentric distance is the coordinate, and consider the class of solutions which fold themselves back to the solar surface after crossing the M (Mach number) = 1 line. If r_m is the heliocentric distance of the top of the loop, the solution of this class, which becomes sonic at r_m is the critical solution for the case of the loop: in terms of the s coordinate it is continuously growing, becoming sonic at s_m and supersonic in the "descending" branch of the loop.

RADIATIVE HYDRODYNAMICS OF FLARES: PRELIMINARY RESULTS AND NUMERICAL
TREATMENT OF THE TRANSITION REGION

A.N. McClymont and R.C. Canfield
Department of Physics, C-011
University of California, San Diego
La Jolla, California 92093 U.S.A.

We report on a comprehensive numerical simulation of flare dynamics,
encompassing the corona, transition region and chromosphere. A coronal
loop geometry, whose magnetic pressure dominates gas pressure, is assumed.
These preliminary results assume optically thin radiation; we are currently incorporating optically thick radiative transfer in the chromosphere. We discuss difficulties in modelling the transition region under
flare conditions, and suggest tentative solutions.

The hydrodynamic equations, incorporating the effects of thermal
conduction, viscosity and radiation, are described by Craig and McClymont
(1976) while the probabilistic method for radiative transfer is discussed
by Canfield and Ricchiazzi (1979). We write the continuity, momentum and
energy equations in Lagrangian form using column mass as the independent
variable and consider the atomic rate equations for the fractional populations rather than absolute number densities. Thus we eliminate all
convective terms. The equations are then written in fully implicit
finite difference form and are solved by simultaneous iteration at each
timestep.

We have carried out preliminary calculations neglecting radiative
transfer effects. The simple model atmosphere used is not claimed to be
a faithful replica of a loop in the solar atmosphere, rather it is intended to illustrate global features. Figure 1 shows the velocity
response of a loop to the impulsive injection of energy at its apex,
raising the coronal temperature from 1.6×10^6K to 5×10^6K. Coronal
material is driven down the legs of the loop and collides with the dense
transition region after 1 minute. The chromosphere is then compressed
downwards while coronal material is reflected upwards again. Thereafter
the atmosphere undergoes oscillations, heavily damped by conduction and
radiation, of period ~ 4 minutes. Global oscillations of this type (and
of greater amplitude for stronger flare heating) appear in all our calculations but, as far as we know, have not been reported observationally.
Figure 2 shows the results of injecting the same amount of energy at a
height of 1000 km in the chromosphere; in this case a surge-like ejection

313

M. Dryer and E. Tandberg-Hanssen (eds.), Solar and Interplanetary Dynamics, 313-316.
Copyright © 1980 by the IAU.

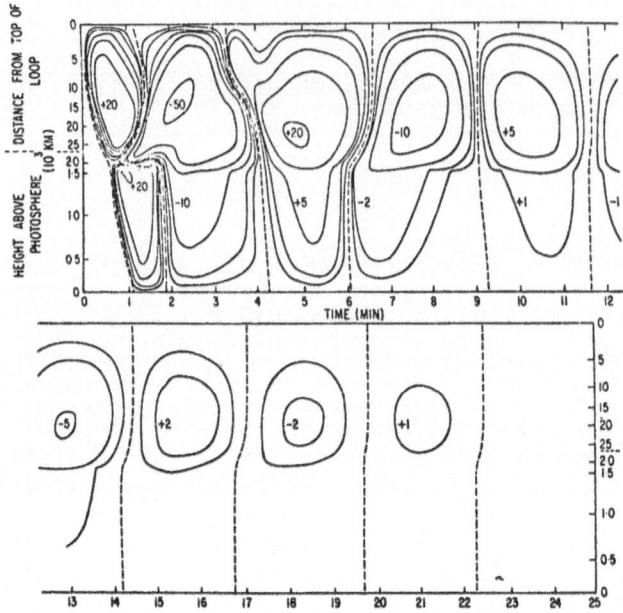

Fig. 1.   Velocity as a function of height and time following the injection of impulse of energy at the top of the loop.   Contours are labelled
in units of km s^{-1} (positive velocities <u>downward</u>).   Note that the vertical axis is a <u>Lagrangian</u> coordinate, heights and distances correspond to
the <u>initial</u> position of gas elements.   The scale is distorted non-
linearly to show both the corona and chromosphere, which extends to a
height of ∿ 2000 km.

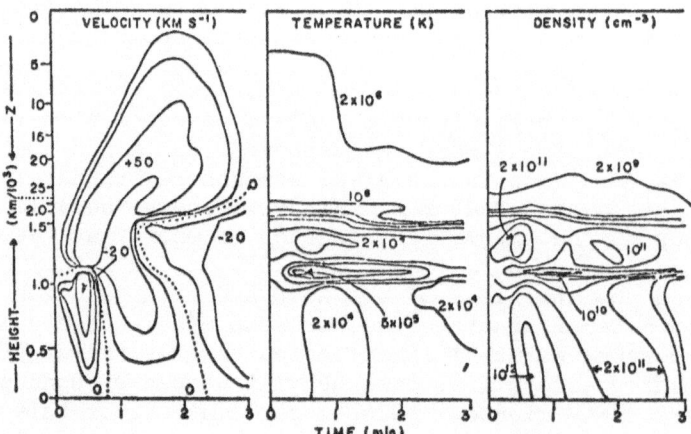

Fig. 2.   Velocity (positive <u>upwards</u>), temperature and density as a
function of column mass and time in a model surge.   Note cool dense
core of surge material above 1000 km, the initial energy release
height.   See Figure 1 caption regarding non-linear vertical axis.

of chromospheric material into the corona occurs. The Lagrangian vertical
axis in Figure 2 does not explicitly show the motion, but integration of
the velocity over time shows that this small surge reaches a height of
only 5000 km. The most interesting feature of the calculation is the
lack of heating of the surge material by the energy release; in fact the
material is compressed as it is driven upwards and cools rapidly through
the enhanced radiative loss rate. Thus we have a possible explanation
for the appearance of cool, dense surges in the corona.

   While investigating the dynamic formation of an atmosphere with a
corona-transition-zone-chromosphere structure from an initially uniform
plasma, Craig and McClymont (1979) found that incorrect results were
obtained for high conductive fluxes through the transition region. It
is now clear that only fluxes for which the scale height of temperature
variation is greater than the finite difference grid spacing can be reproduced accurately. That is, the flux F is limited to $p\kappa(T)/(k\ \Delta N)$,
where p is pressure, $\kappa(T)$ is the conductivity, k is Boltzmann's constant
and ΔN is the Lagrangian grid spacing. With the grid spacing typically
used in numerical modelling of this type (e.g. Kostyuk and Pikel'ner,
1975; Kostyuk, 1976; Somov et al., 1978; Craig and McClymont, 1979) this
criterion is marginally satisfied in the quiet solar atmosphere but
grossly violated under flare conditions. Under dynamic flare conditions
there is an "evaporative" conduction front moving through the plasma at
a "velocity" $u\ (cm^{-2}\ s^{-1})$. In order to reproduce the temporal behavior of
the atmosphere satisfactorily, conditions at a grid point must change
slowly compared to the timestep Δt, i.e. we require $u\ \Delta T << L(T) =$
$p\kappa/(2\ kF)$, where $L(T)$ is the characteristic scale of T variation,
$[(T)(dT/dN)]^{-1}$. Under flare conditions this is very severe restriction
on the timestep; we must depart from conventional finite difference
techniques to handle the transition region. We suggest the following
methods, none of which have been investigated in detail:
   (1) Within the transition zone, formulate the equations for a set
of grid points which convect through the plasma with the conduction front.
   (2) Introduce a "pseudo-conductivity" term in analogy with the
pseudo-viscosity term, with suitable modification of the energy equation.
   (3) Approximate the transition region as a thin interface between
an upper temperature T_1 in the corona and a lower temperature T_2 in the
chromosphere.

REFERENCES

Craig, I.J.D., and McClymont, A.N.: 1976, Solar Phys., 50, p. 133.
Canfield, R.C., and Ricchiazzi, P.: 1979, submitted to Astrophys.J.
Craig, I.J.D., and McClymont, A.N.: 1979, submitted to Solar Phys.
Kostyuk, N.D., and Pikel'ner, S.R.: 1975, Sov.Astron., 18, p. 590.
Kostyuk, N.D.: 1976, Sov.Astron., 19, p. 458.
Somov, B.V., Spektor, A.R., and Syrovatskii, S.I.: 1978, Izv.Acad.Sci.
     USSR. Phys.Ser., 41, p. 273.

DISCUSSION

Uchida: (Comment) I would like to mention that a similar and
elaborated calculation was done recently by Dr. Nagai, one of our
graduate students. He obtained solutions which explain many of the
observed properties of hot loop formation nicely. The oscillatory
response which you found was also shown to occur if the heat input
has a short duration.

Kahler: Do you expect to get an oscillation of the chromospheric
material for an arbitrary energy release profile in the corona or is it
limited to an impulsive input?

McClymont: For the results shown, energy is injected during the
first minute. Any timescale of energy injection shorter than the
oscillation period should result in oscillations. With a more realistic
initial atmosphere, the oscillation probably won't penetrate so deeply-
at present the density of our lower chromosphere is too low.

Emslie: (1) About the Canfield and Ricciazzi (Ap. J., submitted)
radiative losses you use in your calculations. Although I agree, it
results in much shorter computation times. It is important to realize
the effort of Lα backwarming on the chromospheric energy budget, which
is not included in the Canfield and Ricciazzi calculations (c.f., the
large discrepancy between the Canfield and Ricciazzi and Canfield and
Athay (1974) curve at Lα temperatures). This backwarming has been
shown by Machado and Emslie (1979, Ap. J., *232*) and Machado *et al.*
(1980, Ap. J., submitted) to be a strong source of energy to the flare
chromosphere, and it therefore may have a significant effect on your
results.
(2) With regard to the steep temperature gradient you encounter in
the transition zone, it is worth remembering that for very high energy
inputs the collision non-free path might well be greater than the
characteristic transition zone thickness, so that Spitzer (1962)
conductivity is no longer valid (c.f., Spicer, 1979, *Solar Phys., 62*).
Are there possible deviations from Spitzer conductivity considered in
your calculations?

McClymont: We realize that backwarming is the major source of error
in the probabalistic calculations done so far. The error is very
acceptable in view of the fact that the probabalistic method makes
dynamical calculations with radiative transfer possible on a routine
basis. We don't know yet how significant the error will be under all
circumstances.
We expect that classical conductivity will break down under some flare
conditions due to both the mean free path and flux saturation. For these
preliminary results we assumed the Spitzer conductivity. We are currently
investigating the whole problem of thermal conduction in the transition
region.

RECONNECTION DRIVEN CORONAL TRANSIENTS

G.W. Pneuman[*]
Max-Planck-Institut für Physik und Astrophysik, Munich

The association of coronal transients with two-ribbon solar flares is well established. During the Skylab period, every two-ribbon flare when observed close enough to the limb was accompanied by a coronal transient. Flares do not occur with all transients, however many of these transients are associated with soft X-ray enhancements in the corona similar to but less energetic than the intense X-ray loops that occur with two-ribbon flares [cf. MacCombie and Rust (1979) for a review]. The eruption of a filament seems to be the ingredient common to all these events – more so than flares. For these reasons, we consider this class of phenomena, regardless of whether a flare occurs or not, to be exhibiting a common physical process. To produce chromospheric emission requires a substantial amount of energy. Hence, one should expect chromospheric flares to be associated with only the most energetic phenomena. Nevertheless, the most comprehensive observations covering a wide range of wavelengths (Hα, EUV, X-ray, radio, white light) are available for the large two-ribbon flares, and the study of these events sheds the most light on the mechanism which produces coronal transients.

The most striking feature of two-ribbon solar flares are the large flare loop systems which rise slowly upward to great heights in the corona beginning at the very onset of the flare and persisting for some 10-20 hours following the flash phase. The loops are seen in Hα, EUV, and soft X-rays with the cooler Hα loops nested beneath the hot X-ray loops. The footpoints of the loops are rooted in the chromospheric flare ribbons which expand slowly outward from the magnetic neutral line during the course of the flare. The Hα loops are rooted to the insides of the ribbons and the X-ray loops to the middle and outside. A careful examination of the observations shows that the system is <u>not</u> comprised of rising fluxtubes but, rather, of discrete stationary new loops being formed at progressively higher levels in the corona. All of these observations are consistent with the concept that the loops are formed via <u>magnetic</u>

---

[*] On leave from the High Altitude Observatory, National Center for Atmospheric Research, Boulder, Colorado, U.S.A.

M. Dryer and E. Tandberg-Hanssen (eds.), Solar and Interplanetary Dynamics, 317-321.
Copyright © 1980 by the IAU.

reconnection of field lines distended outwards by the eruption of the
pre-flare filament [cf. Pneuman (1979)] for a review .

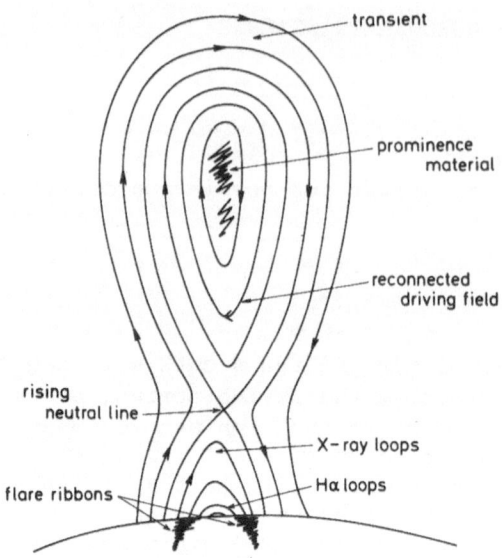

Fig. 1   Schematic of Transient and Flare Loop Configuration

If this concept of the two-ribbon flare (and similar enhancements
without flares) is correct, then the propulsion mechanism for the coronal transient can be understood as a natural consequence of the reconnection process. When two oppositely directed field lines reconnect, a
lower loop is formed rooted to the solar surface (the flare loop). In
addition, an upper loop disconnected from the surface is produced. We
suggest that this upper disconnected new flux provides the driving force
for the transient and prominence material. A schematic of the geometrical configuration of the transient and flare loop system is shown in
Figure 1.

The feasibility of this mechanism can be demonstrated by a crude
mathematical example. Using the same formulation of the MHD equations as
well as the same geometrical model for the transient geometry (an arcade
with a semi-circular top and radial legs) as employed by Pneuman (1979a),
the equation of motion for the transient can be written in the form

$$V \frac{dV}{dS} = \left(\frac{\alpha F^2}{8\pi\rho_o D_o S_o^4} - \frac{B_o^2}{4\pi\beta\rho_o S_o} - \frac{GM_\odot}{S_o^2} \right) \left(\frac{S_o}{S} \right)^2 \tag{1}$$

where V is the velocity, S the displacement from the solar center, F
the reconnected flux, ρ_o the reference density*, D_o and S_o the initial

*Here, the density should be interpreted as an average over the trans-
ient and prominence material.

width and displacement of the arcade, B_0 the initial field strengh in the transient at its top, $\beta = \tan\theta/(1+\tan\theta)$, θ being the half angle between the transient's legs, and α is a constant relating the field strength behind the transient to the reconnected flux. In deriving Equation (1), we have made the assumption $D \propto S$ (D being the width). This is borne out by the observations and by previous solutions of equations of this type (Pneuman, 1980). In order to relate F to the displacement, we note two properties of the reconnection process. Firstly, the strength of the fields that reconnect early in the event is expected to be much larger than of those that reconnect later — since these are lower down in the corona. Secondly, as the process proceeds upward into the corona, the velocity of the neutral point decreases. Hence, the rate of reconnection is faster during the early stages. For these two reasons, we expect the reconnected flux to increase sharply in the beginning but quickly approach a constant value when the increments to the total flux become negligible. This dependence of the flux on time should also be reflected in its dependence upon S since S is a monotonically increasing function of time. These considerations lead us to parameterize F as $F = K[1-(S_0/S)^k]$ where K is the limiting value of the flux and k determines how quickly the flux achieves this limiting value.

Now, the solution of Equation (1) is straightforward. Assuming $V = 0$ at $S = S_0$, the terminal velocity (at $S = \infty$) is given by

$$V^2_{terminal} = \frac{2k^2}{(k+1)(2k+1)}\left(\frac{\alpha K^2}{4\pi\rho_0 D_0 S_0^3}\right) - \frac{B_0^2}{2\pi\beta\rho_0} - \frac{2GM_\odot}{S_0} \; .$$

Thus, the terminal speed of the transient is increased, not only by larger magnitudes of reconnected flux (large K), but also by adding a given amount of flux at a faster rate (large k). A more complete and precise model of this process is currently under preparation (Anzer and Pneuman, 1979).

Finally, we touch on two additional flare associated phenomena which we believe can be understood within the context of the present model — moving Type IV radio bursts and proton events. It is generally believed that the Type IV burst consists of synchrotron radiation produced by relativistic electrons. We suggest that these electrons are energized by the reconnection process and injected into the upper closed fields where they remain trapped. Similarly, protons are injected into this same region behind the transient's leading edge. The unusually late arrival time of energetic protons at 1 AU after major flares could be explained as due to the inability of the protons to escape across the field lines of this closed geometry of the transient. Hence, the protons cannot reach the earth until the transient does.

REFERENCES

Anzer, U. and Pneuman, G.W.: 1979 (in preparation).
MacCombie, W.J. and Rust, D.M.: 1979, Solar Phys. 61, pp. 69-89.
Pneuman, G.W. 1979, Solar Flare Magnetohydrodynamics, (ed. E. Priest),
     Gordon and Preach Pub. Co., New York, London. Paris, (in press).
Pneuman, G.W. 1980, Solar Phys. 65, pp. 369.

DISCUSSION

Kahler: Which comes first: do you first have reconnection which then drives the filament upwards or does the filament first erupt, setting up the reconnection process?

Pneuman: For this mechanism to work, it does not really matter. One could visualize the prominence lifting allowing field lines underneath to collapse inward and reconnect. Alternatively, the reconnection could start first, providing additional magnetic flux beneath the filament which pushes it outward. It may be difficult to distinguish between these two possibilities observationally, but I tend to favor the latter explanation.

Kuperus: I am interested to know whether the prominence eruption is because of reconnection or just the opposite.

Pneuman: This is essentially the same question that Dr. Kahler asked. It is the fundamental question. I favor the reconnection starting first only for the reason that only one mechanism need be evoked for the whole process rather than having to explain the destabilization of the filament as well. The other possibility is that both occur together and are really a manifestation of the same process.

Newkirk: Your model appears to leave the corona in a state after the transient which is just the opposite from that observed. One of the outstanding characteristics of loop-like coronal transients is that the corona in the range 1.5 to 6 $R_\odot$ appears to be <u>open</u> after the passage of the transient with the "legs" of the loop visible for many hours. The final configuration of your model is one with closed fields or a coronal stream.

Pneuman: That's right. A coronal streamer is formed during this process and the field lines above the helmet (in the coronagraph's field of view) are indeed open, being held open by the solar wind. As the X-ray observations suggest , the reconnection rate and rise rate of the neutral point slows down, and the neutral point comes to rest at some height in the corona determined by the equilibrium conditions. Hence, the closed loops rooted to the surface never reach the field of view of the coronagraph and one only sees open field lines there as you said.

Sheeley: In their study of the Skylab/ATM observations of coronal transients, the HAO team emphasized that they found no evidence for the expulsion of magnetic flux from the Sun. In particular, although the curved front of each loop-like structure passed through the coronograph field of view, the ends of the "loop" always seemed to remain tied to the photosphere. Assuming that your reconnection-driven model is correct, why do you suppose the HAO team never saw the back end of a closed loop move outward across the field of view?

Pneuman: In the schematic I showed you can see that, in this model, the outer legs of the transient <u>do</u> remain rooted in the photosphere. It is only the very inner part that is disconnected. I don't think the white light photographs rule that picture out. The inner part of the transient appears rather confusing in the pictures.

I believe that you do see these isolated inner loops in the moving Type IV emission. This emission is of a globular shape and, since it is produced by relativistic electrons, they must be confined in a closed field geometry. Otherwise, the emission volume would expand dramatically.

Michels: During the observations of a white light transient, the frontal loops are visible throughout a long period, while the volume expands greatly - where does the mass come from that is necessary to maintain this luminosity?

Pneuman: This model, of course, does not consider the origin of material for either the flare loops or transient. I'm not sure that additional material is actually needed for the transient, but I can't give you a more definite answer than that.

Anzer: (Comment) There is enough mass already contained in the pre-event helmet.

PROPAGATION OF AN 'HD SHOCK IN THE VICINITY OF A MAGNETIC NEUTRAL SHEET

D. J. Mullan
Bartol Research Foundation of the Franklin Institute, University of Delaware, Newark, DE 19711, and
R. S. Steinolfson
University of Alabama, PO Box 1247, Huntsville, AL 35807

The acceleration of solar cosmic rays in association with certain solar flares is known to be highly correlated with the propagation of an MHD shock through the solar corona (Svestka, 1976). The spatial structure of the sources of solar cosmic rays will be determined by those regions of the corona which are accessible to the flare-induced shock. The regions to which the flare shock is permitted to propagate are determined by the large scale magnetic field structure in the corona. McIntosh (1972, 1979) has demonstrated that quiescent filaments form a single continuous feature (a "baseball stitch") around the surface of the sun. It is known that helmet streamers overlie quiescent filaments (Pneuman, 1975), and these helmet streamers contain large magnetic neutral sheets which are oriented essentially radially. Hence the magnetic field structure in the low solar corona is characterized by a large-scale radial neutral sheet which weaves around the entire sun following the "baseball stitch". There is therefore a high probability that as a shock propagates away from a flare, it will eventually encounter this large neutral sheet.

In a study of small-amplitude MHD waves, Uchida (1973) used ray-tracing techniques to show that such waves are refracted away from (towards) regions of high (low) Alfven speed. Although these results cannot be immediately applied to the present problem (in which the thickness of the neutral sheet is too small to satisfy the ray-tracing criterion) they suggest that a neutral sheet should act as a wave-guide, allowing essentially no transmission across the neutral sheet to the other.

In our work, we studied the encounter between a finite amplitude wave and a neutral sheet, using the two dimensional MHD code described by Steinolfson et al. (1978) in which the solar atmosphere is treated as a single fluid of polytropic index γ. Radiation is neglected, and there is no dissipation (except at shocks). As an initial state, we neglected solar gravity. For purposes of calibration, we first impose a uniform radial magnetic field over the entire computational grid. A pressure pulse is introduced at time t=0 at the equator, on the solar surface, and a shock wave sweeps out in two dimensions. After a time of order 500 seconds, we take a radial cut of the pressure variation at two lat-

M. Dryer and E. Tandberg-Hanssen (eds.), Solar and Interplanetary Dynamics, 323-326.

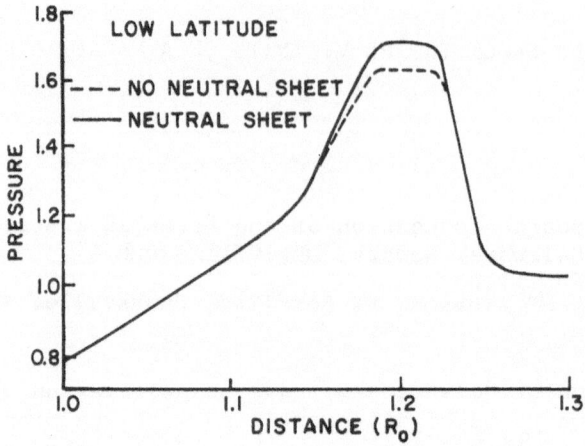

Figure 1

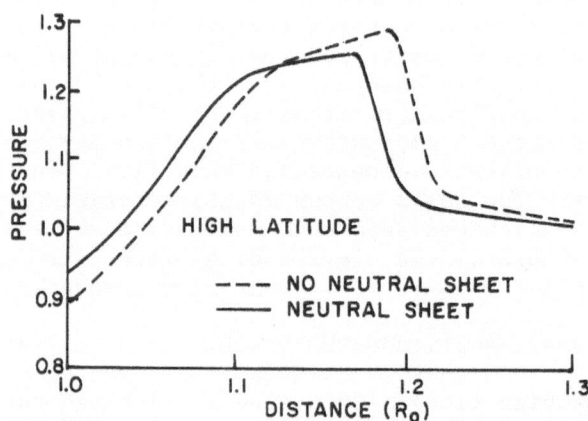

Figure 2

Figures 1 and 2.   Variation of gas pressure as a function of radial distance at two latitudes.   These curves have been derived
by taking constant-latitude cuts through a two-dimensional flow field at a time t $\sim$ 500 seconds after the "flare".
The flare occurred on the surface of the sun (1.0 R_0) at
low latitudes.   Figure 1 represents a cut at a latitude
which lies between the flare and the neutral sheet: notice that the presence of the neutral sheet strengthens
the shock in this region.   Figure 2 is a cut at a latitude which lies beyond the neutral sheet, remote from
the flare site: notice that the shock in this region
is weakened when the neutral sheet is present.

itudes, one at "low" latitudes, the other at "high" latitudes (dashed lines in Figures). Here, "low" and "high" are taken with reference to latitude θ_0 where the radial neutral sheet is to be located. We then inserted a neutral sheet at constant latitude θ_0. The magnetic field was taken to be uniform and directed outward from the sun at low latitudes, and uniform and inward at high latitudes. A smooth variation of field through zero was chosen, spread out over 12 grid points in latitude. To preserve equilibrium in the neutral sheet, the gas density was allowed to increase over the same grid points. Temperature was assumed to remain uniform at all points. Again, the same pressure pulse was inserted, and the pressure wave was followed. Radial cuts at the same two latitudes as before showed that the shock is strengthened at low latitudes, and weakened at high latitudes (solid lines in Figs. 1 and 2). The weakening of the shock at high latitudes is due partly to reflection of the shock off the density enhancement associated with the neutral sheet, and partly to dissipation of the shock in the neutral sheet. An examination of the temperature contours shows significant heating within the neutral sheet compared with the temperature which prevails at the same locations in the absence of the neutral sheet.

The weakening of the transmitted shock partially confirms our expectations based on Uchida's earlier work, although the neutral sheet does not block the transmission of the shock completely. Quantitative evaluation of transmission efficiencies in different regions of parameter space is currently underway.

The work of RSS was supported by NASA under Contract NAS8-33216 and by NOAA under Contract NOAA/04-78-B01-6. The work of DJM is partially supported by the National Science Foundation under Grant ATM-7820936. Acknowledgment is made to the National Center for Atmospheric Research, which is sponsored by the National Science Foundation, for the use of its computing facilities.

REFERENCES

McIntosh, P. A. 1972, Rev. Geophys. Spa. Sci. 10, p. 837.
McIntosh, P. A. 1979, paper presented at IAU Symp. No. 91, Cambridge Mass., August.
Pneuman, G. W. 1975, in IAU Symp. No. 57 (ed. G. Newkirk) [Dordrecht: Reidel] p. 35.
Steinolfson, R. S., Wu, S. T., Dryer, M., Tandberg-Hanssen, E. 1978, Ap. J. 225, p. 259.
Svestka, Z. 1976, Solar Flares (Dordrecht: Reidel).
Uchida, Y. 1973, in High Energy Phenomena in the Sun (eds. R. Ramaty and R. G. Stone) [Washington: NASA SP-342), p. 577.

DISCUSSION

Rosenau: (Comment) The effect that you have described is not a new one and in fact it is merely a composition of two well known phenomena: (a) shock moving in a decreasing magnetic field (like your case) is decelerated; (b) shock moving into increased density zone is weakened. Therefore, this work must be evaluated as a quantative treatment of these effects.

TWO-FLUID THEORY OF INTERPLANETARY SHOCK WAVES

P. Rosenau
Department of Mechanical Engineering
Technion, Israel Institute of Technology
Israel

ABSTRACT

A 2-fluid time-dependent analytical model of the perturbed solar
wind is presented. The expansion of newly emitted material, caused,
for instance, by the outburst of a solar flare, is simulated by a
spherical piston. For a given thermal conductivity in the limit of
strong coupling, one fluid flow in a thermally conducting medium is
recovered. A pattern of flow which resembles one-fluid flow in
adiabatic medium may be recovered if heat is removed from the perturbed
plasma into the propelling plasma. The perturbed flow consists of a
thermal precursor which is followed by a shock across which electrons
are isothermal while protons are compressed and heated. Finally, we
show that the post-shock rise and fall of density cannot be used to
distinguish piston-driven waves from blast waves.

I believe that I do not have to convince anyone in the audience
why at least a two-fluid description of the perturbed solar wind is
needed. For a Quiet Solar Wind (Q.S.W.) this was recognized more than
twelve years ago. However, construction of a time-dependent multi-
fluid theory of the solar wind is a matter infinitely more difficult.
What impinges upon our physical understanding are mathematical
difficulties. Not only is the process time dependent (and thus
essentially described by Partial Differential Eqs.) but to make things
worse its mathematical nature is not yet well understood. Psycho-
logically, this is probably a barrier that stopped many workers in the
field from learning the safe ground of ideal MHD theory which is
described by Symmetric Hyperbolic Eqs.; a well tamed mathematical beast.

The specific problem to be discussed here is: Consider material
emitted through the outburst of a solar flare that propagates into
the Q.S.W. and perturbs it. A simple model to describe the dynamics
of the perturbed medium is that of a piston expanding into the Q.S.W.
(modelled after Parker). The piston simulates the interface between
the newly emitted and compressed "old" material.

M. Dryer and E. Tandberg-Hanssen (eds.), Solar and Interplanetary Dynamics, 327-331.
Copyright © 1980 by the IAU.

Applying group invariant methods (in the scientific folklore known also as similarity transformations) we have constructed analytical solutions to the above problem taking into account the ambient inhomogeneity, magnetic field, finite thermal conductivity and ion-electron coupling. As the full analysis of the problem will be published (J.G.R., Vol. 84, p. 5897) we shall present here only a general overview of our results.

A.   The equations we employ are   (standard notation):

$$\frac{dn}{dt} + \frac{n}{r^2} \frac{\partial}{\partial r} (r^2 V) = 0; \quad m_i n \frac{dV}{dt} + \frac{\partial}{\partial r} n(T_e + T_i) = 0$$

$$\frac{n}{\gamma-1} \frac{dT_e}{dt} - T_e \frac{dn}{dt} = -\frac{1}{r^2} \frac{\partial}{\partial r} (r^2 q_e) + Q$$

$$\frac{n}{\gamma-1} \frac{dT_i}{dt} - T_i \frac{dn}{dt} = -Q$$

$$\frac{d}{dt} = \frac{\partial}{\partial t} + V \frac{\partial}{\partial r}; \quad P_e = nT_e; \quad P_i = nT_i; \quad \gamma = \frac{5}{3}.$$

Q is the rate of heat exchange between ions and electrons, and q_e is the radial electron heat flux. In a simple, monatomic, fully ionized plasma they are given by

$$Q = Q_0 n^2 T_e^{-3/2} (T_i - T_e)$$

$$q_e = -\lambda_e \frac{\partial}{\partial r} T_e \qquad \lambda_e = K_0 T_e^{2.5}.$$

The classical expression for Q and λ_e are applicable only near the sun. To account for the continuous decrease in τ_e, the electron collision time, we introduce a mathematical artifact, namely $\tau_e \sim (T_e^{3/2}/n) \left[\frac{\alpha R_\odot}{r}\right]^y$, α, y = const. which effectively inhibits thermal conductivity and increases Q. As for the motion of the disturbing piston we assume that its position is given as

$$r_p(t) = At^N, \quad A = const.$$

The ensuing wave is assumed to propagate into a cold Q.S.W. with expansion velocity V and density $n = n_o(R_\odot/r^2)(n_o = \text{const.})$.

B.   One can show that the disturbed region contains a shock that for electron is isothermal while the density, ion temperature, and two pressures may jump across the shock which satisfies jump conditions appropriate to two-fluid theory. On the piston in addition to its velocity one has to specify electron temperature or (electron) heat flux via the piston. It is important to note that in our theory the shock is preceeded by a non-linear electron thermal wave.

C.   Main results. The following figures were obtained as a result of numerical integration of ordinary differential equations which are the mathematical consequence of the invariance of our equations of motion under one parameter group of (stretching) transformations. Asymptotic analysis was used to bypass singular points.

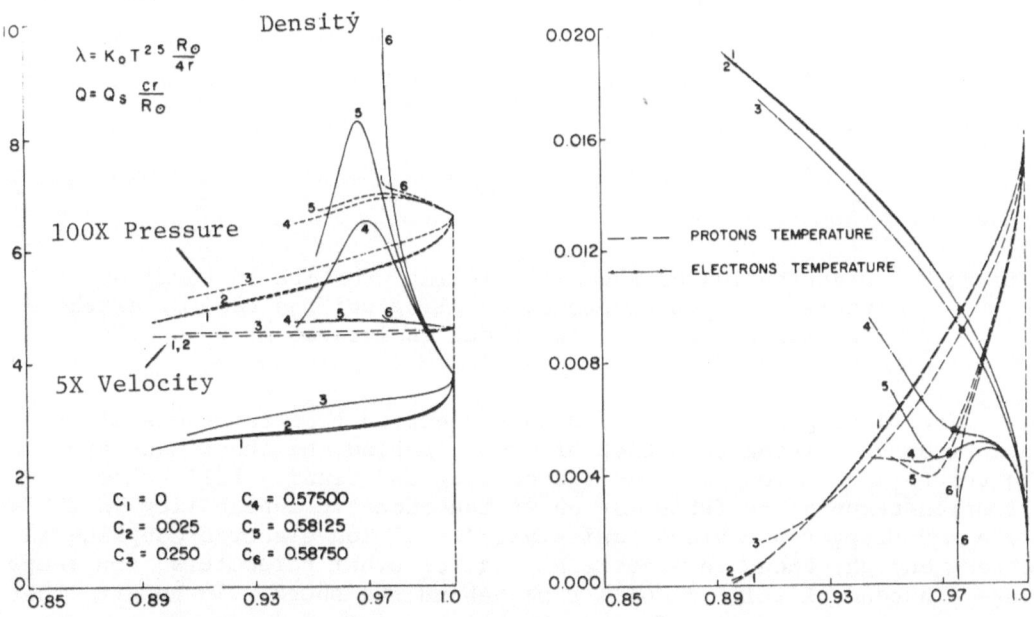

Figure 1.   Distribution of the self-similar profiles in terms of variable electron-proton coupling. Note the non-monotone change of density as the coupling is varied. The front of the disturbance is normalized to $\xi = 1$ (1 AU) and the shock is located at $\xi_s = 0.9999$.

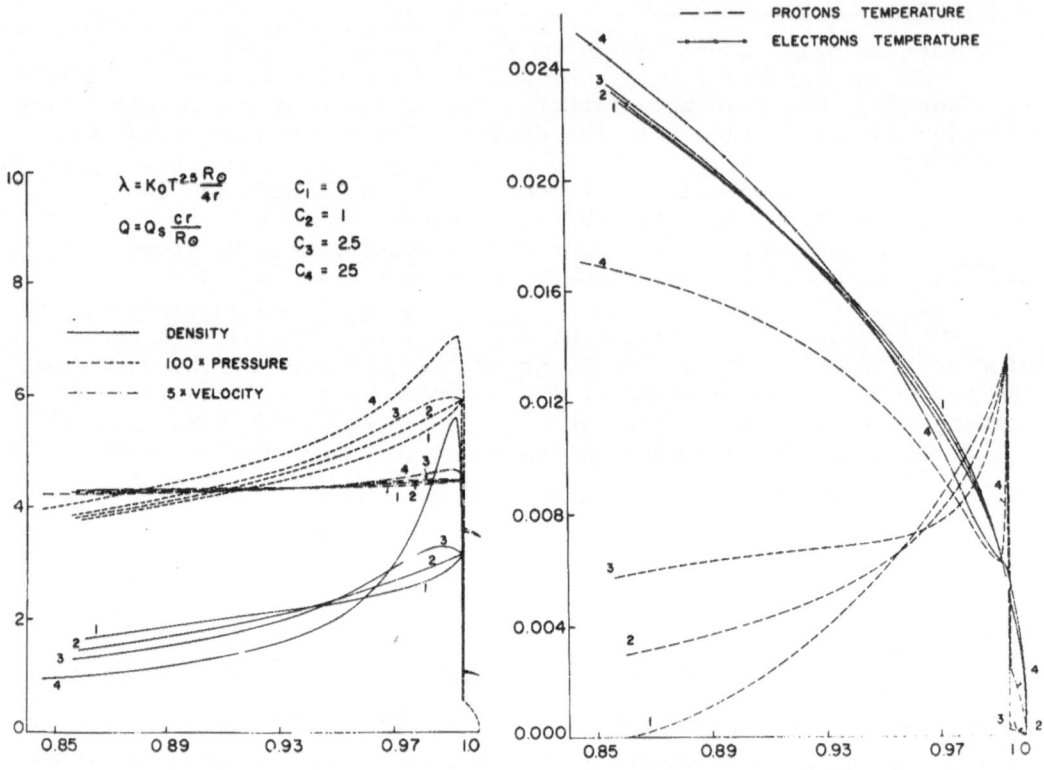

Figure 2. Distribution of the self-similar profiles in terms of variable electron-proton coupling for the case of extended thermal precursor. The shock is located at ξ_s = 0.995.

The first general result is that electron temperature is always higher on the piston than that of ions. Behind the shock ions are hotter than electrons but undergo cooling and finally fall below electron temperature (this may onset ion-acoustic instability which in turn may trigger anomalous resistivity). If ion-electron coupling is strong enough, then, in a certain range of other parameters, ion temperature (instead of falling) will rise behind the shock. In general there is a very large variety of post shock flows. This is in clear contradistinction to the one fluid theory where only one pattern was observed: density increases (decreases) and the temperature decreases (increases) behind the shock. (The second case is realized in non-adiabatic one-fluid theory; Rosenau and Frankenthal, Ap. J., 1976). Since these phenomena are caused by an expanding piston R and F events cannot be used to distinguish between piston and blast events. Also a change from one pattern into another depends continuously on the value of the different (four in number) parameters that enter into the problem and thus support the observation by Burlaga that each of the experimentally observed shock waves today has a unique pattern of behavior.

Finally, to obtain in the limit of strong coupling the one-fluid models, the heat flux via the piston must be properly treated. The adiabatic one-fluid model may be obtained only if heat is removed from the flow via the piston (the propelling stream is cooler than the propelled one). If heat is added to the flow, one obtains in the limit a one-fluid flow in a thermally conducting medium. The fact that we do observe adiabatic-like patterns in solar wind which is a heat-conducting medium, could thus be merely a result of heat extraction from the perturbed flow.

DISCUSSION

Ivanov: What could you say about the structure of the shock front?
Rosenau: It is very similar to the steady state shock structure in the 2-fluid description of the plasma. (See, for instance, the book by Zeldovich and Raizer.) There is however, an important difference, namely, there the downstate does not relax to a uniform state. Indeed we are able to prove that on the piston, electrons are hotter than protons though immediately behind the shock the inverse is true in all but the weakest shocks.

TRANSIENT DISTURBANCES OF THE OUTER CORONA

R. T. STEWART
Division of Radiophysics, CSIRO, Sydney, Australia

1. INTRODUCTION

I consider it a great honour to give this talk today in place of my friend and colleague, the late Dr. S. F. Smerd, who died last December. All who knew Steve miss his enthusiasm and keen insight into solar physics.

For recent review papers on observations of coronal transient disturbances I refer you to the works of Hildner (1977), Dulk (1979) and MacQueen (1979). Today I will give a brief historical review of this subject, followed by a detailed description of three events observed at radio and whitelight wavelengths, concentrating mainly on the radio evidence. I will not discuss the beautiful soft X-ray, EUV and green-line observations of the lower corona, since they are outside the scope of this review.

2. EARLY OBSERVATIONS

Perhaps the first recognition of a coronal transient was the eclipse drawing by G. Tempel on 18 July 1860 (Eddy, 1974). Next came the detailed observations of eruptive Hα prominences by Secchi and Respighi in the 1870s (Abetti, 1934). From the early 1900s onwards many photographs of eruptive prominences were taken from various observations, including the outstanding coronograph observations of Lyot. Then in the 1940s Bartels and others suggested that geomagnetic sudden commencements which follow large solar flares by about two days could be due to the arrival at the Earth of solar ejecta.

Shock Waves

The first clear indications that geomagnetic disturbances were caused by shock waves came from the metre-wavelength observations of Type II bursts (Wild et al., 1953), which were interpreted as being due to the passage of a shock wave through the corona. An example is shown in Figure 1(a). There is a clear fundamental second-harmonic structure in these bursts. Theoretical work by Ginzburg and others

333

M. Dryer and E. Tandberg-Hanssen (eds.), Solar and Interplanetary Dynamics, 333-355.

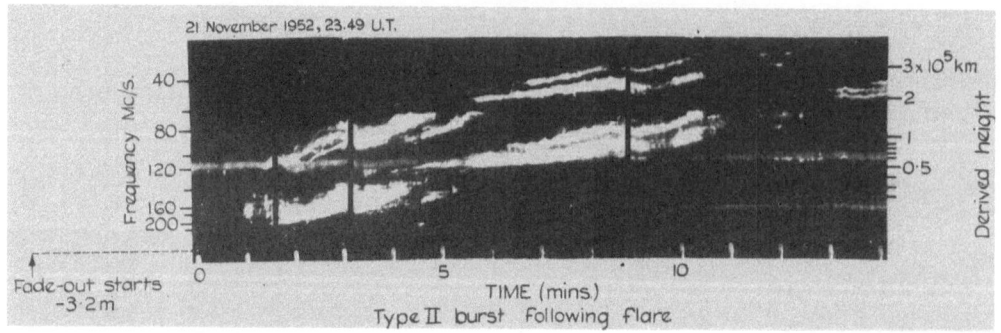

Fig. 1(a) - One of the first dynamic spectral records of a Type II
   solar radio burst. Note the splitband and fundamental-harmonic
   structure (Wild et al., 1953).

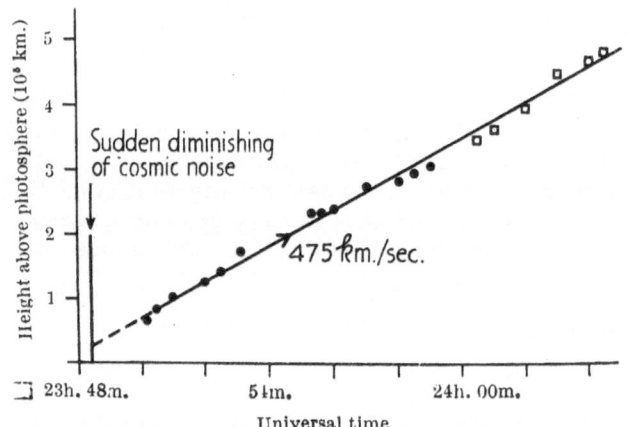

Fig. 1(b) - Derived height-time plot of the Type II burst of Figure 1(a).
   Fundamental frequencies were converted to heights using a standard
   plasma density model (Wild et al., 1953).

showed that the emission was plasma radiation from mildly relativistic
electrons (10-100 keV), presumably accelerated locally in the corona
by the passage of the shock. An early height-time plot of one of these
early events (Wild et al., 1953) is given in Figure 1(b), where it can
be seen that the slope of the Type II burst extrapolates back nicely to
the start of the solar flare.

   Later two-dimensional radioheliograph pictures at 80 MHz (Kai,
1970) showed the extent of an advancing front (or bow wave) of a shock-
wave disturbance in the corona (Fig. 2a). Note that this advancing
front is travelling at a slightly lower projected velocity than that of
the Type II shock wave (Fig. 2b). From the 1960s interplanetary shock
waves have been observed from spacecraft (Hundhausen, 1972). Kilometre

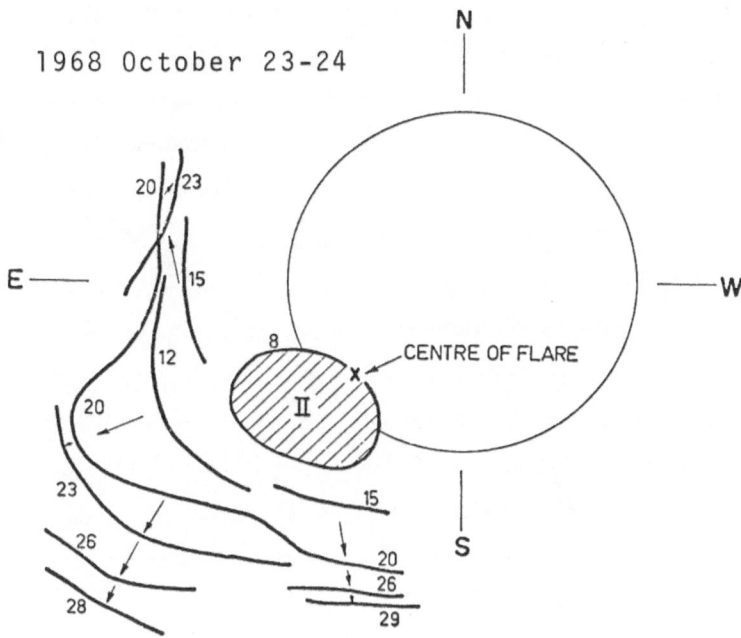

Fig. 2(a) - 80 MHz Culgoora radioheliograph observations on 1968 October 23-24 of an "advancing front" moving Type IV burst (heavy outlines) and a Type II burst (hatched region). Figures on the diagram indicate the elapsed time in minutes since the start of the flare (Kai, 1970).

observations of a Type II burst from IMP-6 (Malitson et al., 1973) showed that the Type II shock wave reached the Earth's orbit (Fig. 3a), thus confirming that the Type II burst is closely associated with the interplanetary shock wave.

We now know that large flares produce not only Type II bursts and interplanetary shock waves but also fluxes of energetic electrons and protons which sometimes escape to the Earth's orbit (Fig. 3b). These cause cosmic ray decreases and auroras.

Expanding Magnetic Arches

As well as the shock-wave-associated disturbances we sometimes see evidence for slower, expanding magnetic arches in the corona. Note the beautiful helical structure in the Hα threads of the eruptive prominence "Grandpa" reproduced in Figure 4(a) (Akasofu and Chapman, 1972). Radio evidence also exists for expanding magnetic arches. For example, the event of 1968 November 22 (Wild, 1969) showed two oppositely polarized (in the circular sense) sources on the legs of an expanding arch and another unpolarized source at the top (Fig. 4b). The leg sources were attributed by Wild to plasma radiation at the 80 MHz plasma level and the top source to gyrosynchrotron emission.

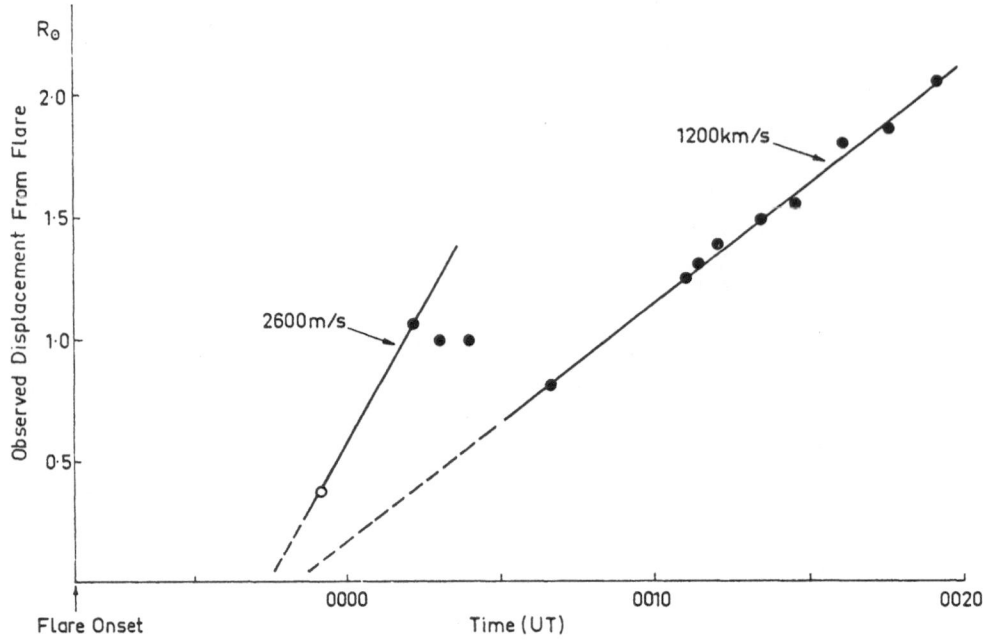

Fig. 2(b) - Plots of the Type II and IVM source displacements with time
of the event of Figure 2(a). Note the Type II plot has the steeper
slope and precedes the IVM plot (Kai, 1970).

Another majestic expanding arch which might have been magnetically
controlled is shown in Figure 5(a). This transient event was observed
in white light by the HAO coronagraph aboard Skylab on 1973 June 10
(MacQueen et al., 1974; Hildner et al., 1975). The dark areas indicate
the excess coronal mass moving slowly outwards through the corona to
heights ⁓6 $R_\odot$, leaving behind an underlying region of depleted coronal
material.

Evidence for fast changes in magnetic field during coronal
disturbances can be seen in Figure 5(b). This figure shows changes
due to Faraday rotation of signals from Pioneer-6 as they pass
tangential to the solar limb at heights ranging from 6 to 11 $R_\odot$ (Levy
et al., 1969). Such changes may be associated with the expansion of a
magnetic arch during the transient. The active region producing the
solar flares associated with these transients was probably the same one
which produced the expanding arch event of Figure 4(b).

Ejected Plasmoids

The first moving radio sources were detected by the Nancay
interferometer (Boischot, 1958). Since then many examples of isolated
moving Type IV sources have been observed by the Culgoora radioheliograph (Smerd and Dulk, 1971). One of the most spectacular
occurred on 1969 March 1-2. Nicknamed "Westward Ho", it moved out to

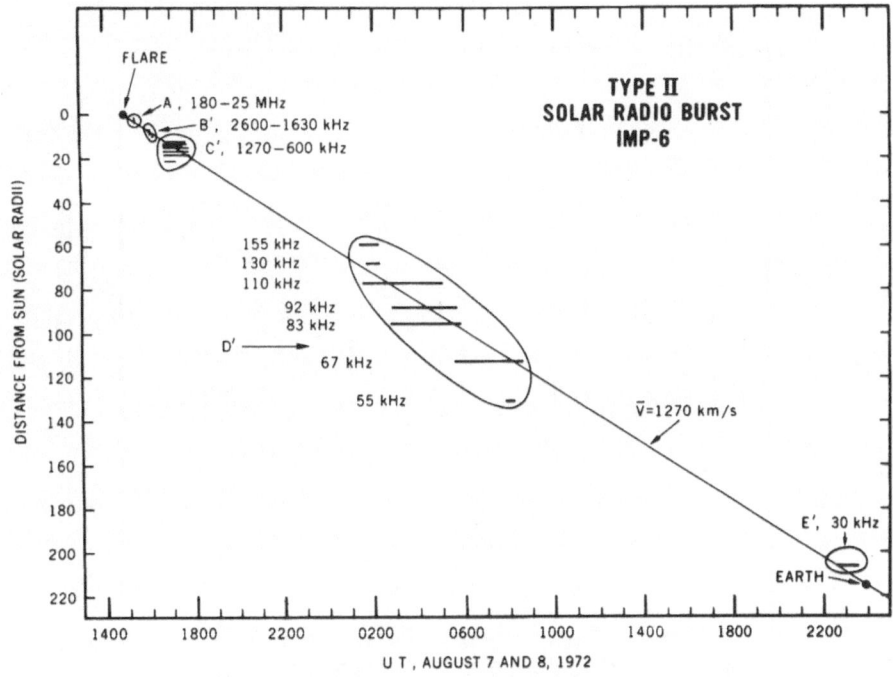

Fig. 3(a) - Derived height-time plot of a Type II burst observed at
  kilometre wavelengths (Malitson et al., 1973).

a height ∿6 R$_\odot$ before it finally faded after separating into two
oppositely polarized sources (A and B of Fig. 6b) (Riddle, 1970). The
speed of the moving sources was fairly constant at ∿270 km s^{-1}; it
was considerably slower than that of the Type II burst and Hα spray
which preceded it (Fig. 6a). To explain the high degree of circular
polarization observed in moving Type IV sources of this kind by
gyrosynchrotron emission mechanism one must assume that the (plasmoid)
source contains a magnetic field of several gauss (Dulk, 1973).

3. SOME RECENT OBSERVATIONS

    Probably the most complete observation of a coronal disturbance to
date is the event of 1973 Jan. 11 (Stewart et al., 1974a) in which an
Hα flare-spray, a K-coronameter transient, a moving Type IV source and
a white-light expanding cloud were observed at successive heights in
the corona. The Hα spray, moving Type IV source and the brightest part
of the white-light cloud (or the main part of the ejected mass)
probably constituted a piston moving outwards behind the Type II shock
wave and the leading edge of the white-light cloud (Fig. 7b). Stewart
et al. (1974a) estimated that ∿10^{16} g of excess mass was ejected in the
cloud and that only ∿10% of this mass could be accounted for by the
K-coronal depletion observed at heights ⩽1.5 R$_\odot$ (Fig. 7a). Comparable

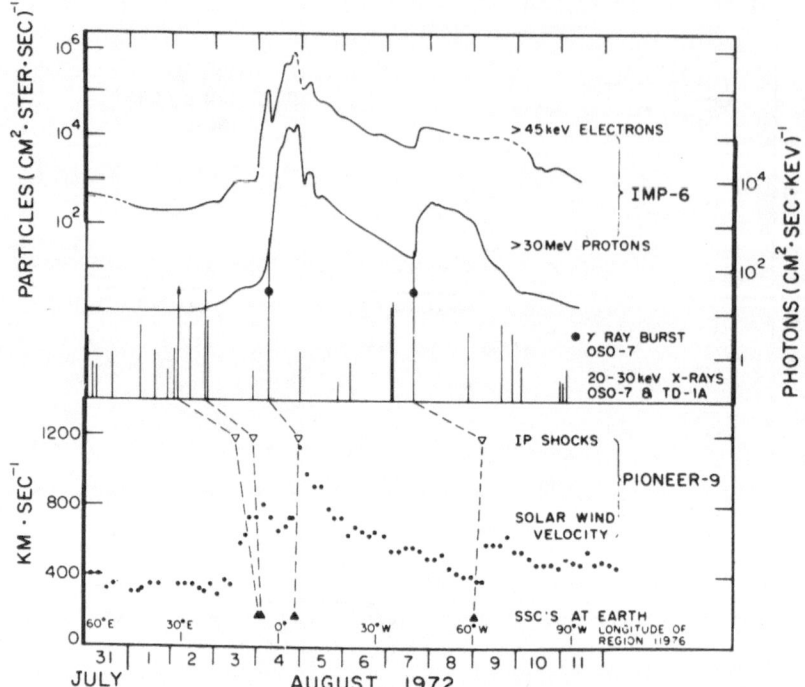

Fig. 3(b) - A summary of solar non-thermal phenomena for the 1972
July 31 to August 11 period, taken from the World Data Center "A"
report UAG-28 and Solar Geophysical Data monthly reports. Except for
the Pioneer 9 plasma observations at 0.78 AU all the data are from
Earth-orbiting spacecraft. Note the correspondence between energetic
particle injections, shock waves, γ-ray bursts, and intense hard
X-ray bursts (Lin, 1977).

or slightly higher masses are measured behind interplanetary shock
waves (Hundhausen, 1972; Hirshberg et al., 1972).

    One of the outstanding achievements of the HAO coronagraph on
Skylab was the detection of loop transients. A beautiful example of a
white-light loop transient is shown in Figure 8 (MacQueen et al., 1974;
Gosling et al., 1974). Another such event which occurred on
1973 September 14/15 was associated with a slow-drift continuum radio
event (SDC) observed by the Culgoora radioheliograph. The radio continuum sources (A and B of Fig. 9c) were stationary and occurred at
progressively greater heights at lower frequencies. A third source C
at 160 and 80 MHz was associated with a Type I burst storm in the later
stages of the event. From the height-time plot of Figure 10 it can be
seen that the radio continuum occurs at a particular height (or
frequency) soon after the passage of the leading edge or front of the
white-light disturbance and ends with the passage of the white-light
loop.

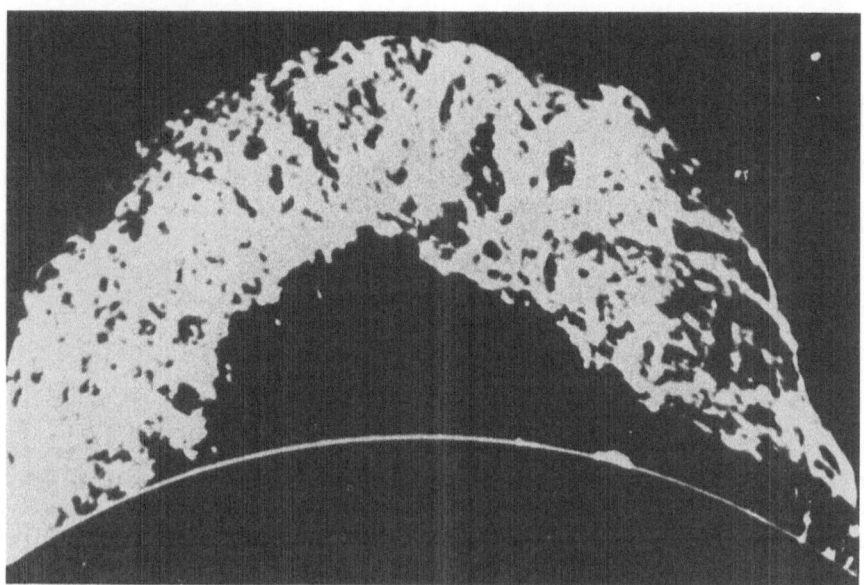

Fig. 4(a) – Eruptive prominence photographed by W. Roberts at Climax in Hα light with a polarizing filter, 1946 June 4^d,69 UT.

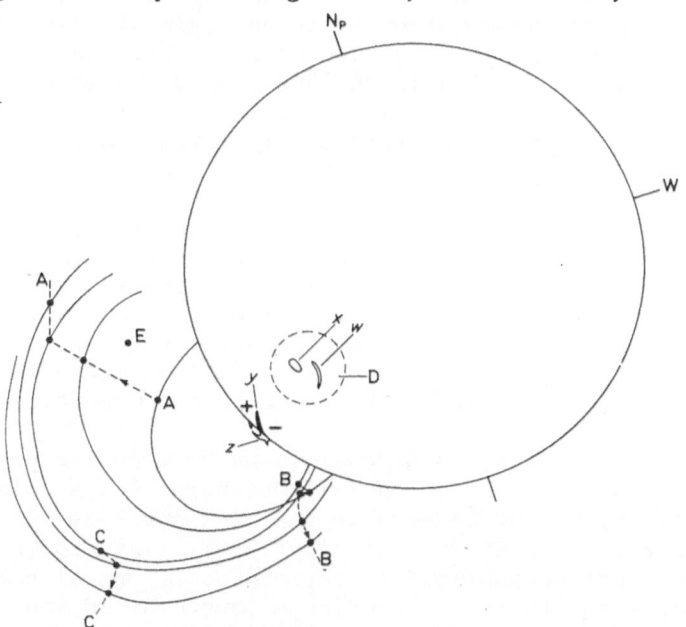

Fig. 4(b) – Observed geometry of the outburst of 1968 November 22. The loops beyond the limb indicate the evolution of the expanding magnetic structure of the moving type IV burst inferred from 80 MHz Culgoora radioheliograms. A,B,C,D and E refer to radio sources, w and x are Hα flares, y is a dark filament which formed during the outburst and z is an active prominence. The + and - signs indicate magnetic polarity (Wild, 1969).

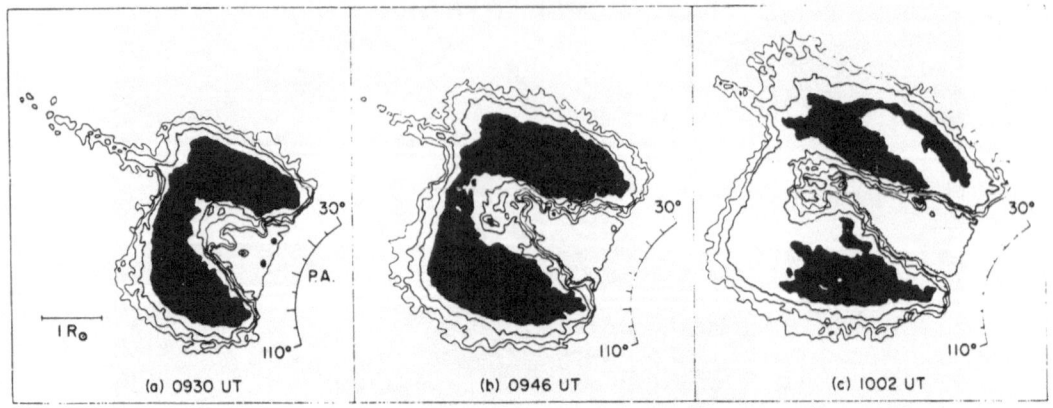

Fig. 5(a) - Contour plots of excess mass (darker colour indicating
higher mass) of an eruptive prominence event observed in white light
by the HAO coronagraph on Skylab (Hildner et al., 1975). See also
white light picture of Fig. 8 which shows several loop structures.

    Dulk et al. (1976) have suggested a model for this event in which
the energetic electrons emitting the radio continuum event are
accelerated locally in the corona by the passage of a shock wave - the
particles emitting gyrosynchrotron or second harmonic plasma radiation
in the region between the leading edge of the disturbance and the white-
light loop (Fig. 10). No clearly defined Type II burst was observed to
confirm the existence of a shock wave. However, this is not surprising,
because the active region which (presumably) produced this flare event
was ~30° behind the western limb of the visible disk.

    Dulk et al. (1976) were able to estimate the kinetic and magnetic
energy densities at various positions through this coronal transient
event. At most places the magnetic energy density was about 10 times
greater than the kinetic energy density except at the fastest moving
material, where the two were comparable. This suggested that the whole
coronal transient event was probably *magnetically controlled* (see later
discussion) even after allowing for potential and magnetic energy input.

    The third event I want to discuss is an Hα eruptive prominence and
a moving Type IV burst observed on 1977 October 4-5. A coloured
illustration of this event appears on the back cover of the publication
containing the reference Stewart et al. (1978). This coloured photograph shows radioheliograph source centroid positions of both a RH
circularly polarized stationary continuum source at 80 and 160 MHz and
a LH polarized moving Type IV source at 160, 80 and 43 MHz. The moving
source progressed outwards with time along the same direction as the
Hα ejecta. Figure 11 shows that for a period ~30 m the fastest moving
Hα ejecta were *coincident* with the moving Type IV source.

    Stewart et al. (1978) have estimated that the magnetic energy
density obtained by assuming a plausible radio emission mechanism for
the moving Type IV burst (such as second-harmonic plasma or low-harmonic

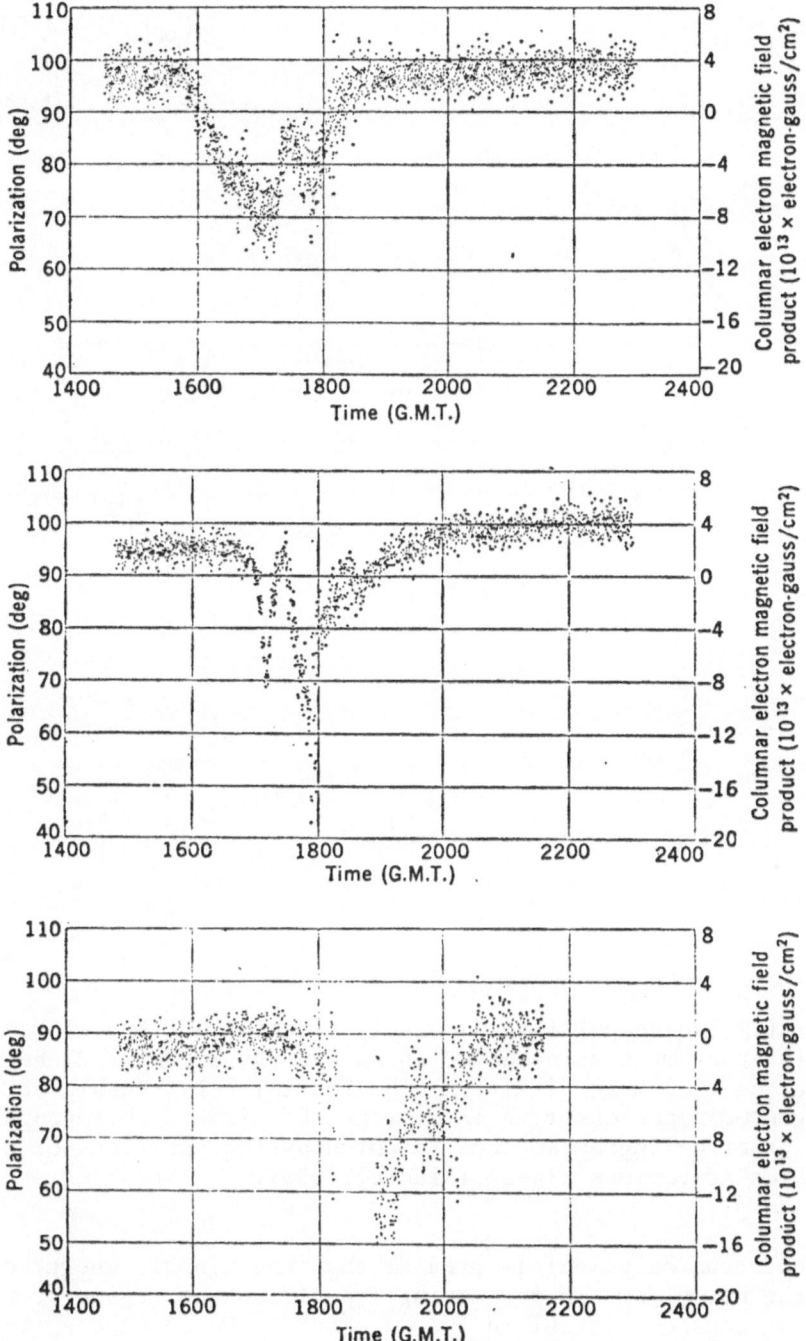

Fig. 5(b) – Plots of the polarization changes (due mainly to Faraday rotation in the corona) of a radio signal from Pioneer-6 passing tangentially to the limb with closest approach of 10.9 $R_\odot$ on 1968 November 4 (top figure), 8.6 $R_\odot$ on 1968 November 8 (middle figure), and 6.2 $R_\odot$ on 1968 November 12 (bottom figure). After Levy et al. (1969).

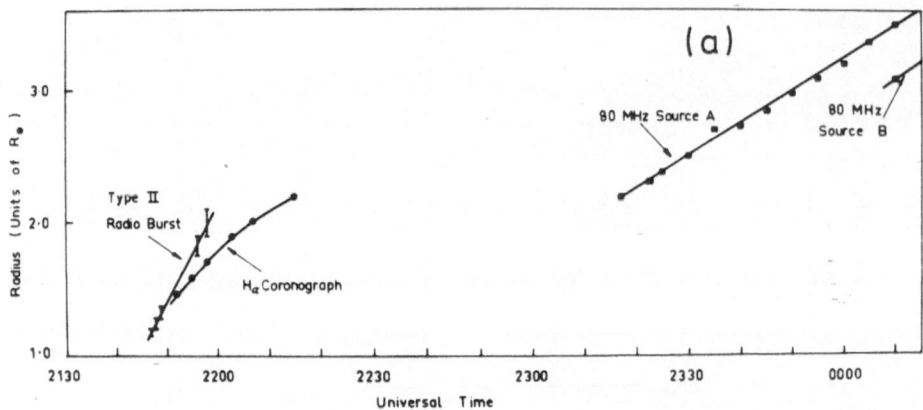

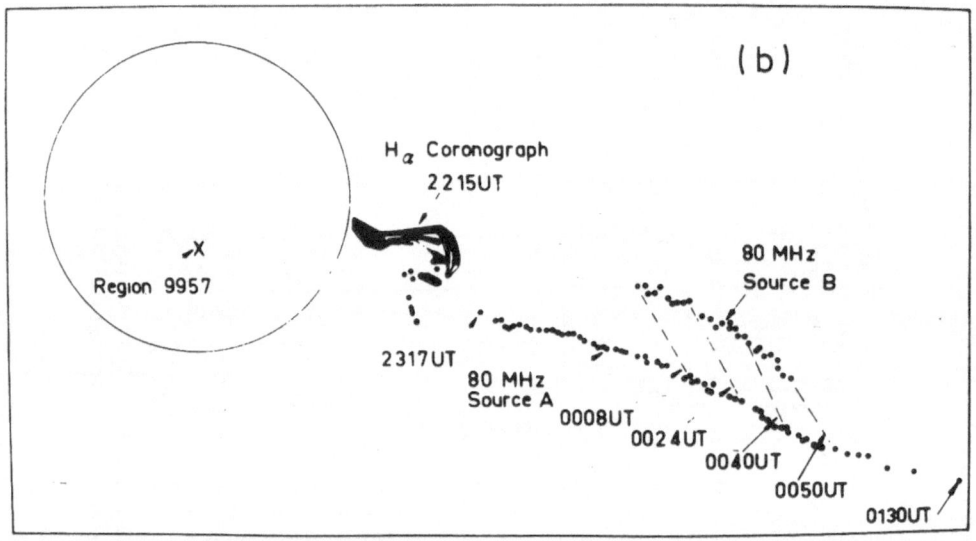

Fig. 6 - (a) The temporal relationship between the Type II radio burst,
   the rising prominence observed in Hα and the 80 MHz radioheliogram
   sources on 1969 March 1-2.   (b) The spatial relationship between the
   rising prominence observed in Hα (sketched from a photograph) and the
   80 MHz radioheliogram sources (denoted by the positions of the
   centroids at various times) (Riddle, 1970).

gyrosynchrotron emission) is greater than the kinetic energy density of
either the hot plasma in the moving Type IV source region or the
cooler and denser material in the Hα ejecta.   Hence once again it appears
that the coronal transient is *magnetically controlled*.   Unfortunately
there were no white-light observations during this event, but it seems
likely from previous evidence that a white-light transient occurred.

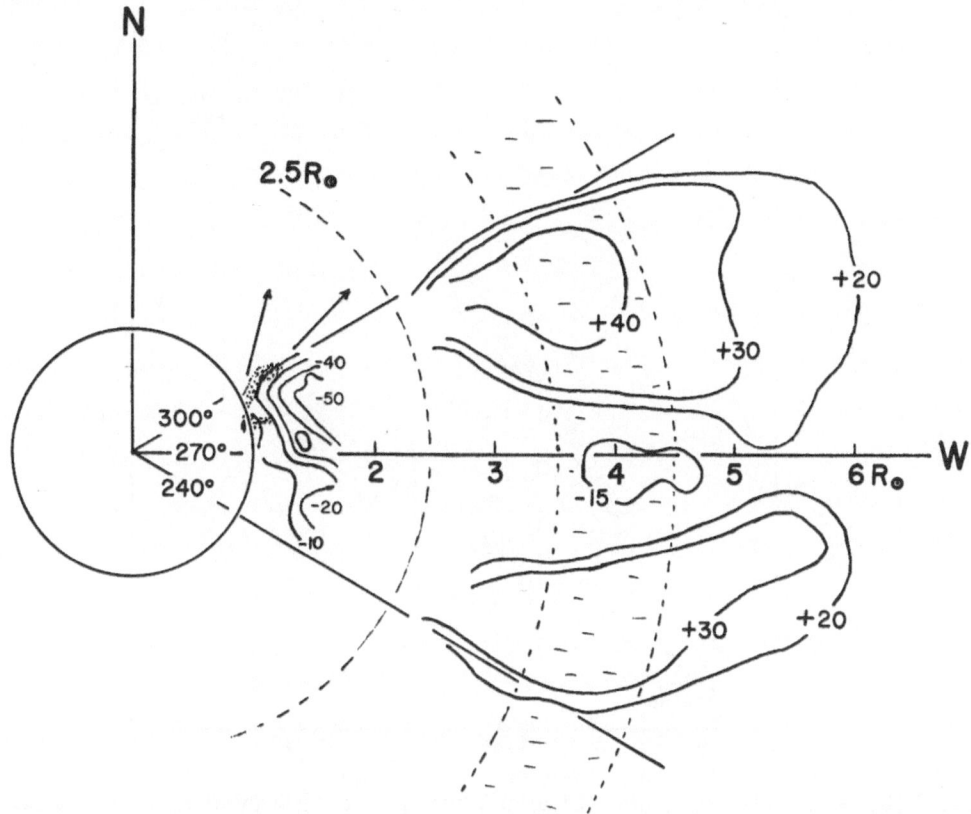

Fig. 7(a) - Composite plots of the percentage change of the brightness
of the K-corona (full lines) during the coronal disturbance of
1973 January 11. The contours below 1.6 $R_\odot$ were obtained by subtracting fixed-height scans taken by HAO coronal activity monitor at
different times before and after the transient event. The contours
between 2.5 and 6.0 $R_\odot$ were obtained by subtracting the OSO-7
coronagraph picture at 00^h14^m from the one at 02^h28^m UT. The
stippled region and the arrows indicate the Hα spray and ejection
cone. The dashed arc at 2.5 $R_\odot$ indicates the outer edge of the OSO-7
coronagraph occulting disk. The contours show the tangentially
polarized component; the polarization of the OSO-7 picture was
tangential everywhere except in the hatched region between 3.5 and
4.5 $R_\odot$, where it was radial. The direction of the occulter support
of the coronagraph was along the west axis.

4. PROPERTIES OF CORONAL TRANSIENTS

Speed of Transients

     The observed speeds of moving Type IV bursts and most white-light
excess mass transients vary from ~100 km s^{-1} to $\geq$1000 km s^{-1} (see e.g.
Smerd and Dulk, 1971; Gosling et al., 1976; Hildner, 1977) - i.e.
from about the sound speed in the corona (~150 km s^{-1}) to greater than
the Alfvén speed (~400 km s^{-1}).

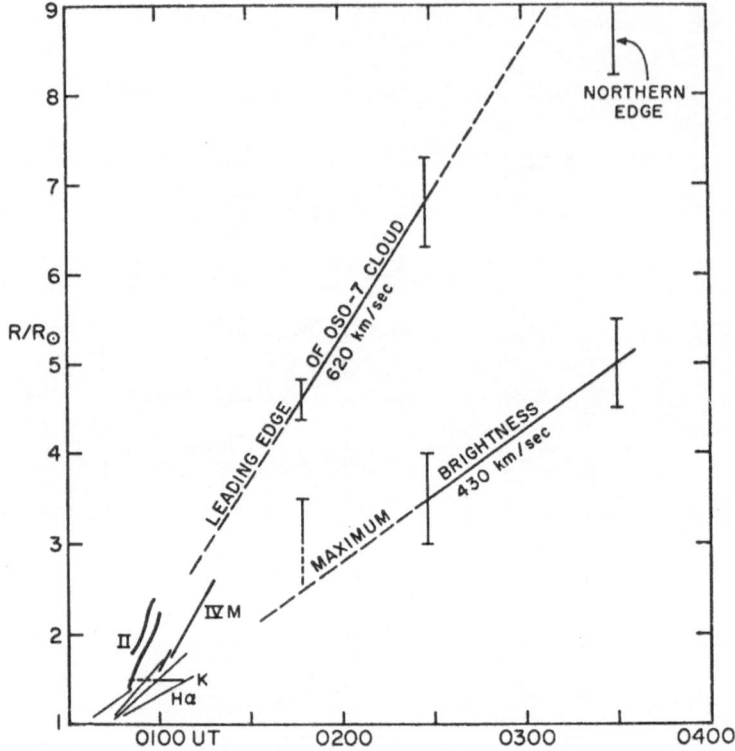

Fig. 7(b) - Combined height-time plots of the observed moving sources
in the coronal disturbance of 1973 January 11. The height of the
leading edge of the white-light cloud as well as the height of the
maximum excess brightness of the cloud is labelled accordingly. The
times and errors of measurement for the OSO-7 cloud are indicated.
At 03^h30^m only the northern edge of the cloud was in the field of
view of the coronagraph. Also shown are the trajectories of the Hα
spray material (light lines), the Type II and moving Type IV bursts
(labelled II and IVM), and the duration of the K-corona transient at
1.5 $R_\odot$ (labelled K). The projected radial velocities are 430 km s^{-1}
(the region of maximum brightness in the OSO-7 white-light cloud),
620 km s^{-1} (the leading edge of the OSO-7 cloud) and 800 to 1200 km s^{-1}
(the Type II burst) (Stewart et al. 1974a).

        According to Gosling et al. (1976) most flare-associated
transients (FLA) have speeds >500 km s^{-1} and these are associated with
Type II or IV radio bursts. On the other hand, non-flare events such
as eruptive prominences (EPL) have velocities <500 km s^{-1} and are not
associated with Type II/IV bursts. Hence it is reasonable to assume
that the fastest transients are associated with coronal shock waves.
In one event, 1973 September 7, Gosling et al. (1975) were able to
show that the ejected mass seen in white light and the inferred
mechanical energy input were sufficient to explain an interplanetary
shock wave event detected by Pioneer 9.

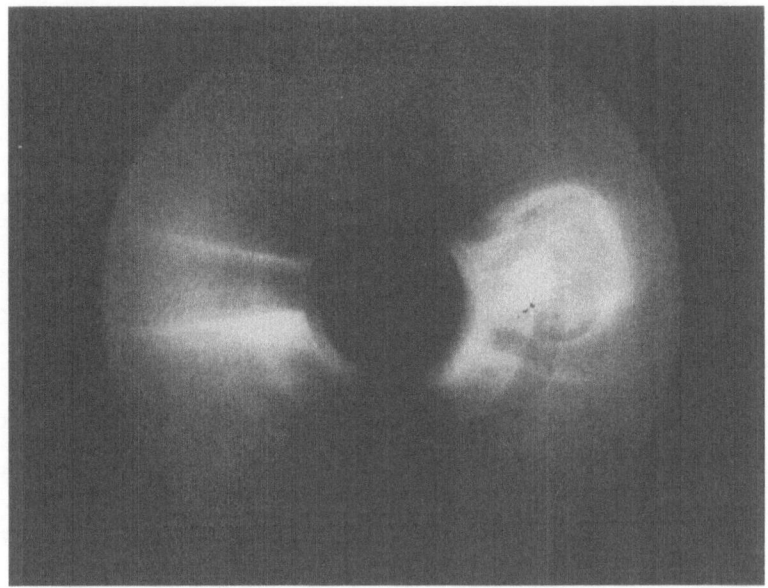

Fig. 8 - HAO white-light coronograph picture of a loop transient event on
1973 June 10 resulting from an eruptive prominence on the east limb of
the Sun (MacQueen et al., 1974).

Excess Mass and Mechanical Energy

Table 1 lists the excess mass and mechanical energy input for five
FLA events and five EPL events detected by OSO-7 or Skylab. It can be
seen that the FLA events are generally more energetic ($E \approx 10^{31-32}$ erg)
by an order of magnitude than the EPL events ($E \approx 10^{30-31}$ erg) (with the
exception of the EPL event of 1973 August 10), but the two types of
events contain about the same excess mass, $M \approx 10^{15-16}$ g. Essentially the
difference is due to the faster speed of the FLA events. According to
Hildner (1977), about 3% of the mass escaping to 1 AU from the Sun during
Skylab period was in the form of coronal transients, and during solar
maximum the amount could be as high as 10%. Transients are often preceded by a fore-runner (Jackson and Hildner, 1978) which contains about
10% of the transient mass.

Magnetic Energy Release

Dulk (1973) has estimated that ~10^{31-32} erg of magnetic energy are required to explain the properties of a moving Type IV burst source by gyrosynchrotron emission. A similar argument for the 1977 October 4-5 event described above requires ~10^{31} erg (Stewart et al., 1978). For two white-
light and radio transient events, Dulk et al. (1976) and Gergeley et al. (1979)
require 10^{31-32} erg of magnetic energy. McLean and Dulk (1978) calculated that
the speed and duration of Type II bursts imply that >10^{31} erg of
(mainly) transported magnetic energy is involved in a coronal disturbance.

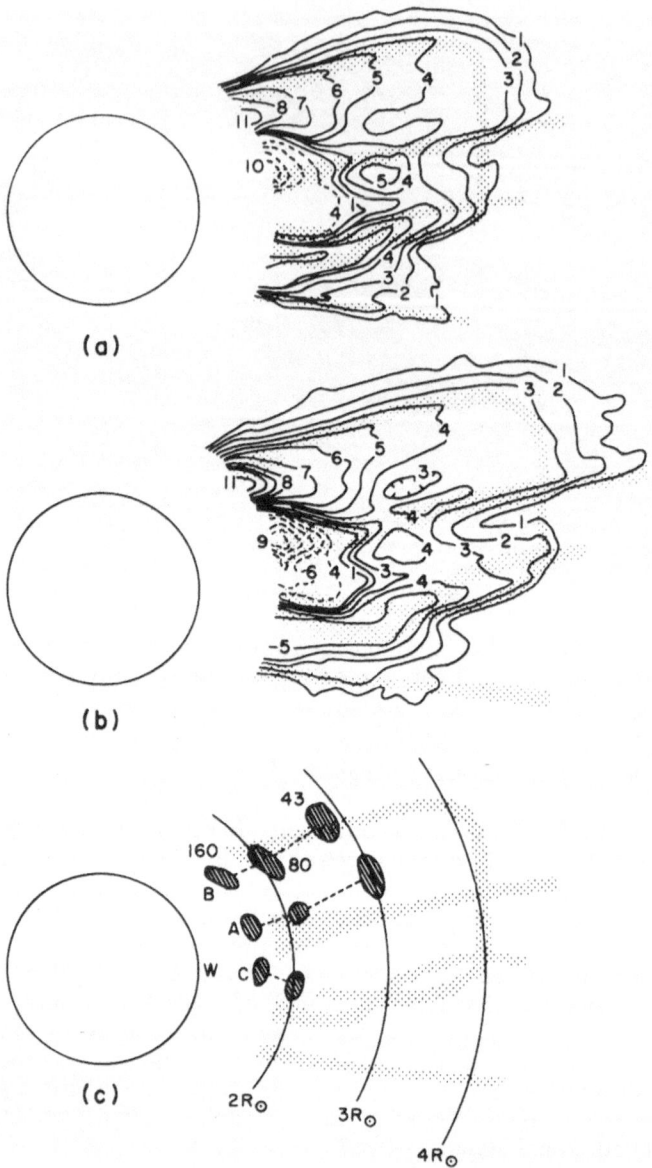

Fig. 9 – Mass (or density) contours of the excess material in the corona on
1973 September 15 at 00^h32^m (a) and 00^h41^m (b). The stippled areas show the
location of the principal visible structures at the two times. The hatched
areas in (c) represent the positions of the Culgooraradioheliograph sources;
all but the source "C" had disappeared before 00^h32^m, the time of the
visible structures. Crosses show the positions of the active regions, on
and beyond the limb. Dashed contours indicate rarefaction regions where
the material was *depleted* relative to the pre-event corona. Sample contour
levels in terms of mass, electron density and plasma frequency (the latter
two obtained by assuming a disturbance depth of 0.5 $R_\odot$) are given in the
Table of Dulk et al. (1976).

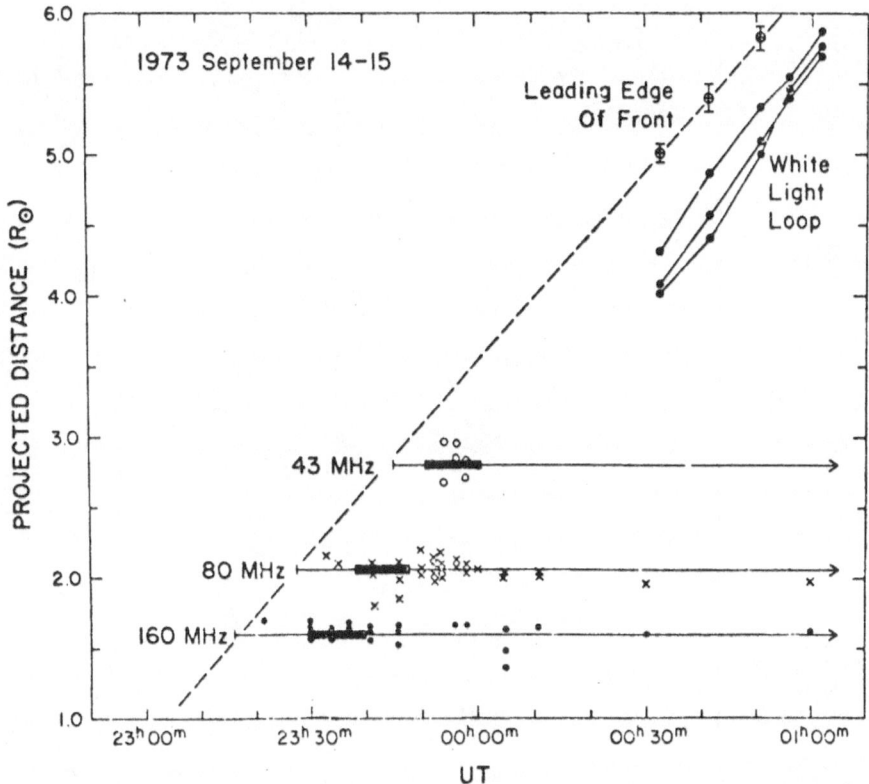

Fig. 10 - Height-time diagram showing the radio and white-light data. Individual points show the radio source heights as measured from radioheliograms. Light lines show the burst durations measured from spectrograms and heavy bars emphasize the high-intensity portions. The height of the white-light loop at three position angles was determined from five coronagraph pictures, and the white-light front was determined from microdensitometer scans of three photographs (Dulk et al., 1976).

We note that these estimated magnetic energies are similar to the total energy required to explain interplanetary shock waves (Hundhausen, 1972). By contrast, the impulsive energy release of mildly relativistic particles in a solar flare is usually $\leqslant 10^{32}$ erg and just possibly approaches this number in the very largest flares (Lin, 1977). Hence it seems that a major component of the energy in a (flare-associated) coronal transient is in the form of transported magnetic energy behind a shock wave disturbance. There is also sufficient energy in the shock wave to accelerate secondary particles such as MeV electrons and protons (see Fig. 3b) during large solar flares (Lin, 1977).

TABLE 1. PROPERTIES OF CORONAL TRANSIENTS *†

(1) Type	(2) Height range ($R_\odot$)	(3) Outward Velocity (km s^{-1})	(4) Excess mass (g)	(5) Mechanical energy input (erg)	(6) Class	(7) Reference
1971 December 14 ≥2.5^h UT						
Hα flare spray	1–1.5	~1700				
Type II burst	1.5–3	900–1200				Kosugi (1976) Fig. 6
Type IVM burst	1.6–3	~700				
WL leading edge	5–8	~1000				
WL compact clouds	7–9		3 x 10^{15}	2 x 10^{31}	FLA	Brueckner (1973)
WL excess mass	7–9	≤700	4 x 10	10^{32}		
1973 January 11 >01^h UT						
Hα flare spray	1–1.8	300–600		≤10^{30}		
Type II burst	1.4–2.4	800–1200			FLA	Stewart et al. (1974c) Fig.6
Type IVM burst	1.5–2.5	600–700		10^{27}		(Dulk, 1973)
WL leading edge	3–9	620				
WL excess mass	3–5	430	3 x 10^{16}	2 x 10^{31}		
1973 January 11 ≥18^h UT						
Hα flare spray	1–2	200–400				
Type II burst	1.7–2.4	800–1200			FLA	Stewart et al. (1974b) Fig. 8
Type IVM burst	1.7–3	500–400				
WL leading edge	3.3–5.3	750				
WL excess mass	2.7–3.2	230	10^{16}	10^{31}		

(1)	(2)	(3)	(4)	(5)	(6)	(7)
1973 September 7 ≥11.4h UT						
Hα flare surge	Disk event	?			FLA	Gosling et al. (1975)
Type II burst		>960				
WL leading edge						
WL excess mass			2.4×10^{16}	1.1×10^{32}		
Interplanetary shock wave	~1 AU	722				
Interplanetary excess mass		600	4.2×10^{16}	1.2×10^{32}		
1973 September 14 ≥23.5h UT						
No Hα flare observed	Behind limb?				FLA	Dulk et al. (1976)
WL leading edge	4–6	720				
WL excess mass	4–6	350	4×10^{15}	$\sim10^{31}$		
Radio continuum source	1.5–3.0	Nil				
Coronal depletion	≤3 $R_\odot$	~120				
1972 September 18 ≥17h UT						
Hα eruptive prominence					EPL	Tousey (1973)
WL excess mass	4–6	250–450	7×10^{15}	1.7×10^{31}		
1973 June 10						
WL leading edge	3.6–5.0	~500			EPL	MacQueen et al. (1974)
WL excess mass			1.8×10^{16}	$\sim3 \times 10^{31}$		
Coronal depletion	2–4					Hildner et al. (1975)

(1)	(2)	(3)	(4)	(5)	(6)	(7)
1973 August 10						
Hα eruptive prominence					EPL	Gosling et al. (1974)
WL leading edge	2–6	~400				
WL excess mass			4×10^{15}	8.4×10^{31}		
WL loop			$\sim1 \times 10^{15}$			Anzer and Poland (1978)
1973 August 13						
Hα eruptive prominence					EPL	Rust and Hildner (1976)
Soft X-ray loop	1–2	Constant acceleration $\sim12.5\ \mathrm{m\,s^{-1}}$ $v<2.6\ \mathrm{km\,s^{-1}}$ max.	1.3×10^{15}			
WL excess mass	2–5		1.9×10^{15}	3×10^{30}		
1973 December 19						
Hα EUV eruptive prominence	1–3 $R_\odot$	Slow rise $\sim14\ \mathrm{km\,s^{-1}}$ then	2×10^{14}	8×10^{29}	EPL	Schmahl and Hildner (1977)
WL cloud	3–6	rapid expansion $<100\ \mathrm{km\,s^{-1}}$	1.8×10^{15}	2.7×10^{30}		

*WL = white light; IVM = moving Type IV burst

†Notes on events are given on the following page.

1971 December 14: Disrupting WL streamer may have initiated event. 1-D radio positions suggest that three IVM sources may have been associated with the three compact WL clouds. The latter appeared later and at much greater heights than the IVM sources. OSO-7 event.

1973 January 11: K-coronal depletion at heights ≤ 1.5 $R_\odot$ accounts for only 10% of excess mass in WL cloud (3-9 $R_\odot$). OSO-7 event.

1973 January 11: Homologous event to 1973 January 11 >02^h UT; K-coronal brightening observed at 2.7 $R_\odot$. OSO-7 event.

1973 September 7: Only interplanetary shock wave observed so far to be associated with a WL coronal transient. Skylab event.

1973 September 14: Event appears to be magnetically controlled. Radio emission appears to be associated with the region between the leading edge and the loop of the white-light transient. Skylab event.

1972 February 18: Circular WL cloud diameter ~2 $R_\odot$ and density ~2.5×10^6 cm^{-3} at 5 $R_\odot$ located radially above eruptive prominence. OSO-7 event.

1973 June 10: Skylab event.

1973 August 10: Skylab event.

1973 August 13: There was probably enough mass in soft X-ray loop to explain the WL excess mass at greater heights. Skylab event.

1973 December 19: Only ~10% of Hα prominence mass was expelled from Sun. Skylab event.

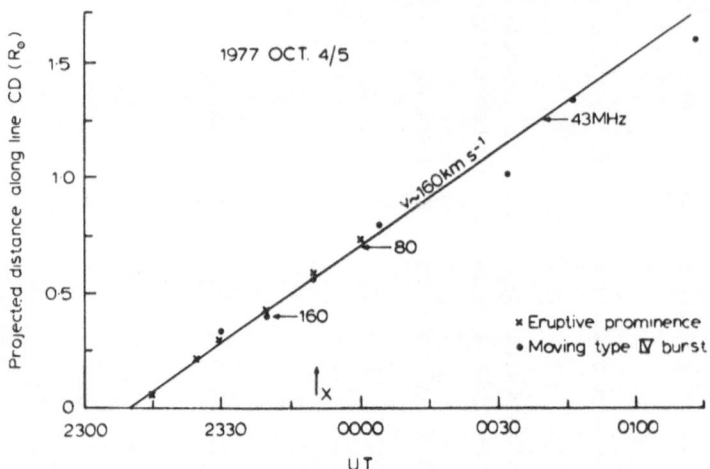

Fig. 11 - Distance versus time plots of the moving fronts of the Hα prominence material and the main type IV sources for the two events. Vertical
arrows labelled X indicate times of maximum soft X-ray fluxes. Horizontal
arrows indicate the times of maximum radio flux at 160, 80 and 43 MHz
(Stewart et al., 1978).

Other Properties of Coronal Transients Suggesting Magnetic Control

Coronal transients appear to be restricted to solar latitudes $\pm 40^\circ$
and to be associated with active regions (Hildner et al., 1976). During
Skylab about one transient occurred every two days, but the frequency
of occurrence will probably increase during solar maximum because flares,
EPLs, Type II and moving Type IV bursts all show a solar cycle dependence
(see Fig. 2 of Hildner (1977)). The frequency of white-light transients
increases with sunspot number, suggesting that transients occur above
strong photospheric magnetic field regions. Hildner et al. (1976) argue
that this pattern of occurrence is consistent with their belief that
the forces propelling transient material outward are primarily magnetic.

Transients appear to be restricted in latitude, having sides that
are slightly inclined away from the radial direction (see Fig. 3 of
Hildner (1977)). Even loop transient events do not expand uniformly in
all directions but have rather straight sides. The most obvious reason
for the transient shape is that the sides of the transient are contained
by radially directed magnetic fields (Anzer, 1978).

5.   CONCLUSION

All the evidence collected to date can be interpreted as indicating
that coronal transients are driven outwards by magnetic forces. The
loop and arch-like shapes of white-light events suggest expanding
magnetic loops or arches which may have been stable magnetic structures
in the lower corona prior to the event.

How the magnetic forces are released is not at all clear from observations, but several mechanisms have been proposed (Dulk et al., 1976; Mouschovias and Poland, 1978; Anzer, 1978. See also papers by Pneuman, Steinolfson and Wu, Maxwell and Dryer, and Nakagawa et al., in this issue). It would be beyond the scope of this paper to discuss these models.

What is clear from the observations is that there is sufficient magnetic energy available in the transient ($\sim 10^{32}$ erg) to explain the energetics of the coronal disturbance.

It is anticipated that the forthcoming solar maximum experiments will greatly enhance our knowledge of this complex and fascinating subject.

REFERENCES

Abetti, G. A: 1934, *The Sun*. English transl., by J. B. Sidgwick, 1957 (1st edn.), Faber and Faber, London.
Akasofu, S., and Chapman, S.: 1972, *Solar Terrestrial Physics*, p. 470, Oxford University Press.
Anzer, U.: 1978, *Solar Phys.* 57, p. 111.
Anzer, U., and Poland, A. I.: 1978, HAO preprint.
Boischot, A.: 1958, *Ann. Astrophys.* 21, p. 273.
Brueckner, G. F.: 1973, *Coronal Disturbances*, IAU Symposium 57, (ed. G. Newkirk), p. 333.
Dulk, G. A.: 1973, *Solar Phys.* 32, p. 491.
Dulk, G. A.: 1979, IAU Symposium 86, S. F. Smerd Memorial Symposium, Maryland, August 1979.
Dulk, G. A., Smerd, S. F., MacQueen, R. M., Gosling, J. T., Magun, A., Stewart, R. T., Sheridan, K. V., Robinson, R. D., and Jacques, S.: 1976, *Solar Phys.* 49, p. 369.
Eddy, J. A.: 1974, *Astron. Astrophys.* 34, p. 235.
Gergely, T. E., Kundu, M. R., Munro, R. H., and Poland, A. I.: 1979, *Solar Phys.* (in press).
Gosling, J. T., Hildner, E., MacQueen, R. M., Munro, R. H., Poland, A. I., and Ross, C. L.: 1974, *J. Geophys. Res.* 79, p. 4581.
Gosling, J. T., Hildner, E., MacQueen, R. M., Munro, R. H., Poland, A. I., and Ross, C. L.: 1975, *Solar Phys.* 40, p. 439.
Gosling, J. T., Hildner, E., MacQueen, R. M., Munro, R. H., Poland, A. I., and Ross, C. L.: 1976, *Solar Phys.* 48, p. 389.
Hildner, E.: 1977, in *Study of Travelling Interplanetary Phenomena 1977* (Proc. L. D. De Feiter Memorial Symposium, Tel Aviv, June 7-10, 1977) (eds. M. A. Shea, D. F. Smart and S. T. Wu), p. 3.
Hildner, E., Gosling, J. T., MacQueen, R. M., Munro, R. H., Poland, A. I., and Ross, C. L.: 1975, *Solar Phys.* 42, p. 163.
Hildner, E., Gosling, J. T., MacQueen, R. M., Munro, R. H., Poland, A. I., and Ross, C. L.: 1976, *Solar Phys.* 48, p. 127.
Hirshberg, J., Bame, S. J., and Robbins, D. E.: 1972, *Solar Phys.* 23, p. 467.

Hundhausen, A. J.: 1972, *Coronal Expansion and the Solar Wind,* Springer-
    Verlag, New York.
Jackson, B. V., and Hildner, E.: 1978, *Solar Phys.* 60, 155.
Kai, K.: 1970, *Solar Phys.* 11, p. 310.
Kosugi, T.: 1976, *Solar Phys.* 48, p. 339.
Levy, G. S., Sato, T., Seidel, B. L., Stelzried, C. T., Ohlson, J. E.,
    and Rusch, W. V. T.: 1969, *Science* 166, p. 597.
Lin, R. P.: 1977, in *Study of Travelling Interplanetary Phenomena 1977*
    (Proc. L. D. De Feiter Memorial Symposium, Tel Aviv, June 7-10,
    1977) (eds. M. A. Shea, D. F. Smart and S. T. Wu), p. 23.
MacQueen, R. M.: 1979, *Phil. Trans, R. Soc. Lond.,* in press.
MacQueen, R. M., Eddy, J. A., Gosling, J. T., Hildner, E., Munro, R. H.,
    Newkirk, Jr., G. A., Poland, A. I., and Ross, C. L.: 1974, *Astro-
    phys. J.* 187, p. L85.
MacQueen, R. M., Gosling, J. T., Hildner, E., Munro, R. H., Poland, A.
    I., and Ross, C. L.: 1974, *Proc. Soc. Photo-Opt. Instrum. Eng.* 44,
    p. 207.
Malitson, H. H., Fainberg, J., and Stone, R. G.: 1973, *Astrophys. Lett.*
    14, p. 111.
McLean, D. J., and Dulk, G. A.: 1978, *Proc. Astron. Soc. Aust. 3,*
    p. 251.
Mouschovias, T. C., and Poland, A. I.: 1978, *Astrophys. J.* 220, p. 675.
Riddle, A. C.: 1970, *Solar Phys.* 13, p. 448.
Rust, D. M., and Hildner, E.: 1976, *Solar Phys.* 48, p. 381.
Schmahl, E. J., and Hildner, E.: 1977, *Solar Phys.,* 55, p. 473.
Smerd, S. F., and Dulk, G. A.: 1971, in *Solar Magnetic Fields* (IAU
    Symposium 43) (ed. R. Howard), p. 616, Reidel, Dordrecht.
Stewart, R. T., Hansen, R. T., and Sheridan, K. V.: 1978, in *Physics of
    Solar Prominences* (IAU Colloquium 44, Oslo, Aug. 14-18, 1978)
    (eds. E. Jensen, P. Maltby, and F. Q. Orra), p. 315, Institute of
    Theoretical Astrophysics, Blindern, Oslo.
Stewart, R. T., McCabe, M. K., Koomen, M. J., Hansen, R. T., and Dulk,
    G. A.: 1974(a), *Solar Phys.* 36, p. 203.
Stewart, R. T., Howard, R. A., Hansen, F., Gergely, T., and Kundu, M.:
    1974(b), *Solar Phys.* 36, p. 219.
Tousey, R.: 1973, *Space Res.* 13, p. 713.
Wild, J. P.: 1969, *Solar Phys.* 9, p. 260.
Wild, J. P., Murray, J. D., and Rowe, W.C.: 1953, *Nature* 172, p. 533.

DISCUSSION

Newkirk: We should be very cautious in identifying the forerunners
of transients as a shock front. This phenomenon appears to accompany
transients regardless of whether the main transient is supersonic or
not. If we were dealing with a shock, the density enhancement would
depend upon the Mach number. The observations show no evidence for
such a relationship.
Stewart: I did not say the forerunner was a shock front.

Dryer: (Comment) The observations are inconclusive concerning the question of identifying forerunners with shocks. The small density enhancements detected in front of transients are entirely consistent with the notion that a very weak shock (hence, a very small density increase) could be responsible. During the time-dependent mass motion observations by Skylab, OSO-7 or P78-1, we have no evidence concerning the question as to whether the local plasma velocities (within or outside the transients) are supersonic or not. I suggest, therefore, that this question should be considered to be an open one. We must keep in mind, also, the fact that even non-shocked MHD wave motion (with little or even no bulk motion) can produce a moving compression wave, followed by an expansion wave. This could even precede-due to preflare activity-the main type II shock.

Rosenau: (Comment) While the propagation speed of 100 km s^{-1} is indeed too small for a generation of transonic shock (sonic speed is ~ 150 km s^{-1}) a slow shock can be generated. Therefore generation of a shock cannot and should not be precluded.

Moore: In terms of Jerry Pneuman's magnetic configuration sketch for coronal transients, do we usually view transients from the same aspect as in his sketch, or from the side?
Stewart: The same aspect because white-light loops are sometimes observed.

Pneuman: (Comment) The slide that Dr. Stewart showed is, I think, approximately the same view of the transient as in the slide I showed yesterday - since you see a loop-like structure in the white-light photograph.

Sheeley: Why are 10^{33} electrons such a large number? It is a relatively negligible fraction of the number that occurs in a typical coronal transient.
Stewart: There are about 10^{40} thermal (~1keV) electrons in a mass transient. We require $\geq 10^{33}$ electrons with energies $\geq$ 1 MeV. If these are part of a non-thermal distribution with, say, E^{-3} the number of electrons with energies > 10keV would be > 10^{37} which is an appreciable fraction of the total number of electrons in the transient.

Ivanov: (1) Could you estimate the Alfvenic-Mach number of the shock front? (2) What is characteristical length between shock front and piston front?
Stewart: See Dulk et al. (1976) for specific values but I think the Mach number in this and similar events is quite small ~2-4 while the characteristic length is $\leq$ 0.5 $R_\odot$ at a height of say 5 $R_\odot$.

TRANSIENT PHENOMENA ORIGINATING AT THE SUN - AN INTERPLANETARY VIEW

Devrie S. Intriligator
University of Southern California
Los Angeles, California 90007, U.S.A.

ABSTRACT

An overview is given of various phenomena observed by numerous spacecraft in the interplanetary medium. These phenomena are related to transient solar events such as flares and coronal holes. The effects of such transient solar events are extensive. At times, a solar event can affect the interplanetary medium out to distances as far as 17.2 AU and over a wide range of azimuthal angles. Also some phenomena, such as high frequency fluctuations (precursors of shocks) in the interplanetary medium, appear beyond 1 AU. Thus, transient phenomena are significantly modified by their passage through this medium. The conclusion is reached that transient events originating on the sun and passing through the interplanetary medium represent complex and significant physical problems.

I. INTRODUCTION

An understanding of transient events originating at the sun is important because it provides insight into solar and interplanetary particle and field phenomena. Analyses of these events associated with solar phenomena can provide information on various processes and energy release mechanisms on the sun or in the solar corona. Comparisons of associated observations made at different points in the solar system provide information on the dynamics and evolution of plasma phenomena as they propagate through the interplanetary medium.

Initially highly localized phenomena can propagate identifiably near the ecliptic plane to large heliocentric distances and expand over a wide range of azimuthal (longitudinal) angles. Simultaneous particle and field measurements by the same spacecraft frequently show significant changes associated with the passage of transient phenomena in the interplanetary medium. At present a number of active spacecraft located near the ecliptic plane are providing in situ observations that can be used to study the spatial extent and the development of

357

M. Dryer and E. Tandberg-Hanssen (eds.), Solar and Interplanetary Dynamics, 357-374.

transient phenomena originating at the sun. An overall evaluation of
the latitude or three-dimensional solar system effects of these transient phenomena must await in situ observations out of the ecliptic, such
as those planned on the International Solar Polar Mission.

   In this paper analyses of several transient phenomena are presented. Several of the interplanetary phenomena studied are apparently
associated with solar flares, others appear to be associated with
coronal holes. All of the events studied are recent, occurring in 1978
or 1979. Analyses of both the radial and azimuthal propagation of several events based on in situ observations indicate that they can drastically affect the structure of solar system particle and field regimes
at least to heliocentric distances of 17.2 AU and over a wide range of
($\geq 90^{\circ}$) of azimuthal angles.

II. OBSERVATIONS

   Daily values of the solar wind proton number density and speed at
Venus during six solar rotations are shown in Figure 1. These parameters
were obtained at Venus by the NASA Ames Research Center plasma analyzer
(Intriligator et al., 1979a,b; Wolfe et al., 1979) on the Pioneer Venus
Orbiter spacecraft, after orbit insertion on December 4, 1978. The
variations of the parameters are indicative of the gusty nature of the
solar wind, particularly during the rising portion of the solar cycle.

   The current solar cycle in terms of sunspot numbers is shown in
Figure 2. This figure is adapted from the Solar Geophysical Data and
indicates that to date this solar cycle has been exceptionally active
as compared with the average for the last thirteen cycles.

   The arrows in Figure 1 indicate two of the peaks in solar wind

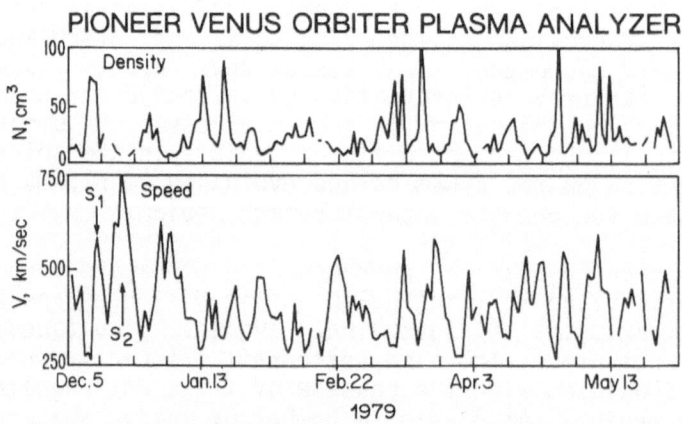

*Figure 1. Daily solar wind speeds and densities measured at
Venus. Values were obtained near noon UT for each orbit, upstream from
the bow shock.*

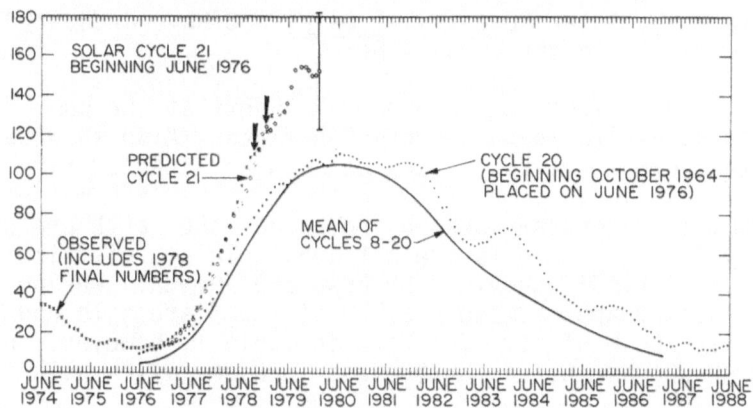

Figure 2. Observed and predicted sunspot numbers adapted from Solar Geophysical Data. The two arrows above the September 1978 and December 1978 values refer to the times of the two flares giving rise to transient phenomena discussed in this paper. Transient phenomena associated with a coronal hole in December 1978 are also presented.

speed that will be studied in greater detail below. The first event (S_1) we have tentatively associated with a 2N flare at S13 E29 on December 10th at 2332 UT. The event marked S_2 in Figure 1 appears to be a stream from a coronal hole, and forms an interesting contrast to the flare stream S_1. S_2 appears to represent a reemergence of stream activity near the region where a large coronal hole was located in previous solar rotations. In Figure 3 we have adapted the Carrington

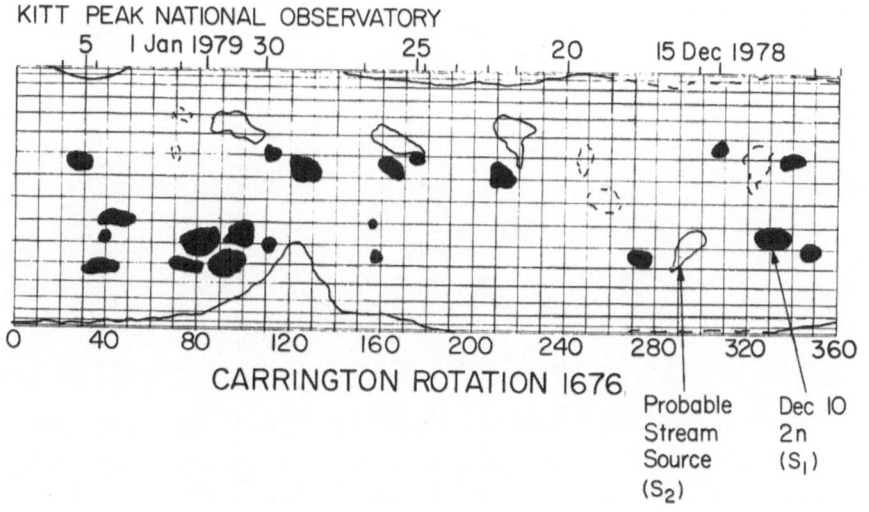

Figure 3. Carrington Rotation adapted from Solar Geophysical Data.

Rotation from the <u>Solar Geophysical Data</u>, and have indicated the location of the flare and the coronal hole.

   Table I summarizes the sequence of events at the sun, Venus, and
Earth associated with the December 10th flare. Also shown are the
Pioneer Venus Orbiter (PVO) plasma data on an expanded time scale for
December 13th indicating the forward shock. The ISEE magnetometer data
indicating the subsequent forward shock near the Earth are shown, unfortunately ISEE plasma data are not available at this time. The ISEE
data shown were kindly provided by Edward J. Smith. During this time
Venus was located approximately 50° W of the flare site and Earth,
approximately 30° W of it. The data in Table I, however, indicate that
a forward shock was seen at both sites in association with this event.

TABLE I

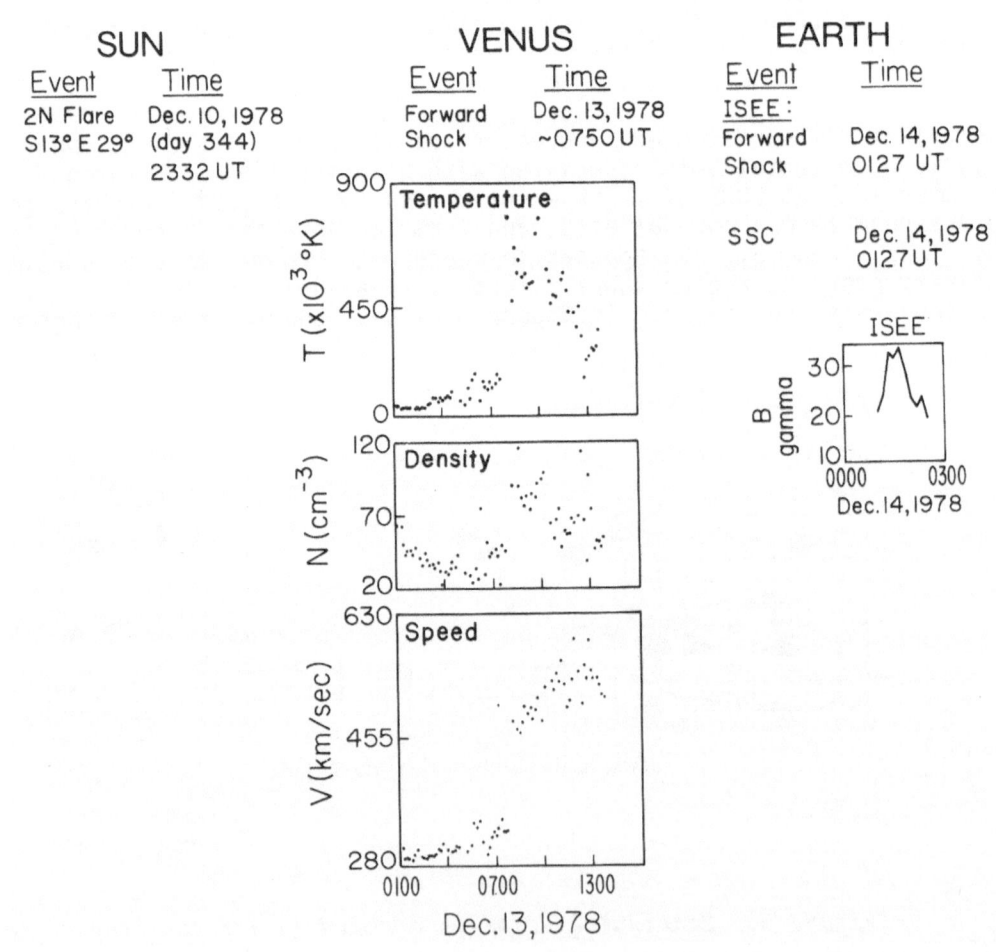

Table II indicates the calculated local shock speed obtained at Venus using the flux conservation equation (Intriligator 1976):

$$V_{shock} <V,N> = \frac{V_2 N_{p2} - V_1 N_{p1}}{N_{p2} - N_{p1}} \qquad (1)$$

for this event in December. This calculated local shock speed is compared with the average speeds of transit in Table III, where $V_{s \text{ to } V}$ refers to the average transit speed from the sun to Venus and $V_{s \text{ to } E}$ refers to the average transit speed from the sun to Earth.

The sequence of events summarized in Table I indicates that the flare associated plasma arrived at PVO at Venus on December 13, 1978 at 0750 UT and at Earth on December 14, 1979 at 0127 UT. Since, as noted above, at this time Venus (at 0.7 AU) was located approximately 50° W of the flare site and the Earth was located approximately 30° W of the flare site, it is particularly interesting to note the earlier time of arrival of the shock front at Venus. The comparison of the calculated shock speeds (Table III) indicates that the average transit speeds from the sun to Venus and from the sun to Earth were similar even though Venus was located approximately 20° further west of the flare site than the Earth. It is tempting to conclude that these analyses imply a spherically symmetric shock front over large ($\gtrsim 50^{\circ}$) azimuthal ranges. Alternatively, perhaps the shock front was not spherically symmetric and it propagated radially outward more rapidly within a few degrees ($\lesssim 20^{\circ}$) of the flare site and propagated radially outward more slowly at other azimuthal locations ($>30^{\circ}$) from the flare

TABLE II

LOCAL SHOCK SPEED USING EQUATION (1)

EVENT	DAY	TIME, UT	V_1 km/sec	N_{p1} protons/cm^3	V_2 km/sec	N_{p2} protons/cm^3	$V_{SHOCK <V,N>}$ km/sec
$S_1(V)$	D 347 1978	0750	330	46	465	137	533

TABLE III

COMPARISON OF CALCULATED SHOCK SPEEDS

EVENT	DAY	TIME, UT	$V_{s \text{ to } V}$ km/sec	$V_{SHOCK <V,N>}$ km/sec	$V_{s \text{ to } E}$ km/sec
$S_1(V)$	D 347 1978	0750	533	533	563

site. In the case of the December 10th flare it is not possible to
unambiguously select one of these alternatives since at this time no
spacecraft was located within a few degrees of the flare site.

In comparing Table II and Table III, it is interesting that the
local shock speed at Venus in the solar wind and the average transit
speed from the sun to Venus are the same for this event. Previous
observations, most notably of the August 1972 events (Intriligator 1976,
1977a; Smith 1976), have indicated essentially constant average speed of
propagation beyond 1 AU, but higher speeds near the sun.

It is interesting to speculate that if, as is the case beyond 1 AU,
there is an essentially constant average speed of propagation near the
sun, it would imply for many previous flare studies (Intriligator 1976,
1977a; Smith 1976) that before the onset of the H-alpha flare there was
a local increase in the solar wind flux, and the solar wind speed. In
the case of the first large flare of the August 1972 events this would
imply that the enhanced solar wind flux began at approximately 2000 UT
July 29, 1972 or more than 100 hours earlier than the onset of the
H-alpha flare.

There is a wide range of fluctutions in the solar wind associated
with these events. Figure 4 shows the low frequency power in the PVO
solar wind speed fluctuations associated with the December 10th event.
The variances (σ_V^2) shown are indicative of the levels of power in the
10^{-3} to 10^{-4} Hz frequency range. These variances indicate that upon

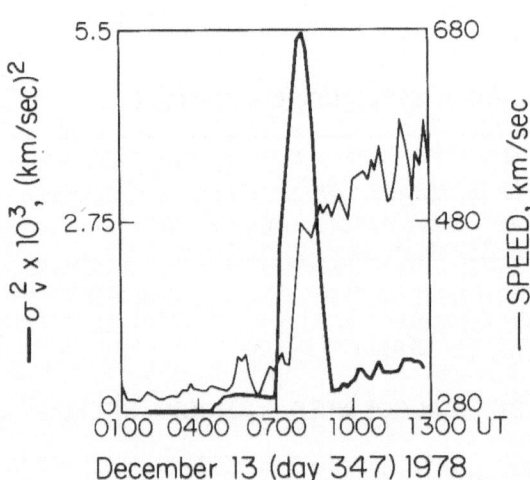

Figure 4. Solar wind streaming speed and the variances (σ_V^2)
in the frequency range of $\sim 10^{-3}$ to $\sim 10^{-4}$ Hz associated with the fluctuations of the hourly values of the speed at PVO.

the arrival of this event at PVO there are enhanced levels of low frequency power associated with the fluctuations of the hourly values of the solar wind speed. Figure 5 is from ISEE 3 and shows the higher frequency power also observed at times. As indicated in the figure, this higher frequency power occurs at frequencies six orders of magnitude higher than the fluctuations shown in the previous figure. These data were kindly provided by F. L. Scarf, who has identified these fluctuations as interplanetary ion sound waves.

While many fluctuations are associated with high speed streams in the solar wind (Intriligator, 1979), there are also other phenomena associated with the propagation of high speed streams in the solar wind. Figure 6 shows a high speed stream observed at 1 AU (IMP) and the same stream as seen at Pioneer 11. As indicated in the figure, there is a similarity in the overall speeds and duration of the stream at both spacecraft but there are some differences particularly associated with the peak of the stream. In this interval Pioneer 11 was at 1.6 AU and was radially aligned with IMP. The histogram of each of these streams is shown in Figure 7. Figure 7 shows that, as reported in Intriligator (1977b), there is a narrowing of the solar wind speed histograms with increasing radial distance. In addition, there is an erosion of the highest speeds as the stream propagates radially outward.

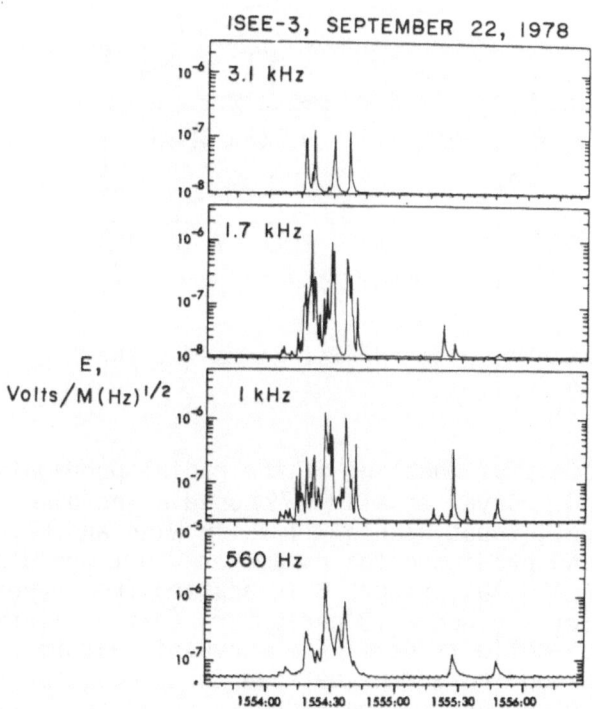

Figure 5. ISEE 3 plasma wave observations of the interplanetary ion sound waves at frequencies of ~10+3 Hz.

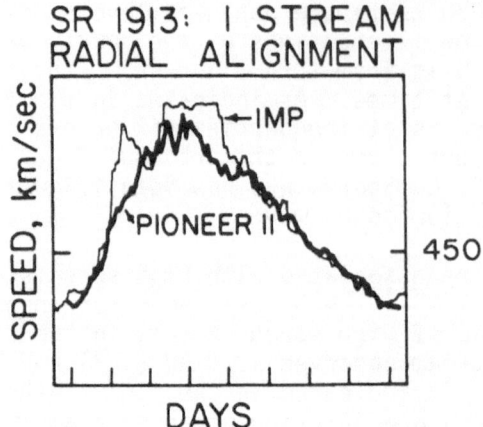

Figure 6. Superposition of solar wind speeds associated with a high speed stream at Pioneer 11 and at IMP.

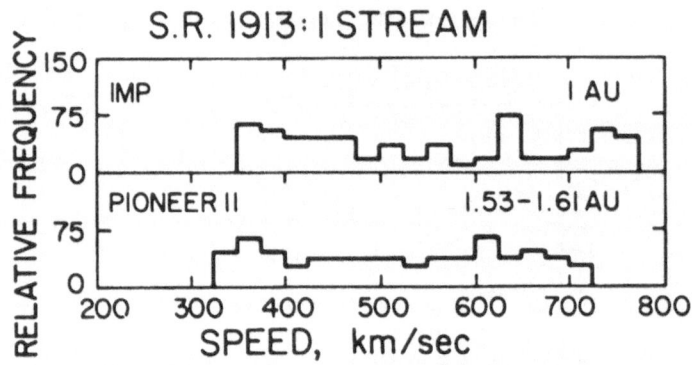

Figure 7. Solar wind speed histograms for the high speed stream shown in Figure 6.

There has been successful modeling of the radial propagation of events in the solar wind. Dryer et al. (1979) used a one-dimensional magnetohydrodynamic model, input Pioneer 11 solar wind and magnetic field data at 2.8 AU, and predicted the parameters that would be observed at Pioneer 10 at 4.9 AU. Figure 8 is adapted from Dryer et al. and indicates the predicted Pioneer 10 parameters (dashed lines) and the observed parameters (solid lines). The curves in Figure 8 indicate that generally there is good agreement between the predictions of the model and the observed parameters. The temperature is the parameter that is not predicted well by this model and this could be the result of employing a single fluid model that does not allow for the thermal exchange between protons and electrons.

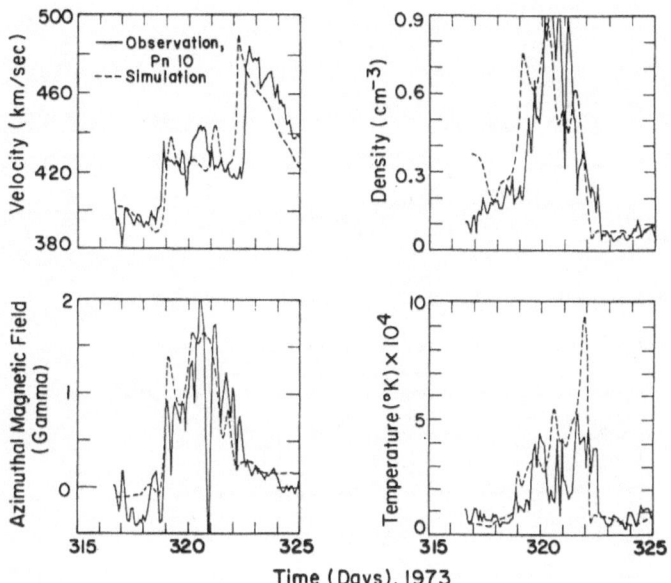

Figure 8. Comparison between 1-D MHD simulation and Pioneer 10 observations at 4.9 AU. The numerical simulation inputs Pioneer 11 observations at 2.8 AU, (Dryer et al., 1979).

At extended heliocentric distances, corotating interaction regions (CIR's) are formed in the solar wind (Smith and Wolfe, 1976). Figure 9 is from Smith and Wolfe and shows some of the characteristics in the plasma and field data that are indicative of the CIR's. At the leading edge of the CIR there is a forward shock that is characterized by a sharp increase in the solar wind speed, V, and the magnetic field magnitude, B. At the trailing edge of a CIR there is usually a reverse shock that is characterized by another sharp increase in the solar wind speed, V, and a distinct decrease in the magnitude of the magnetic field strength, B. In Figure 10, from Barnes and Simpson (1976), the Jovian electron (3-6 MeV) and the low energy proton data are shown. As indicated in the figure, the CIR's, which are denoted by the cross-hatched boxes above the abscissa, typically exclude the Jovian electrons and meanwhile there is an apparent increase in intensity of the 0.5-1.8 MeV protons. It is tempting to associate these proton observations and those of Marshall and Stone (1977, 1978); Pesses et al. (1978); McDonald et al. (1976), and Van Hollebeke et al. (1979) with some local interplanetary acceleration process which may, in fact, be accelerating the protons in the vicinity of CIR's.

Figure 11, which was kindly provided by F. L. Scarf, shows the high frequency (e.g., 1.78 kHz) precursors associated with an interplanetary shock observed at 2.25 AU in February 1978 by Voyager 2. These high frequency fluctuations were apparently not observed at 1 AU in association with this same shock (Scarf, private communication). Therefore,

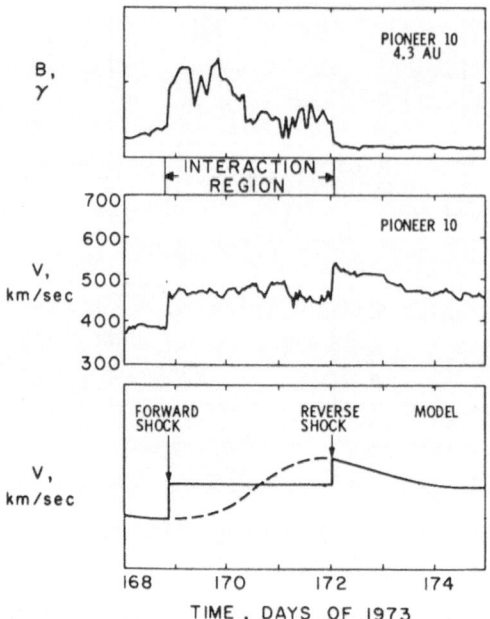

Figure 9. *Corotating Interaction Region (CIR), (Smith and Wolfe, 1976).*

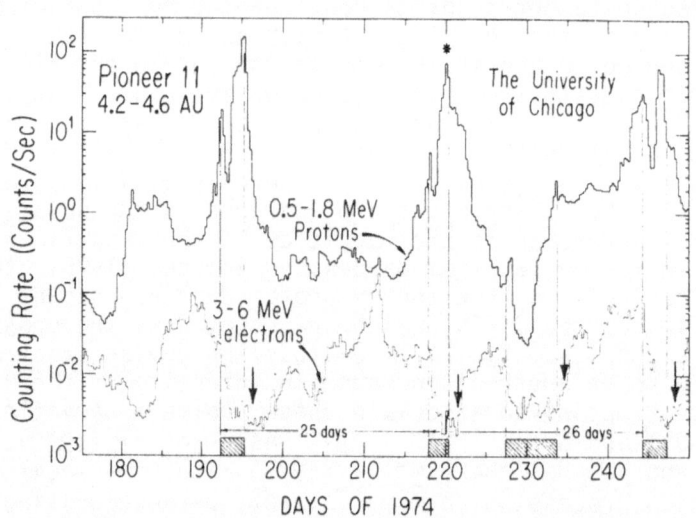

Figure 10. *The count rates for ~1 MeV protons and 3-6 MeV electrons are shown for ≈3 solar rotations. Well-identified CIR's occurring during this interval are indicated by cross-hatched bars along the abscissa. CIR's appear to exclude Jovian electrons and to produce MeV protons at forward-reverse shock pairs, (Barnes and Simpson, 1976).*

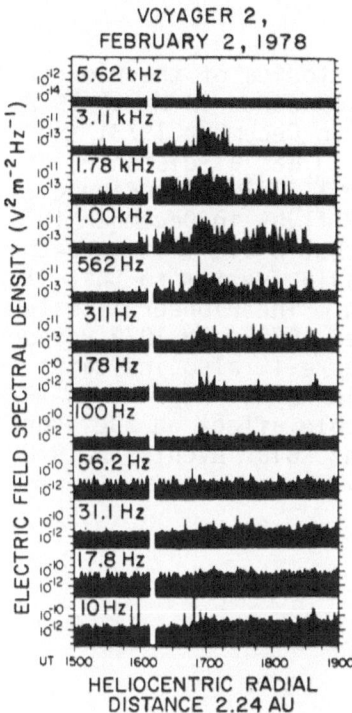

Figure 11. Precursors (e.g., 1.78 kHz) associated with an interplanetary shock at 2.24 AU as observed by the plasma wave experiment on Voyager 2. The shock occurred at approximately 1700 UT on February 2, 1978. High frequency precursor activity in several (e.g., 1.78 kHz, 1.00 kHz) channels began to be observed almost one hour before the shock. These data were kindly provided by F. L. Scarf.

at least in the case of this event, there are high frequency fluctuations that appear to be generated in the interplanetary medium as the shock propagates beyond 1 AU.

It is tempting to associate these high frequency fluctuations with the interplanetary acceleration processes that may give rise (e.g., through wave-particle interactions) to the observed increases in 1 MeV proton intensities in the vicinity of the CIR's. The identification of these interplanetary acceleration mechanisms is, in my opinion, one of the most important areas of interest in interplanetary physics today.

At times a transient phenomenon originating at the sun can drastically affect the overall structure of the particles and field environment in the solar system, over a wide range of heliocentric distances and azimuthal angles. In 1978, for example, a series of transient solar events occurred which significantly affected the solar system environment out to heliocentric distances as far as 17.2 AU and over a wide (>90°) range of azimuthal angles.

Figure 12 shows the outbound trajectories of the Pioneer 10 and Pioneer 11 spacecraft. Pioneer 10 was launched in March 1972 and was the first spacecraft to flyby Jupiter (in December 1973), and now is on an escape trajectory. In 1978 Pioneer 10 was at distances of ~15 to ~17 AU. Pioneer 11 was launched in April 1973. By using gravity assist at Jupiter during the Pioneer 11 flyby in December 1974, the spacecraft was placed in a trajectory that would lift it out of the ecliptic and carry it back across the solar system so that it would be the first spacecraft to flyby Saturn. The Pioneer 11 Saturn flyby successfully occurred in early September 1979. In 1978 Pioneer 11 was at distances of ~6.5 to ~8 AU. Figure 12 also indicates the wide azimuthal separation between the two Pioneer spacecraft in 1978. The next two figures illustrate the remarkable affect on the overall struc-ture of the heliosphere of the transient solar events in 1978.

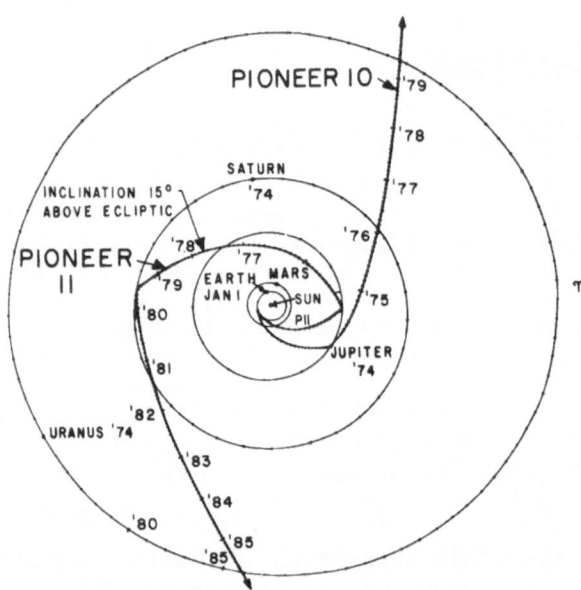

Figure 12. The Pioneer 10 and Pioneer 11 trajectories. The location of Pioneer 10 in September 1978 is indicated by the arrow to the right of the "Pioneer 10" label in the figure. Similarly, the location of Pioneer 11 in September 1978 is denoted by the arrow to the right of the "Pioneer 11" label in the figure. This figure indi-cates the large heliocentric and azimuthal separation of the two space-craft at this time.

Figure 13 is from Pyle <u>et al.</u> (1979) and shows that transients
originating at the sun can affect the galactic cosmic ray intensities
at distances of 15 to 17 AU. The curves in the top panel of Figure 13
indicate the galactic (energetic protons) cosmic ray intensities from
late 1977 through 1978 observed on Pioneer 10. The anomalous helium
intensities observed at Pioneer 10 are shown in the second panel in
Figure 13. The solar wind velocities measured by the NASA Ames Research
Center plasma analyzer on Pioneer 10 are shown in the third panel. The
bottom panel in Figure 13 shows the Climax Neutron Monitor data for this
time interval.

The Pioneer 10 observations indicate that for many months there
were depressed levels of galactic cosmic ray intensities over a wide

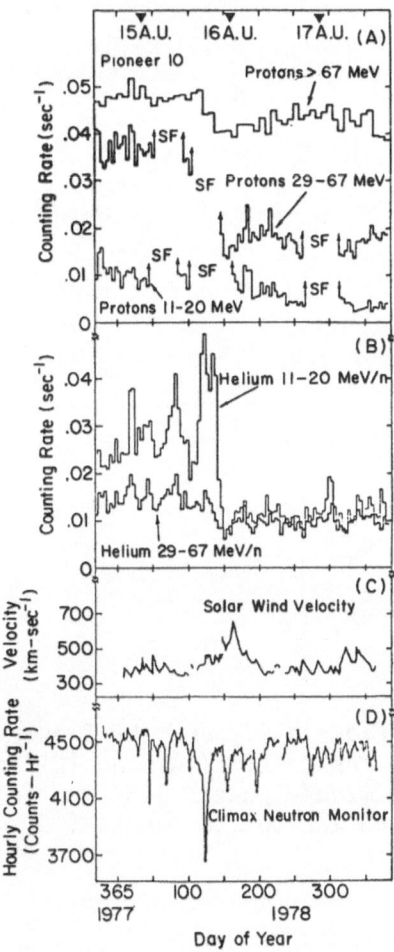

*Figure 13. Pioneer 10 observations from ~15 AU to ~17 AU from
Pyle <u>et al.</u> (1979). See text for a description of the observations.*

range of heliocentric distances and angles. The onset of these depressed levels was associated with a high speed stream in the solar
wind as indicated by the increase in the solar wind velocity data shown
in the figure. It is interesting to note that the high speed stream
persisted at these distances for several solar rotations.

Figure 14 is from Pyle et al. (1979) and depicts the 11-20 MeV and
low energy (0.5-1.8 MeV) protons, and the solar wind velocity at Pioneer
10 and 11, and IMP 8 during part of 1978. Pyle et al. state that the
higher energy protons observed at the spacecraft appear to be generated at the flare. They also interpret the times of arrival and patterns
of intensity buildup and decay in terms of expected effects on the particles as they propagate away from the sun.

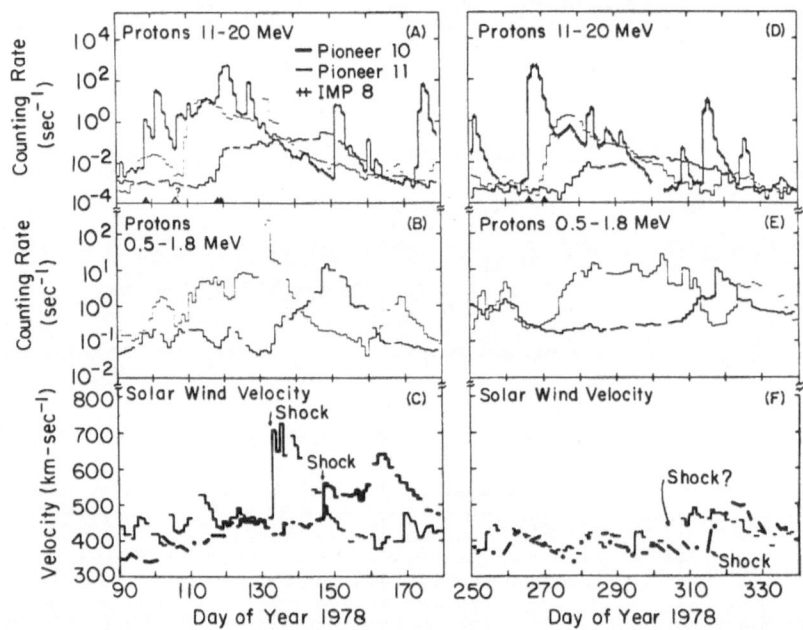

Figure 14. Pioneer 10, Pioneer 11, and IMP 8 observations in
1978 from Pyle et al. (1979). The top panels ((A) and (D)) show the
11-20 MeV proton observations, the middle panels ((B) and (E)) show the
0.5-1.8 MeV proton observations, and the bottom panels ((C) and (F)),
the solar wind speed.

III. DISCUSSION

The observations presented here give some indication of the wide range of complex and significant physical problems that are associated with an interplanetary view of transient phenomena originating at the sun. The observations show that phenomena originating at the sun affect the structure of the particles and fields regimes throughout the explored solar system, extending out to at least 17-20 AU in the vicinity of the ecliptic plane.

It is evident that further observational and theoretical studies along the lines of thoSe presented here are necessary for future progress in this field. In particular, an understanding of the radial evolution of the interplanetary events and their associated phenomena (e.g., fluctuations) would require more multipoint spacecraft studies, including those among radially aligned spacecraft as emphasized in Intriligator (1979).

Additional studies between azimuthally separated spacecraft, however, can also significantly contribute to our understanding of solar and interplanetary physics. The discussion of the December 10th flare-associated phenomena presented above, serves as an appropriate illustration of this point.

Finally, in the next few years there is the prospect of making a great deal of progress in our understanding of the fundamental physical processes that govern these phenomena. We will have a number of well-instrumented spacecraft strategically located throughout the solar system obtaining important data. To make appropriate use of these observations we will need complementary theoretical studies including those in plasma physics and we will need to take an interdisciplinary approach interrelating the various aspects of the solar phenomena, interplanetary phenomena, and the basic physics and plasma physics phenomena. Our understanding of the solar and the interplanetary plasma processes associated with these phenomena is, in my opinion, one of the most significant scientific challenges today.

ACKNOWLEDGMENTS

The author is indebted to Drs. F. L. Scarf and E. J. Smith for generously providing their data. The author has benefitted from helpful discussions with Drs. Dryer, Scarf, and Smith. W. David Miller participated in the analyses of the data. This work was performed at the University of Southern California and was supported by NASA contracts NAS2-7969 and NAS2-9478.

REFERENCES

Barnes, C.W., and Simpson, J.A.: 1976, Ap. J., 210, pp. L91-L96.
Dryer, M., Smith, Z.K., Smith, E.J., Mihalov, J.D., Wolfe, J.H.,
    Steinolfson, R.S., and Wu, S.T.: 1979, J. Geophys. Res., 83,
    pp. 4347-4352.
Intriligator, D.S.: 1976, Space Sci. Rev., 19, pp. 629-660.
Intriligator, D.S.: 1977a, J. Geophys. Res., 82, pp. 606-617.
Intriligator, D.S. in Shea, M.A., Smart, D.F., and Wu, S.T. (eds):
    1977b, Study of Travelling Interplanetary Phenomena, D. Reidel,
    pp. 195-225.
Intriligator, D.S.: 1979, Solar Wind 4, Springer-Verlag (in press).
Intriligator, D.S., Collard, H.R., Mihalov, J.D., Whitten, R.C.,
    Wolfe, J.H.: 1979a, Science, 205, pp. 116-119.
Intriligator, D.S., Wolfe, J.H., and Mihalov, J.D.: 1979b, IEEE
    Transactions on Geoscience Electronics, (in press).
Marshall, F.E., and Stone, E.C.: 1977, Geophys. Lett., 4, pp 57-60.
Marshall, F.E., and Stone, E.C.: 1978, J. Geophys. Res., 83, pp. 3289-
    3298.
McDonald, F.B., Teegarden, B.J., Trainor, J.H., and von Rosenvinge,
    T.T.: 1976, Ap. J., 203, pp. L149-L154.
Pesses, M.E., Van Allen, J.A., and Goertz, C.K.: 1978, J. Geophys.,
    Res., 83, pp. 553-562.
Pyle, K.R., Simpson, J.A., Mihalov, J.D., Wolfe, J.H.: 1979, 16th
    International Cosmic Ray Conference Papers, Paper SP 7-12.
Smith, E.J.: 1976, Space Sci. Rev., 19, pp. 661-686.
Smith, E.J., and Wolfe, J.H.: 1976, Geophys. Res. Lett., 3, 137-140.
Van Hollebeke, M.A.I., McDonald, F.B., Trainor, J.H., von Rosenvinge,
    T.T.: 1978, J. Geophys. Res., 83, 4723-4731.
Wolfe, J., Intriligator, D.S., Mihalov, J., Collard, H., McKibbin, D.,
    Whitten, R., Barnes, A.: 1979, Science, 203, pp. 750-752.

DISCUSSION

Sheeley: Can you confirm Murray's (Dryer) suggestion (for the
Pioneer Venus quick look data) that there was a shock apparently
associated with the May 8, 1979 coronal transient?

Intriligator: Available quicklook data suggest rapid changes in a
few hours, but do not clearly indicate a shock. We will have more
complete information on this event when data tapes are available.

Bhatnagar: Have you tried to look into the correspondence with
coronal transients and shocks observed at Saturn (etc.) distances?

Intriligator: We have just started to look into the correspondence
of interplanetary phenomena and coronal transients. For example, the
coronal transients in May of this year were brought to our attention
by Dr. Dryer, and we have started analyses of our Pioneer Venus Orbiter
plasma analyzer data in the vicinity of Venus to ascertain the
correspondence between coronal transients and our observations. To date
we have not performed any studies of coronal transients and plasma

observations at extended heliocentric distances such as in the vicinity of Saturn. In the case of interplanetary shocks, Dryer et al., (*JGR*, *83*, 1165, 1978) have associated several flare generated shocks observed in conjunction with Type II radio bursts on 1976 March 20 (during STIP Interval II) with Pioneer 10 observations of a possible shock at 9.7 AU on 1976 April 9.

Dryer (Commet to Bhatnagar): I would like to comment on Dr. Bhatnagar's question concerning the interplanetary identification of coronal transients. There are two published cases in the literature. Wu et al. (*Solar Phys.*, *49*, 187, 1976) discuss the OSO-7 white light transient on 15-16 June 1972; the Adelaide IPS signature of an interplanetary shock; a terrestrial sudden commencement; and the Pioneer 10 plasma observations of a possible shock followed by its disturbed flow. They used a 1-D hydrodynamic code to simulate the velocity and density profiles as observed at 1.6 AU by Pioneer 10. The observed (simulated) mass and energy in the interplanetary transient (assumed to subtend π steradians) were 7.7×10^{15}g (5.0×10^{15}g) and 1.4×10^{31} erg (4.5×10^{31} erg). The second published case is described by Gosling et al. (*Solar Phys.*, *40*, 439, 1975) who discuss the Skylab white light transient on 7 September 1973 and the Pioneer 9 plasma observations at 0.98 AU of a possible shock that was followed by its disturbed flow. The observed mass and energy in the interplanetary transient (again assuming π steradians) were 2.4×10^{16}g and 1.1×10^{32} erg.

The mass and energy for the two cases noted above are in reasonable agreement with the estimates given for coronal transients observed by Skylab (Rust et al., *Solar Flare Workshop*; P. Sturrock, Ed., pp. 314-315, Colo. Univ. Press, 1979). Phenomenologically, then, we seem to have several good cases for identifying interplanetary "signatures" with the solar-observed coronal transient.

A third case is currently suspected. Michels et al. (this Proceedings) observed a coronal transient on 8 May 1979 with the P78-1 coronograph (Solwind). I have suggested, on the basis of daily samples of Pioneer-Venus-Orbiter plasma data, that this latter spacecraft at 0.78 AU may have detected the interplanetary "signature".

Sheeley: Did you say there was a solar event on the east limb 3-4 days ago?

Intriligator: McMath Region 16239 flared on the east limb on August 18th and 20th, and near the central meridian on August 26th. Region 16252 had a 98 minute 1B flare at 2137 UT on August 27th near the east limb, and minor flares on August 26th and early on the 27th that produced SID's.

Benz: The observed presence of ion acoustic waves in shocks could easily be explained by an unstable current. It represents enhanced resistivity and, by Ohm's law, an increase in the electric field. This can accelerate charged particles above a critical velocity (runaway process). Thus the observed presence of energized protons in shocks is probably not a coincidence. What is the ratio of electron to proton temperature?

Intriligator: As indicated in my presentation, I agree with you
that it is tempting to associate the enhanced intensities of 0.5 to
1.8 MeV protons at CIR's with wave-particle processes or other
acceleration processes. As to the ratio of electron to proton temperatures, on Pioneer 10 and 11 generally the intensity of electrons was
not high enough beyond ~2 AU for us to measure them except during
special events such as the August 1972 events. Since, on Pioneer 10
and 11, CIR's were generally observed beyond ~3 AU we do not have
simultaneous plasma electron temperatures to compare with our plasma
proton temperatures.

Kahler: What is the interplanetary signature of a low-speed mass
ejection?
Intriligator: This is one of the areas we are studying at the
present time and hopefully in the near future we will have some specific
information on this.

MEASUREMENTS OF MASS FLOW IN THE TRANSITION REGION AND INNER CORONA

Gary J. Rottman
The Laboratory for Atmospheric and Space Physics
University of Colorado
Boulder, Colorado U.S.A.

ABSTRACT A recent sounding rocket experiment has provided high spectral resolution line profiles across the solar disk. The objective of this experiment is to provide information on the systematic velocity fields at the base of the corona by observing the displacement, width and shape of EUV emission lines.

A new solar rocket experiment was launched on June 5, 1979. The instrument consists of a small telescope providing moderate spatial resolution ($\lesssim$ 1 arc minute) and a spectrograph with stability and high spectroscopic resolution ($\simeq$ .028A at 600A). The CODACON detector system developed by G.M. Lawrence of this laboratory uses a microchannel plate array to provide 1024 simultaneous measurements spanning approximately 30A of the solar spectrum. The spectrograph uses a 3600 1/mm grating and the spectral range 609-634A is recorded in second order overlapping 1218-1268A in first order. Figure 1 is a typical limb spectra obtained during this experiment.

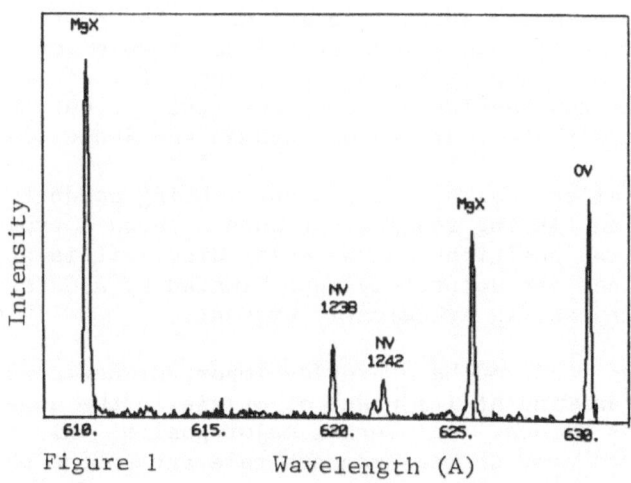

Figure 1   Wavelength (A)

375

M. Dryer and E. Tandberg-Hanssen (eds.), Solar and Interplanetary Dynamics, 375-378.

The instrument does not contain an internal wavelength standard
and it is not possible to assign an absolute wavelength scale to the
observations. In future experiments a platinum hollow cathode source
will provide this standard. In flight the wavelength scale of the
spectrograph is extremely stable and the relative position of the lines
can be accurately measured. Velocity shifts of the emission lines can
then be analyzed two ways. First, a relative velocity field is es-
tabished by recording the position of a particular line as the instrument field of view sweeps from limb to limb across the solar disc. In
this way a systematic flow within a particular feature, for example a
coronal hole, can be measured relative to the quiet corona. A second
analysis will compare the positions of high temperature coronal lines
with positions of certain low temperature lines, for example SiII.
Such low temperature lines have been studied extensively by OSO-8 to
provide information of the velocity fields in the chromosphere.

The launch opportunity requires a coronal hole near disc center and
the most reliable information on such features is the Kitt Peak 10830
spectroheliogram. Examination of these data in mid-May revealed the
development of a low latitude hole which was projected forward one solar
rotation to establish a launch date near June 5th. This coronal hole
was again identified at the east limb from the Kitt Peak data of May
31st and subsequent spectroheliograms through June 3rd enabled an
accurate projection forward to the launch time on June 5th. The roll
angle of the pointed rocket instrument was set to give a scan plane
inclined $\sim 15^{o}$ to the solar equator passing through the coronal hole.
Analysis of the intensity of the MgX lines verifies that data were
obtained in the coronal hole, in the quiet corona and in an active
flaring region at the limb.

A preliminary analysis of the data has been completed. The data
were corrected for instrumental effects and a background signal was
removed. The position of the emission lines was determined by a "center-
of-mass" routine and velocity shifts are evident. The routine is
somewhat inaccurate due to weak blinds in the solar emission lines as
well as noise in the data and background. Because of these deficiencies
in the first analysis the results are only qualitative, but they are
in general agreement with the results of Cushman and Rense (1976).

A more detailed analysis is presently being conducted using a
least-squares Gaussian fitting routine. Figure 2 shows a typical fit to
the OV 629A line at two positions on the solar disc. It is apparent
that the emission lines are accurately approximated by a Gaussian profile and that velocity shifts are clearly present.

The observations reported here provide important new constraints
for studying the solar wind at its place of origin in the upper transition region and inner corona. An overall major goal of this rocket
research program is to provide measurements relevant to the physical
processes by which both momentum and energy are deposited into the
expanding corona.

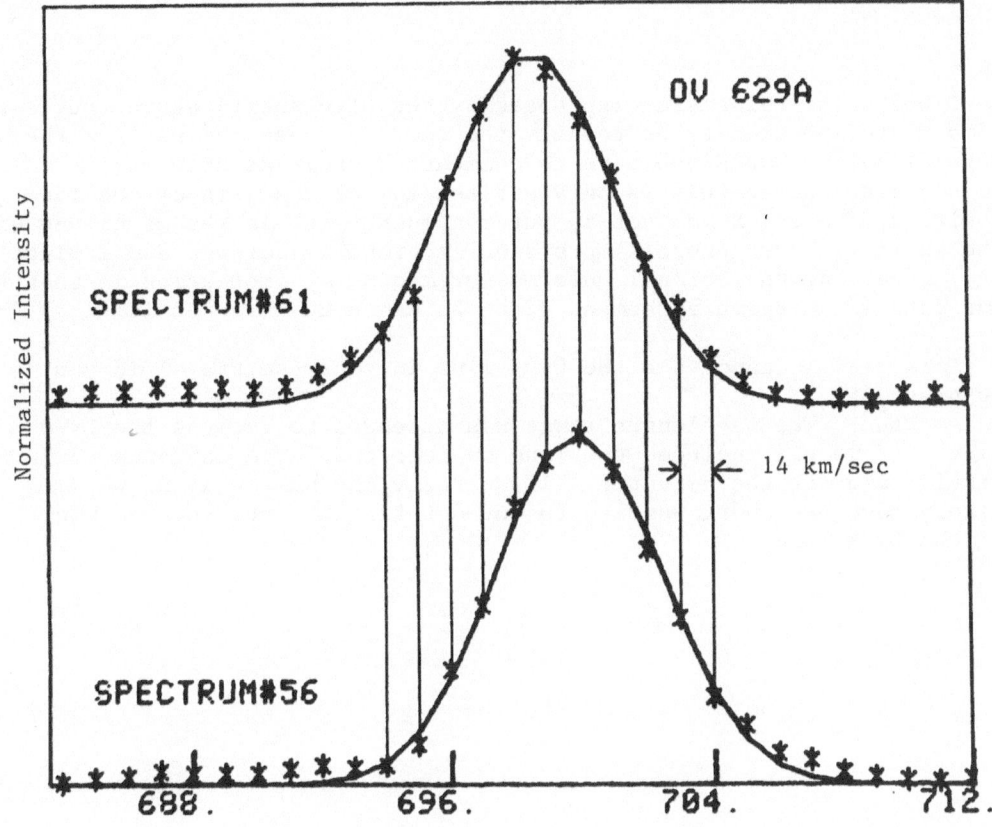

Figure 2

ACKNOWLEDGEMENTS

    The assistance and guidance of Frank Orrall, a co-investigator on this rocket research program, are gratefully acknowledged. This work is supported by NASA contract NAS-5178.

REFERENCE

Cushman, G.W., and Rense, W.A.: 1976, Astrophys. J. Letters 207, No. 1, Part II, L-61.

DISCUSSION

Dryer: The velocities (as noted by the blue shift) above the west limb flare appeared to be roughly the same as those coming from the coronal hole. Would you care to comment on their magnitudes?

Rottman: Since this is only a first moment analysis of the line position, I caution you not to put too much faith in the magnitude of the shifts. Your general impression is probably correct and I might add, regarding the coronal hole measurements, my impression is that our data is in basic agreement with the Rense-Cushman result.

Porsche: I cannot see the Ly α line in your spectra. Did you suppress it?

Rottman: The wavelength range was selected to exclude the Ly-α line. If we had included the line we felt the large radiance would likely saturate the detector. If we refly the same grating we will likely move the short wavelength cut-off further into the red wing of the Ly-α line.

THE ASSOCIATION OF TYPE III BURSTS AND CORONAL TRANSIENT ACTIVITY

B. V. Jackson* and G. A. Dulk
University of Colorado at Boulder, U.S.A.

K. V. Sheridan
Division of Radiophysics, CSIRO, Sydney, Australia

Metric Type III radio bursts are nearly always associated with solar active regions. In particular, the occurrence of Type III bursts within a few minutes of the onset of the solar flares (Wild et al. 1954) is well known. During the Skylab period, a broad peak in the number of isolated Type III bursts occurred about 5 h prior to large Hα solar flares (Jackson and Sheridan 1979).

Here we follow up on a report by Jackson et al. (1978) that a peak in the number of isolated Type III bursts occurs prior to mass ejection transients observed from Skylab. Histograms show a broad peak approximately 8 hours prior to the time that the most massive portions of the transients reach 3 R_O. Further, those Type III bursts whose positions were measured with the Culgoora heliograph generally emanated from the vicinity of the eventual mass ejections. Figure 1 shows the temporal distribution of the bursts observed to lie within 20° of the solar position angle of the eventual transient; there is a pronounced peak in the number of bursts some hours prior to the transient.

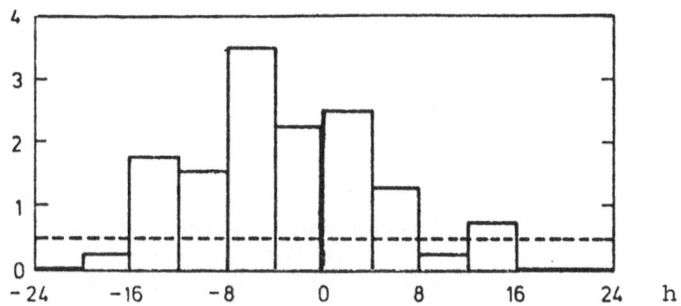

FIGURE 1. Number of isolated Type III bursts or burst groups per hour prior to and after mass ejection transients. The dashed line gives the mean number of Type III bursts per hour over ±126 h, excluding the interval of the histogram. 17 transients are represented.

M. Dryer and E. Tandberg-Hanssen (eds.), Solar and Interplanetary Dynamics, 379-380.

The lead time of the Type III bursts is not the same from one event to another, but depends on the speed of the eventual transient: the slower transients lag the Type III bursts much more than do the faster transients. Most of the Type III bursts of individual events occur near the time of initial motion of the outermost material of the mass ejections, i.e., the onset of motion of the transient forerunners described by Jackson and Hildner (1978).

From Hα, X-ray, and magnetogram data, there is no evidence for an extraordinary input of energy into the corona at the time of the peak in Type III activity prior to the transients. However, there is evidence that the transients arise from active regions in which the areas of strong magnetic field gradually increase in size over a period of days. Thus, the data indicate that the corona slowly stores the energy that is quickly released at the time of the mass ejection. We suggest that the increase and subsequent decrease in Type III activity indicates a new phase in the coronal adjustment to the long-term addition of energy, with some of the energy going into lifting the plasma and perhaps some into the kinetic energy of moving plasma. If this is correct, the peak in Type III activity indicates the beginning of outward coronal plasma motion and can be studied for clues to the energy input mechanism(s) responsible for mass ejection transients.

*Present address: University of California, San Diego, La Jolla, California, U.S.A.

REFERENCES

Jackson, B. V., Sheridan, K. V., Dulk, G. A., and McLean, D. J.: 1978, Proc. Astron. Soc. Australia, 31, 241.
Jackson, B. V., and Hildner, E.: 1978, Solar Phys. 60, 155.
Jackson, B. V., and Sheridan, K. V.: 1979, Proc. Astron. Soc. Australia, in press.
Wild, J. P., Roberts, J. A., and Murray, J. D.: 1959, Nature, 173, 532.

DISCUSSION

Webb: Did you observe the distribution of Type IIIs on the disc, and if so, do they give the same type of peak in Type III number prior to solar flares as is observed prior to limb transients?

Jackson: During the Skylab period a peak in Type III burst number was found approximately 5 h prior to solar flares of importance 1 or greater. The Type III bursts of this peak are generally associated with the active region where the flare occurs. The often-reported peak in Type III activity at the onset of flares and subflares is well observed for approximately 11% of the ∿ 1300 Type III bursts observed during Skylab. However, there is no increase in the number of observed subflares 5 hr prior to the flares during Skylab.

RECENT VERY BRIGHT TYPE IV SOLAR METRE-WAVE RADIO EMISSIONS

R.A. Duncan, R.T. Stewart and G.J. Nelson
Division of Radiophysics, CSIRO, Sydney, Australia

INTRODUCTION

Stewart et al. (1978) have reported moving Type IV solar metre-wave radio outbursts with brightness temperatures between 10^8 and 10^{10} K. We now report Culgoora radioheliograph observations of four more Type IV radio sources, some moving, some stationary, but all with brightness temperatures above 10^9 K, and one with a brightness temperature above 10^{13} K. We also describe one of the previously reported events (that of 1977 September 20) in more detail. The interest of these events is that their high brightnesses place great strain upon the gyro-synchrotron theory of radio emission.

OBSERVATIONS

Details of the outbursts are shown in Table I. Brightness and

TABLE I

Brightness temperature (T_b), polarization, and motion of the Type II and Type IV phases of four recent intense solar metre-wave radio outbursts.

Date	Type II max. phase at 43 MHz		Type IV					Moving or Stationary
			Max. phase at 80 MHz			Max. phase at 43 MHz		
	UT	$\log_{10} T_b$	UT	$\log_{10} T_b$	Pol. %	UT	$\log_{10} T_b$	
1977 Sept.7	2241	10.7	2259	9.2	40 R	2310	9.3	Slowly M
1977 Sept.16	2244	10.2	2307	10.3	60 R	2333	9.9	S
1978 May 7	-	<12.7	0341	12.0	20 R	0334	13.7	M
1979 Feb.16	0156	11.8	0214	9.5	0 L	0230	9.9	M

M. Dryer and E. Tandberg-Hanssen (eds.), Solar and Interplanetary Dynamics, 381-385.

polarization are tabulated for the time of maximum brightness at the stated frequency. No Type II was seen to precede the Type IV of 1978 May 7; a Type II during the Type IV would not have been seen unless it had been brighter than about $10^{12.7}$ K. We see that in all events the brightness temperatures at 43 and 80 MHz exceeded 10^9 K and that in one event, 1978 May 7, the brightness temperature at 43 MHz exceeded 10^{13} K.

Spectral plots for these four events are given in Figure 1. The

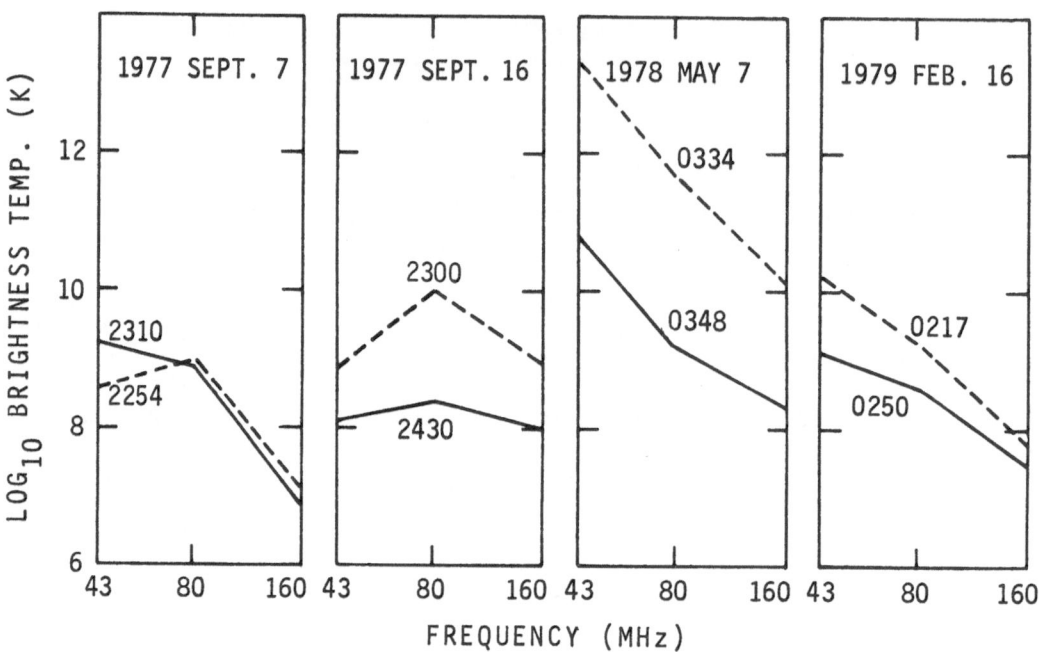

Fig. 1 - Brightness temperature versus observing frequency. Dashed and solid lines respectively refer to the earliest and latest times that the sources were visible at all three radio frequencies. The latest time was usually set by the disappearance of the source at 160 MHz. The apparent coronal height of the sources was often different at different frequencies.

stationary source seen on 1977 September 16 showed a moderately flat spectrum. The moving sources seen on the other dates had steeper spectra, with spectral indices between -4 and -5.

Table I shows also the circular polarization of the Type IV radio sources at 80 MHz at the time of maximum brightness at 80 MHz (the heliograph cannot measure polarization at 43 MHz). These four events showed polarizations of from 0% to 60%; another event seen on 1977 September 20 and described later, showed a polarization, at the time of maximum brightness, of 100%. The polarization increased after the time of maximum brightness in all events except those of 1978 May 8,

where the polarization remained steady, and 1977 September 20, where
the polarization, having reached 100%, could rise no higher.

Type IV sources, that is sources which emit broad-band long-lived
continuum, may be divided into two broad classes: those continuum
sources which remain stationary at normal coronal heights and those
continuum sources which move progressively outwards to very great
coronal heights. The latter have often been considered to be isolated
plasmoids carrying their own magnetic field and energetic electrons,
and emitting by gyro-synchrotron emission. The four events listed in
Table I almost certainly include examples from both classes.

However, very high moving Type IV sources can be identified with
confidence only if the source lies near the limb and has a component
of velocity radially outwards at right angles to the line of sight.
Such circumstances prevailed during the event of 1977 September 20
(Stewart et al., 1978; Duncan and Dixon, 1979). This event began
with two adjacent and oppositely polarized sources on the limb. The
southern source was about 55% left-hand circularly polarized throughout
the event. The northern source was about 60% right-hand circularly
polarized at the start (0325 UT) of the event, but it rapidly increased
in polarization and was approximately 100% polarized at all times after
0329 UT. As shown in Figure 2, the southern partially left-hand-polarized
source remained stationary, but the northern totally right-hand-polarized
source moved outward at a projected velocity of 710 km s^{-1} and reached
a projected radial distance of 3.4 $R_\odot$ before fading (Fig. 2). Thus in
this event we have a clear example of both a stationary and a moving
Type IV radio source, each with a brightness temperature exceeding
10^9 K.

The moving component of the very bright event observed on
1978 May 7 showed only about 20% polarization at the time of maximum
brightness. It differed from all the other events listed in Table I
in that its polarization did not increase after the time of maximum
brightness; in fact, at many times its polarization fell well below
20%. Its projected velocity carried it across the face of the disk,
so that we have no way of knowing whether or not it rose to great
coronal heights.

DISCUSSION

Moving Type IV sources have often been attributed to gyro-
synchrotron emission from sub-relativistic electrons trapped in an
isolated moving plasmoid (Dulk, 1973; Kai, 1978). However, the
observations reported here and elsewhere (Kai, 1969, 1978; Stewart
et al., 1978) throw doubt on this interpretation (cf. McLean, 1974).
Because moving Type IV radio emission is usually strongest at low
frequencies (Fig. 1) and because it is often substantially circularly
polarized, we cannot invoke electrons with energies much above 100 keV
(Kai, 1969; Dulk, 1973; Stewart et al., 1978). However, 100 keV
electrons are incapable of giving gyro-synchrotron brightness

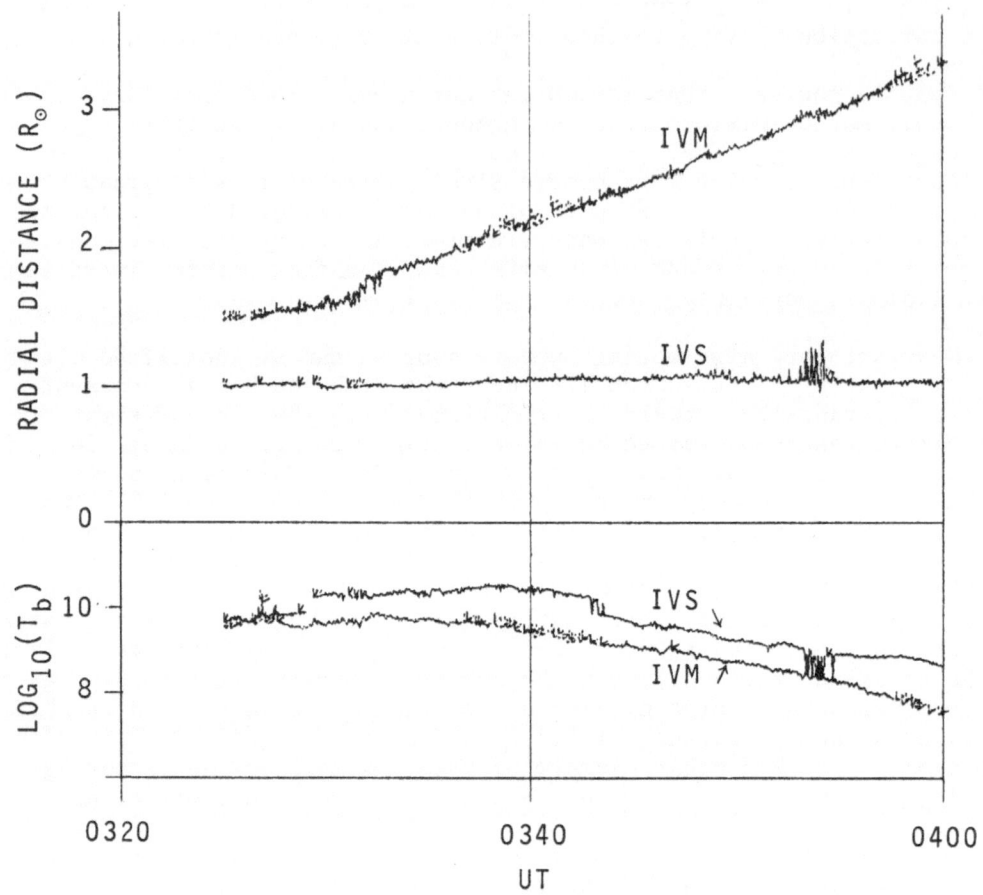

Fig. 2 - Plots of radial distance (top) and brightness temperature
(bottom) of the two sources seen on 1977 September 20, both measured
at 80 MHz.

temperatures much greater than 10^8 K (Stewart et al., 1978); even if
they radiated as a perfect black body they could, from Boltzmann's
relation, give a brightness of no more than 10^9 K. Thus, in summary,
on the incoherent gyro-synchrotron model the observed spectra and
polarization of Type IV radio sources are difficult to reconcile with
their observed brightness. The very high brightnesses suggest amplified,
or coherent, emission.

A feature of the event of 1977 September 7 suggested coherent
plasma emission. During this event Type III emission was seen to come
from the same high moving source as Type IV emission. As Type III
emission is universally believed to arise through plasma emission
this observation implies that - notwithstanding the great height of the
source - plasma densities were appropriate for plasma emission. Hence
the Type IV also probably arose through plasma emission (Duncan, 1978).

References

Dulk, G.A.: 1973, *Solar Phys.* 32, 491.
Duncan, R.A.: 1978, *Proc. Astron. Soc. Aust.* 3, 253.
Duncan, R.A., and Dixon, J.M.: 1979, UAG Report, World Data Center A, Boulder, Colorado (in press).
Kai, K.: 1969, *Proc. Astron. Soc. Aust.* 1, 189.
Kai, K.: 1978, *Solar Phys.* 56, 417.
McLean, D.J.: 1974, in *Coronal Disturbances*, Proc. IAU Symp. 57 (G. Newkirk, ed.) D. Reidel, Dordrecht, p. 301.
Stewart, R.T., Duncan, R.A., Suzuki, S., and Nelson, G.J.: 1978, *Proc. Astron. Soc. Aust.* 3, 247.

THE SOLAR MASS EJECTION OF 8 MAY 1979

D. J. Michels, R. A. Howard, M. J. Koomen, & N. R. Sheeley, Jr.
E. O. Hulburt Center for Space Research
Naval Research Laboratory
Washington, D. C. 20375, USA
and
B. Rompolt
Astronomical Observatory of the Wroclaw University
51-622 Wroclaw UL. Kopernika 11, Poland

Abstract: This paper describes the main features of the 8 May 1979 solar mass ejection, including the eruption of a polar crown filament to 1.5 R$_\odot$ during 0810-1036 UT and the passage of material through the outer corona, from 2.6 to 10.0 R$_\odot$, during 1028-1246 UT.

An earlier paper in this Symposium (Sheeley et al., 1979) described the NRL SOLWIND coronagraph. Briefly, it is optically similar to the NRL coronagraph on NASA's OSO-7 satellite (Koomen, et al., 1975), but it has greatly improved temporal resolution. With this instrument in orbit on the P78-1 satellite, images of the entire outer corona, from 2.6 to 10.0 R$_\odot$, are obtained every ten minutes.

Figure 1(a) is a composite, showing the appearance of the sun in Hα at 0701 UT on 8 May. As is typical at epochs near solar maximum, polar crown filaments encircle the north and south poles. One of these filaments, is visible as a prominence at S55W90 (prominence photo at

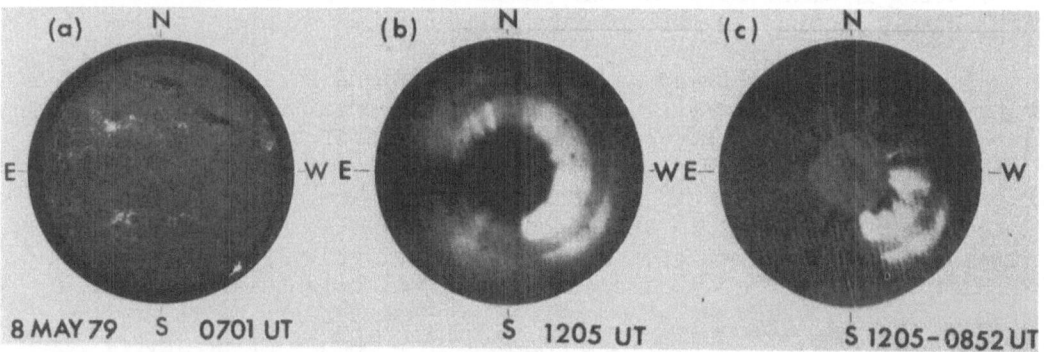

Fig. 1: (a) Composite, showing Hα limb and disk observations of the south polar crown filament prior to its eruption at S55W90. (b) Outer corona at 1205UT (c) Same data as (b), but with pre-event image at 0852 UT subtracted.

M. Dryer and E. Tandberg-Hanssen (eds.), Solar and Interplanetary Dynamics, 387-391.

0704 UT). This filament later erupted, (Figure 2), beginning at about 0810 UT and reaching maximum distance of 1.5 R$_\odot$ from the center of the sun at 1021 UT. Subsequently the prominence faded, and some of the material fell back to the solar surface. By 1120 UT, vestiges of the H α prominence had disappeared. The bulk of the prominence material displayed radial velocities averaging 40 km s^{-1}; the maximum velocity observed was 165 km s^{-1}.

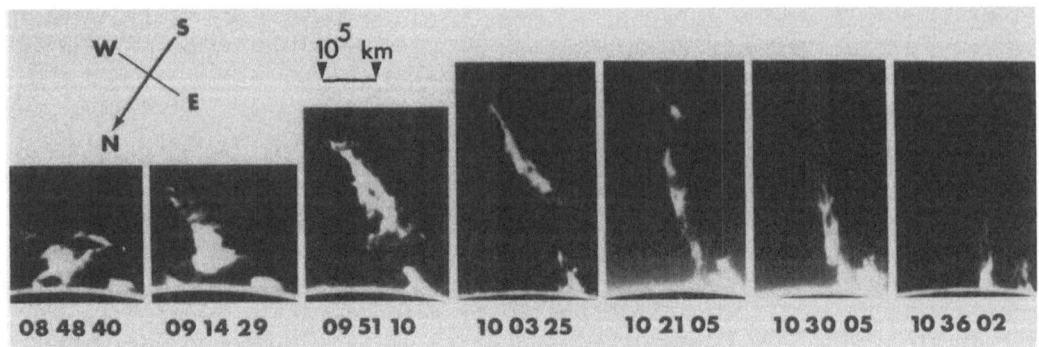

Fig. 2: Sequence showing eruptive prominence in Hα .

Figure 1(b) is an image from the orbiting coronagraph, recorded at 1205 UT. Here the transient is seen in the SE quadrant, midway through the instrumental field of view. The roughly circular brightening measures approximately 6 R$_\odot$ in diameter at this time. Note that the instrument's occulting disk forms a shadow that extends to 2.6 R$_\odot$. The outer edge of the field of view has been masked down to about 8 R$_\odot$. The slightly off-center ring at ~5 R$_\odot$, darker than its surroundings, is caused by a focal plane polarizer with its axis everywhere approximately radial; the remainder of the focal plane contains a concentric polarizer with its axis everywhere at right angles to the solar radius, thus discriminating in favor of K-coronal radiation. There is a second annular ring, barely visible in this picture, at ~8 R$_\odot$. (Koomen et al., 1975).

Figure 1(c) illustrates a powerful technique for studying temporal changes of coronal intensity. Here, the data contained in the 256 x 256-element picture matrix of Figure 1(b) has had subtracted from it a similar picture matrix corresponding to an observation at 0852 UT, prior to onset of the transient. The resulting gray-scale presentation then represents only changes in the corona that have taken place between the times of these two pictures. Areas of neutral gray signify no change; brighter areas indicate an increase in coronal brightness and darker areas indicate a decrease. The subtraction has clearly revealed the coronal mass ejection as well as changes in a number of pre-event streamers, even at great distances from the transient.

In Figure 3 one can trace the development of the transient from 1028 UT until 1246 UT, when the leading edge of the ejecta had almost reached the limit of the instrumental field of view at 10 R$_\odot$. The leading edge moves outward with a constant radial velocity of approximately 500 km s^{-1}. Projected back to the solar limb, this would indicate an onset near 0950 UT. Figure 2 of the accompanying paper (Sheeley et al., 1979) shows the pre-event corona at 0852 UT, which has been subtracted to produce these images. Comparison of the 0852 UT image with that at 1028 UT shows that the first manifestation occurred in the region of the pre-event streamers at S15-60W90. Also, there developed a brighter and more sharply-defined feature at S65. This

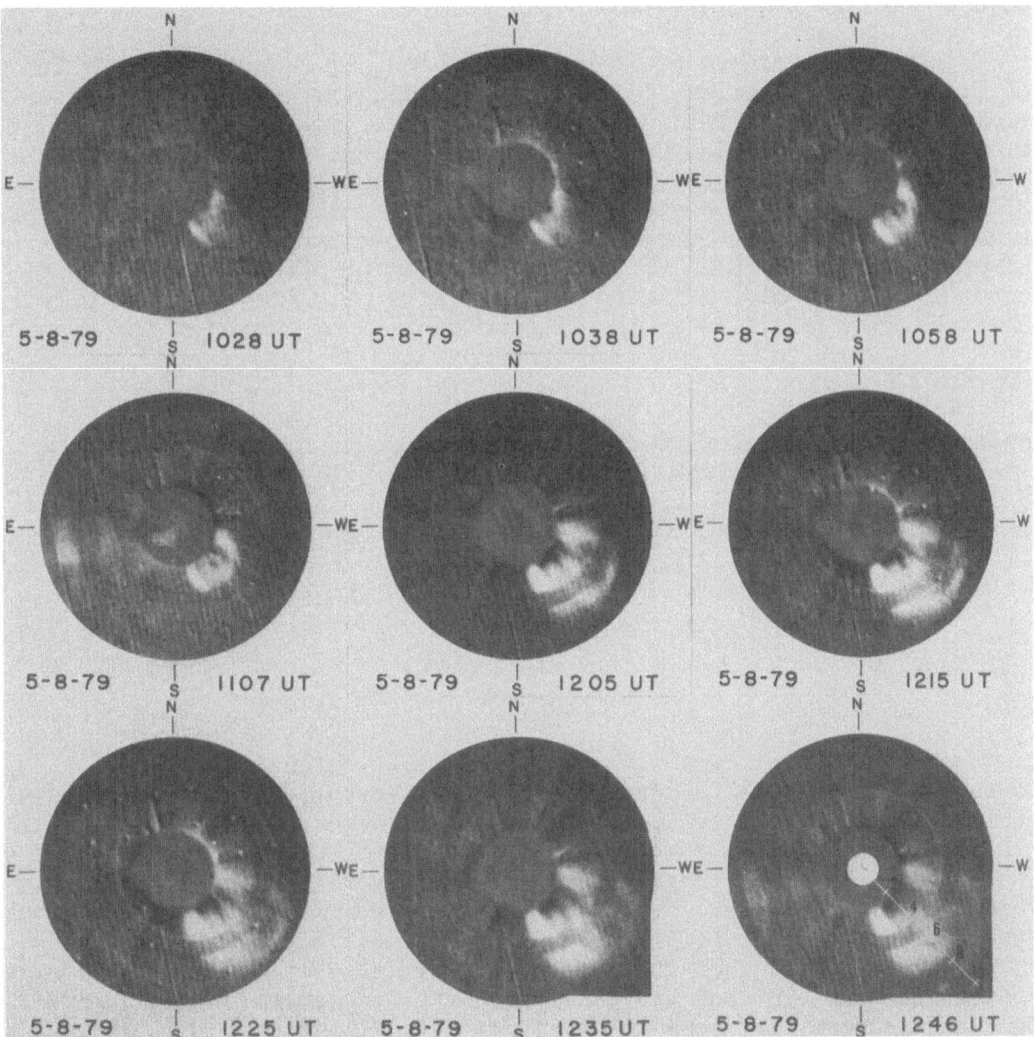

Fig. 3: Difference images, showing coronal changes from the time of a reference image at 0852 UT to the times indicated. In the last two images, the mask has been cut to show expansion of the transient to the instrumental limit of 10 R$_\odot$. The scale marks distance from sun center in solar radii; the white spot indicates the size of the photospheric disk.

continued to be the brightest part of the transient. In the images starting with 1038 UT, there is a neutral gray spot (signifying no brightening) imbedded in the outward moving plasma, suggestive of a cavity within the ejecta. A small, denser region may be seen within this cavity, particularly in the 1058 and 1107 UT images. The spacecraft passed into the Earth's shadow after recording the 1107 UT image. In the next observation, at 1205 UT, there is evidence for disappearance of the pre-event streamer cluster, as indicated by the darkened area adjacent to the occulter shadow.

These observations could not have been made without the assistance of D. Roberts, F. Harlow, R. Seal, and R. Chaimson, who provided essential technical support at NRL. Dr. I. Garczynska, Dr. J. Paciorek, and P. Majer, of the Wroclaw Observatory assisted in the Hα prominence observations. Dr. P. Simon, University of Paris Observatory at Meudon, kindly supplied the Hα disk photograph. We are indebted to the U. S. Department of Defense Space Test Program for integration and launch support, and to the NASA Office of Solar Physics which provided spare coronagraph and solar-pointing hardware from its OSO-7 program, and also assisted in acquisition of the P78-1 data used in preparing this paper.

REFERENCES

Koomen, M. J., Detwiler, C. R., Brueckner, G. E., Cooper, H. W., and Tousey, R.: 1975, Appl. Opt. 14, 743.
Sheeley, N. R., Jr., Howard, R. A., Michels, D. J., and Koomen, M. J.: 1979, in M. Dryer and E. Tandberg-Hanssen (eds.), "Solar and Inter-planetary Dynamics", I. A. U. Symposium No. 91, Cambridge, MA, 27-31 August 1979.

DISCUSSION

Moore: In comparison with the white-light coronal transients observed from Skylab for quiescent filament eruptions, was there anything unusual or unique about this transient and to what degree was it average or typical?

Michels: The observation is unique in several respects: first, in continuity of coverage, because we have coronal images every ten minutes whenever the spacecraft is in sunlight - it was not possible to show more than a sampling of the data here; then too, the event itself appears to be different in that no vestiges of the transient, or of the pre-event streamers, are seen the next day (cf. Fig. 2 of the accompanying paper (Sheeley et al., 1979)). Anzer and Poland have said that, in those Skylab events studied, residual "legs" of the transient remained for one to two days afterward. (Anzer, U. and Poland, A. I.: 1979, Solar Phys., 61, 95.)

Sheeley: It may also be mentioned that this event took place near solar maximum. No Skylab events involved polar crown filaments because there were none at that time.

Martres: In this session we learn that the three dimensional spatial extension of the coronal transients is mainly similar to a bubble. Now we try to associate these coronal transients to destablize filaments which have a quasi-planar extension: how do you explain this apparent contradiction?

Michels: The form of the ejected material is not made completely clear from this observation. For example, if the transient is associated with a pre-existing arcade of loops following the sinuous course of a typical filament channel, then loops would be seen at many aspect angles and these, when viewed in superposition, could well present a bubble-like appearance.

Newkirk: The question of the 3-dimensional structure of coronal transients is not completely answered. However, the polarization measures in white light indicate that this gross structure resembles a loop rather than a bubble.

Martres: If the large transient arches seen in the corona in the sky plane are to be associated to filaments, their "curvature" would be directly correlated to the angle of the main direction of the chromospheric filament with the line-of-sight up to the limit of a radial perturbation for a filament perpendicular to the limb. It seems to me that the coronal arches showed here are about of the same shape (?). Is it right?

Michels: More complete analysis, particularly if we have a number of events, may help us to discriminate between the different possibilities.

Kahler: I don't really see a problem with the three-dimensional aspect of the coronal transients. If we imagine an arcade of loops over-lying the filament and running along the neutral line, then it shouldn't matter whether the neutral line lies along the line of sight or perpendicular to it, because in either case there will be a sub-stantial angular extent to the loop structures.

McIntosh: (Comment) The occurrence of coronal transients above polar-crown erupting filaments warns us that the cause of such activity cannot be emerging magnetic flux or flare-like triggering phenomena. I think we must consider large-scale, slow processes, such as the shear between moving large-scale magnetic structures.

Engvold: (Comment) With reference to the discussion on how coronal transients may appear in 3-dimensions, I would like to mention what we have learned about eruptive prominences (ascending prominences and flare sprays) in this regard. To my knowledge, the cases which are recorded both spectroscopically (motion in the line-of-sight) and on narrow band filtergrams (motion in the plane of the sky), are suggestive of rising, expanding "bubbles" of material.

VARIATIONS OF INTERPLANETARY PARAMETERS AND COSMIC-RAY INTENSITIES

A. Geranios
University of Athens, Nuclear Physics Laboratory
Cosmic Ray Group
Solonos str. 104, Athens 144
Greece

ABSTRACT

Observations of cosmic ray intensity depressions by earth bound neutron monitors and measurements of interplanetary parameter's variations aboard geocentric satellites in the period January 1972-July 1974 are analysed and grouped according to their correlation among them. From this analysis of about 30 cases it came out that the majority of the depressions correlates with the average propagation speed of interplanetary shocks as well as with the amplitude of the interplanetary magnetic field after the eruption of a solar flare. About one fourth of the events correlates with corotating fast solar wind streams. As the recovery time of the shock-related depressions depends strongly on the heliographic longitude of the causitive solar flare, it seems that the cosmic ray modulation region has a corotative-like feature.

I. INTRODUCTION

Many authors, in order to find the mechanism of cosmic ray decreases, investigated any possible relations between cosmic ray intensities and interplanetary and solar parameters (Conforto, 1973; Barnden, 1973; Hedgecock, 1975; Bland, 1976; Barouch and Burlaga, 1975; Kane, 1977; Iucci et al., 1973, 1975; Lockwood and Webber, 1977). In many of these studies only the pronounced CR decreases, the so-called Forbush Decreases (Forbush, 1966; Lockwood, 1971) were used and their characteristics as the amplitude and the recovery time, were related to solar flares, IMFs, solar wind velocities, and hydromagnetic shock waves (Dryer, 1975). Barouch and Burlaga found inverse correlations between almost all kinds of CR intensity fluctuations and IMF amplitude, observations which are also supported by Lockwood and Webber (1977).

In the past most times the well-known models of Gold (1960), and Parker (1963) have been used to explain the CR decreases. However, unique conclusions could not always be drawn because of the lack of

393

M. Dryer and E. Tandberg-Hanssen (eds.), Solar and Interplanetary Dynamics, 393-398.
Copyright © 1980 by the IAU.

knowledge on the state of the interplanetary plasma and the processes
going on there. Therefore, in this work we continue correlative
studies of earthbound CR observations and in situ plasma observations.
Using low rigidity neutron monitor stations cosmic ray data we identify
CR decreases and compare them to simultaneously occurring enhanced
solar activity, IMF amplitude and solar wind velocity, in order to
improve our understanding of the macroscopic conditions and mechanisms
responsible for these decreases.

II. DATA PRESENTATION

Neutron monitor data from Alert and Deep River stations of the
period January 1972 to July 1974 showing CR decreases are analysed together with interplanetary parameters measured by HEOS 2, or taken from
multiple spacecraft measurements (Grunwaldt, 1975; King, 1975). For
each CR decrease greater than 2% the hourly averages of neutron
monitor data corrected for atmospheric pressure and normalized to the
counting rate just before the depression were plotted. Long-term
variation corrections were not applied because of the relatively short
time period under consideration. The diurnal variation effect was
taken into account for the estimation of the maximum depression. Three-
hour averages of the solar wind speed and the IMF amplitude in the
vicinity of the earth were plotted for the same periods of time. An
example of a moderate CRD falling into a high speed solar wind stream
and not related with any solar flare or a Storm Sudden Commencement
(ssc) is shown in Figure 1.

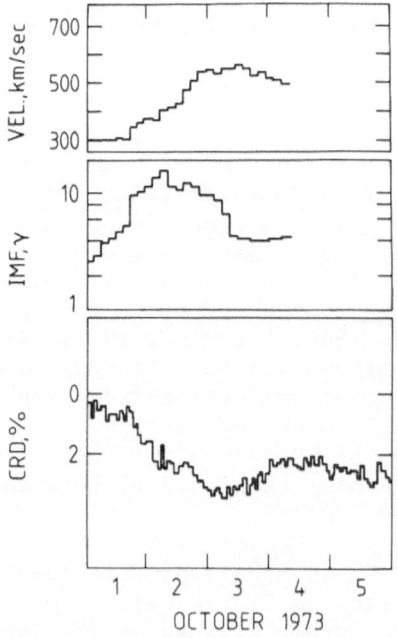

Figure 1. A high speed solar wind stream falling into a CR decrease.

The travelling time of the solar wind from the point of observation to the magnetosphere could be neglected. We used the time delay between the flare onset and the observed ssc in order to calculate the mean transient velocities of the interplanetary disturbances which were suspicious of having caused a CRD, and we tried to identify the solar flares which possibly could have generated these disturbances. The selection criteria of the flares were that their optical importance was greater than 1 N and the time delay was no longer than six days.

3.   DISCUSSION-CONCLUSION

Assuming radial expansion of shock waves, the correlation between their mean propagation speed and the amplitude of the related CRD came out to be positive, Figure 2.

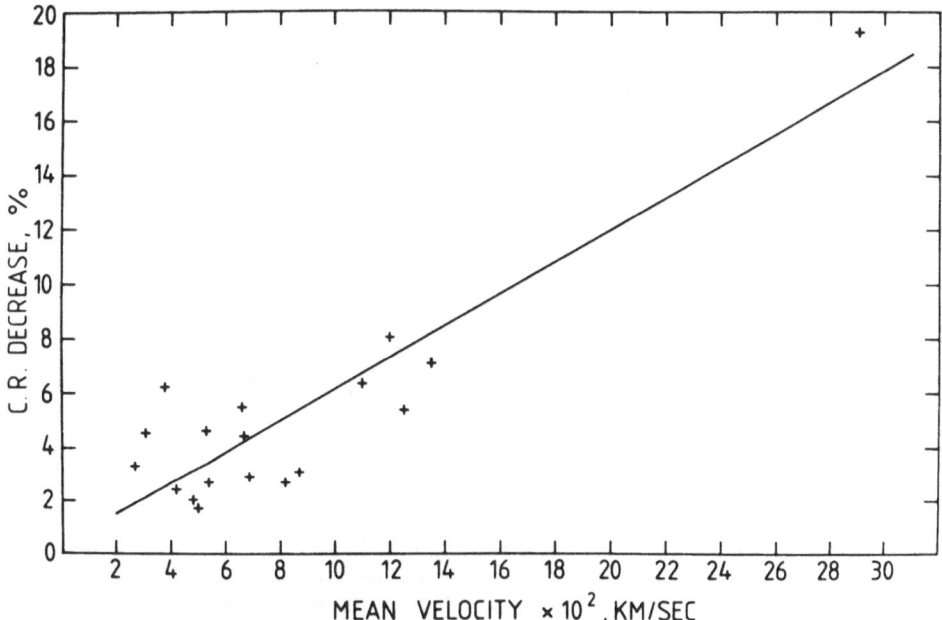

Figure 2.   Correlation between cosmic ray depressions and the mean
            propagation speeds of the related shocks.

The same correlation of solar wind stream-related CRDs showed to be only qualitative, where now the maximum speed of the stream during the CRD is used. For the same cases we correlated the maximum amplitude of the IMF observed at 1 AU with the amplitude of the CR depressions, and we got a similar relation (Figure 3).

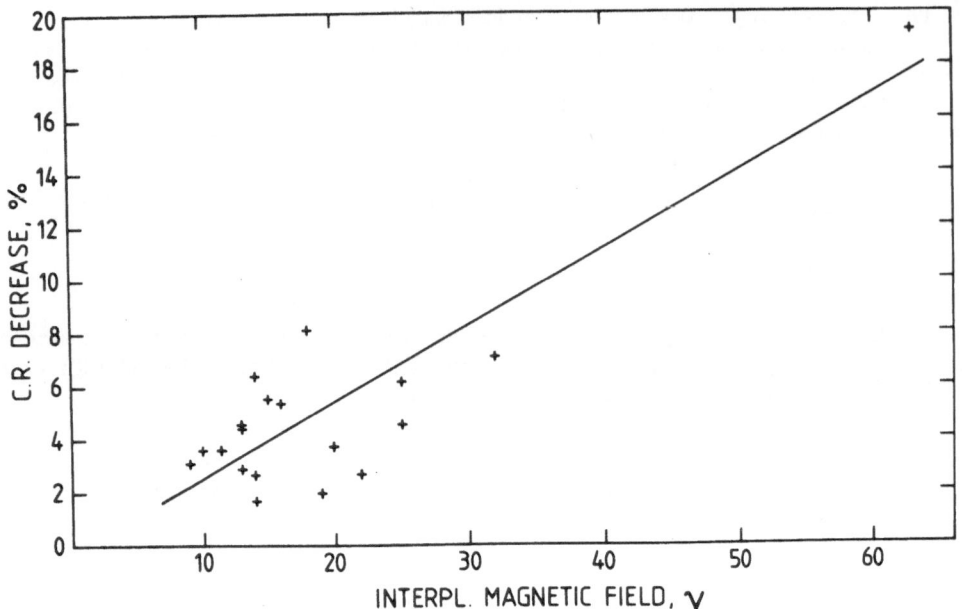

Figure 3. Correlation between cosmic ray depressions and IMF amplitude
for shock related events.

        The correlation between CR depressions and fast solar wind streams,
where a shock wave was absent, was almost zero. Again, for the shock
related cases we calculated the momentary speed (W) of the shock
assuming a radially symmetric motion and mass conservation

$$W = (V_2N_2 - N_1V_1)/(N_2 - N_1)$$

where V and N are the measured speeds and densities of the solar wind,
and subscripts 1 and 2 refer to values before and after the shock
front passage, respectively. Correlation between W and CRD amplitude
failed. This was due perhaps to the fact that W characterizes the
shock only at the time of the observation, where the mean propagation
speed characterizes the motion of the shock from the sun up to 1 AU.
In order to estimate whether or not the modulation region in which
CR are depressed corotates with the sun, we plotted the recovery time
for each CRD (i.e., the time which is needed for the CR to reach its
predecrease value) with the heliographic longitude of the related flare,
Figure 4. It is evident that long-recovering CRDs correspond on the
average to solar flares generated in the eastern part of the solar
disc, while short recovering decreases correspond to flares in the
western part. This "asymmetry" could reveal a westward corotative
motion of the modulated region and therefore, for eastern situated
flares, the earth-bound neutron monitors remain longer inside the
modulation region, while for western flares this time is considerably
shorter. The positive correlation between CRD and mean propagation

speed of the shock, shown in Figure 2, could emphasize the "frozen-in" characteristics of the IMF in the solar wind.

Finally, due to the positive correlation between the CRD amplitude and the IMF amplitude, as well as mean speed of the related shock, and due to the negative correlation between the recovery time of the heliographic longitude of the causitive flare we suggest that the cosmic ray modulation region has to corotate with the sun as it expands radially with the characteristics of a shock front.

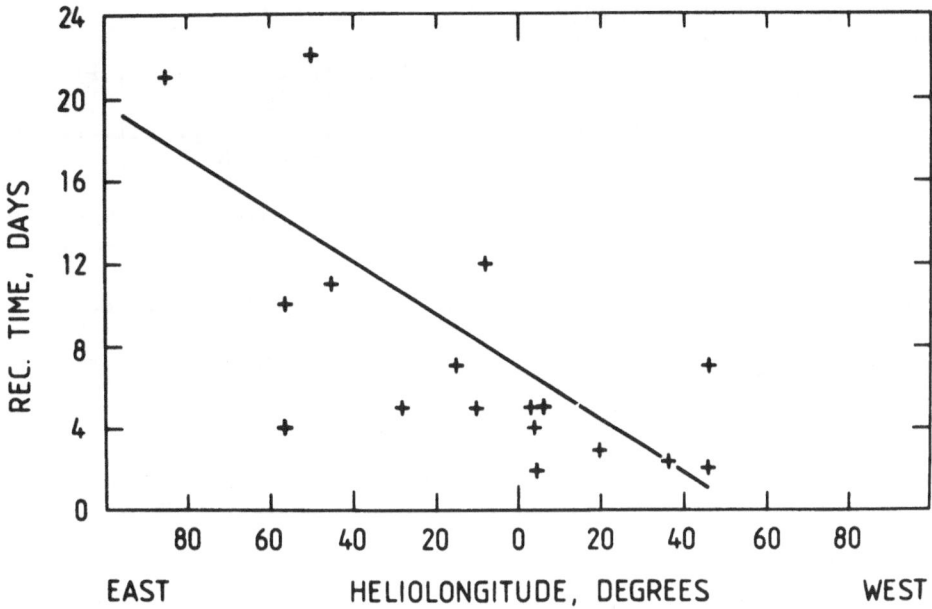

Figure 4.   Correlation between the recovery time of CRD and the heliographic longitude of the related flare.

REFERENCES

Barnden, L. R.:   1976, *Proc. 13th Int. Conf. Cosmic Rays, 2,* pp. 1271-
    1276.
Bland, C. J.: 1976, *J. Geophys. Res., 81,* pp. 1807-1811.
Barouch, E., and Burlaga, L. F.: 1975, *J. Geophys. Res., 80,* pp. 449-
    456.
Conforto, A. M.: 1973, *N. Cimento 17B,* pp. 300-312.
Dryer, M.: 1975, *Sp. Sci. Rev., 17,* pp. 277-326.
Forbush, S. E.: 1966, *Encyclopedia of Physics, Vol. XLIX/I,* ed. S.
    Flugge, Springer, New York.
Gold, T.: 1960, *Astrophys. J., Suppl. Ser. 4,* pp. 406-413.
Grunwaldt, H.: 1975, Private communication, Max-Planck-Institut für
    Extraterrestrische Physik, 8046 Garching, F.R.G.
Hedgecock, P. C.: 1975, *Solar Phys., 42,* pp. 497-527.
Iucci, N., Parisi, M., Storini, M., and Villoresi, G.: 1973, *Proc.
    Int. Conf. Cosmic Rays, 13th, 2,* pp. 1261-1265.
Iucci, N., Parisi, M., Storini, M., Villoresi, and Zangrilli, L.: 1975,
    Proc. Int. Conf. Cosmic Rays 14th, pp. 1064-1068.
Kane, R. P.: 1977, *J. Geophys. Res., 82,* pp. 561-577.
King, J. H., 1975, *Interplanetary Magnetic Field Data Book,* NASA/NSSDC
    75-04, Greenbelt, Maryland.
Lockwood, J. A.: 1971, *Sp. Sci. Rev., 12,* pp. 658-715.
Lockwood, J. A., and Webber, W. R.: 1977, *J. Geophys. Res., 82,* pp.
    1906-1914.
Parker, E. N.: 1963, *Interplanetary Dynamical Processes,* Inters.

TWO CLASSES OF FAST SOLAR WIND STREAMS: THEIR ORIGIN AND INFLUENCE ON THE GALACTIC COSMIC RAY INTENSITY

N.Iucci, M.Parisi, M.Storini and G.Villoresi
Istituto di Fisica dell'Università
Laboratorio Plasma Spazio del C.N.R.
Piazzale delle Scienze, 5 - 00185 Roma, Italy.

The analysis of solar wind (bulk velocity v, proton density n, proton temperature T) and magnetic field ($\vec{B}$) data in the years 1964-1974 makes possible to identify two main classes of high-speed streams (ΔV = (vm - vo)$\geqslant$ 100 km/sec, vm being the maximum daily mean speed and vo the mean value between the speeds immediately preceding and following the stream; duration $\Delta t \geqslant$ 2 days):

a) Regular high-speed streams (RHSS's) with the following behaviours in the interplanetary parameters: n shows a peak immediately before the increase in v, sharp decrease towards low values during the v increase phase, recovery towards normal values during the v decrease phase; field magnitude B shows a peak terminating at the end of the v increase where it decreases rapidly towards a level which remains almost constant, constant field polarity except for minor fluctuations lasting a few hours; T increases with v, decreases slightly during the B descent phase and subsequently tends to follow the v behaviour. The initial phase of these streams is consistent with a stream interface (e.g. Burlaga, 1974) and the behaviours of the various parameters during the stream are consistent with a solar wind from a quiet solar region with unipolar diverging magnetic field.

b) Complex high-speed streams (CHSS's) showing at least one of the following features differentiating them from the RHSS's: simultaneous increases in v, n, B and T (radially outmoving shocks), large fluctuations in v, n, B and T in the high-speed period, fluctuations in field polarity lasting more than 3-4 hours, T behaviour departing from v behaviour, periods of very low T denoting flare ejecta pistons (e.g. Gosling et al., 1973).

The period during which the solar wind was emitted from the Sun is computed for each stream, by using speeds measured at the Earth assuming a constant Sun-Earth solar wind speed. The following features are obtained: nearly all (96%) the RHSS's are emitted during periods in which no "strong active regions" (AR's) exist at the central meridian and no solar flares accompanied by type IV radioemission fairly well extended in time and frequency (type IV SF's) occur; nearly all (92%) the CHSS's are associable with AR's and with type IV SF's;

M. Dryer and E. Tandberg-Hanssen (eds.), Solar and Interplanetary Dynamics, 399-401.

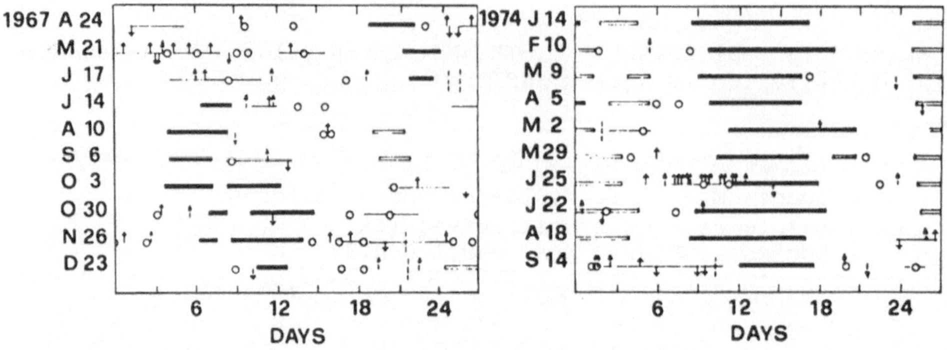

Fig. 1. Bartels display of dates of emission of RHSS's (▭ : positive polarity; ▬ : negative polarity) and CHSS's (—); central meridian passage of strong AR's (o) and occurrence of type IV SF's (↑ between E55-W55; ↓ E55-E90, W55-W90; | undefined longitude; ↑↓¦ idem for type IV SF's either in decametric waves only or lasting $<$10 minutes).

the RHSS's tend to recur several times. On the other hand only 60% of CHSS's recur 3-4 times at the most; the RHSS's tend to be longer and more frequent during the years 1973-1974; the CHSS frequency tends to follow the solar activity cycle. Moreover, practically all the observed near equatorial coronal holes (Nolte et al., 1976; Sheeley et al., 1976; Broussard et al., 1978) correspond to RHSS's even in respect of magnetic field polarity; the few instances of disagreement generally occur for small streams or to small coronal holes. Therefore we deem it plausible that the RHSS's are an almost stationary emission from coronal holes, while the CHSS's are generally emitted from the vicinities of AR's producing type IV SF's. See in Fig. 1 examples of such periods.

In order to study the cosmic ray (CR) modulation during the RHSS's, superposed epoch analyses of neutron monitor intensity (I) and v are performed; the events happening during the recovery of Forbush decreases (FD's) or followed by FD's and those with gaps longer than one day in the v data are not taken into account. The 88 RHSS's are divided into 5 groups according to duration as shown in Fig. 2; for each group the CR intensity follows the sign-changed v behaviour: $\Delta I/I \approx - K.\Delta v$, where K = 0.00005 sec/km. The factors causing such small CR modulation might be the following: increase of CR convection and decrease of diffusion in the stream interaction region which at great distances from the Sun may cover all magnetic field lines contained in the RHSS; due to magnetic field regularity in RHSS, CR's can diffuse freely parallel to the field lines, thus CR's with particularly small pitch angle can be absorbed by the Sun atmosphere. A glance at the CR intensity during CHSS's shows that most FD's with amplitude $\geqslant$ 1.5% occur during such periods. It also appears that for most individual events, v and I behaviours are not correlated; in Figure 3 the correlation plot between

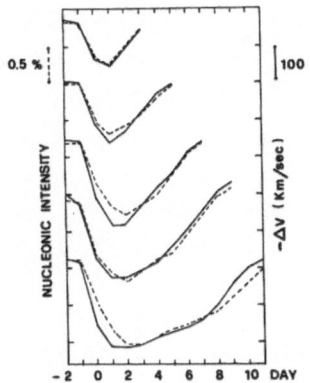 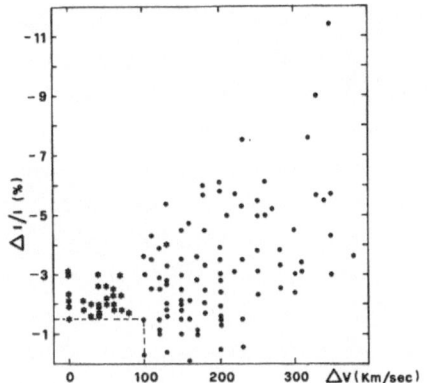

Fig. 2 (to the left). Superposed epoch analyses of high-latitude nucleonic intensity (dashed lines) and sign-changed solar wind speed (full lines) for 88 RHSS's divided in 5 groups according to their duration.

Fig. 3 (to the right). Nucleonic intensity maximum depression versus the maximum speed increase for 88 CHSS's (·) and for 29 FD's without CHSS's (*).

the maximum CR depression and ΔV is shown; besides the FD's associated with CHSS's there are 29 FD's where either $\Delta V < 100$ km/sec or $\Delta t < 2$ days. Out of these 29 FD's, 26 are associated with type IV SF's occurring near the Eastern limb (E50-E90) or are not flare-associated, i.e. are probably produced by type IV SF's occurring behind the East limb (Iucci et al., 1979), that is when the Earth enters into the modulated region from the Western edge where the speed increase is likely to be lower than in the radially advancing front. Therefore it can be stated that the CHSS's coming from the vicinities of AR's generating type IV SF's are generally associated with FD's whose time behaviours and maximum amplitudes are little correlated with the v at the Earth.

REFERENCES

Broussard, R.M., Sheeley, Jr., N.R., Tousey, R., and Underwood, J.H.: 1978, Solar Phys. 56, pp 161-183.
Burlaga, L.F.: 1974, J. Geophys. Res. 79, pp. 3717-3725.
Gosling, J.T., Pizzo, V., and Bame, S.J.: 1973, J. Geophys. Res. 78, pp. 2001-2009.
Iucci, N., Parisi, M., Storini, M., and Villoresi, G.: 1979, Nuovo Cimento 2C, pp. 1-52.
Nolte, J.T., Krieger, A.S., Timothy, A.F., Vaiana, G.S., and Zombeck, M.V.: 1976, Solar Phys. 46, pp. 291-301.
Sheeley, Jr., N.R., Harvey, J.W., and Feldman, W.C.: 1976, Solar Phys. 49, pp. 271-278.

A LARGE DECAMETRIC WAVELENGTH ANTENNA ARRAY FOR IPS OBSERVATIONS OF RADIO SOURCES

CH. V. SASTRY
Indian Institute of Astrophysics
and
Raman Research Institute, Bangalore, India

Most observations of interplanetary scintillations of radio sources are made at frequencies around 80 MHz. These observations are limited to regions close to the sun, where the scintillations are maximum at this frequency. It is possible to extend these observations to the weakly scattering regions beyond 1 A.U. by making measurements at low frequencies. We have built a low frequency antenna system at Gauribidanur, India (Lat. 13° 36' N and Long. 5 hrs. 10 min.), which can be used for this purpose. Although this system will not be dedicated to IPS, we intend to use it exclusively for solar wind observations during periods of interest.

The antenna system consists of two broadband (25 to 35 MHz) arrays arranged in the form of the letter 'T'. One of the arrays is oriented in the EW direction with dimensions 1.6 km/25 M and the other one is along the NS direction with dimensions of 0.5 km/ 40 M. By correlating the outputs of the two arrays a beam of 30 arc minutes is produced at 30 MHz. The physical area exceeds 50,000 sq. metres. It will be possible to observe above 50 sources around the sun for solar wind mapping. A detailed description of the antenna and receiver system and the programs undertaken will be published elsewhere.

DISCUSSION

Dryer: Do you have any plans for cooperative studies with radio observations at other longitudes (such as those at Clark Lake, Nancy, etc.), as well as at the new and complementary IPS station at Ahmedabed? Such studies would be extremely useful especially during SMY/STIP periods when spacecraft tracking would be especially requested.
Sastry: Yes. We are planning to have cooperative studies with Ahmedabad IPS array and Cocoa Cross Operations at Clark Lake.

M. Dryer and E. Tandberg-Hanssen (eds.), Solar and Interplanetary Dynamics, 403.

INTERPLANETARY SCINTILLATION - PRELIMINARY OBSERVATIONS AT 103 MHz

S. K. ALURKAR and R. V. BHONSLE
Physical Research Laboratory
Ahmedabad-380009, India

ABSTRACT

A 3-station interplanetary scintillation (IPS) observatory is being developed mainly with a view to study the solar wind plasma. The first IPS telescope operating at 103 MHz at Thaltej near Ahmedabad has been put into regular operation since April 1979. With only half the antenna aperture (~2500 m^2) presently in use, observations of 8-10 sources are being made to calculate scintillation index, temporal spectrum of intensity fluctuations and scale size of density irregularities.

INTRODUCTION

Interplanetary Scintillation (IPS) of small diameter radio sources (less than 1") has been proven to be very effective in studying the microscale (~100 km) structure and dynamics of the interplanetary medium in the plane of ecliptic as well as outside it. IPS measurements have also been used to derive structure of radio sources up to a resolution limit of 0.02" (Hewish et al., 1964; Cohen et al., 1967; Armstrong and Coles, 1972; Kakinuma and Watanabe, 1976). In this note we present preliminary observations made from the IPS station at Thaltej near Ahmedabad.

In Figure 1 is shown the locations of the three IPS stations being developed in India. The radio telescope near Ahmedabad has become operational since April 1979 and the remaining two telescopes are expected to be completed by mid-1980.

OBSERVATIONS AND ANALYSIS

The radio telescope with which the present IPS observations were made consists of a filled aperture dipole antenna array with an aperture of ~5000 m^2 at an operating frequency of 103 MHz. The obser-

M. Dryer and E. Tandberg-Hanssen (eds.), Solar and Interplanetary Dynamics, 405-408.

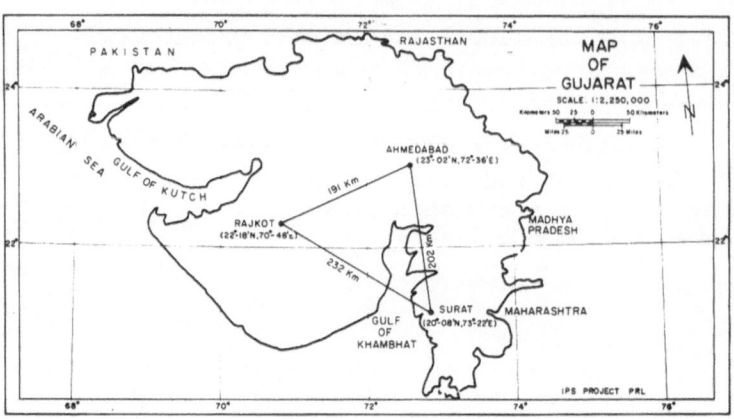

Figure 1.   Geometry of IPS stations in India.

vations reported here were taken using half the antenna aperture. The
RF signals after being stepped up by low noise preamps are combined in
a multibeam forming matrix yielding a 32-beam pattern, with individual
beams having a size of (8° EW x 2^O NS). A selected beam is connected to
a total power receiver having a predetection bandwidth of 2 MHz and the
intensity fluctuations are recorded on a strip-chart with an overall time-
constant of 0.1 sec. After A/D conversion at a sampling frequency of
20 Hz the data are recorded on a digital magnetic tape and processed on
an IBM 360/44 computer. Normally, 20-25 min of useful data per day for
each source are available. So far observations on 3C 48, 3C 144, 3C 147,
3C 161, 3C 196, 3C 237, 3C 273, 3C 298 and 3C 459 have been made.

The IPS power spectra were computed using the fast Fourier trans-
form algorithm. Autocorrelation functions and power spectra were
computed from successive 50 sec of data and after editing, the selected
spectra were added to get an average spectrum with a spectral resolution
of 0.04 Hz. Usually each source was recorded for about 30 min followed
by an off-source recording of at least 5 min. A few examples of IPS
spectra of some sources are shown in Figure 2. The straight lines on
the spectra are exponential fits to the data in the frequency range
0.3 - 3 Hz for strong and 0.3-1.5 Hz for weak scintillating sources.

The first and (square root) second moments f_1 and f_2 of the
intensity spectra were computed from the relations

$$f_1 = \int_o^{f_c} f\, P(f)df / \int_o^{f_c} P(f)\, df, \text{ and}$$

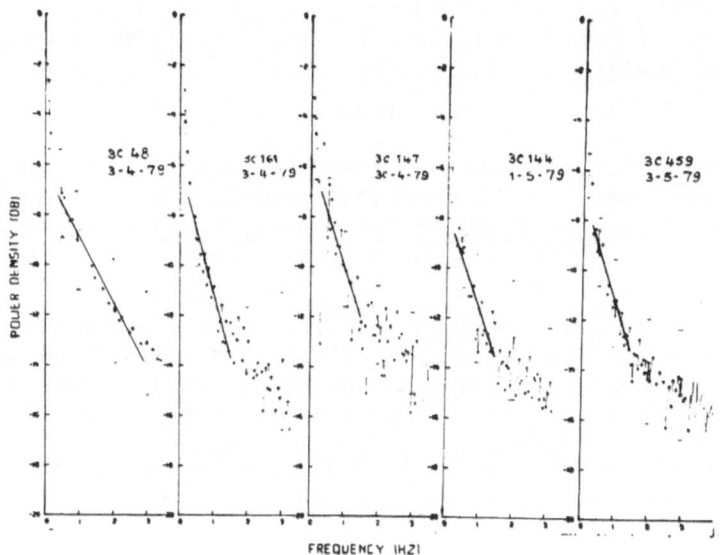

Figure 2. IPS temporal spectra.

$$f_2 = \int_0^{f_c} f^2 \, P(f) \, df \Big/ \int_0^{f_c} P(f) \, df$$

where $P(f)$ is the spectrum and f_c is the cut-off frequency, giving the upper limit of frequency up to which the spectrum can be accurately determined. f_1 and f_2 are a measure of the width of the spectrum, the former being less affected than the latter by the noise in the tail of the spectrum. Also the first moment gives more weight to large scale components in which case the accuracy is higher due to increased signal/noise ratio (Readhead et al., 1978).

Table I gives the summary of the observed parameters. Only a few specimen days were selected for preliminary analysis. The values of scintillation index (S.I.), which is the rms fluctuation of intensity about its mean, were derived using the relation $(S.I.)^2 = (\langle I^2 \rangle - \langle I \rangle^2)/\langle I \rangle^2$.

DISCUSSION

The temporal power spectrum of intensity is best described by a power law $P \propto f^{-n}$ with index n of about 2.6 for a strongly scintillating source, like 3C 48 at a solar elongation of $31°$ in the frequency range

TABLE I.   Summary of Observed IPS Parameters

Date	Source	Elong-ation	S.I.	f_1 (H_z)	f_2 (H_z)	f_2/f_1	Power law index, n
3-4-79	3C 48	31°	0.31	1.38	1.59	1.2	2.6
3-4-79	3C 161	86°	0.03	0.66	0.76	1.1	4.8
30-4-79	3C 147	53°	0.01	0.84	0.67	0.8	4.1
1-5-79	3C 144	44°	0.34	0.55	0.62	1.1	3.9
3-5-79	3C 459	50°	0.03	0.76	0.77	1.0	4.2

0.3-3 Hz. This is in good agreement with the mean value of the index of 2.4 ± 0.2 for solar elongation larger than 10° obtained by Milne (1976). For weakly scintillating sources the power law index is about 4 in the frequency interval 0.3 - 1.5 Hz, since their solar elongations are also greater. The ratio f_2/f_1 is nearly unity; while for a noise-free exponential spectrum it is equal to $\sqrt{2}$.

ACKNOWLEDGMENT

We express our sincere thanks to Professors D. Lal and S. P. Pandya for showing keen interest in this work. We are grateful to all our colleagues in the IPS group for their dedicated efforts in developing the IPS station. Financial support for this project came from the Departments of Space and Science & Technology, Government of India.

REFERENCES

Armstrong, J. W., and Coles, W. A.: 1972, *J. Geophys. Res., 77,* pp. 4602-4610.

Cohen, M. H., Gundermann, E. J., Hardebeck, H. E., and Sharp, L. E.: 1967, *Astrophys. J., 147,* pp. 449-466.

Hewish, A., Scott, P. F., and Wills, D.: 1964, *Nature, 203,* pp. 1214-1217.

Kakinuma, T., and Watanabe, T.: 1976, *Space Sci. Rev., 19,* p. 611.

Milne, R. G.: 1976, *Aust. J. Phys., 29,* pp. 201-209.

Readhead, A. C. S., Kemp, M. C. and Hewish, A.: 1978, *Mon. Not. R. Astr. Soc., 185,* pp. 207-225.

IPS OBSERVATIONS OF FLARE-GENERATED DISTURBANCES

Takashi Watanabe[*]
Space Environment Laboratory, NOAA/ERL
Boulder, Colorado 80303, U.S.A.

ABSTRACT

Dynamical behaviour of a flare-generated disturbance can be followed by IPS observations through a comparison between the IPS observations and a model of the disturbance. The highly turbulent post-shock
plasma can be attributed to the contact surface of the disturbance.
Two examples of IPS observations of flare-generated disturbances will
be discussed.

It has been proved that the IPS (interplanetary scintillation)
technique has an ability to determine large scale properties of flare-
generated disturbances (e.g., Kakinuma and Watanabe, 1976). Since we
have no direct method to estimate the position of a disturbance on a
line of sight, there exists an ambiguity in interpretation of the IPS
observations. Watanabe (1977) has proposed a method to improve this
situation depending on a fact that IPS observations provide the transverse component of the flow vector of highly disturbed post-shock
plasma. The change of observed flow speed with time reflects a combination of the projection effect, deceleration and anisotropic expansion
of the disturbance. Dynamical characteristic and, consequently, the
position of the disturbance at a given time can be estimated through
a comparison between the IPS observations and a suitable model of the
disturbance.

Several examples show that the highly turbulent post-shock plasma
which produces high level scintillations is not distributed in the
region immediately behind the shock front, but it appears at 2-6 hours
after the shock-front passage at around 1 AU from the sun (Watanabe,
1977). This region seems to correspond to the high-density post-shock
plasma which might be attributed to the contact surface, or the piston,
of the disturbance (e.g., Dryer, 1972). Consequently, we can determine dynamical behaviour of the piston of the disturbance by the IPS
observations.

___

* NRC-NAS Associate. On leave from Res. Inst. Atmospherics, Nagoya Univ.

M. Dryer and E. Tandberg-Hanssen (eds.), Solar and Interplanetary Dynamics, 409-412.
Copyright © 1980 by the IAU.

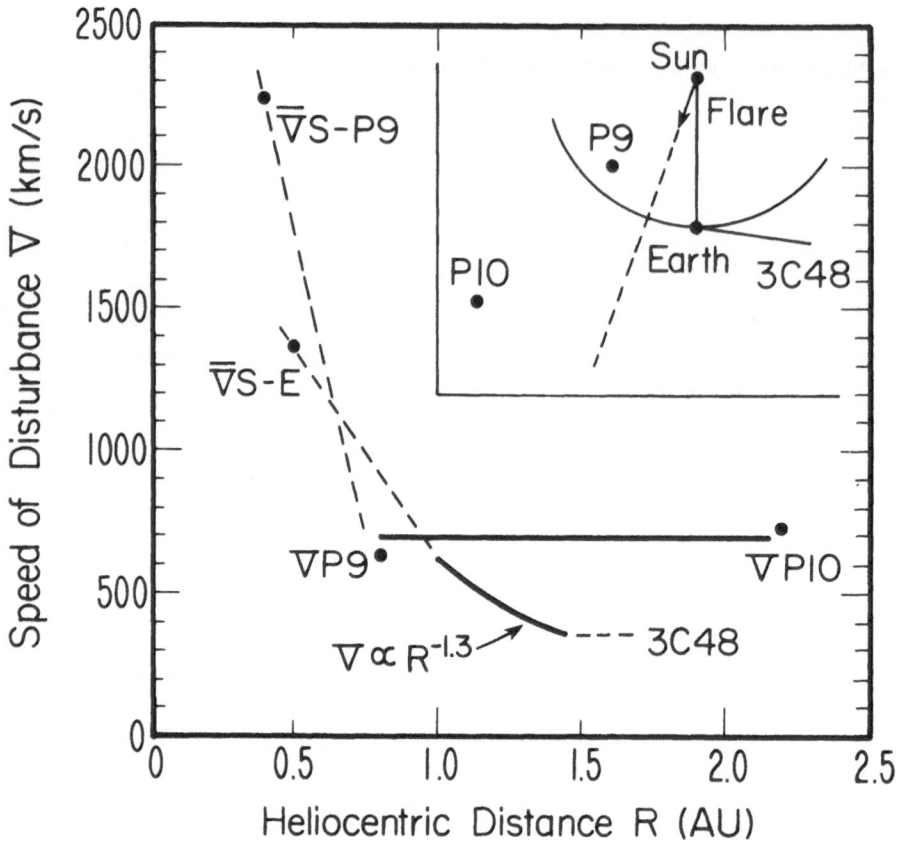

Fig 1.   Speeds of the disturbance generated by the solar flare at
2035 UT on Aug. 2, 1972 as a function of heliocentric distance, R.
VP9 and VP10 indicate the local shock speeds observed at Pioneer 9
and at Pioneer 10 respectively. $\overline{V}$S-P9 and $\overline{V}$S-E are mean speeds in
the region between the sun and two points of Pioneer 9 and the Earth
respectively.

The disturbance generated by the solar flare at 2035 UT on Aug. 2,
1972 has been well studied by many authors (e.g., review given by Smith,
1976) depending on observations made by Pioneer 9 (0.8 AU) and Pioneer
10 (2.2 AU) which were in radial alignment at about 20° east of the
flare normal. Watanabe (1977) also discussed the dynamical nature of
this disturbance depending on the IPS observations of 3C48 whose line
of sight was in the region to the west of the flare normal. Figure 1
shows the dynamical behaviour of this disturbance determined by above-
mentioned observations. In the direction of 20° east of the flare
normal (Pioneer 9-Pioneer 10) the shock front   propagated with approximate constant speed of about 700 km/sec in the region between Pioneer 9

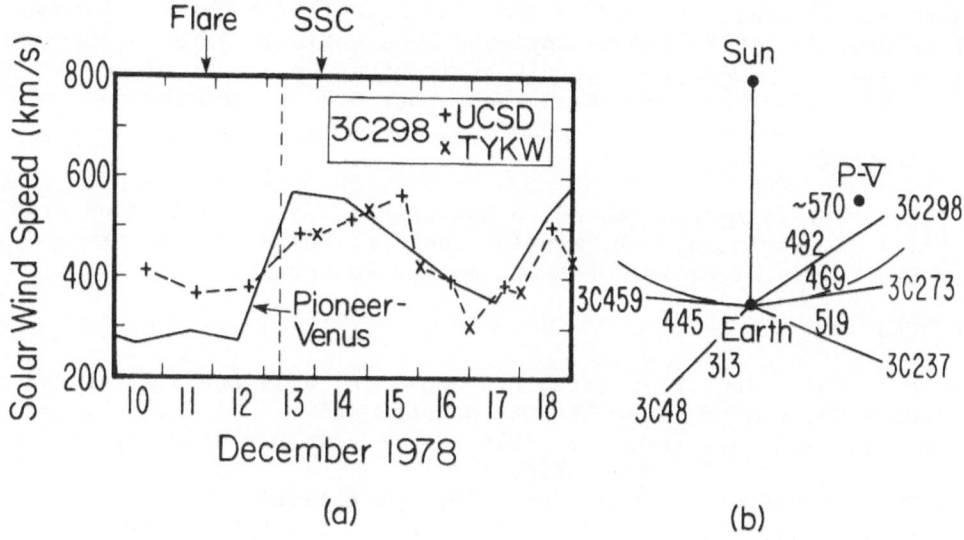

(a)   (b)

Fig. 2a. Solar wind speeds observed by Pioneer-Venus and IPS of 3C298.
Approximate arrival time of the shock front at Pioneer-Venus is indi-
cated by a vertical broken line (Mihalov, 1979). The arrival of the
shock front at the Earth was known by SSC of geomagnetic storm.
Fig. 2b. A geometry of Pioneer-Venus (P-V) and the lines of sight.
Numbers are observed flow speeds of the disturbance in km/sec.

and Pioneer 10 after very strong deceleration in the region between
the sun and Pioneer 9 which is inferred from very high mean speed
(VS-P9) of 2220 km/sec. Small deceleration of the piston in the region
between Pioneer 9 and Pioneer 10 is suggested. IPS observations of
3C48 showed that the western part of the disturbance was decelerated
in the region between the sun and 1.4 AU from the sun until the expanding speed was slowed down to the ambient wind speed of 350 km/sec
at about 70° west of the flare normal. These observations indicate an
anisotropic nature of this disturbance.

        Shortly after the Venus orbit insertion of Pioneer-Venus (PV), a
flare-generated disturbance was detected by this space probe on Dec.
13, 1978 (Wolfe et al., 1979). The IPS observations made at UCSD and
Toyokawa (TYKW) also detected this disturbance. Figure 2a shows the
solar wind speeds observed by PV and IPS of 3C298 whose line of sight
was situated in the region fairly close to PV. Figure 2b shows a
geometry of PV and the lines of sight. Observed flow speeds are also
indicated. All of the observed flow speeds by western radio sources
(3C298, 3C273 and 3C237) can be explained by a model of the disturbance with constant speed of 520 km/sec in the region between 0.8 AU
and 1.2 AU from the sun. The mean speed between the sun and PV

(0.8 AU) is about 820 km/sec (Mihalov, 1979) when we consider this disturbance was generated by one of the two solar flares observed around
18 UT on Dec. 11, 1978. Above-mentioned observations again suggest the
small deceleration near the Earth's orbit after an amount of deceleration in the region between the sun and about 0.7 AU from the sun.

ACKNOWLEDGMENTS

    The author is much indebted to unpublished data provided by Drs.
W.A. Coles, T. Kakinuma, J.D. Mihalov and J.H. Wolfe. Thanks are
due to Dr. Murray Dryer for comments on the manuscript.

REFERENCES

Dryer, M.: 1972, in C.P. Sonett, P.J. Coleman, Jr., and J.M. Wilcox
    (eds.) Solar Wind, NASA SP-308, pp. 453-465.
Kakinuma, T., and Watanabe, T.: 1976, Space Sci. Rev., 19, pp. 611-627.
Mihalov, J.D.: 1979, Communicated.
Smith, E.J.: 1976, Space Sci. Rev., 19, pp. 661-686.
Watanabe, T.: 1977, in M.A. Shea, D.F. Smart, and S.T. Wu (eds.)
    Contributed Papers to the Study of Travelling Interplanetary
    Phenomena/1977 (Proceedings of COSPAR Symposium B, Tel Aviv,
    Israel, June 1977), AFGL-TR-77-0309, pp. 139-149.
Wolfe, J., Intriligator, D.S., Mihalov, J., Collard, H., McKibbin, D.,
    and Whitten, R.: 1979, Science, 203, pp. 750-752.

DISCUSSION

    Ivanov: Eroshenko *et al.* (Innsbruk, 1978) have interpreted the
structure of interplanetary stream of August 4, 1972 in the following
manner: near $\sim 06^h$ UT the boundary of flare ejecta was observed. Where
is the high turbulence region: in the shock layer or in the ejecta?
    Watanabe: IPS observations of 3C48 during August 4-5, 1972 seem
to be well explained if we assume that the highly-turbulent post-shock
plasma followed the shock front with a time-lag of 2-4 hours (namely at
$\sim 06^h$ UT on August 4, 1972) in the region near the Earth's orbit. I
have suggested that this turbulent region corresponds to the contact
surface dividing the flare-ejecta from the shocked gas (Watanabe, 1977).
The conclusion given by Eroshenko *et al.* (Innsbruk, 1978) seems to
support my idea.

MAGALERT: AUGUST 27, 1978

JoAnn Joselyn
Space Environment Services Center
NOAA/ERL/SEL

Joseph F. Bryson, Jr.
Capt, USAF, OIC, OL-B, AFGWC

On August 27, 1978, a major geomagnetic storm began which eventually resulted in short period geomagnetic fluctuations of over 500 gammas in Boulder, and sightings of aurora as far south as Santa Fe, New Mexico. This storm was not obviously precipitated by flare or coronal hole solar plasma, but was apparently associated with a large solar filament which abruptly disappeared on August 23, 1978. Preliminary results of a study inspired by this storm are that 16 of the 59 geomagnetic storms which have occurred since the beginning of the current 11 year solar cycle can only be traced to disappearing filaments and some of the other storms which have been blamed on flares or coronal holes are also associated with disappearing filaments. Filament eruptions have been identified with coronal mass ejections, especially those observed with the Skylab white-light coronograph. However, there are some points of difference between typical coronal transients and geoactive coronal transients which may suggest fruitful research.

The key word in the title, Magalert, is taken from a series of international warning words and implies that a major geomagnetic disturbance is in progress or is expected. A Magalert was issued on August 27, 1978, by The Space Environment Services Center in response to a truly significant geomagnetic storm. Figures 1-4 show the three components of the Boulder magnetic field for the days of interest. Figure 1 shows the extremely quiet conditions which existed before the storm onset; Figure 2 shows the sudden commencement at 0247 UT signaling the arrival of a solar shock; Figure 3 shows the most disturbed day of the storm (over 500 gammas of fluctuation of the magnetic field and coincident with sightings of aurora in Boulder and Santa Fe, New Mexico); and Figure 4 shows the decay and end of the storm. This event is significant not only for its geophysical effects principally on communications systems, but also because it was a surprise - no solar optical flare or appropriate coronal hole heralded its arrival. Since the SESC (operated jointly by the Space Environment Laboratory and the Air Weather Service of the U.S. Air Force) is the only U.S. facility dedicated to the real-time monitoring and prediction of the space environment, we have done some considerable postmortem work on this event.

M. Dryer and E. Tandberg-Hanssen (eds.), Solar and Interplanetary Dynamics, 413-420.

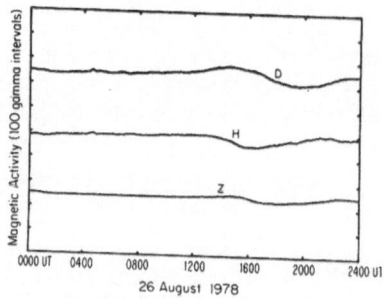

Figure 1. Pre-storm quiet conditions on 26 August 1978.

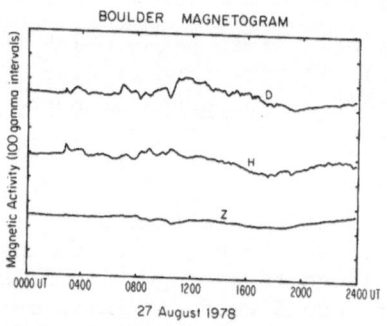

Figure 2. SSC at 0247 UT on 27 August 1978.

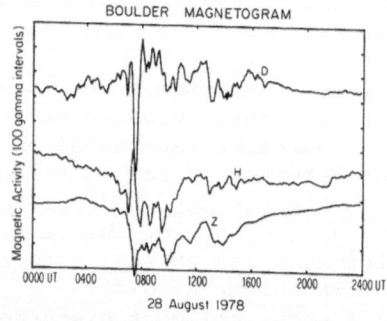

Figure 3. Peak of magnetic storm conditions on 28 August 1978.

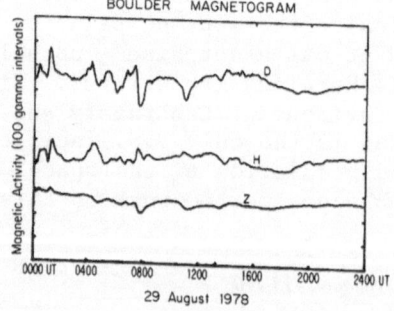

Figure 4. Storm decay on 29 August 1978.

First, we rechecked x-ray and solar radio data as well as flare reports and coronal hole maps. Only subflares were reported between August 22 and 27, and only 4 of those were associated with x-ray bursts above background. Importantly, those x-ray bursts were all well more than an order of magnitude below the signatures expected for potential geomagnetic effects. All radio bursts for a week preceding the storm were quite small and typical of normal quiescent solar behavior, and the only low solar latitude coronal hole was just approaching central meridian on August 27, fully four days before any geomagnetic effects are usually expected. However, an inspection of the daily H-alpha photos revealed that a filament disappeared between August 22 and 23 (Figures 5 and 6). The filament was still present at local Boulder sunset early on August 23, but Palahua Observatory, Hawaii, verified that the main filament disappeared totally between 0114 and 0128 UT on 23 August. If this filament disruption is indeed the source of the magnetic storm, the disturbance took approximately 4 days and 90 minutes to arrive at earth for an average speed of 427 km/s. This is in agreement with the "zero-order" near real-time solar wind speeds which were supplied by the University of California at San Diego using interplanetary scintillation techniques. Recently, preliminary ISEE-3 solar wind data supplied by S. Bame and published by Domingo et. al. (1979) shows a preshock velocity of approximately 300 km/s and a post shock velocity of approximately 450 km/s. Approximate pre and post shock solar wind densities are less than 10 cm^{-3} and greater than 40 cm^{-3}, respectively.

That filament eruptions could show up as geomagnetic storms really shouldn't have been surprising. The Skylab white-light coronograph experiment observed an association between sudden mass ejections from the sun with active and eruptive prominences and surges and significant mass and energy input into the solar wind (Gosling et. al., 1974). Gosling et. al. (1975) showed that eruptive prominence coronal events typically traveled out at speeds near 330 km/s, accelerated with height above the solar surface, and were not associated with metric wavelength type II and IV radio bursts which are correlated with faster moving flare-ejecta. However, obvious large-scale disappearing filaments such as the one shown here are not all that common in contrast with an estimate by Hildner et. al. (1976) that an average of 30 coronal transients per month occurred during the May 1973 to February 1974 Skylab period (although the transients did prefer helio longitudes where solar activity was high). In a study of a specific slowly ascending prominence and a more rapid accompanying loop-shaped coronal transient, Hildner et al. (1975) found that the bulk of the ejected material did not originate in the ascending prominence but must have come from the low corona above the prominence. They reported that the total event was far larger, more energetic, and longer lasting than would be inferred from the prominence observations alone. H-alpha filament eruptions have also been linked with long-delay enhancements in the x-ray emission or transient coronal holes (Rust, 1979). However, for this event, no

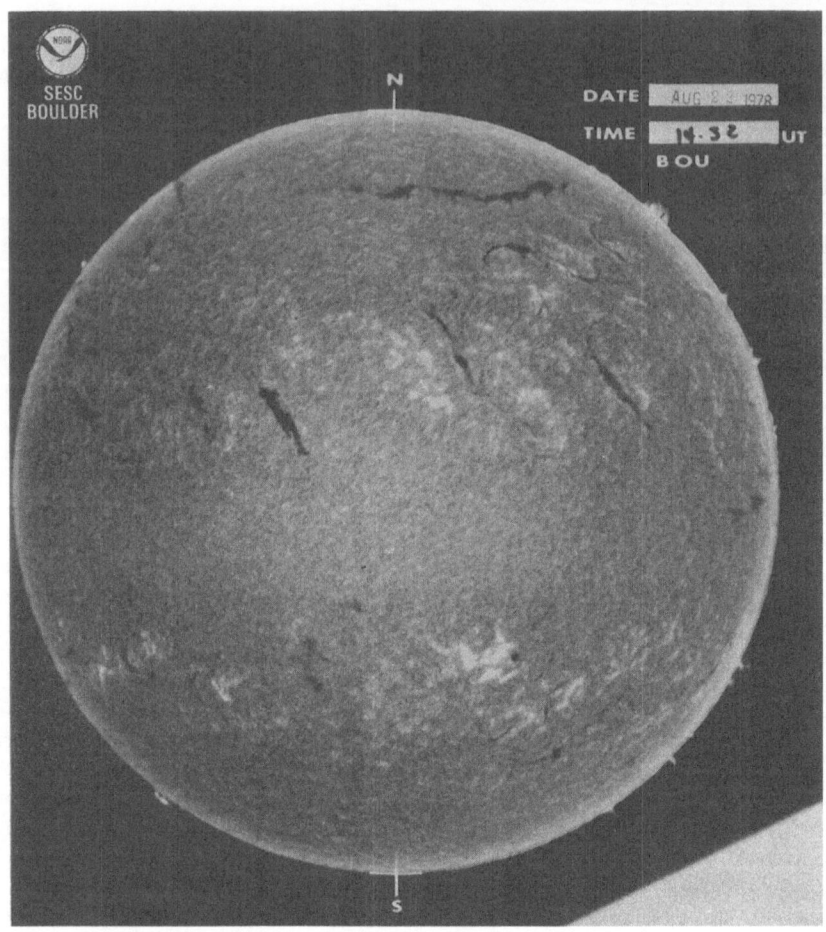

Figure 5.   Low latitude Hα filament near central meridian at 1432 UT,
            22 August 1978.

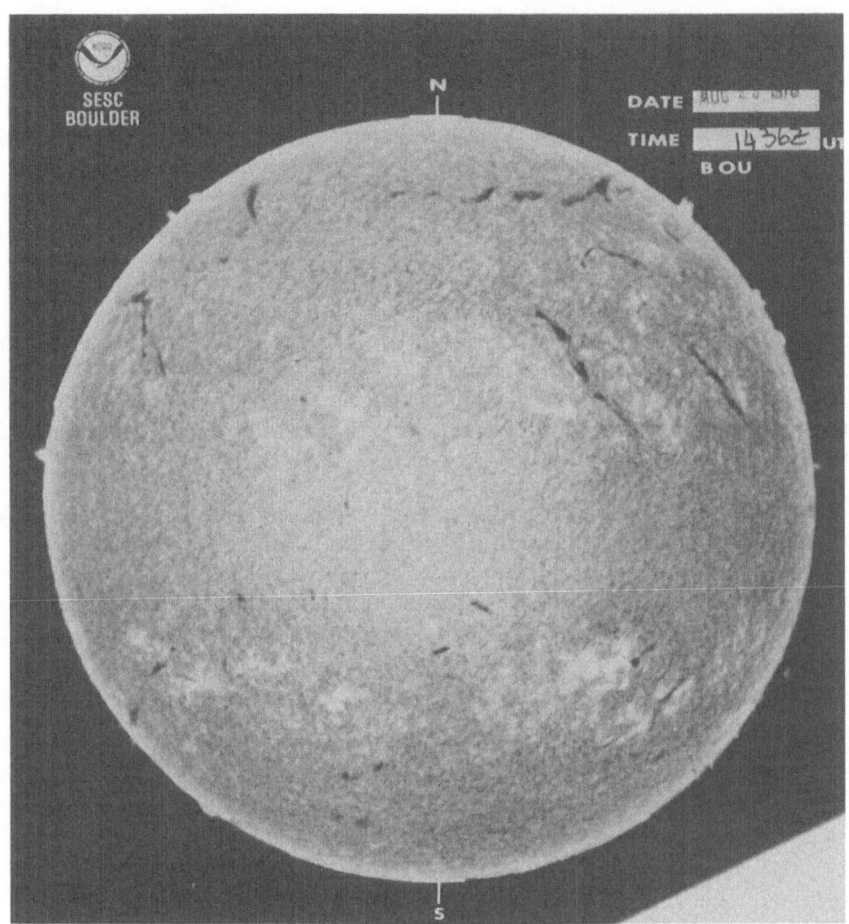

Figure 6.   Patrol film at 1438 UT, 23 August 1978. Filament disappeared entirely between 0114 and 0128 UT.

well-defined x-ray signature was observed and the solar wind speed is slower and the density much higher than that generally associated with coronal hole high speed streams.

Since this particular geomagnetic storm almost exactly one year ago, 9 other storms have been linked to disappearing filaments. And the preliminary results of a study in progress shows that of the 59 geomagnetic storms that have been recorded since the beginning of this solar cycle, 16 - including the 2 largest storms - can only be explained as the effects associated with a disappearing filament. The remaining storms find explanations in flares or coronal holes although some of those also occur in conjunction with disappearing filaments. Of course, there are disappearing filaments which are not connected with geomagnetic storms. We have not yet cataloged the differentiating factors between those filaments which link to geomagnetic storms and those which do not. I suspect the answers may not be obvious, especially in H-alpha photos.

We conclude that for the SESC, large disappearing filaments are surely worthy of note as harbingers of significant magnetic activity. We also submit that these storms and the interplanetary data between the coronal events and the earth may offer useful clues in understanding coronal dynamics and the underlying solar physics associated with coronal transients.

REFERENCES

Domingo, V., Hynds, R. J., and Stevens, G.: 1979, the Tokyo "Cosmic Ray Conference".

Gosling, J. T., Hildner, E., MacQueen, R. M., Munro, R. H., Poland, A. I., and Ross, C. L.: 1974, J. Geophys. Res., 79, p.4581.

Gosling, J. T., Hildner, E., MacQueen, R. M., Munro, R. H., Poland, A. I., and Ross, C. L.: 1976, Solar Phys., 48, p.389.

Hildner, E., Gosling, J. T., Hansen, R. T., and Bohlin, J. D.: 1975, Solar Phys., 45, p.363.

Hildner, E., Gosling, J. T., MacQueen, R. M., Munro, R. H., Poland, A, I., and Ross, C. L.: 1976, Solar Phys., 48, p.127.

Rust, David M.: 1979, Solar Phys., in press.

DISCUSSION

   Martres: We must not forget that: (1) a D.B. (disparition brusque) has always a cause (flare or new emerging flux), and this cause is very often to be found far from the filament (between some heliographic degrees to 60° or more) and we observe only one solar hemisphere. (2) A D.B. is a slow event (duration one, two days) in the case of a quiescent prominence as your example is, and it never produces a shock into the coronal gas as geomagnetic storms ask.
   Joselyn: (1) Since we do not really understand why (how) quiescent prominences exist, it seems that we also do not understand why they disappear. Flares and new emerging flux regions are daily occurrences, and yet quiescent filaments often exist for several rotations before they disappear. Of course, it is possible that the origin of this particular event may have been on the invisible solar hemisphere, but the statistics are building up. For 16 non-flare, non-coronal hole associated storms since January 1977, a large filament disappeared 3 to 5 days before the storm began. For another 4 storms, the explanation is not obvious and perhaps these might indicate that their source is on the backside of the sun.
   (2) This D.B. was not a particularly slow event (14 minutes), but many do take hours to disappear completely and are apparently "shock-less." However, the relationships between solar events and geophysical effects, i.e., the necessary and/or sufficient conditions for geo-physical effects, are not at all clear. Geometry is part of the answer and so is the southward component of the convected magnetic field, but there is much more to know about the physics of coronal transients and their interplanetary propagation.

   McIntosh: (Comment) The distribution of disk positions of disappearing filaments followed by magnetic storms is too broad to permit a simple model of propagation of the transient density wave. It appears from a preliminary study that magnetic storm filament disruptions often occur on large-scale neutral lines that appear to be the solar source of an interplanetary sector boundary. The disappearing filaments may affect the Earth if they occur anywhere in a sector boundary that intersects the ecliptic plane near the sun-earth connection longitude; that is, the position of the disappearing filament may not be as important as the equator-crossing portion of its underlying neutral line.

   Webb: I am surprised that this event had no soft X-ray signature. I have two questions: (1) Did you check both GEOS and Solrad X-ray data for the event; and (2) Did you check to see if the event had a microwave signature, which the Skylab results have shown in a characteristic of the thermal nature of filament eruptions?
   Joselyn: I have consulted all the data available in the EDIS Solar-Geophysical Data publication. In addition, I have examined the 1 minute resolution X-ray plots from both of the NOAA satellites, GOES-2 and GOES-3. (The X-ray channels from these satellites are received in

real-time at the Space Environment Services Center).   There are no reports of microwave bursts for several days surrounding this event.

Ivanov   (Comment): Now there are methods of determination of the shock normal from interplanetary data. You can determine the normal, if there are the corresponding data.   And the normal shows the Sun's region from which the shock arrived.

Dryer (Comment):   I want to repeat my earlier comment that the important fact is the magnitude of the energy release which triggered the eruption.   This magnitude can be at the very low side of "release" spectrum which is insufficient to produce X-rays, Hα emission, radio spectral data, etc.   The release of even such small amounts of energy perturbs the corona and solar wind and, provided the ΔV exceeds the appropriate characteristic speeds, eventually produces a finite amplitude MHD wave which may steepen into a shock wave (possibly a slow shock as suggested by Rosenau in his comment after Dr. Stewart's review paper).

CLASSIFICATION AND INVESTIGATION OF SOLAR FLARE SITUATIONS CONFORMABLY TO INTERPLANETARY AND MAGNETOSPHERIC DISTURBANCES

K. G. IVANOV, N. V. MIKERINA, and L. V. EVDOKIMOVA
Institute of Terrestrial Magnetism, Ionosphere and Radio Wave
  Propagation
Academy of Sciences
Moscow Region, Troytsk, USSR

1. CLASSIFICATION

A new classification of large solar flares is presented. If the time spacings between the flares are over 2.5 days, the flares and the corresponding interplanetary streams are classified as isolated ones. The concepts of the nearest, intermediate and distant flare zones (N, I and D zones) are introduced. The limits of the zones are determined at 30° longitude intervals and at 15° latitude ones. The classification is applied to the flares of 1966–1974 (Ivanov et al., 1979). It allows one to study the interplanetary and the magnetospherical disturbances more systematically.

2. OCCURRENCES OF SSC'S AFTER THE ISOLATED FLARES

Table 1 gives the number of isolated flares N and the number of SSC's: N_{ssc}. One can see that the occurrences of SSC's are nearly independent of the flare positions.

TABLE 1

Zones	N	N_{ssc}	$N_{ssc}/N, \%$
N	13	8	60
I	54	36	65
D	50	25	50
Total	117	69	60

It is especially strange that, in the N zone, 40% of the flares did not generate an SSC.

M. Dryer and E. Tandberg-Hanssen (eds.), Solar and Interplanetary Dynamics, 421-424.

3.   RELATION OF OCCURRENCES OF SSC'S AFTER ISOLATED FLARES WITH
      MAGNETIC AXES OF BIPOLAR GROUPS OF SOLAR SPOTS

Figure 1 shows the flares of the nearest zone (the asterisks) and
the magnetic axes NS of the bipolar groups.

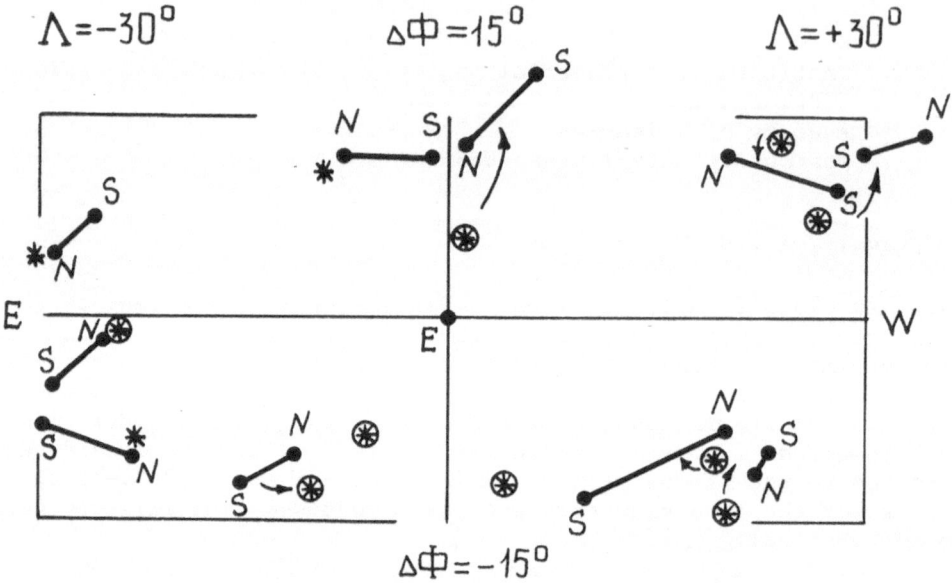

Figure 1.   Flares and magnetic axes of N zone. Note the helioprojec-
             tion of the Earth onto the solar disk, E.

We now establish the following rule: If a straight line, passing
through a flare in parallel to the corresponding NS, cuts the parallel
of helioprojection of the Earth E at a point which is removed from the
E less than 30°, then the flare generates a SSC. The rule is
statistically performed on the Sun (Table II).

TABLE II

Zones	$\|\Lambda^*\| \leq 30^\circ$			$\|\Lambda^*\| > 30^\circ$		
	N	N_{ssc}	$\dfrac{N_{ssc}}{N}$, %	N	N_{ssc}	$\dfrac{N_{ssc}}{N}$, %
N	5	5	100	5	5	100
I	10	9	90	15	10	65
D	2	2	100	14	11	80
Total	17	16	95	34	26	75

The relation can be interpreted in the following manner. Let an
isolated stream be strongly flattened to the plane of a large circle
which passes through the flare and is parallel to the NS. Then the

favourable positions of the NS occur when the angle $\Lambda^* \lesssim 30^\circ$, and the forward shock wave envelops the Earth and generates a SSC.

4. SOLAR PROTON EVENTS IN CONNECTION WITH MAGNETIC AXES OF BIPOLAR GROUPS

According to Tverskoy, acceleration of solar protons takes place in the forward shock layer. The strongly flattened shock wave must have an effect on the solar proton events (SPE) near the Earth. The favourable positions of NS (in this case) cut the parallel of Earth's helioprojection at the points of $0^\circ \div 60^\circ W$. The consequence of the model is confirmed by data (Table III).

TABLE III

$\Lambda = 0^\circ - 60^\circ W$			$\Lambda \neq 0^\circ - 60^\circ W$		
N	N_{SPE}	$N_{SPE}/N,\%$	N	N_{no}	$N_{no}/N,\%$
15	10	65	42	36	85

Table III gives the numbers of isolated flares N, the number of SPE, and the number of flares without SPE – N_{no}.

REFERENCES

Ivanov, K. G., Mikerina, N. V., Zovoykina, A. I., and Treschotkina, V. M., Catalogue of Flare Situations of 1966–1974, IZMIRAN, Moscow, 1979.

DISCUSSION

McIntosh: Your result is very interesting and will be tested immediately at the NOAA Space Environment Services Center in Boulder. Can you relate your classification to the production of a B_z interplanetary field favorable for geomagnetic disturbance (i.e., strong southward component)?

Ivanov: Yes. If, after an isolated flare and for the favourable position of the magnetic axis of a bipolar group, the Earth was enveloped by flare ejecta, then the sign of B_z-component of IMF was in a strong accordance with the point of view with the direction of NS. However, we have analysed only the flares of nearest zone.

Simon: Could you describe more in detail the criteria used to select the "active" and "unactive" flares?

Ivanov: Let us consider the following situation. There are a bipolar

group of solar spots and a large flare near the group. The line
connecting the spots is the magnetic axis - NS. There is also the
projection of the Earth onto the solar disk for the time of the flare,
E. We draw a line passing through the flare in parallel to the NS.
The line crosses the heliographical parallel of E in a point P. If
the angle distance $|PE|$ is less than 30° of longitude, then SSC is
generated by the flare. In the other case, SSC is absent.

Tandon: Have you also considered flares associated with complex
groups in your analysis?
  Ivanov: No, we have analysed only isolated flares in bipolar groups.

PHYSICAL DRIVING FORCES AND MODELS OF CORONAL RESPONSES
(Invited Paper)

S.I. Syrovatskii and B.V. Somov
Lebedev Physical Institute,
Academy of Sciences of the USSR, Moscow

The reasons for nonstationary hydrodynamic flows in solar atmospere are
reviewed. It is emphasized that a rapid local heating of corona or upper
chromosphere can scarcely provide a very large mass of solar plasma
ejections observed in corona and interplanetary space. We suggest that
coronal transients and interplanetary ejections are produced by magnetic
field evolving in solar atmosphere. Magnetic reconnection in current
sheets can play essential role in this process. The suitable approximation of strong magnetic field is formulated. Some solutions of MHD equations in this approximation are demonstrated. Their applications to coronal conditions are discussed.

I. INTRODUCTION

During solar flares a significant fraction, if not the majority of the
total energy, can release in the form of plasma motions in chromosphere
and corona (see monograph by Svestka, 1976). For large flares these motions initiate in interplanetary space strong shocks with energy comparable with the total energy output. Webb et al. (1978a) have obtained
accurate estimates of the mechanical energy budget in the 5 September
1973 flare, based on detailed observations of the different forms of motions. The mechanical energy appeared to be comparable with the change
of magnetic energy in corona. This is not a surprising result because
the magnetic energy in the corona is well known to be enough for all forms
of flare energy release.

The other essentially nonsteady phenomena in solar atmosphere (e.g.
coronal transients and flarelike events) differ from flares not only by
scale (size, power) but also by their spatial structure and by the relative role of different energy release channels. Nevertheless, taking
into account that magnetic field is the main energetic factor in active
regions, we suggest that all those rapid phenomena have principally the
same origin. Namely, we suppose that all those processes are due to magnetic forces.

M. Dryer and E. Tandberg-Hanssen (eds.), Solar and Interplanetary Dynamics, 425-441.
Copyright © 1980 by the IAU.

According to modern concepts (see Heyvaerts et al., 1977; Syrovat-skii, 1977; Baum et al., 1978) the conversion of magnetic energy into other forms, particularly nonthermal, can be connected with the appearance and explosive distruction ('rupture') of current sheets (including the sheets which may appear in nonevolutionary force-free configurations, e.g. Burnes and Sturrock, 1972; Low and Nakagawa, 1972; Jockers, 1978; Bobrova and Syrovatskii, 1979). In order to interpret fast motions in the chromosphere and corona two different regimes of magnetic reconnection in current sheet are appropriate. First of them is the rupture of the laminar (nonturbulent) quasi-steady sheet as a result of some instability and excitation of anomalous resistivity. This regime provides the most powerful energy release and corresponds to the flash phase of flares. The second regime is the rapid reconnection in quasi-steady turbulent sheet. We think that this process can correspond to such coronal phenomena as rapid ascents of X-ray arches and transients.

We shall not discuss here the questions concerning the interpretation and simulation of large amplitude disturbances in the interplanetary space. These very important problems have been developed in detail by American scientists (for a review, see Dryer, 1979). It is more close to our consideration the magnetic field influence on the propagation of disturbances initiated at the coronal base (Steinolfson et al., 1978; Wu et al., 1978; Dryer et al., 1979). The channeling and blocking effects by the magnetic field were studied in those papers. Opposite to that passive role of the coronal magnetic field, we consider the last as a primary cause for plasma motions. For example, an emerging flux (like a magnetic piston) can cause the plasma to flow upward near the loop tops and downward near the loop legs (Nakagawa et al., 1976; Steinolfson et al., 1979). On the other hand, the rapid changes of magnetic field are necessary for explanation of fast motions. To our mind, those changes could result from the reconnection in current sheets. Before discussing these problems, let us consider briefly nonmagnetic mechanisms of plasma ejections.

2. THE MASS OF A PLASMA EJECTED BY FLARE HEATING OF SOLAR ATMOSPHERE

Energy release in the form of heating, particle acceleration etc. results finally in the local heating of the solar atmosphere to high temperatures (see Somov and Syrovatskii, 1976). Large gradient of pressure appears on the boundary of a heated region and can excite fast motions including shocks and plasma ejections.

Consider the at first sight most efficient process of ejections when the chromosphere is impulsively heated by flare electrons with energies $E_e \geqslant E_o \sim (10 - 30)$ keV. This process is really accompanied by the upward high-speed flow of heated plasma (Somov et al., 1977). Note here that the ejected mass is determined essentially by the position of the flare transition layer but not by the 'critical level' where the direct electron heating can be compensated by radiative cooling, as Brown (1973) affirms. Really, the true approach to the problem must include

the heat conduction (Shmeleva and Syrovatskii, 1973) that can provide
an effective transport of a flare energy even for very large energy
fluxes (Somov, 1979a).

The depth ξ (in atoms cm^{-2}) above the flare transition layer depends mainly on the boundary energy flux F_o (erg $cm^{-2}s^{-1}$). For maximum
values of F_o during the flash phase of flares, the depth ξ equals about
$(1-3) \times 10^{19}$ cm^{-2} (Somov et al., 1979). Hence the mass $\sim (2-5) \times 10^{-5}$ g cm^{-2}
can be ejected from unit area. Let S_I be the area heated by energetic
electrons during one 'elementary flare burst' (a single spike of hard
X-rays). Bearing in mind some 'mean' impulsive flare, we suggest that S_I
is equal to the area of fast (5-10 s) small (1-3") chromospheric flash
which coincides in time with hard X-ray burst (see Zirin and Tanaka,
1973). Then $S_I \sim 5 \times (10^{16}-10^{17})$ cm^2 and the ejected mass equals $M_I \sim 10^{12}-$
3×10^{13} g. Even if flare as whole ejects the total mass $M \sim 10$ M_I, this
is too small compared with the observed values $M_{obs} \sim 10^{16}-10^{17}$ g (e.g.
Dryer et al.,1976a). Thus, the impulsive heating of the chromosphere by
energetic electrons can not provide the observed mass. Note here that
the fast heating by large thermal fluxes with the same F_o yields the same or somewhat less ejected mass (Sermulinia et al., 1979).

The ejected mass is even smaller if the heating occurs not inside
the chromosphere but in the corona. The numerical study of the dynamic
response of a static corona on the finite-amplitude disturbances shows
that the needed mass can be ejected upward only from beneath the boundary placed at the coronal base (Nakagawa et al., 1975). It means, for
example, that for large flare spray the mass of order 10^{16} g is ejected
from the chromosphere, but the physical reason for such ejection remains
unspecified. In the numerical simulation for the fast-mode MHD shock
propagation through the corona during the flare of 1973 September 5, the
ejected mass is about 6×10^{16} g (Dryer and Maxwell, 1979). The mass of the
same order is assumed for the model of interplanetary disturbances in the
solar wind (Dryer et al., 1976b; Wu et al., 1976). All these approaches
assume tacitly that the needed mass is ejected from the chromosphere just
as a response to the flare energy release. Is really the large mass observed in the coronal transients and interplanetary disturbances taken
directly from the chromosphere?

This question is especially critical one if we take into account
that the most of transients ($\sim 60\%$) are not accompanied by observable
chromospheric flares (Munro et al., 1979). This implies that the observed mass ought to be stored in the corona before a transient or raked
up during it. Coronal transients are often associated with eruptions of
prominences. For this reason, one can assume that the transients are
caused by sudden heating and eruption of a previously cooler material,
e.g. cold dense filament. Another possibility is that the needed mass is
expelled from the low corona (Rust and Hildner, 1976). Then, the large
area $S \sim 10^{21} - 10^{22}$ cm^2 is needed to gather the observed mass.

One more source of mass is conceivable. The current sheets could,
in principle, accumulate coronal plasma. Magnetic field moves into the

sheet and annihilates there whereas plasma carried with the field is accumulating in the sheet or in the neighbourhood. The current sheets are most effective in strong magnetic fields. For this reason, we consider in Section 2 the approximation of strong field.

3. THE MAGNETOHYDRODYNAMIC APPROXIMATION OF A STRONG FIELD

For a weak field, magnetic effects are only small corrections to hydrodynamic ones. Contrary to it, the strong field approximation (Syrovatskii, 1966) implies that a magnetic force dominates over all other forces: gas pressure gradient, inertia etc. This means formally the following. The complete set of the ideal MHD equations can be written in the next dimensionless form (e.g. Somov and Syrovatskii, 1974):

$$\varepsilon^2(\delta^{-1}\,\partial\vec{v}/\partial t + (\vec{v}\cdot\text{grad})\vec{v}) = -\,\gamma^2(\text{grad }p)/\varrho - (\vec{B}\times\text{curl }\vec{B})/\varrho, \quad (1)$$

$$\partial\vec{B}/\partial t = \delta \text{ curl }(\vec{v}\times\vec{B}), \quad (2)$$

$$\partial\varrho/\partial t + \delta \text{div }\varrho\vec{v} = 0, \quad (3)$$

$$\partial S/\partial t + \delta\,(\vec{v}\cdot\text{grad})S = 0, \quad (4)$$

$$\text{div }\vec{B} = 0, \quad (5)$$

$$p = p(\varrho, S). \quad (6)$$

These equations contain three dimensionless parameters

$$\delta = VT/L, \qquad \varepsilon = V/V_A, \qquad \gamma^2 = p_0/\varrho_0 V_A^2, \quad (7)$$

where L, T, V, ϱ_0, p_0 and B_0 are the scales of length, time, velocity, density, pressure and field strength, respectively. One can consider a magnetic field as a strong one if ["I" represents the integer "one" – Eds.]

$$\gamma^2 \ll 1 \text{ and } \varepsilon^2 \ll 1. \quad (8)$$

It follows from Equation (1) that in zeroth order in small parameters (8) the magnetic field is a force-free one:

$$\vec{B}\times\text{curl }\vec{B} = 0, \quad (9)$$

or that is simply a potential field in the absence of currents.

In the first order, we neglect the gas pressure gradient in comparison with the inertia (Somov and Syrovatskii, 1974):

$$\varepsilon^2(\delta^{-1}\,\partial\vec{v}/\partial t + (\vec{v}\cdot\text{grad})\vec{v}) = -(\vec{B}\times\text{curl }\vec{B})/\varrho, \quad (10)$$

which implies the approximation of a strong field and cold plasma:

$$\gamma^2 \ll \varepsilon^2 \ll 1. \quad (11)$$

This approximation is especially appropriate for fast motions caused by changes of a strong field. Just this case is of interest for us.

The parameter δ rates the local derivative $\partial/\partial t$ relative to the convective term $\vec{v}\cdot\text{grad}$. $\delta \gg 1$ for quasi-steady flows, $\delta \ll 1$ for small disturbances. Generally, $\delta = 1$, then the equations are

$$\varepsilon^2\,d\vec{v}/dt = -(\vec{B}\times\text{curl }\vec{B})/\varrho, \quad (12)$$

$$\delta\vec{B}/\delta t = \text{curl } (\vec{v}\times\vec{B}),$$ (I3)

$$\delta\varrho/\delta t + \text{div } \varrho\vec{v} = 0.$$ (I4)

Expand the solution in the small parameter ε^2 (i.e. $f = f^{(o)} +$ $+\varepsilon^2 f^{(I)} + \ldots$). In zeroth order, magnetic field is determined by equation

$$\vec{B}^{(o)}\times\text{curl } \vec{B}^{(o)} = 0$$ (I5)

with time-dependent boundary conditions. Following these conditions, the field changes and drives a plasma. Kinematics of this motion is determined by two equations. Equation (I2) means that the acceleration is perpendicular to magnetic field lines, namely

$$\vec{B}^{(o)}\cdot d\vec{v}^{(o)}/dt = 0.$$ (I6)

The second one coincides with the 'freezing-in' Equation (I3):

$$\delta\vec{B}^{(o)}/\delta t = \text{curl } (\vec{v}^{(o)}\times\vec{B}^{(o)}).$$ (I7)

The plasma density can be found from the continuity Equation (I4):

$$d(\ln \varrho^{(o)})/dt = -\text{ div } \vec{v}^{(o)}.$$ (I8)

Thus, the Equations (I5)-(I8) determine all unknown zeroth order variables. The same procedure can be continued to any order in the small parameter ε^2. For our aim, it is enough to consider only zeroth order, e.g. to neglect departures of the field from force-free (or potential) one.

To solve the problem we proceed as follows. At first, we find the force-free (or potential) field $\vec{B}^{(o)}(\vec{r},t)$ at any instant from the Equation (I5) and time-dependent boundary conditions. Then, the velocity $\vec{v}^{(o)}(\vec{r},t)$ follows from the Equations (I6) and (I7) and from the initial condition for velocity $\vec{v}^{(o)}(\vec{r},0)$ along the magnetic field. Finally, the continuity Equation (I8) and an initial plasma distribution $\varrho^{(o)}(\vec{r},0)$ yield the density $\varrho^{(o)}(\vec{r},t)$. Let us demonstrate this procedure by some examples.

4. HYDRODYNAMIC FLOWS NEAR A STEADY CURRENT SHEET

Let the steady current sheet be of the width 2b along the x axis (Figure I). The uniform electric field $\vec{E}_o$ is parallel to the z axis.

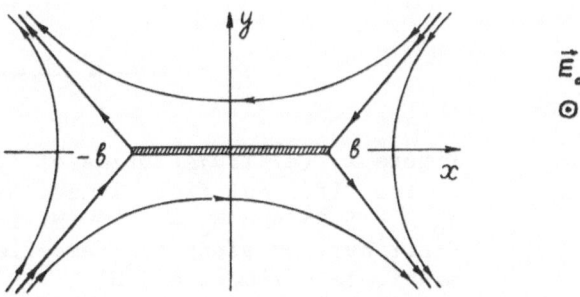

Figure I. The field lines near the steady current sheet.

Two-dimentional motions near such a sheet can be described by the Equations (I5)-(I8) formulated in terms of the vector potential $\vec{A} = (0, 0, A(x,y))$:

$$\Delta A = 0, \tag{I9}$$

$$d\vec{v}/dt \times \text{grad } A = 0, \tag{20}$$

$$dA/dt = 0, \tag{2I}$$

$$\partial\rho/\partial t + \text{div } \rho\vec{v} = 0. \tag{22}$$

According to Equation (I9) the 'potential' $A(x,y)$ is a harmonic function of a complex variable $z = x + iy$. For this reason, it is convenient to introduce the complex potential as the analytic function

$$F(z,t) = A(x,y,t) + iB(x,y,t). \tag{23}$$

In the case under consideration the complex potential is of the form (Syrovatskii, I97I):

$$F(z,t) = \tfrac{1}{2}h_o z(z^2 - b^2)^{\frac{1}{2}} - \tfrac{1}{2}h_o b^2 \ln\frac{z + (z^2 - b^2)^{\frac{1}{2}}}{b} + A(t). \tag{24}$$

The coefficient h_o is the magnetic gradient in the absence of current sheet (when $b = 0$). The term $A(t)$ is introduced formally to take into account the magnetic flux dissipation in the sheet.

The Equations (20)-(22) can be written in Lagrange variables as the set of four ordinary differential equations for the trajectories of 'fluid particles' $x(t)$ and $y(t)$ and the continuity equation for density

$$\rho(x,y,t)/\rho_o(x_o,y_o) = D(x_o,y_o)/D(x,y). \tag{25}$$

Here $\rho_o(x_o,y_o)$ is the initial density distribution. The Jacobian on the right side can be found, for example, by the simultaneous calculations of three neighbouring trajectories.

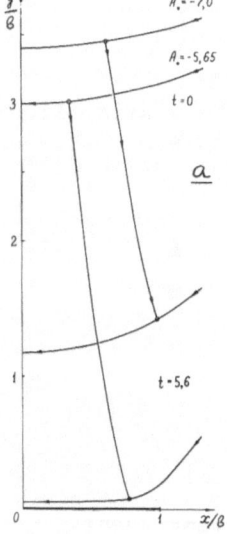

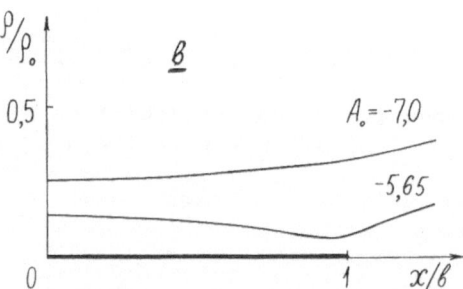

Figure 2. (a) Trajectories of fluid particles placed on two different field lines: $A(x,y,t) = A_o$ with $A_o = -5.65$ and $A_o = -7$. (b) The relative density distributions along the same field lines at the dimensionless time $t = 5.6$.

The results of numerical solution (Somov and Syrovatskii, 1974) show that for considered supersonic ($\gamma^2 < \varepsilon^2$) flows near the steady current sheet there is no stationary solution for density. The last decreases monotonously on the field lines flowed to the sheet (Figure 2). As a final result, the current sheet could be surrounded by a very low density plasma if the electric field $\vec{E}_o$ acts long enough. This conclusion is important for the problem of stabilization and rupture of current sheets in laboratory and cosmic plasmas (e.g. Bulanov et al., 1977; Bulanov and Sasorov, 1978).

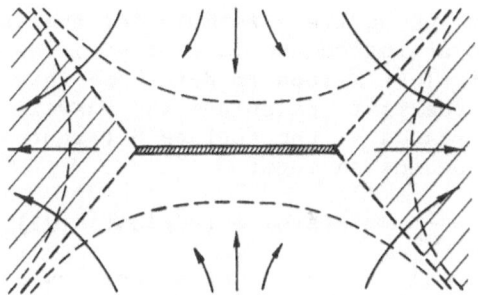

Figure 3. The regions of contraction (two shaded sections) and of depletion (the rest) near the steady current sheet. The motion trajectories of plasma are shown schematically by arrows.

On the field lines which have undergone magnetic reconnection in the sheet a plasma is raked-up and compressed (Figure 3). Together with a plasma flowed out from the sheet, this compressed plasma moves rapidly in opposite directions along the x axis. Fast plasma ejections are met in the laboratory and numerical simulation of the current sheet development and rupture (Kirii et al., 1979; Podgornii and Syrovatskii, 1979). Ivanov and Platov (1977) consider a similar process (the raking-up near a neutral line, i.e. b=0) as a model for loop prominence formation.

5. THE RUPTURE OF A CURRENT SHEET AND THE FORCES ACTING ON ITS EDGES

Magnetic field is weak or fully absent inside a current sheet. However the external strong field can penetrate in one or more points (gaps) when and where the sheet ruptures. In this case, magnetic forces begin to act on the edges of appearing gap and will enlarge it (Somov and Syrovatskii, 1975).

In the simplest case one can imagine that before the rupture the current sheet is a uniform current in the y = 0 plane (Figure 4a) flowing parallel to the z axis and creating the magnetic field jump from B_o at y < 0 to $- B_o$ at y > 0. Let us assume that the gap is formed parallel to the z axis as a result of some instability of the sheet. 2a(t) is the width of this gap (Figure 4b). In the approximation of ideal conductivity, the field lines at the surface of the current sheet remain the same as before rupture. The magnetic field is potential one outside currents and, if a new current does not arise along X-type neutral line at the gap (see more general case in Somov and Syrovatskii, 1975), the complex potential is of the particularly simple form

$$F(z,t) = iB_o(z^2 - a^2(t))^{\frac{1}{2}}.$$ (26)

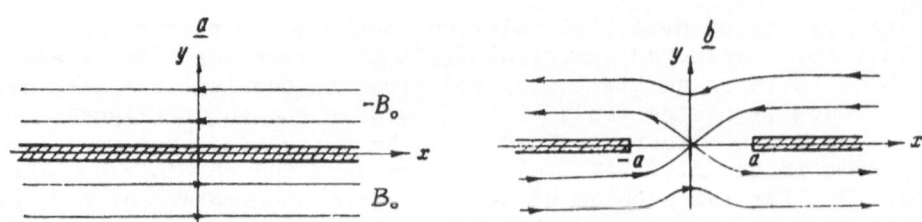

Figure 4. Field lines near the primary sheet (a) and ruptured one (b).

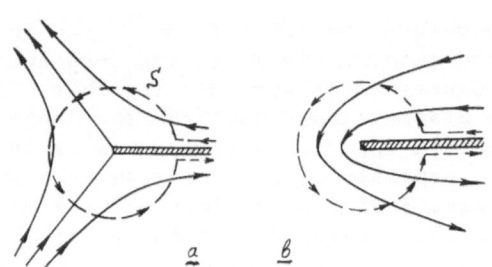

Figure 5. Field structure near free edge of the sheet (a) and the edge of the gap.

Let us derive a formula for the force acting on the edge of gap. The simplest method to determine this force is to calculate the Maxwell stresses at the surface S (Figure 5) bounding an edge:

$$\vec{F} = \frac{1}{4\pi} \oint_{S} \left[-\vec{B}(\vec{B}\cdot\vec{n}) + \tfrac{1}{2}B^2\vec{n} \right] dS. \quad (27)$$

A real current sheet always has a finite width. But according to (27) there is no magnetic force on the free edge ($F \sim z^{3/2}$, Figure 5a) of the sheet. On the other hand, at the edge of the gap ($F \sim z^{\frac{1}{2}}$, Figure 5b) the magnetic force acts along the current sheet (the x axis):

$$F_x = \pm \tfrac{1}{8}B_o^2 a \quad \text{at} \quad x = \pm a. \quad (28)$$

This result can be understood easily in another way, namely by calculation of the free magnetic energy for the system of currents

$$W = - \tfrac{1}{8}B_o^2 a^2. \quad (29)$$

Since W equals the work done by the magnetic field on the expansion of the gap, the force acting on the gap edge is

$$F_x = \partial(\tfrac{1}{2}W)/\partial a = \pm \tfrac{1}{8}B_o^2 a \quad \text{at} \quad x = \pm a. \quad (30)$$

Thus, the edges of the gap are subjected to magnetic forces which are proportional to the width of the gap and tend to enlarge it.

We can draw two conclutions from this analysis. First of them is a condition of sheet rupture. If the sheet has a finite thickness d and if the gas pressure inside the sheet is balanced by external magnetic pressure, then the sheet can be ruptured in the case of $d < a$. This conclusion follows from a comparison of the magnetic force in (30) with the opposing gas pressure force $pd = (B_o^2/8\pi)d$. The second conclusion is the exponential growth of small perturbations in the initial stage of the rupture.

At nonlinear stage of the rupture, in order to write an equation of motion for the edge, it is necessary to take into account the mass of a plasma involved in motion and also thermal effects (Bulanov and Sasorov, 1978). The gas mass can be estimated as

$$M(a) = n_s m_i ad \quad (g\ cm^{-1})\tag{31}$$

where n_s is plasma concentration inside the sheet. Thermal effects being disregarded,

$$d(Mda/dt)/dt = F_x(a) = \tfrac{1}{8}B_o^2 a.\tag{32}$$

From this it follows that the edge moves with the constant acceleration

$$\ddot{a} = 4\pi V_A^2/3d \quad \text{where} \quad V_A = B_o/(4\pi n_s m_i)^{\tfrac{1}{2}}\tag{33}$$

is the characteristic value of Alfven velocity.

To take into account thermal effects it is necessary to solve hydrodynamic equations $y = 0$, $|x| \lesssim a$. The boundary conditions for pressure at the point $x = a$ describes the action of the magnetic force:

$$p(a) = F_x(a)/d.\tag{34}$$

The solution of this self-similar problem (Bulanov and Sasorov, 1978) does not change the conclusion about the character of the current sheet rupture affected by magnetic force. It is essential here that after the rupture the plasma motion velocity becomes larger than the velocity (33).

6. PLASMA ACCUMULATION INSIDE CURRENT SHEET

Let us estimate the plasma mass inside a quasi-steady laminar current sheet. Following Syrovatskii (1976), we use the momentum and continuity equations and Ohm's law from which there follow the order of magnitude relations (see Figure 6):

$$n_o V_d b = n_s V_x a, \quad B_s^2/8\pi = 2n_s kT = \tfrac{1}{2}n_s m_i V_x^2,$$
$$cB_s/4\pi a = \measuredangle E_o.\tag{35}$$

Here n_o is plasma concentration near the sheet, a and b are half-thickness and half-width of the sheet, $V_d = cE_o/B_s$ is the drift velocity near the sheet, E_o is the electric field along the zeroth line of primary field. From (35) it follows that the velocity of plasma outflow from the laminar sheet equals the Alfven velocity inside the sheet (cf. Sweet, 1958)

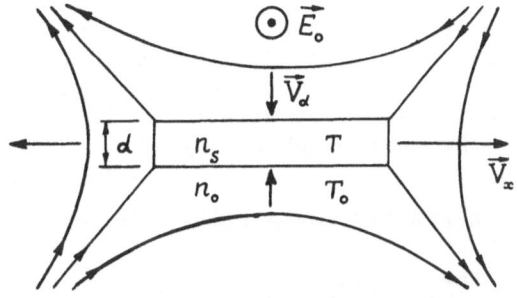

Figure 6. Steady model of the sheet.

$$V_x = B_s/(4\pi n_s m_i)^{\tfrac{1}{2}},\tag{36}$$

and the other quantities are expressed through the external parameters n_o, E_o and h_o and through the sheet temperature T (Syrovatskii, 1976). Specifically, in the current sheet being of the length 1 the plasma mass is

$$M = 4ablm_i \approx 5 \times 10^{-21}(E_o^5 n_o^4/h_o^7)^{1/3}T^{1/6}l \quad (g).\tag{37}$$

Let, for example, $n_o = 5 \times 10^8$ cm^{-3}, $E_o = 4 \times 10^{-4}$ CGSE, $h_o = 5 \times 10^{-7}$ G/cm and $T = 8 \times 10^4$ K (cf. Syrovatskii, 1976). Then $M \approx 141$g. If $1 = 10^{10}$ cm, the current sheet mass has to be $M \sim 10^{11}$g. This value is small compared to the observed ejected mass (Section 2).

Thus, the mass of plasma accumulated inside the current sheet is insignificant. However,   another point is essential. Firstly, an emerging strong magnetic field can lift plasma from the chromosphere and low corona. Some part of the ascendant plasma flows into the current sheet and is ejected from it with near Alfven velocity. This ejected plasma moves together with and along the magnetic field lines which have undergone the reconnection inside the sheet. Partially those fast motions of plasma are directed upward through corona. Secondly, during a fast reconnection the above-mentioned raking-up mechanism can also give rise to upward fast ejection. The total mass involved in both processes can be estimated roughly if one takes into account that the emerging and reconnecting magnetic field pushes out a plasma practically from the whole coronal volume placed below the current sheet. For the sheet at the height $h \sim 3 \times 10^{10}$ cm, this mass is of about $m_i n h^3 \sim 10^{16}$ g if $n \sim 3 \times 10^8$ cm^{-3} is the coronal plasma concentration.

In our opinion, many observations give an evidence for the hypothesis that current sheets lie at the tops of loops or arcades of loops. In particular, X-ray and EUV observations of flares and flarelike events (for a review, see Somov, 1978) together with observations and calculations of magnetic field confirm that many active processes in the corona are conditioned by emergence of a new magnetic flux. The front boundary of the flux is observed as an arcade of interrelated loops whose density is two-three orders of magnitude larger than that in the surrounding corona. Observed temperature distributions and plasma motions agree qualitatively with the assumption that at the tops of loops there are current sheets where the magnetic reconnection occurs. Optical observations (e.g. Mahmudov et al., 1979; Ostapenko, 1979) also confirm that at the tops of the loops plasma condensations are formed from which plasma flows out with the initial acceleration (due to magnetic forces in the current sheet model).

7. APPEARANCE OF A CURRENT SHEET IN A PLASMA MOVING IN DIPOLAR FIELD

In the presence in plasma of singular lines of magnetic field (zeroth lines $\vec{B} = 0$, $\vec{E} \neq 0$ in the simplest case) a continuous flow becomes impossible because the freezing-in Equation (2) is violated at these lines (Syrovatskii, 1978). As has been shown by Syrovatskii (1971), current sheets appear where the singular lines would take place in the absence of plasma (Figure 7). The general solution including the reversed currents at the sheet edges is shown in the Figure 7b. In this case, there exists magnetic force acting on the current sheet edges (cf. Figure 5b). Besides, magnetic field strength becomes infinite value here in the approximation of infinitely thin sheet. For the quasi-steady sheet the structure of magnetic field near the edges is shown in Figure 5a.

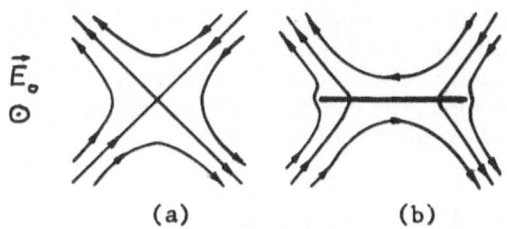

Figure 7. Formation of current sheet in the place of a singular zeroth point (line along the z axes).
(a) Field lines near the primary
zeroth point, (b) the potential
field outside the sheet.

We consider below an idealized mo--
del when a current sheet appears
as a result of capture and extension of a dipolar magnetic field by
a plasma flow (Somov and Syrovatskii, 1972). Let the two-dimensional
dipole be placed in the base of a
semicylindrical region (Figure 8)
inside of which the conditions (II)
are satisfied. We assume that the
magnetic flux is conserved on the
boundary R of the region expanding
in accordance with a specified law
R = R(t). For the corona, such a
boundary is the simplest model of

transition region from the low corona where the strong field conditions
are satisfied to the upper corona in which the kinetic energy of solar
wind dominates.

In zeroth order in small parameter ε^2, the 'potential' $A(x,y,t)$ is
defined by the Laplace equation (I9) with boundary conditions

$$A(x,y,t) = A(r,\varphi,t) = \begin{cases} 0 & \text{if } y = 0 \\ (\mathcal{L}m/r)\sin\varphi & \text{if } r = R(t) \end{cases} \quad (38)$$

and singularity of the dipolar type at $r = 0$. Here r and φ are polar coordinates, $\mathcal{L}$ is a fraction of the magnetic flux penetrating through the
boundary at the initial time ($\mathcal{L} = I$ in Figure 8a), $R_0 = R(0)$. The solution in terms of the complex potential is obvious:

$$F(z,t) = im/z - im(\mathcal{L}R - R_0)z/(R^2 R_0). \quad (39)$$

Pattern of magnetic field lines is shown in Figure 8b. If the radius $R(t)$
of the boundary increases to a value larger than $2R_0/\mathcal{L}$, then a zeroth
point of X-type appears inside the region. This point coordinate is

$$z_0 = iR(R_0/(\mathcal{L}R - R_0))^{\frac{1}{2}}. \quad (40)$$

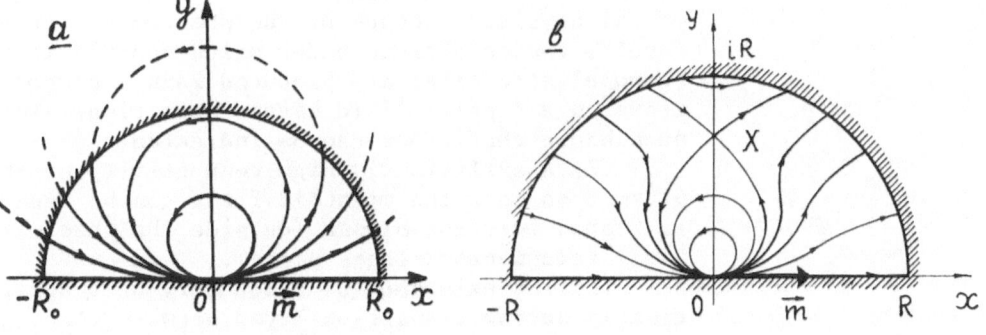

Figure 8. The magnetic field of two-dimensional dipole m with field lines frozen into the boundary R(t). (a) The dipolar field has penetrated
through boundary at t = 0. (b) Magnetic field in the absence of plasma.

One can see that electric field at the X-type point is nonzero. Therefore, that point is singular in the sense of the freezing-in condition and should be eliminated from the region where the analytic function $F(z,t)$ is defined. This can be done by the cut on the complex plane along the y axis from the boundary to a certain height h (Figure 9).

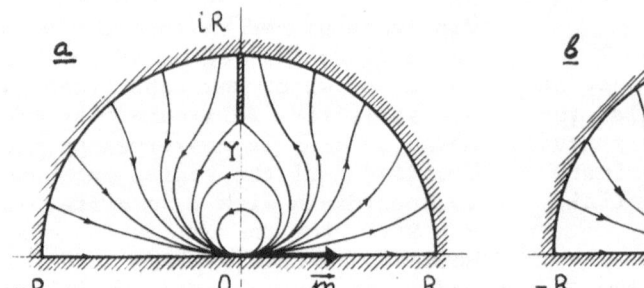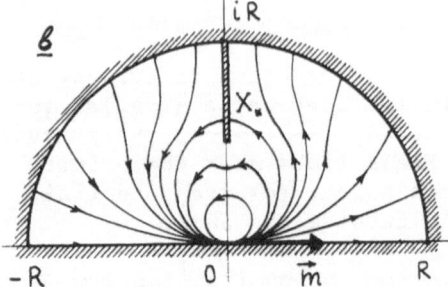

Figure 9. Field lines corresponding to the solution with the current sheet. (a) The length of the cut is minimal, (b) the cut length is larger than the minimal one (there is a reversed current at the sheet edge).

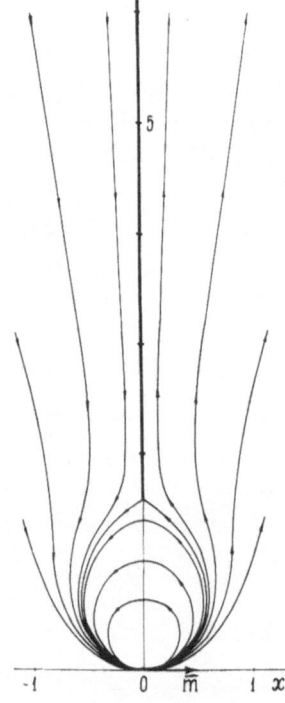

Figure IO. Magnetic field corresponding to the asymptotic solution at $\alpha R/R_0 \to \infty$.

The introduction of the cut takes into account that at the time when the zeroth point appears ($R = 2R_0/\alpha$) a current sheet begins to develop. The width of this sheet increases with increasing $R(t)$. The general solution of the problem with the sheet (Somov and Syrovatskii, I972) is shown schematically in Figure 9. If $\alpha R/R_0 \to \infty$, the solution rapidly reaches its asymptotic form which is shown in Figure IO for the case when the reversed current is absent. Near the dipole, the field has its usual structure. At large distances, the field lines tend to become radial straight lines.

The model considered has the advantage that it can be fully elaborated conserving the fundamental physical essence of the phenomenon. As a result, the conditions under which neutral layers (coronal streamers) are produced when a plasma flows in a dipolar field become more clear. Three such basic conditions can be indicated:

(I) A sufficiently high conductivity of the plasma, so that the magnetic field can be regarded as frozen into the plasma (outside the sheet itself where reconnection takes place).

(2) The existence of a boundary or a sufficiently narrow transition layer between the region where magnetic forces dominate (magnetic cavity) and the region where the plasma kinetic energy dominates (solar wind).

(3) A penetration of the magnetic field from

the magnetic cavity into the region of wind, i.e. the 'capture' of the field by the solar wind.

8. PLASMA MOTIONS INITIATED BY RAPID CHANGES OF MAGNETIC FIELD

Current sheets screen magnetic fluxes from different sources. For example, if a new emergent flux differs in direction from magnetic field of an active region, then the new and old fluxes are separated by the surface of screening current (Figure IIa). The magnetic reconnection in this current sheet changes the topology of field: the fluxes are redistributed between different sources (Figure IIb). This can illustrate the slow (quasi-steady) evolution of magnetic fields in an active region.

 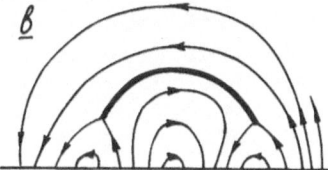

Figure II. Current sheet between the new and old fluxes.

Another situation is more interesting in connection with nonsteady phenomena like flares, transients etc. During the slow evolution described above some critical state can be reached. Starting from it the continuous deformation of magnetic field does not lead to a new equilibrium, and fast dynamic phase of evolution begins. For the case under consideration the critical state corresponds to an approximate equality of effective dipolar moments for the new emerging and old magnetic fields. The sudden transition to the eruptive phase is accompanied by the fast change of magnetic topology.

To illustrate this affirmation let us consider the potential field for four magnetic regions of interchanging polarities on the photosphere. If the normal component is uniform in each region it is easy to calculate the magnetic field in the whole space over the photosphere (Syrovatskii, 1979). For symmetrical case the neutral (zeroth) point of magnetic field appears on the symmetry axis (Figure I2a). The neutral point height is determined by the effective moments of the weak (background) field (S_0, N_0) and of the evolving internal bipolar group (N, S). When the magnetic moment of the internal group increases, the neutral point ascends with increasing velocity upto an infinite height if the effective moments of the emerging and background fields become nearly equal. In this time the 'closed' magnetic configuration turns into the 'open' one (Figure I2b). Note that the background field can be very weak provided its total flux (or, more exactly, the effective magnetic moment) is comparable with the emerging flux. It is important that the ascent velocity of the neutral point increases infinitely near the critical value of the emerging magnetic moment.

In real conditions, in a plasma of high conductivity, a current
sheet is formed in the place where the neutral point should be in the
absence of the plasma (Figure 7). The reconnection in the current sheet
provides the transition from a 'closed' magnetic field to an 'open' one.
The reconnection rate determines the real velocity for the 'neutral point' ascent. It is known that maximum rate can be near (about a few ten-
thes) to the Alfven velocity in the undisturbed plasma. For the corona
it corresponds to values of order thousand km/s. The coronal reconnection
is especially effective if the current sheet is turbulent after a flare.
Under such a rapid reconnection the real picture of magnetic field lines
differs only a little from the ideal picture of potential field with the
neutral point (Figure I2).

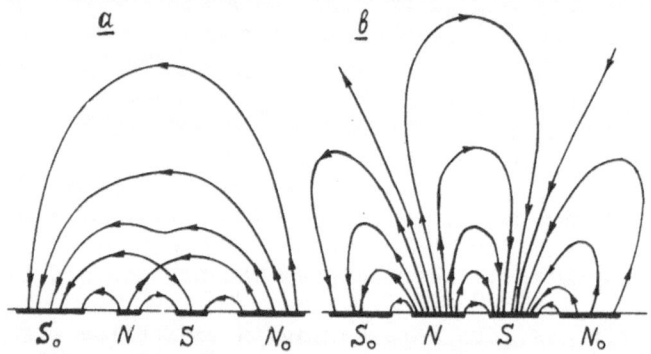

Figure I2. Potential field of four magnetic regions. (a) The
close configuration with the neutral point. (b) The critical
instant when the magnetic field opens.

Thus, the emergence of a new magnetic field can result in very fast
coronal motions accompanied by the reconnection and appeerence of regions
with an 'open' field (coronal holes). To our mind, it is very likely
that such changes of strong magnetic field can cause coronal transients.

9. CONCLUSIONS

For the interpretation of fast motions in the corona (under condition of
relatively slow photospheric motions) the concept of current sheet is of
principial weight. Current sheets can store magnetic energy during a long
slow evolution. This pre-eruption phase can be ended with a sudden explosion (a fast dynamic phase) by transition into turbulent state. The explosion is accompanied by direct ejections from a current sheet and by
fast plasma motions in the surrounding chromosphere and corona due to
rebuilding of total magnetic structure. One example of such a motion is
considered by Somov and Syrovatskii (I979).

On the other hand, under certain conditions, the turbulent current
sheet over ascending magnetic flux can penetrate rapidly (with Alfven
velocity of order one) into the upper corona and rebuild the whole mag-

netic field from a 'closed' topology to an open one. These processes create the fast plasma motions and the direct conversion of magnetic energy into heat and radiation.

Note here that Webb et al. (1978b) found a tendency for transients to occur on the borders of growing coronal holes. Rust (1978) presented 'the observational evidence for transient field openings' in the transient coronal holes. According to the model considered above (contrary to the model by Pneuman, 1979) this process is result from the magnetic reconnection in the ascending current sheet (the 'neutral point') on the tops of rising loops.

For more detailed investigation of disturbances in the upper corona and interplanetary space and for comparison of them with definite active processes on the Sun it is necessary, to our mind, to develop further the nonsteady MHD theory for radiative plasma flows in the low corona and upper chromosphere. What should future theory take into account (see also Somov, 1979b)?

First of all, this is an inhomogeneity of magnetic field and an existence of the singular (neutral in the simplest case) lines. Models with homogeneous or simple dipolar field do not have those lines and are not satisfactory for this reason. The singular lines appear, for example, if besides the emerging 'dipolar' magnetic flux there is an 'old' background field in the corona. This field, even if it is very weak, occupies large areas and can have a sufficient magnetic flux to interact with local emerging flux efficiently.

The second important factor is the solar wind. It has a strong influence on the magnetic field in upper corona. Development of solar wind theory (especially near the Alfven surface) is necessary to interprete coronal streamers and holes. In the internal corona the magnetic field structure is determined mainly by photospheric sources, but this is not true for the upper corona. The approximation of a strong field is not applicable here. The investigation of a transition region between the internal corona with a strong magnetic field and the upper corona with a weak field trailed by solar wind is an important problem to be solved.

REFERENCES

Baum, P.J., Bratenahl, A., and Kamin, G.: 1978, Astrophys. J. 226, pp. 286-300.
Bobrova, N.A. and Syrovatskii, S.I.: 1979, Solar Phys. 6I, pp. 379-387.
Brown, J.C.: 1973, Solar Phys. 3I, pp. I43-I69.
Bulanov, S.V., Sasorov, P.V., and Syrovatskii, S.I.: 1977, Soviet Phys. (Letters) JETP 26, pp. 565-568.
Bulanov, S.V. and Sasorov, P.V.: 1978, Fizika Plasmy 4, pp. 746-757.
Burnes, C.W. and Sturrock, P.A.: 1972, Astrophys. J. I74, pp. 629-638.
Dryer, M.: 1979, Invited Rev. Paper, Solar Wind 4 Conf. (H. Rosenbauer, Ed.), Springer-Verlag, Berlin, in press.

Dryer, M. and Maxwell, A.: I979, Astrophys. J., 23I, pp. 945-959.

Dryer, M., Smith, Z.K., Steinolfson, R.S., Mihalov, J.D., Wolfe, J.H., and Chao, J.K.: I976a, Geophys. Res. 8I, pp. 465I-4660.

Dryer, M., Steinolfson, R. S., and Wu, S. T.: I976b, Space Research XVI (M. J. Rycroft, Ed.), Akademie-Verlag, Berlin, pp. 685-69I.

Dryer, M., Wu, S.T., Steinolfson, R.S., and Wilson, R.M.: I979, Astrophys. J. 227, pp. I059-I07I.

Heyvaerts, J., Priest, E.R., and Rust, D.M.: I977, Astrophys. J. 2I6, pp. I23-I37.

Ivanov, L.N. and Platov, Yu.V.: I977, Solar Phys. 54, pp. 35-44.

Jockers, K.: I978, Solar Phys. 56, pp. 37-53.

Kirii, N.P., Markov, V.S., Syrovatskii, S.I., Frank, A.G., and Hodzgaev, A.Z.: I979, in 'Flare processes in plasmas', Lebedev Phys. Inst. Proc. IIO, pp. I2I-I6I (in Russian).

Low, R.P. and Nakagawa, Y.: I972, Astrophys. J. I99, pp. 237-246.

Mahmudov, M.O., Nikolskii, G.M., and Zhugzhda, Yu.D.: I979, Solar Phys., in press.

Nakagawa, Y., Steinolfson, R.S., and Wu, S.T.: I976, Solar Phys. 47, pp. I93-203.

Nakagawa, Y., Wu, S.T., and Tandberg-Hanssen, E.: I975, Solar Phys. 4I, pp. 387-396.

Ostapenko, V.A.: I979, Soviet Astron. AJ, in press.

Pneuman, G.W.: I979, On reconnection driven coronal transients, (these priceedings).

Podgornii, A.I. and Syrovatskii, S.I.: I979, in 'Flare processes in plasmas', Lebedev Phys. Inst. Proc. IIO, pp. 33-56 (in Russian).

Rust, D.M.: I978, Transient coronal holes and magnetic reconnection. Preprint AS&E-43I8.

Rust, D.M. and Hildner, E.: I976, Solar Phys. 48, pp. 38I-388.

Sermulina, B.Ja., Somov, B.V., Spektor, A.R., and Syrovatskii, S.I.: I979, Gasodynamics of impulsive heated solar plasma, (these proc.).

Shmeleva, O.P. and Syrovatskii, S.I.: I973, Solar Phys. 33, pp. 34I-362.

Somov, B.V.: I978, in 'Solar cosmic rays', Proc IX Leningrad Seminar on Space Phys., pp. I85-2IO (in Russian).

Somov, B.V.: I979a, Bull. Acad. Sci. USSR, Phys. Series 43, No. 4 (Allerton press Inc., New York).

Somov, B.V., Spektor, A.R., and Syrovatskii, S.I.: I977, Bull. Acad. Sci. USSR, Phys. Series 4I, pp. 32-43.

Somov, B.V., Spektor, A.R., and Syrovatskii, S.I.: I979, in 'Flare processes in plasmas', Lebedev Phys. Inst. Proc. IIO, pp. 73-94.

Somov, B.V.: I979b, in 'Flare processes in plasmas', Lebedev Phys. Inst. Proc. IIO, pp. 57-72 (in Russian).

Somov, B.V. and Syrovatskii, S.I.: I972, Soviet Phys. JETP 34, pp. 992-997.

Somov, B.V. and Syrovatskii, S.I.: I974, in 'Neutral current sheets in plasmas', Lebedev Phys. Inst. Proc. 74, pp. I4-72 (in Russian).

Somov, B.V. and Syrovatskii, S.I.: I975, Bull. Acad. Sci. USSR, Phys. Series 39, pp. I09-III (Allerton Press Inc., New York).

Somov, B.V. and Syrovatskii, S.I.: I976, Soviet Phys. Uspekhi I9, pp. 8I3-835 (American Inst. Phys.).

Somov, B.V. and Syrovatskii, S.I.: 1979, Magnetically driven motions in
    solar corona, (these proceedings).
Steinolfson, R.S., Wu, S.T., Dryer, M., and Tandberg-Hanssen,E.: 1978,
    Astrophys. J. 225, pp. 259-274.
Steinolfson, R.S., Wu, S.T., Dryer, M., and Tandberg-Hanssen, E.: 1979,
    Solar Wind 4 Conf. (H. Rosenbauer, Ed.), Springer-Verlag, Berlin,
    in press.
Svestka, Z.: 1976, Solar Flares. D.Reidel Publ. Co., Dordrecht, Boston.
Sweet, P.A.: 1958, in 'Electromagnetic Phenomena in Cosmic Physics',
    Proc. IAU Symp. No. 6, pp. 123-132.
Syrovatskii, S.I.: 1966, Soviet Astron. AJ 10, pp. 270-279.
Syrovatskii, S.I.: 1969, Astrophys. Space Sci. 4, pp. 246-250.
Syrovatskii, S.I.: 1971, Soviet Phys. JETP 60, pp. 1727-1736.
Syrovatskii, S.I.: 1976, Soviet Astron. (Letters) AJ 2.
Syrovatskii, S.I.: 1977, Solar Phys. 53, p. 247.
Syrovatskii, S.I.: 1978, Astrophys. Space Sci. 56, pp. 3-12.
Syrovatskii, S.I.: 1979, Lebedev Phys. Inst. Preprint, in press.
Webb, D.F., Cheng, C.C., Dulk, G.A., Edberg, S.J., Matrin, S.F., McKenna-
    Lawlor, S., and McLean, D.J.: 1978a, Preprint AS&E-4364.
Webb, D.F., McIntosh, P.S., Nolte, J.T., and Solodyna, C.V.: 1978b, Solar
    Phys. 58, pp. 389-396.
Wu, S.T., Dryer, M., and Han, S.M.: 1976, Solar Phys. 49, pp. 187-204.
Wu, S.T., Dryer, M., Nakagawa, Y., and Han, S.M.: 1978, Astrophys. J.
    219, pp. 324-335.
Zirin, H. and Tanaka, K.: 1973, Solar Phys. 32, pp. 173-207.

DISCUSSION

Moore: What is your opinion of the magnetic field configuration
proposed by Jerry Pneuman as being appropriate for magnetic field
reconnection in a filament eruption?

Somov: Contrary to the model by Dr. Pneuman, the opening of
magnetic field in our model is result from magnetic reconnection in
the ascending current sheet.

Nakagawa: (Comment) Those models presented represent schematic
illustrations at best. I do not believe these figures or models, unless
realistic physically self-consistent computations are carried out and
numerical results are obtained to compare with observations.

Callebaut: Have you thought of or tried to introduce some of the
features of the Uchida-Sakurai model? I think in particular of the
interchange in stability which allows the reconnecting surface and
volume to be increased very much.

Somov: I think that the interchange instability considered in the
Uchida-Sakurai model is very essential for the case when gas pressure
gradient is not disregarded. But this instability can hardly be
effective in a strong magnetic field in the internal (low) corona,
especially if a small field component exists along a current sheet.

THEORETICAL INTERPRETATION OF TRAVELING INTERPLANETARY PHENOMENA AND
THEIR SOLAR ORIGINS

S. T. Wu
The University of Alabama in Huntsville
Huntsville, Alabama 35807, U.S.A.

ABSTRACT

Recent theoretical studies on Traveling Interplanetary Phenomena
(TIP) and their relation or presumed relation to their solar origins
will be reviewed. An attempt is made to outline the theoretical
studies in the context of mathematical methods and physical processes.
The following alternative approaches are examined: analytical vs.
numerical methods; magnetohydrodynamics vs. hydrodynamics; processes
with or without dissipation; continuum (macroscopic) vs. the kinetic
(microscopic) approach. In particular, the flare-generated interplanetary shocks are used as examples to illustrate these theoretical
studies within the context of TIP. Some emphasis will be placed on MHD
wave propagation through the inner corona and its maturity to a fully-
developed interplanetary shock. Further, their propagation and the disturbing effects on the solar wind will be considered. Cases concerning
the classification and characteristics of blast-produced shocks and long-
lasting ejecta are also discussed in the context of numerical simulations.

In this review, it has been revealed that: (i) sophisticated
numerical simulations are significant for the progress of hydrodynamical
and magnetohydrodynamical studies; (ii) these numerical simulation
studies have improved significantly the understanding of non-linear mode-
coupled wave interactions from the lower corona to interplanetary space;
and (iii) lack of emphasis on the kinetic (microscopic) approach limits
our understanding on microscopic interactions. We suggest, therefore,
that future directions should emphasize the physical processes of the
continuum approach (i.e., hydrodynamics and MHD theory) and the kinetic
approach to reveal further understanding of microscopic interactions.

I.   INTRODUCTION

A large amount of data about transient phenomena in the corona and
its extension into interplanetary space has been accumulated by the
space program during the past decades. In order to gain basic physical

M. Dryer and E. Tandberg-Hanssen (eds.), Solar and Interplanetary Dynamics, 443-458.

insight from these data, theoretical interpretation of them has become a
necessity. An attempt is made in this paper to summarize the theoretical
studies in the context of mathematical methods and physical processes.

Significant progress in our understanding of traveling interplanetary phenomena has been made. In particular, the study concerning
solar flare generated interplanetary shocks has been reported in several
recent reviews by Hundhausen (1972a,b), Burlaga (1974), Dryer (1974,
1975) and Wu et al. (1977). Although the present discussion will refer
to these earlier papers, consideration of the theoretical interpretation
since that time will be emphasized. The purpose of this paper, then is
to outline some fundamental developments in the last few years. Accordingly, it will be assumed that the reader has some familiarity with
earlier works which are referred to in the reviews noted above. An
attempt will be made to categorize these recent developments. Thus, the
following approaches are examined: analytical vis-à-vis numerical methods;
magnetohydrodynamic (MHD) vis-à-vis hydrodynamic description; macroscopic
(continuum) vis-à-vis microscopic (kinetic) approach. Thus, we shall
begin our discussion with the macroscopic theory (i.e., the MHD and
hydrodynamic descriptions) in Section II with those models using
analytical methods in which the discussion of physical regions (i.e.,
corona/corona-interplanetary) are included. In Section III, we shall
discuss the recent developments of microscopic theory in this area. In
the final Section IV, current research and future directions of this
line of research are discussed.

II. MACROSCOPIC (CONTINUUM) THEORY

In this approach, the coronal and the corona-interplanetary media
have been represented by fluid models. Thus, the hydrodynamical and
magnetohydrodynamical formalisms are applied. The problems can be
classified into two categories: (i) corona and (ii) corona-interplanetary space in order to distinguish two cases of basic physical
behavior. For example, in the corona, the initial steady-state can be
approximately represented by an isothermal and hydrostatic equilibrium
atmosphere where the low subsonic and sub-Alfvénic velocities may be
neglected. However, in the corona-interplanetary case, the initial
steady-state atmosphere must include the characteristics of the solar
wind together with its imbedded magnetic field. That is, the initial
state is an atmosphere in hydrodynamic equilibrium instead of in isothermal and hydrostatic equilibrium as in the case of the corona. The
mathematical methods used to solve these problems are either analytical
or numerical in nature, with each approach complimentary to the other.

II.1. Analytical Analysis

II.1.A Corona. It is well known that the most spectacularly-
observed traveling phenomenon in the corona is the so-called "Coronal
Transient." These transient phenomena are seen in white light and, in
some cases, in X-ray and radio wavelengths (MacQueen et al., 1974);

Stewart et al., 1974; Rust and Hildner, 1976). Some progress has been made toward the theoretical interpretation of these phenomena. In essence, these theoretical studies can be summarized into two categories. The first theoretical approach views coronal transients as a global wave phenomenon (Nakagawa, Wu and Han, 1978; Wu et al., 1978; Steinolfson et al., 1978; Dryer et al., 1979). The method used for this approach utilizes numerical analysis of the complete nonlinear MHD equations and will be discussed later. An alternative interpretation suggested by Mouschovias and Poland (1978) views coronal transients as expanding flux tubes in the corona (1.6 ~ 6 $R_\odot$; $R_\odot$ being the solar radius). In their

model, the latter workers assume that a white-light loop-like transient density enhancement seen by the coronograph is a magnetic flux tube which originates below the occulting disk of the coronograph (1.6 $R_\odot$).

The flux tube is assumed to expand through a background coronal plasma and magnetic field. This global background atmosphere remains unaffected by the transient. In this model, shown in Figure 1, the material and field within the loop does not bear any relation to the surrounding coronal material and field. The force responsible for the outward expansion of the loop is the magnetic buoyancy force which is local in

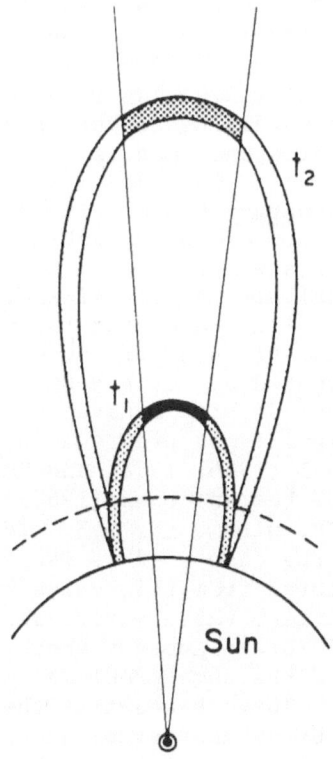

Figure 1.   Schematic representation of a coronal loop transient at two different times, t_1 and t_2 (Mouchovias and Poland, 1978).

nature. Mouchovias and Poland (1978) made a comparison between the observed 1973 August 10 coronal transient event and their model, with some reasonably good phenomenological agreement. The main deficiency of this model is its lack of interaction between the loop and coronal background. Nevertheless, it will be interesting to pursue models such as this one further (Anzer and Poland, 1979). Another example of a model of traveling coronal phenomena which deserves attention is the magnetic reconnection model given by Kopp and Pneuman (1976). In this model, they examined the theoretical consequences during the extended relaxation phase which must follow events such as flares, flare sprays, and eruptive prominences. This phase is characterized by a gradual reconnection of the outward-distended field lines. It is further shown that the enhanced coronal expansion which occurs on open field lines just before they reconnect appears adequate to supply the large downward mass flow observed in the H_α loop prominence systems during the post-transient relaxation phase. In addition, this enhanced flow may produce nonrecurrent high speed streams in the solar wind after such events. Again, the disadvantage of this model is that the lack of systematic approach prevents a complete description of dynamics of the problem which is expected by an analytical method. In fact, a more systematic numerical study has been carried out recently by Steinolfson and Wu (1979), which is being presented in this symposium. The reader is referred to that paper in this Proceedings. The works of Anzer (1979), Syrovatskii and Somov (1979) and Somov and Syrovatskii (1979), as reported in this symposium, should also be noted. They have discussed the driving forces for physically-meaningful coronal response models.

 II.1.B. Corona —Interplanetary Space. The required governing equations to describe the physics of these problems are highly non-linear. Thus, the most appropriate method used to seek an analytical solution is the similarity analysis. A self-similar treatment of a spherical magnetohydrodynamic disturbance for the propagation of inter-planetary shocks limited to the vicinity of the equatorial plane of an axisymmetric geometry is presented by Lee and Chen (1968), for the special case where the upstream density behaves as r^{-2}; r being the radial distance. Recently, significant progress in this area has been made by Rosenau and Frankenthal (1976, 1978) and Rosenau (1977, 1978). They extended the treatment of Lee and Chen (1968) to cases where the ambient density behaves as r^{-w}, with $0 \leq w < 3$. This extension is significant. It reveals that the case $w = 2$ is isolated in the sense that the infinitesimal departures from this value result in qualitative changes in the nature of the flow, thereby revealing significant physical meaning for interpretation of interplanetary shock structures. Rosenau and Frankenthal (1978) studied the same problems further by considering a thermally conducting medium. They showed that the motion consists of a thermal precursor followed by an isothermal shock. The magnetic field plays a fundamental role. These workers showed that a very modest transverse magnetic field depresses the peak density, blocks the heat flow, and widens the perturbed domain. Figure 2 (plasma parameters) and 3 (magnetic field) show a comparison of observed data (Heos-1 data on

March 25, 1969) with theoretical predictions based on an adiabatic
magnetohydrodynamic model (Dryer, 1974) and an MHD model which incor-
porates heat conduction. In these results, a significant improvement of
the theoretical predictions is demonstrated by including heat conduction.
Rosenau and Frankenthal (1978), also noted that the heat conduction is
a variable parameter which can be used together with a variable length
of the thermal precursor to improve the prediction of the shock structure.

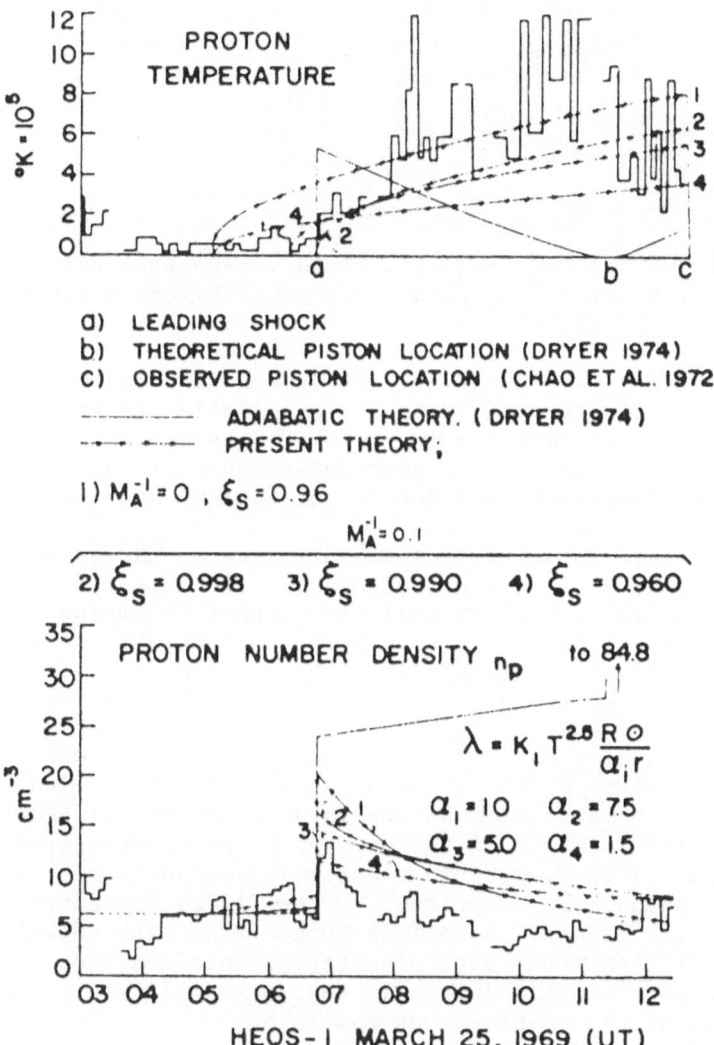

Figure 2.   A comparison of observed data (Heos-1 on March 25, 1969) with
            theoretical predictions based on adiabatic magnetohydrodynamic
            model and on the model which incorporates heat conduction, λ,
            as a variable parameter together with a variable length of the
            thermal precursor ξ_s (Rosenau and Frankenthal, 1978).

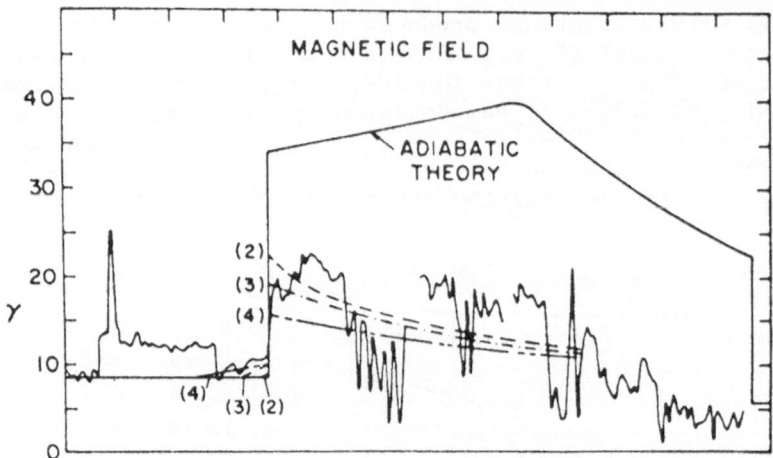

Figure 3.   Display of observed magnetic field versus theoretical
            predictions with same parameters as in Figure 2 (Rosenau
            and Frankenthal, 1978).

In summary, we conclude that these investigations represent substantial
progress on the self-similar theory by its application to the study of
the propagation of an axisymmetric magnetohydrodynamic shock in a
thermally conducting medium. All these solutions are classified as
piston-driven shock solutions, and double pair shocks are revealed.

On the other hand, Summer (1975) investigated an MHD blast-wave
type solution applied to a flare-produced shock in the solar wind by
using similarity analysis. Similar analysis without the magnetic field
but with a gravitational field was done by Rao and Purohit (1973). A
pure gasdynamic model using Lagrange-function approach to investigate
the geometric characteristics of propagation of the interplanetary
shocks was presented by Krimsky and Transky (1973). Dryer (1970) also
used the similarity analysis to study the electrical field effect
(with finite resistivity) on the propagation of solar flare-induced
interplanetary shock waves. He found the largest effects of joule
heating (with implied location of turbulence) to be concentrated within the piston region. However, this class of solution is restricted
to small magnetic Reynold's number flow. Dryer (1972) has extended
this work with finite magnetic Reynold's number flow with anomalous
electric conductivity to study interplanetary double-shock ensembles.
He has shown that even substantial joule heating has little effect on
the gross features of the double-shock ensemble.

II.2. Numerical Analysis

Because of the limitations of similarity analysis on multi-
dimensional, time-dependent problems wherein arbitrary input conditions
are prescribed at the boundary, numerical analysis has become an important tool to interpret and understand the physics of the observed

coronal and corona-interplanetary transient phenomena. Again, we shall divide our discussions into two parts: corona and corona-interplanetary space.

II.2.A. Corona. A recent review concerning numerical modeling of coronal and interplanetary responses to solar events has been given by Wu, Nakagawa and Dryer (1977). In this work, a detailed account of the development of numerical modeling of transient phenomena in these two regions has been presented. Hence, this work will not be repeated.

In a recent development, Steinolfson et al. (1978) presented a boundary perturbation type MHD model in the meridional plane in contrast to the equatorial plane model given by Nakagawa et al. (1978) and Wu et al. (1978), with the later two works having been done in the context of the blast-wave-type solution. Steinolfson and Wu (1979) recently applied this model with Helmet streamer magnetic field configuration to study the coronal response. In the studies mentioned above, the perturbations are considered to be in the nature of a thermodynamic pulse (i.e., changes in temperature, density or both). Recently, Steinolfson et al. (1979) presented a solution with an emerging magnetic flux perturbation. This work shows a distinct result in comparison with a thermodynamic pulse. They found that there exists a high (i.e., β greater than one) region between the shock and contact surface and a low β (less than one) region behind the contact surface (β being the ratio of plasma pressure to magnetic pressure). Dryer et al. (1979) used this model to simulate the 1973 August 21 coronal transient event by using observed plasma parameters (i.e., employing data from S-056 soft X-ray experiment by NASA/MSFC and Aerospace Corporation on board Skylab) as an initial and long lasting pulse. They demonstrated very good agreement with observed data. The outcome of each of these calculations depends strongly on the value of β. Figure 4 shows the β distribution before and after the 1973 August 21 event. The essence of the numerical method for these calculations can be found in the works of Nakagawa and Steinolfson (1976) and Han, Wu and Nakagawa (1979).

As the reader may note, this plane model exhibits the non-linear interaction between fast and slow mode MHD waves; however, the Alfvén mode (transverse wave) is excluded. In order to relax this deficiency, a two-dimensional, time dependent, non-plane, MHD model has been presented recently by Nakagawa et al. (1979), and Wu et al. (1979b). In the work of Nakagawa et al. (1979), a new way of interpreting the energy storage and release in repeated flares was suggested.

II.2.B. Corona-Interplanetary Space. In the previous section, we have summarized briefly the recent development of numerical models for the corona. We will now discuss the current status of modeling in the corona-interplanetary case. Typical results for this problem can be found in the work of Dryer et al. (1976, 1978) and Zakaidakov and Synakh (1977). These workers used a one-dimensional, time-dependent MHD model to study the evolution of the ·structures of the interplanetary shocks. Also, Dryer et al. (1978) utilized this model to simulate

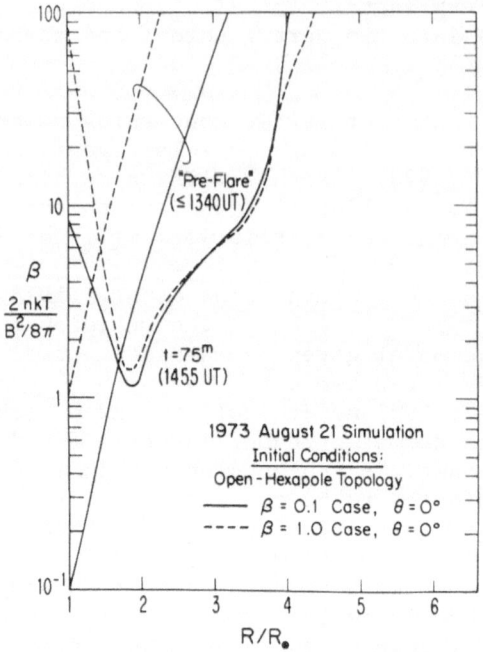

Figure 4.   The distribution of β (the ratio of plasma pressure to
            magnetic pressure) through the solar atmosphere before
            and after the 1973 August 21 event.

essential features of space-probe data. In particular, the shock pair
ensembles are reproduced by these numerical solutions. Wu et al. (1979a)
extended their model (Wu et al., 1978) to the corona-interplanetary
case by including the solar wind characteristics in the initial state.
In this work, they considered the case of stream-stream interactions.
Typical results for the disturbed density and temperature contours from
the lower corona (18 R_s being solar radius) to 1 A.U. (Astronomical
Units) are depicted in Figure 5.   This figure shows development of MHD
shocks in two dimensions. In a subsequent paper by D'Uston et al. (1979),
they utilized this model to study the nonsymmetric properties of the
propagation of flare interplanetary shocks. Figure 6 shows their numerical
results for the evolution of the position of the forward shock and the
reverse shock. The non-spherical nature of the shock front is clearly
exhibited.

III. MICROSCOPIC (KINETIC) THEORY

    In this context, one treats the problem from a particle point of
view, in which the Boltzmann transport equation is used as the basis
of the formulation.   Practically very little work has been done in the

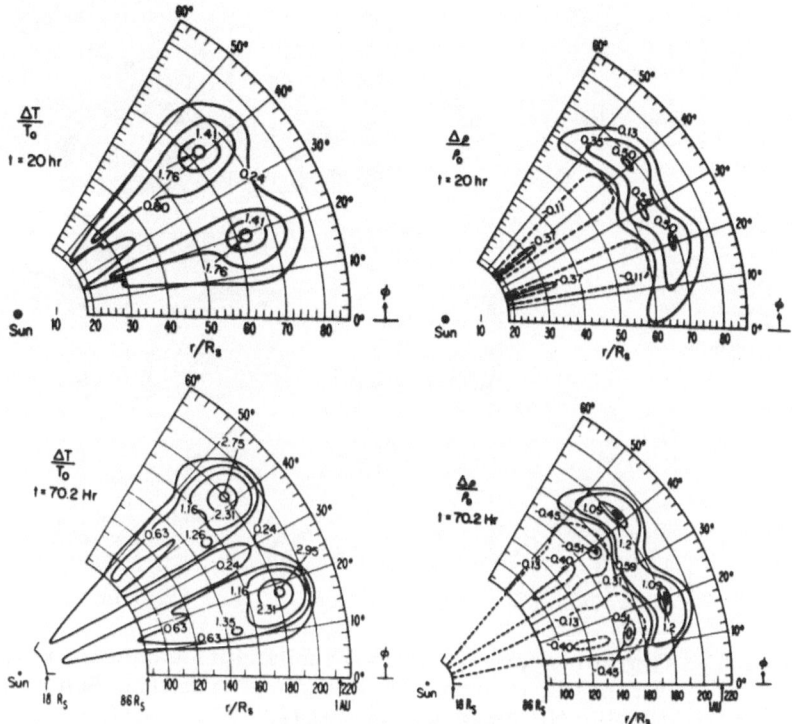

Figure 5. Disturbed temperature enhancement contours $(\frac{\Delta T}{T_0})$ and density enhancement contours $(\frac{\Delta \rho}{\rho_0})$ in the solar equatorial plane at (a) 20 hr and (b) 70 hr after the introduction of disturbance. In the density contours, the solid line represents positive enhancement (i.e., compression), and the broken line represents negative enhancement (i.e., rarefaction) (Wu et al., 1979).

study of transient phenomena in the corona and corona-interplanetary space by using kinetic theory. However, significant advancement of plasma kinetic theory has been made in fusion research in recent years. For example, Liewer and Krall (1973), have used such a method to study the electromagnetic turbulence in fusion plasmas. In the area of solar and interplanetary dynamics, Jockers (1970) presented a solar wind solution by using kinetic theory via the moment method; Fahr, Bird and Ripken (1977) used the moment equations from the collisionless Boltzmann equation to study the solar wind expansion with spherically symmetric magnetic fields. Smith (1971, 1972a,b) used the kinetic theory to study the plasma radiation from collisionless MHD shock waves in the corona and its application to Type II radio bursts. All these results refer to steady state solutions. It is understood that, the classical approach (i.e., Chapman-Enskog method) to seek solution of Boltzmann equation is only limited to the case in which the gradients of the thermodynamical property are small and only particle-particle collision

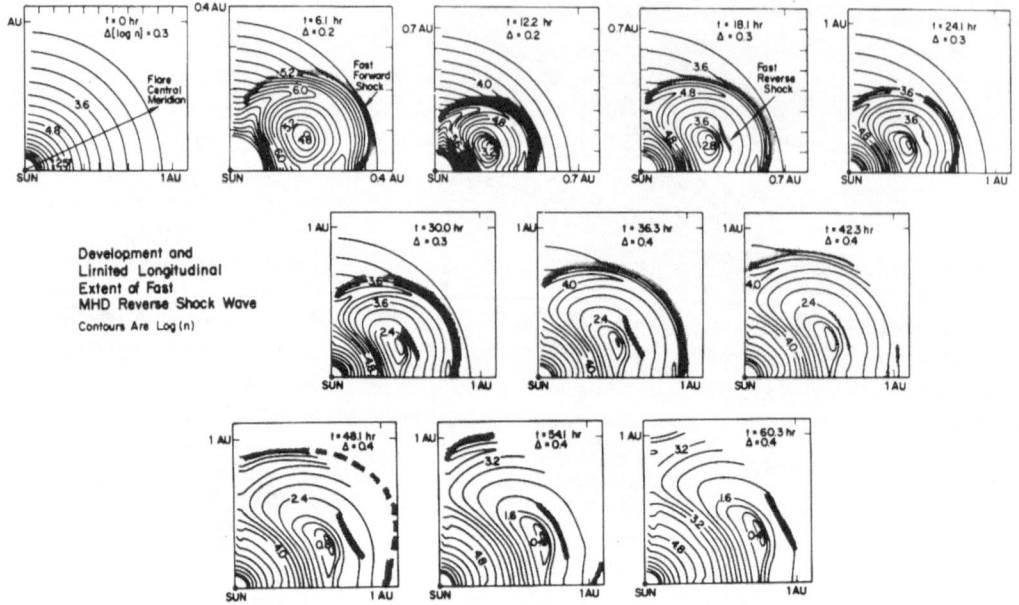

Figure 6.   Contour maps of Log(n) showing the development and the
             extension of the shock fronts from the Sun's surface to
             1 A.U. (D'Uston et al., 1979).

is accounted for. In the heliospheric region, the collective behavior
of the medium depends not only on the particle-particle collision ρ,
but also on the wave-particle interaction.   Therefore, a new approach
is needed to understand the more realistic physical behavior of the
TIP in the corona and corona-interplanetary medium. In this symposium,
Cuperman (1979) has suggested such a theoretical study. In his study,
higher order moments equations (i.e., fluid description) are developed
with more realistic closure conditions and transport coefficients; these
transport coefficients are obtained from quasi-linear kinetic theory.

Concerning numerical analyses of the Boltzmann transport equation
for transient phenomena in the corona and corona-interplanetary
environments, little work has been done because of the mathematical
complexity involved. Again, significant progress can be found in
fusion research. Also, Wu and Dryer (1972) used a time dependent
Boltzmann equation to study the solar wind interaction with celestial
objects via numerical methods, however, only particle-particle interaction is taken into account in their collision integrals and a
Maxwellian distribution function is also assumed in their computation.
Recently a more advanced numerical analysis of kinetic theory has been
done by Scudder and Olbert (1978) to explain the observed characteristics of the electron distribution function in the solar wind. Their
results are in agreement with observations. All these works indicate
that this approach is a promising one.

The inter-relationship between the macroscopic (fluids) theory and microscopic (kinetic) theory is shown in Table I, in which the various levels of sophistication of both approaches with their corresponding physical interpretation are outlined. It should be noted that the current status of the macroscopic approach is far more advanced than the microscopic approach. However, we note the advantages of using the microscopic approach to interpretate the physical behavior of the TIP.

IV. CONCLUDING REMARKS

In summary, this study has revealed that:

(i) Sophisticated numerical analyses are significant for the progress of hydrodynamical and magnetohydrodynamical studies without dissipative processes.

(ii) These numerical studies have improved significantly the understanding of non-linear mode-coupled wave interactions from the corona to interplanetary space.

(iii) The similarity analysis brought significant progress to the understanding of the MHD shock-disturbed flow. However, due to its limitation to a single spatial dimensional configuration and inability to treat realistic boundary conditions, only simple geometry and asymptotic solutions can be studied. Nevertheless, similarity theory is complementary to the numerical studies.

(iv) Lack of emphasis on the microscopic approach limits our understanding of detailed microscopic interactions which have essential effects on macroscopic dynamics, such as the plasma turbulent structures observed in the corona and heliospheric space.

In conclusion, we suggest that future research directions should emphasize the physical processes of the macroscopic (continuum) approach together with the kinetic approach to reveal further understanding of the microscopic interactions. Consequently, a hybrid model for the TIP in the corona and corona-interplanetary space should be constructed.

Acknowledgment. The author would like to thank Drs. M. Dryer, Y. Nakagawa and E. Tandberg-Hanssen for helpful discussions and reading of the manuscript. The work was done under a grant from NSF-ATM77-22482 and Contract NAS8-28097, from NASA/Marshall Space Flight Center, Alabama.

TABLE 1.   PARALLELISM BETWEEN MACROSCOPIC AND MICROSCOPIC DESCRIPTIONS
UNDER VARIOUS LEVELS OF SOPHISTICATION

Order of the Level of Sophistication	Physical Interpretation	Model Description	
		Macroscopic	Microscopic
Zeroth	System at complete thermodynamic equilibrium is steady, homogeneous.	System is in hydrostatic equilibrium with no motion.	Distribution function f is Maxwellian with constant uniform temperature $f^{(o)}$.
First	System at local-thermodynamic equilibrium.	System can be represented by a single fluid. Hydrodynamic and magnetohydrodynamic (MHD) description applied, no dissipation process.	Distribution function f is local Maxwellian $f^{(o)} = f^{(o)}(n, \vec{u}, T)$ where $n = n(\vec{r}, t)$, $\vec{u} = \vec{u}(\vec{r}, t)$, $T = T(\vec{r}, t)$.
Second	Multiple component system at local thermodynamic equilibrium.	System should be represented by multi-fluids. Hydrodynamic and MHD description applied.	Distribution function f is local Maxwellian for each specified species; $f_s^{(o)} = f_s^{(o)}(n_s, \vec{u}_s, T_s)$; where $n_s = n_s(\vec{r}, t)$; $\vec{u}_s = \vec{u}_s(\vec{r}, t)$; $T_s = T_s(\vec{r}_s, t)$.
Higher Order	Multiple component system at non-local-thermodynamic equilibrium.	System will be represented by dissipative multi-fluids system. Hydrodynamic and MHD description applied.	Distribution function for each species becomes highly non-Maxwellian $f_s = f_s^{(o)} + f_s^{(1)}$.

REFERENCES

Anzer, C. F., and Poland, A. C., *Solar Phys.*, *61*, 95.

Anzer, C. F., 1979, in Dryer, M., and Tandberg-Hanssen, E. (eds.),
    Proc. IAU Symposium 91, D. Reidel Publishing Co., Dordrecht,
    Holland.

Bulaga, L., 1974, *Proc. Conf. on Flare-Produced Shock Wave in the
    Corona and Interplanetary Space*, 11-14 September 1972, National
    Center for Atmospheric Res./HAO, p. 123.

Cuperman, S., 1979, in Dryer, M. and Tandberg-Hanssen, E. (eds), *Proc.
    IAU Symposium 91*, D. Reidel Publishing Co., Dordrecht, Holland.

Dryer, M., 1970, *Cosmic Electrodynamics*, *11*, 348.

_____, 1972, in C. P. Sonett, P. J. Coleman, Jr., and J. M. Wilcox
    (eds), *Solar Wind, NASA SP-308*, Supt. of Documents, Washington,
    D.C., p. 453.

_____, 1974, *Space Science Reviews*, *15*, p. 403.

_____, 1975, *Space Science Reviews*, *17*, p. 277.

Dryer, M., Smith, Z. K., Steinolfson, R. S., Mihalov, J. D., Wolfe,
    J. H., and Chao, J.-K., 1976, *J. Geophys. Res.*, *81*, p. 4651.

Dryer, M., Smith, Z. K., Smith, E. J., Mihalov, J. D., Wolfe, J. H.,
    Steinolfson, R. S., and Wu, S. T., 1978, *J. Geophys. Res.*, *83*,
    p. 4347.

Dryer, M., Wu, S. T., Steinolfson, R. S. and Wilson, R. M., 1979,
    Ap. J., *227*, p. 1059.

D'Uston, C., Dryer, M., Han, S. M., and Wu, S. T., 1979, *Space
    Environment Lab. Report, NOAA*, Boulder, CO.

Fahr, H. J., Bird, M. K., and Ripken, H. W., 1977, *Astron. and
    Astrophys.*, *58*, p. 339.

Han, S. M., Wu, S. T., and Nakagawa, Y., 1979, *Computers and Fluids*, *7*,
    p. 97.

Hundhausen, A. J., 1972a, *Coronal Expansion and Solar Wind*, Springer-
    Verlag, New York.

_____, 1972b, in E. R. Dyer (General Ed.), *Solar Terrestrial
    Physics/1970, Part II*, D. Reidel Publishing Co., Dordrecht-
    Holland, p. 1.

Jocker, K., 1970, *Astron. and Astrophys.*, *6*, 219.

Krimsky, G. F., and Transky, I. A., 1973, *Geomagnatizmi Aeronomyia*, *13*,
    p. 777.

Kopp, R. A., and Pneuman, G. W., 1976, *Solar Phys.*, *50*, p. 85.

Lee, T. S., and Chen, T., 1968, *Planet. Space Sci.*, *16*, p. 1483.

Liewer, P. C., and Krall, N. A., 1973, *Phys. Fluids, 16,* p. 1953.

MacQueen, R. M., Eddy, J. A., Gosling, J. T., Hildner, E., Munro,
    R. H., Newkirk, G. A., Jr., Poland, A. C., and Ross, C. L., 1974,
    Ap. J. (Letters), 187, L85.

Mouschovias, T. Ch., and Poland, A. C., 1978, *Ap. J., 220,* p. 675.

Nakagawa, Y., and Steinolfson, R. S., 1976, *Ap. J., 207,* p. 296.

Nakagawa, Y., Wu, S. T., and Han, S. M., 1978, *Ap. J., 219,* p. 314.

_____, _____, and _____, 1979, in Dryer, M., and
    Tandberg-Hanssen, E. (eds), *Proc. IAU Symposium 91,* D. Reidel
    Publishing Co., Dordrecht, Holland.

Rao, Ranga Melam P. and Purohit, Sharad, C., 1973, *Astron. and
    Astrophys., 23,* p. 155.

Rosenau, Philip, 1977, *Phys. Fluids, 20,* p. 1097.

_____, 1978, *Phys. Fluids, 21,* p. 1455.

Rosenau, Philip and Frankenthal, Shimshon, 1976, *Phys. Fluids, 19,*
    p. 1889.

_____ and _____, 1978, *Phys. Fluids, 21,*
    p. 559.

Rust, D. M., and Hildner, E., 1976, *Solar Phys., 48,* p. 381.

Scudder, J. D., and Olbert, S., 1978, *EOS, 59,* No. 12, p. 1175.

Smith, Dean F., 1971, *Ap. J., 170,* p. 559.

_____, 1972a, *Ap. J., 174,* p. 121.

_____, 1972b, *Ap. J., 174* p. 643.

Somov, B. V., and Syrovatskii, S. I., 1979, in Dryer, M., and
    Tandberg-Hanssen, E. (eds), *IAU Symposium 91,* D. Reidel
    Publishing Co., Dordrecht, Holland.

Steinolfson, R. S., Wu, S. T., Dryer, M., and Tandberg-Hanssen, E.,
    1979a, in Rosenbauer, H. (ed), *Solar Wind IV,* Springer-Verlag,
    (in press).

Steinolfson, R. S., and Wu, S. T., 1979, in Dryer, M., and Tandberg-
    Hanssen, E. (eds.), *Proc. IAU Symposium 91,* D. Reidel Publishing
    Co., Dordrecht, Holland.

Stewart, R. T., McCabe, M. K., Koomen, M. J., Hansen, R. T., and Dulk,
    G. A., 1974, *Solar Phys., 36,* p. 206.

Summers, D., 1975, *Astron. and Astrophys., 45,* p. 151.

Syrovatskii, S. I., and Somov, B. V., 1979, in Dryer, M. and Tandberg-
    Hanssen, E. (eds), *Proc. IAU Symposium 91,* D. Reidel Publishing
    Co., Dordrecht, Holland.

Wu, S. T., and Dryer, M., 1972, in Grard, R. J. L. (ed), *Photon and
    Particle Interactions with Surfaces in Space,* D. Reidel Publishing
    Co., Dordrecht-Holland, p. 453.

Wu, S. T., Nakagawa, Y., and Dryer, M., 1977, in Shea, M. A., et al., (eds), *Study of Traveling Interplanetary Phenomena*, D. Reidel Publishing Co., Dordrecht, Holland, pp. 43-62.

Wu, S. T., Dryer, M., Nakagawa, Y., and Han, S. M., 1978, *Ap. J., 219*, p. 324.

Wu, S. T., Han, S. M., and Dryer, M., 1979a, *Planet. Space Sci., 27*, p. 255.

Wu, S. T., Han, S. M., Nakagawa, Y., Tandberg-Hanssen, E., and Dryer, M., 1979b, *Bull. Am. Astron. Soc., 11*, no. 2, p. 410.

Zakaidakov, V. V., and Synakh, У. S., 1977, in Pivovarov, V. G. (ed), *Novosibirsk Matematicheskiye Modeli Blizhnego Kosmosa, Nauka*, USSR Academy of Sciences, Siberian Dept., Computer Center, p. 33.

DISCUSSION

Somov: Magnetic field in the internal (low) corona is a very important energetic factor. Magnetic force can dominate over all other forces. What do you think about the possibility that magnetic field is a primary reason for fast plasma motions in corona?

Wu: Magnetic field is an important force to control the dynamical behavior of the corona. However, the other factor being equally important is the dynamics itself which has been shown in these numerical calculations. These calculations, however, do not consider the *initial* conversion of magnetic energy into kinetic and thermal forms as, for example, you and Prof. Syrovatskii have discussed in various works. Once such conversion takes place, the subsequent mass motion must respond (non-linearly) to the time-dependent interplay of the relative magnitudes of inertial, thermal and magnetic forces. Thus, we may conclude that the essential factor to control the dynamical behavior in the post-flare corona is the plasma motion and magnetic field *interaction*. Consideration of $\beta(t)$ implies the importance of the modulating effect of the magnetic field.

Ivanov: What do you think about including electromagnetic turbulence in hydromagnetic theory?

Wu: To include turbulence in hydromagnetic calculation is a necessary step to be taken to improve our physical interpretation of the observations. However, we have no self-consistant theory to describe hydromagnetic turbulence. This is why I am suggesting the construction of a *hybrid* model. Using kinetic theory to obtain such microscopic turbulent structure is the first essential step. Having done so, one may then put these results into macroscopic theory to construct new models. See Liewer and Krall (1973) for some work along this line.

Unidentified: (Comment) The fact that we see a type II radio burst associated with interplanetary shocks inevitably points to a non-Maxwellian electron distribution. The presence of Langmuir waves in such sources cannot be explained otherwise.

Lemaire: I believe there is much more work done in the kinetic approach than mentioned in your review. For instance: Lemaire and Scherer, *JGR*, 1971 and *Rev. Geophys.*, 1974; Hollweg, *JGR*, 1971; Lemaire (Proceedings of Toulouse Meeting), 1978 March; and Scudder and Olbert, *JGR*, 1979. I consider, for instance, that the recent work by Scudder and Olbert, including collisions as a post-exospheric approximation is the right way to go in future solar wind modeling!

Wu: I think you would agree that it is difficult (nor is it my intention to attempt) to list all publications in this review. I completely agree with you, as I have mentioned in my presentation, that the kinetic approach is *one* of the ways to improve our modeling effort. The work by Scudder and Olbert is one of several approaches (referenced in the text) currently under consideration. Whether it is, as you say, "the right way to go" is a matter of subjective opinion. It should be considered (in my opinion) to be an open question and not one which excludes alternative considerations (see, for example, Prof. Cuperman's review, this Proceedings).

PHYSICAL PROCESSES AND MODELS OF INTERPLANETARY
RESPONSES: SUGGESTED THEORETICAL STUDIES

S. Cuperman
Department of Physics and Astronomy
Tel Aviv University, Ramat Aviv, Israel

ABSTRACT

Three ways to improve the theory and therefore the understanding of
the physical processes in the interplanetary medium (during both quiet
and disturbed periods of solar activity) are suggested. They are: (1)
the development and consequently the use of *higher order moments (fluid)
equations* as well as of more realistic *closure conditions* and *transport
coefficients* for the macroscopic description of the solar wind ; (2)
the undertaking of *computer simulation experiments* on the nonlinear
collective relaxation process through particle-wave-particle interaction
due to the plasma electromagnetic instabilities which may develop under
conditions prevailing in the solar wind; and (3) the consideration of
collective interactions in the evaluation of the transport coefficients,
as deduced from the quasi-linear theory and computer simulation experiments, and their incorporation into the higher order moment equations.

1. INTRODUCTION

It is now recognized that the solar wind phenomenon represents
that part of the solar corona which is not confined by the solar magnetic field, and therefore escapes into interplanetary space. Essentially, the solar wind is a warm, magnetized, almost collisionless
multicomponent plasma whose space and time behaviour depends on the
activity of the sun's surface. The physical processes occurring in
such plasma systems are rather complex.

Since the pioneering work by Parker (1958), a large observational
and theoretical effort has been directed toward the understanding of
the solar wind phenomenon. The theoretical research followed two main
ways of investigation, namely the *macroscopic* (fluid) approach based
on the continuity, momentum and energy equations and the *microscopic*
(kinetic) approach based on Vlasov (correlationless Boltzmann) equation.
Significant progress has been achieved by these two methods; they are
summarized in the comprehensive review papers by Parker (1968, 1969)
Dessler (1967), Scarf (1970), Brandt (1970), Holzer and Axford (1970),

M. Dryer and E. Tandberg-Hanssen (eds.), Solar and Interplanetary Dynamics, 459-474.
Copyright © 1980 by the IAU.

Hundhausen (1972), Dryer and Cuperman (1972), Barnes (1975), Dryer (1975, 1978), Burlaga (1975), Völk (1975), Hollweg (1975, 1978), Holzer (1976), Cuperman (1977, 1979) and Suess (1978).

To further advance the understanding of the solar wind phenomenon, besides additional detailed spacecraft observations, more advanced theoretical methods are required. For example, the familiar fluid theory used so far treats the particle density, streaming velocity and temperature on a equal footing; however, it uses for the heat flux an approximate expression which is invalid over most of the interplanetary medium; also, it neglects higher order moments of the distribution function which is equivalent to taking it to be a maxwellian distribution function, in contradiction to many of the actual observations (c.f. Feldman et al., 1976).

On the other hand, the kinetic (*collisionless*) approach, which correctly accounts for the shape of the particle distribution function was developed for the case of infinite homogenous plasmas. Even in that case, it adequately treats the *linear* stage of evolution and only approximately the later (e.g., quasi-linear) stage. It is not able to predict the dynamical behaviour of an unstable plasma system through its nonlinear stage, as observed in the interplanetary medium (e.g. Abraham-Shrauner et al., 1979).

This paper discusses three ways in which the theory of the solar wind plasma can be advanced. They are: (1) the development and consequently the use of *higher order moments (fluid) equations* as well as of more realistic *closure conditions* and *transport coefficients* for the solar wind; (2) the undertaking of *computer simulation experiments* on the nonlinear collective relaxation process through particle-wave-particle interaction due to the plasma electromagnetic instabilities which may develop under conditions prevailing in the solar wind; and (3) the use of "hybrid-models" in which collective contributions to the transport coefficients as deduced from computer simulation experiments are incorporated into the higher order moments equations in order to describe the actual physical processes in the solar wind.

II. A HIGHER ORDER FLUID THEORY

Recently, a generalized fluid theory which is required for the description of time-dependent, spatially nonhomogeneous, anisotropic, multi-species spherically systems of particles obeying an inverse-square law of interactions was derived by Cuperman et al. (1979).

The generalization consists of the derivation — starting from the Boltzmann equation — of a higher order, closed system of equations for the moments of the particle velocity distribution f_a (a=e, p, α, etc.). Thus, in addition to the familiar equations for the particle density, n_a, streaming velocity, $\underline{<v>}_a$ and temperature, T_a, this *closed* set of equations also includes equations for the heat flux, q_a as well as for the fifth moment, $\zeta_a \equiv \langle(\underline{v}-\langle\underline{v}\rangle)^4\rangle$ which characterizes the particles in the tail of the distribution function.

The particle-particle interaction terms in the Boltzmann's equation were calculated in the way indicated by the Fokker-Planck relaxation theory. Thus, the velocity distribution functions in the collisional terms were represented by expansions in Legendre polynomials corresponding to the particle distribution functions being functionals of all the moments to be considered, i.e.

$$f_a = f_a(n_a, \langle v_r\rangle_a, \langle v_r^2\rangle_a, \langle v_\perp^2\rangle_a, q_{a,r}, \zeta_a, r, t) \tag{1}$$

(r, $\perp$ indicate the radial and tangential directions, respectively).

This set of closed equations is relatively simple and mathematically tractable. Thus, the first five[1] moments of the distribution function f_a are defined as follows ($v_r \equiv u$, $v_\theta \equiv v$, $v_\phi \equiv w$):

$$n_a(r,t) \equiv \int f_a(r,u,v,w,t)\, d^3v$$

$$\langle u(r,t)\rangle_a \equiv n_a^{-1} \int u\, f_a(r,u,v,w,t)\, d^3v$$

$$\alpha_a(r,t) \equiv \langle(u-\langle u\rangle_a)^2\rangle \qquad \rightarrow \quad kT_{a,r}/m_a$$

$$\beta_a(r,t) \equiv \langle v^2\rangle_a = \langle w^2\rangle_a \qquad \rightarrow \quad kT_{a,\perp}/m_a \tag{2}$$

$$\varepsilon_a(r,t) \equiv \langle(u-\langle u\rangle_a)^3\rangle \qquad \rightarrow \quad 1.2\, q_{a,r}/n_a m_a$$

$$\xi_a(r,t) \equiv \langle(u-\langle u\rangle_a)^4\rangle - 3\alpha_a^2 \qquad \rightarrow \quad 0$$

In these definitions, α_a and β_a are the mean squared random velocities in the radial and tangential directions, respectively; ε is related to the radial heat flow; ξ_a represents the excess or deficiency of high velocity particles in the tail of the distribution function relative to a Maxwellian. In the case of a Maxwellian particle distribution function, the quantities (2) take the values indicated by the arrows.

Following the procedure indicated above, Cuperman et al. (1979) obtained the following closure conditions for the "mixed" higher order moments

$$\langle(u-\langle u\rangle)\, v^2\rangle = \varepsilon/3$$

$$\langle(u-\langle u\rangle)^2 v^2\rangle = \zeta/3 - B(\alpha-\beta) \tag{3}$$

$$\langle(u-\langle u\rangle)^5\rangle = 10\, B\varepsilon$$

$$\langle(u-\langle u\rangle)^3 v^2\rangle = 2B\varepsilon$$

where

$$B \equiv (\alpha + 2\beta)/3.$$

Finally, the closed set of moment equations read:

$$\frac{\partial n_a}{\partial t} + \frac{1}{r^2}\frac{\partial}{\partial r}(r^2 n_a \langle u\rangle_a) = 0$$

$$\frac{\partial \langle u\rangle_a}{\partial t} + \langle u\rangle_a \frac{\partial \langle u\rangle_a}{\partial r} + \frac{1}{n_a}\frac{\partial}{\partial r}(n_a \alpha_a) + \frac{2}{r}(\alpha_a - \beta_a) + \frac{\partial \Phi_a}{\partial r} = 0$$

$$\frac{\partial \alpha_a}{\partial t} + \langle u\rangle_a \frac{\partial \alpha_a}{\partial r} + 2\alpha_a \frac{\partial \langle u\rangle_a}{\partial r} + \frac{1}{n_a}\frac{\partial}{\partial r}(n_a \varepsilon_a) + \frac{2}{3}\frac{\varepsilon_a}{r} = \left(\frac{\partial \alpha_a}{\partial t}\right)_c$$

$$\frac{\partial \beta_a}{\partial t} + \langle u\rangle_a \frac{\partial \beta_a}{\partial r} + 2\beta_a \frac{\langle u\rangle_a}{r} + \frac{1}{3n_a}\frac{\partial}{\partial r}(n_a \varepsilon_a) + \frac{4}{3}\frac{\varepsilon_a}{r} = \left(\frac{\partial \beta_a}{\partial r}\right)_c \qquad (4)$$

$$\frac{\partial \varepsilon_a}{\partial t} + \langle u\rangle_a \frac{\partial \varepsilon_a}{\partial r} + \frac{1}{n_a}\frac{\partial}{\partial r}(n_a \xi_a) + 3\alpha_a \frac{\partial \alpha_a}{\partial r} - \frac{4}{r}(\alpha_a - \beta_a)^2 +$$

$$+ 3\varepsilon_a \frac{\partial \langle u\rangle_a}{\partial r} = \left(\frac{\partial \varepsilon_a}{\partial t}\right)_c$$

$$\frac{\partial \xi_a}{\partial t} + \langle u\rangle_a \frac{\partial \xi_a}{\partial r} + 6\varepsilon_a \frac{\partial B_a}{\partial r} - \frac{8}{3}\varepsilon_a \frac{\partial}{\partial r}(\alpha_a - \beta_a) + 4B_a \frac{\partial \varepsilon_a}{\partial r}$$

$$+ 4\xi_a \frac{\partial \langle u\rangle_a}{\partial r} - 4(\alpha_a - \beta_a)\frac{\partial \varepsilon_a}{\partial r} - \frac{20}{3}\frac{\varepsilon_a}{n_a}(\alpha_a - \beta_a)\frac{\partial n_a}{\partial r}$$

$$- \frac{32}{3}\frac{1}{r}(\alpha_a - \beta_a)\varepsilon_a = \left(\frac{\partial \xi_a}{\partial t}\right)_c$$

In Eq. (4)

$$\frac{\partial \Phi_a}{\partial r} = - F_{a,r} = \frac{GM_o}{r^2} - \frac{Z_a E}{m_a} \quad,$$
(5)

where E is the electrostatic field[2], G - the gravitational constant, M_o - the solar mass, Z_a - the particle charge.

The r.h.s. of Eq. (4), representing the rate of change with time of the moments of f_a due to particle-particle interactions (i.e. the transport coefficients) are given by the following expressions[3]:

$$(\frac{\partial \alpha_a}{\partial t})_c = - \frac{8}{15} \frac{k(T_{a,r}-T_{a,\perp})}{m_a \tau_{aa}} \{1 + \sum_{b \neq a} \frac{n_b}{n_a} F_{b,2}\}$$

$$- \frac{2}{m_a} \sum_{b \neq a} \frac{m_{ab}}{m_a+m_b} \frac{k(T_a-T_b)}{\tau_{ab}} + \frac{m_a Q_a}{2^{1/2}kT_a \tau_{aa}} \sum_{b \neq a} \frac{n_b}{n_a} F_{b,0}$$
(6)

$$(\frac{\partial \varepsilon_a}{\partial t})_c = - \frac{174}{400} \frac{q_{a,r}}{n_a m_a \tau_{aa}} \{1 + \sum_{b \neq a} \frac{n_b}{n_a} L_{b,1}\}$$
(7)

$$(\frac{\partial \xi_a}{\partial t})_c = \frac{6}{7} \frac{k^2 T_{a,r}}{m_a^2 \tau_{aa}} (T_{a,r}-T_{a,\perp})\{1+\Delta\xi_{a,2}\} - \frac{6}{15} \frac{Q_a}{\tau_{aa}} \{1+\Delta\xi_{a,0}\}$$

$$+ 12 \sum_{b \neq a} \frac{m_{ab}}{m_a+m_b} \frac{k^2(T_b-T_a)}{m_a^2 \tau_{ab}} (\frac{0.4T_a}{1+\Gamma_b} - T_{a,r})$$
(8)

The expression for $(\partial \beta_a/\partial t)_c$ differs from that for $(\partial \alpha_a/\partial t)_c$ only by the fact that $(-8/15)$ is replaced by $(+4/15)$ in the term proportional to $(T_{a,r}-T_{a,\perp})$. Consequently, the following equation giving the rate of change of the thermal anisotropy $(\alpha_a-\beta_a)$ due to particle-particle interaction is easily obtained:

$$\{\frac{\partial}{\partial t}(\alpha_a- \beta_a)\}_c = -\frac{4}{15} \frac{n_a k}{\tau_{aa}}(T_{a,r}-T_{a,\perp}) \{1+ \sum_{b \neq a} \frac{n_b}{n_a} F_{b,2}\}$$
(9)

In the Eqs. (6)-(9), the "collision" times are defined as ($m_{ab} = m_a m_b/(m_a + m_b)$)

$$\tau_{ab} = \frac{3m_{ab}^2 [(kT_a/m_a)+(kT_b/m_b)]^{3/2}}{4(2\pi)^{1/2} Z_a^2 Z_b^2 n_b \ln\Lambda}$$
(10)

and

$$\tau_{aa} = 3m_a^{1/2}(kT_a)^{3/2}/8\pi^{1/2}Z_a^4 n_a \ln\Lambda . \tag{11}$$

The quantities $F_{b,2}$, $F_{b,o}$, $L_{b,1}$, $\Delta\xi_{a,2}$ and $\Delta\xi_{a,o}$ and Q_i represent simple algebraic expressions and are given in Cuperman et al. (1979).

At this point two remarks are in order:

(i)   In Eqs.(4) there exist three types of deviations from an isotropic Maxwellian which are able to drive the system to a relaxed state namely, the anisotropy factor ($T_{a,r}-T_{a,\perp}$), the heat flow $q_{a,r}$ and the non-thermal tail, ξ_a. To these, one has to add the differences in the temperatures of the various species which act in the same direction (i.e. T_b-T_a).

(ii)   The evolution of a physical system described by such equations exhibits a global non-local behaviour.

For *isotropic plasma components* ($T_{i,r} = T_{i,\perp}$, i = a,b), with the notation $T\equiv(T_r + 2T_\perp)/3=T_r=T_\perp$, the equations read[4]:

$$n_a m_a \frac{\partial\langle u\rangle_a}{\partial t} + n_a m_a \langle u\rangle_a \frac{\partial\langle u\rangle_a}{\partial r} +k\frac{\partial}{\partial r}(n_a T_a)+ \frac{GM_o m_a n_a}{r^2} - n_a Z_a E = 0 \tag{12}$$

$$\frac{3}{2}n_a k \frac{\partial T_a}{\partial t} + \frac{3}{2}n_a \langle u\rangle_a k \frac{\partial T_a}{\partial r} + n_a k T_a \frac{\partial\langle u\rangle_a}{\partial r} + 2n_a k T_a \frac{\langle u\rangle_a}{r}$$

$$+ \frac{1}{r^2}\frac{\partial}{\partial r}(r^2 q_{a,r})= - 3n_a \sum_{b\neq a} \frac{m_{ab}}{m_a+m_b} \frac{k}{\tau_{ab}}(T_a - T_b)$$

$$+ \frac{3}{2^{2/3}} \frac{n_a m_a^2 Q_a}{kT_a \tau_{aa}} \sum_{b\neq a} \frac{n_b}{n_a} F_{b,o} \tag{13}$$

$$\frac{\partial}{\partial t} q_{a,r}+ \frac{\langle u\rangle_a}{r^2}\frac{\partial}{\partial r}(r^2 q_{a,r})+ 4q_{a,r}\frac{\partial\langle u\rangle_a}{\partial r} + \frac{5}{6}m_a\frac{\partial}{\partial r}(n_a\xi_a)$$

$$+ \frac{5}{2}\frac{n_a k^2}{m_a}T_a\frac{\partial T_a}{\partial r} = - \frac{29}{80}\frac{q_{a,r}}{\tau_{aa}}\{1 + \sum_{b\neq a}\frac{n_b}{n_a}L_{b,1}\} \tag{14}$$

$$\frac{\partial\xi_a}{\partial t} +\langle u\rangle_a\frac{\partial\xi_a}{\partial r} + 4\xi_a\frac{\partial\langle u\rangle_a}{\partial r} + \frac{36}{5}\frac{q_{a,r}k}{n_a m_a^2}\frac{\partial T_a}{\partial r}+ \frac{24}{5}\frac{kT_a}{m_a^2}\frac{\partial}{\partial r}(\frac{q_{a,r}}{n_a})$$

$$= - \frac{2}{5} \cdot \frac{Q_a}{\tau_{aa}} \{1+\Delta\xi_{a,o}\} - 12 \frac{T_a}{m_a^2} \sum_{b\neq a} \frac{m_{ab}k^2}{m_a+m_b} \frac{T_b-T_a}{\tau_{ab}} \frac{0.6+\Gamma_b}{1+\Gamma_b} \cdot \quad (15)$$

Finally, to recover the standard three-moment equations case, one has to use the following simplifications:

(i) $\xi_a \equiv n_a^{-1} \int (u-<u>_a)^4 f_a d^3v - 3(kT_{a,r}/m_a)^2 = 0,$

that is neglect the effect of the non-maxwellian tail of the particle velocity distribution. Then, no equation for ξ_a is required;

(ii) in the Eq.(14), *neglect in the l.h.s. all the terms but the fifth*. One obtains

$$q_{a,r} = - \frac{6.9}{1+\sum_{b\neq a} (n_b/n_a)L_{b,1}} \frac{n_a}{m_a} \tau_{aa} k^2 T_{a,r} \frac{\partial T_{a,r}}{\partial r}$$

which is the familiar approximate relation used for the radial heat flux. The simplifications implied by this expression are obvious.

III. COMPUTER SIMULATION EXPERIMENTS OF COLLECTIVE INTERACTIONS IN MAGNETIZED PLASMAS

Computer simulations are able to provide information about the time-evolution and the final state of an initially unstable plasma system. They consist of the simultaneous solution of the Lorentz equation for each one of the many thousands of the plasma particles acted upon by the external and collective electromagnetic fields (see, e.g.: Haber et al., 1970; Morse and Nielson, 1971; Cuperman and Salu, 1972). In the following two examples (relevant for the interplanetary medium) are considered, namely one in which electromagnetic interactions are dominant and another in which electrostatic interactions are dominant.

1. Ion-cyclotron (Alfvén) electromagnetic instability.

For bi-Maxwellian particle distribution functions, the *linearized* dispersion relation for *electromagnetic* ion-cyclotron waves (left-hand polarization) propagating along a static and homogeneous magnetic field $\underline{B}_o$ may be written as (e.g. Cuperman and Landau, 1974):

$$n^2 \equiv c^2 k^2/\omega^2 = 1 + \sum_j \frac{\omega_{pj}^2}{\omega k} \int \frac{f_{\parallel,j}[1+(kv_\parallel/\omega)A_j]}{v_\parallel - s_j} dv_\parallel, \quad (17)$$

where the summation is overall + and −, warm and cold plasma components. The notation used is as follows: complex frequency $\omega_r + i\omega_i$; real wave-number of disturbance k; $v_\parallel$, $v_\perp$ are particle velocity components parallel and perpendicular to $\underline{B}_o$, respectively; plasma frequency ω_p; light velocity c; $f_{\|,j}$ is the particle distribution function defined by

$$f_o \equiv f_\perp f_\| = \frac{1}{2\pi v_{\perp,th}^2} \exp\left(\frac{-v_\perp^2}{2v_{\perp,th}^2}\right) \cdot \frac{1}{(2\pi)^{1/2} v_{\|,th}} \exp\left(\frac{-v_\|^2}{2v_{\|,th}^2}\right) \quad (18)$$

where $v_{\perp,th} \equiv (KT_\perp/m)^{1/2}$ and $v_{\|,th} \equiv (KT_\|/m)^{1/2}$ represent the

thermal velocity in the directions perpendicular and parallel to $\underline{B}_o$, respectively. Other notations are $A_j \equiv (T_\perp/T_\|)_j - 1$ and $s_j = (\omega - \Omega_j)/k$, where Ω_j is the cyclotron frequency defined by $\Omega_j = q_j B_o/m_j c$ and $q_i = Z_j e^j$ is the particle charge. Expressing the integral part in (17) in terms of the plasma dispersion function $Z(\xi_j)$ one may rewrite (17) as

$$c^2 k^2 = \omega^2 + \sum_j \omega_{pj}^2 \{A_j - \frac{1}{\sqrt{2} v_{\|,th,j} k} Z(\xi_j)[(A_j+1)(\Omega_j-\omega)-\Omega_j]\} \quad . \quad (19)$$

For the case of a warm anisotropic plasma consisting of warm ions and cold electrons, assuming (i) $\omega_i \ll \omega_r$, (ii) $v_r \equiv [(\Omega_p-\omega)/k] \gg v_{\|,th,p}$,

(iii) $k^2 c^2/\omega^2 \gg 1$, and (iv) k is given by the 'cold' plasma dispersion relation (Kennel and Petschek, 1966), from (19) one obtains the approximate analytical expressions ($\beta_\| \equiv 8\pi n_p kT_{\|,p}/B_o^2$), v_A- Alfvén speed)

$$\gamma \equiv \omega_i/\Omega_p = \frac{1}{\alpha^{3/2}} \frac{1(1-x)^5}{x^2(2-x)} \frac{^2[(A+1)(1-x)-1]}{(\beta/\pi)^{1/2}} \exp\left\{-\frac{(1-x)^3}{x^2 \beta_{\|,p} \alpha}\right\} \quad (20)$$

$$(kv_A/\Omega_p)^2 = x^2(1-x)^{-1}, \quad x \equiv \omega_r/\Omega_p \quad . \quad (21)$$

The maximum growth rate is (Cuperman et al., 1975)

$$\gamma_m = \left(\frac{\omega_i}{\Omega_p}\right)_m = \left(\frac{\pi}{\beta}\right)^{1/2} \frac{A_p(1-x_1)-x_1}{x_1^2(2-x_1)}(1-x_1)^{5/2} \exp\left[-\frac{(1-x_1)^3}{\beta x_1^2}\right\} \quad , \quad (22)$$

where x_1 and Q_m are simple algebraic expression given in Cuperman et al. (1975). Notice that as a consequence of the assumptions (i)-(iv), the results (20)-(22) only hold for plasma values such that $P_p \equiv \beta_p A_p^2 (A_p+1) \ll 1$.

As it is seen, even though significant information is provided by the approximate theory based on the linearized Eq. (19), its use is rather limited. Thus, it is only for small P_p (i.e., small $T_\perp/T_\|$ and small β) values that the results hold. Moreover, the linear predictions hold only for the initial stage of the instability, which represents a relatively small time period of evolution. Although additional important information can be achieved from a quasi-linear treatment, the only way to relieve the mathematical restrictions and physical approximations required by analytical treatments is by performing computer simulation experiments. This important point has now been established by detailed examination of individually-measured proton velocity distribution functions in space (Abraham-Shrauner et al., 1979) as discussed below.

Figures (1) and (2) give results obtained in the computer simulation of the electromagnetic ion cyclotron instability in homogeneous plasma systems with parameters $kT_{p,\parallel}$ = 25 keV, β=1 and $T_\perp/T_\parallel$ = 100 (Cuperman and Sternlieb, 1975; see also Cuperman et al., 1976). As seen, a strong electromagnetic ion cyclotron instability develops, as expected from theoretical (linear) considerations. The initial low electromagnetic noise ($W_B(0)/W_{tot} \sim 0.0004$) develops into a significant electromagnetic wave energy level which, at its maximum represents about 10% of the total energy in the system; this represents an increase by a factor of 230!

Inspection of $\Delta\overline{W}_{\perp,p}$ and $\overline{W}_B$ curves shows that $\Delta\overline{W}_{\perp,p}$ is always larger than $\overline{W}_B$ by at least a factor of three. This indicates that significant non-linear processes occur starting from the very beginning of the instability. Thus, the electromagnetic field developed during the instability heats (non-resonantly) the protons (mainly) in the plasma system and, consequently, appears to possess less energy than expected from pure 'linear' considerations (which do not take into account such processes). This suggests the true (nonlinear) measure of the instability developed to be the change in the total proton transverse kinetic energy, $W_{\perp,p}(0)$ (which is the actual energy source available in the the system) rather than the enhancement of the electromagnetic ion cyclotron waves.

A natural result of the instability should be the tendency of the system towards thermal equilibrium, $(T_\perp/T_\parallel)_p \rightarrow 1$. This is indeed observed to occur in the experiment as indicated in Figure 2. Thus, after a time about $800\omega_{pp}^{-1}$, the transverse kinetic energy $W_{\perp,p}(0)$ has decreased by about 25% and the parallel kinetic energy $W_{\parallel,p}(0)$ has increased by a factor of about 26! Consequently, the kinetic anisotropy $0.5W_{\perp,p}/W_{\parallel,p}$ which at t = 0 represents the thermal anisotropy, $(T_\perp/T_\parallel)_p$ has decreased from '100' to '2.9'. The fact that the last value is different from '1' (as expected on the grounds of linear theory) may be due to the fact that the quasi-final (from a collective point of view) state in the experiment is one in which a significant amount of electromagnetic wave energy is present. This suggests the establishment of a non-linear quasi-equilibrium state for which the instability criterion may well be different from that predicted by the linear theory.

This result is of special interest for the solar wind. Indeed, recently, Abraham-Shrauner et al. (1979) have analyzed electromagnetic

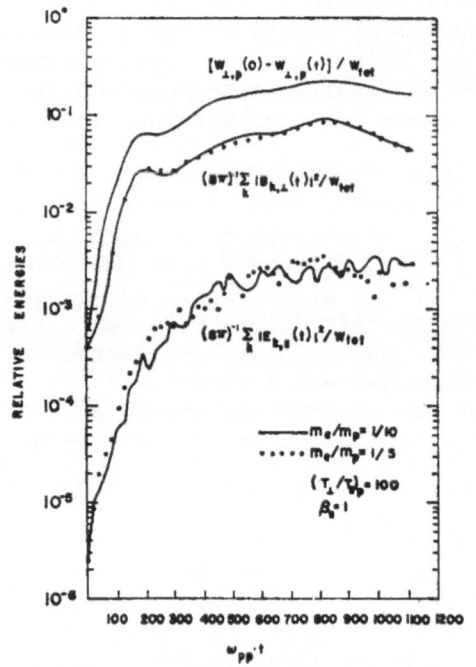
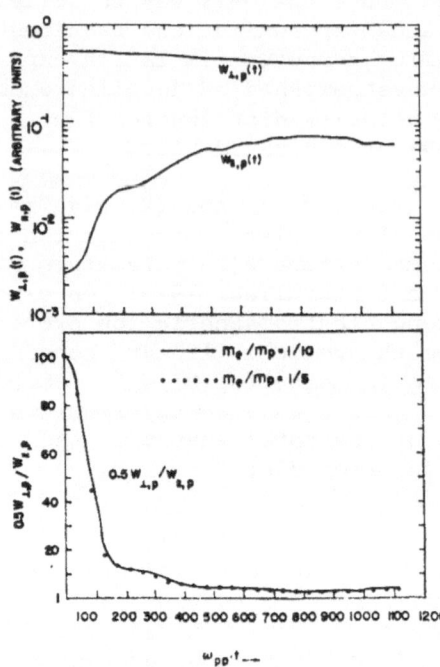

Fig. 1   The evolution in time of $\Delta\overline{W}_{\perp,p}$, the relative change in the total transverse kinetic energy of the warm anisotropic protons, $\overline{W}_B \equiv (8\pi)^{-1}$. $\Sigma_k |B_{k,\perp}(t)|^2/W_{tot}$ the relative transverse magnetic wave energy and $\overline{W}_E \equiv (8\pi)^{-1}\Sigma_k |E_{k,\parallel}(t)|^2/W_{tot}$ the relative longitudinal electrostatic energy.

Fig. 2   The evolution in time of the total proton kinetic energies $W_{\perp,p}$ and $W_{\parallel,p}$ (top) and of the anisotropy ratio $(0.5W_\perp/W_\parallel)_p$ (bottom).   At $t=0$, $(0.5W_\perp/W_\parallel)_p \equiv (T_\perp/T_\parallel)_p = 100$.

instabilities of field-aligned right-hand circularly polarized magneto-sonic waves and left-hand circularly polarized Alfvén waves driven by two drifted proton components for model parameters determined from Imp 7 solar wind data measured during high-speed flow conditions.   The authors found that (i) measured distributions are linearly *unstable with respect to Alfvén waves;* (ii) the characteristic of the proton velocity distributions in the high speed solar wind that is primarily responsible for driving the Alfvén instability is the large thermal anisotropy of the main proton component observed by Feldman et al. (1976) in high speed streams $(T_\perp/T_\parallel = 3.1 + 0.7)$; and (iii) the instability of the Alfvén wave is *inconsistent* with the persistence of the model fits to the measured proton velocity distribution function, since the calculated e-folding times are very short – of the order of 1 min.

To explain the reason why linear stability theory appears to fail for the Alfvén instability the authors adopted the resolution suggested by Cuperman and Sternlieb (1975) to explain the saturation value, ~ 3

in the simulation experiment described above. Namely, the instability criterion of the linear theory ($T_\perp/T_\parallel > 1$) may not apply to the composite wave–particle plasma state. Instead, stabilization may result from the establishment of a nonlinear quasi-equilibrium state which necessarily includes the wave field.

2. Beam–plasma instability.

The *electrostatic* dispersion relation for homogeneous unbounded plasmas in a constant magnetic field $\underline{B} = \underline{B}_{oz}$ and for disturbances with wave vectors $\underline{k}$ along $\underline{B}_o$, may be written in the following form (Montgomery and Tidman, 1964)

$$R_{zz} = -\omega^2 + 2\pi\omega \sum_j \omega_{pj}^2 \int_{-\infty}^{+\infty} v_\parallel dv_\parallel \int_0^\infty v_\perp dv_\perp \frac{\partial f_{oj}/\partial v_\parallel}{kv_\parallel - \omega} = 0, \qquad (24)$$

where the summation is over the plasma components. Here ω is the complex frequency of the disturbance, f_{oj} is the equilibrium distribution function and ω_{pj} is the plasma frequency of the j-plasma component. The notations $\parallel$, $\perp$ refer to parallel and transverse direction with respect to $\underline{B}_o$. The longitudinal electrostatic waves propagating parallel to $\underline{B}_{oz}$ are unaffected by the magnetic field.

In the following we will not elaborate on the linear (or quasi-linear) theories, as done for the case discussed in Section III.1. Rather, we will discuss the non-linear simulation results. Computer simulation experiments of the interaction between electron beams and background plasmas have been carried out by a number of authors (see, e.g. Cuperman et al., 1976). Both electrostatic and electromagnetic interactions were simultaneously considered. The relative beam concentrations considered were $\varepsilon \equiv n_b/n_p = 1, 0.1$ and 0.01 (cases A, B and C, respectively). In all cases, the background plasma (1 eV thermal energy) was penetrated by an electron beam of 1 keV streaming energy and 5% thermal spread in the streaming direction. The following results emerged (see also Fig. 3):

(1) All the systems were strongly unstable against the electrostatic beam–plasma instability; the electromagnetic interaction was negligible. In the linear stage, the measured growth rates were in satisfactory agreement with the linear predictions for cold beam–plasma interacting systems.

(2) The maximum relative electrostatic energy developed, $\overline{W}_{e.s.}^{max} \equiv W_{e.s.}^{max}/W_{tot}$ was 5.7, 8.8 and 6.3% for cases A, B and C, respectively. Thus, although ε varied by a factor of 100, $\overline{W}_{e.s.}^{max}$ was almost unchanged. This indicates an almost linear relationship between $W_{e.s.}^{max}$ and W_{tot}: $W_{e.s.}^{max} \simeq 0.7\, W_{tot}$ for $10^{-2} < \varepsilon < 1$. These $\overline{W}_{e.s.}^{max}$ values were reached after a time $t_s\omega_p \simeq 18$, 30 and 75, respectively.

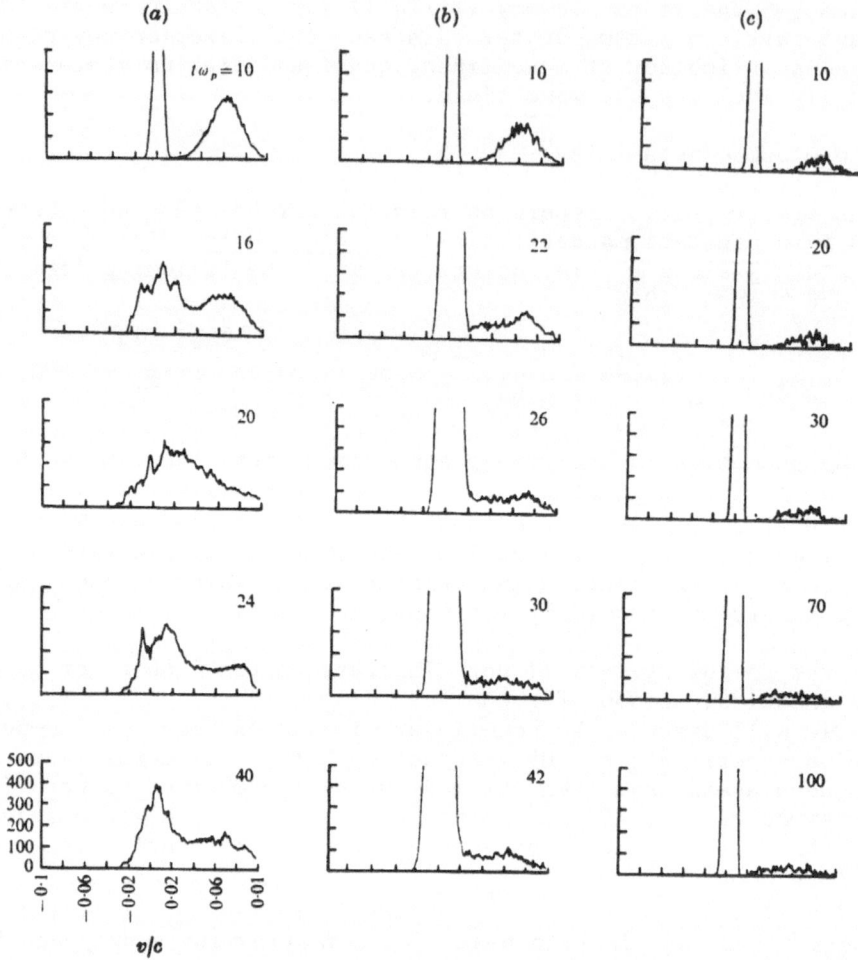

Fig. 3   The velocity distribution functions of the beam-plasma
          systems at several times of evolution.

(3) At the same time t_s, the streaming energy of the beam represented the following fraction of its initial value: 49.5%(A), 67.7%(B)
and 76.7%(C). The parallel thermal energy in the beam increased by a
factor of 4.5, 3 and 1.8, respectively. The largest change occurred
in the parallel thermal energy of the background plasma which increased
by a factor of 279, 136 and 2.1, respectively.

(4) At the end of the run, t_{end} in all three cases the kinetic
states of the beam and plasma electrons were almost the same as at the
saturation time, in spite of periodic changes which occurred between
t_s and t_{end}. (In case A, for example, at the time $t\omega_p \simeq 25$ the

streaming energy in the beam decreased below the plasma thermal energy: $W_{b,s}$ = 330eV and W_p = 430 eV.)

(5) As a result of nonlinear wave-particle and wave-wave inter-actions, electromagnetic waves were generated in the unstable beam-plasma configuration. However, the relative importance of electromagnetic to electrostatic activity was small, $W_{e.m.}/W_{e.w.} \sim 10^{-5}$.

(6) At the end of the run, particle distribution functions which were "nonlinearly" stable were obtained.

These results are relevant for the solar wind. Lemons et al., 1979, performed a linear analysis of plasma configurations measured with the Los Alamos instruments on Imp 7 and 8 during or close to times when *electrostatic* fluctuations have been observed. The authors con-cluded that the *ion-beam instability* is more likely the cause of the observed e.s. fluctuations. Since the time required by the Imp particle analyzers to make a complete measurement of the particle distribution is much larger than the periods of enhanced electrostatic fluctuations, one expects the observed distribution functions to be stable.

The suggestions by Lemons et al. (1979) are consistent with the results of the computer experiments of Cuperman et al., 1976, described above. The experiments provide complete quantitative information on the evolution of the beam-plasma systems through their nonlinear and final stages, which are the ones mostly observed by the instruments; this is so because of the short duration of the linear stages. Indeed, the particle distribution functions in the last row of Figure 3 (cor-responding to the end of the computer simulation experiments) are re-presentative for the "stable" distribution functions discussed by Lemons et al. (1979) in their work.

IV. HYBRID-MODELS

Both approaches indicated above are very useful for theoretical solar wind studies. However, each one has its limitations. Indeed, the fluid approach does not include the collective interactions which could play a significant role. The computer simulations, on the other hand assume specially uniform plasma systems[5]. A hybrid method exploit-ing the capabilities of both approaches is obviously desired, and such a method is being used for fusion plasmas in which electromagnetic tur-bulence plays an important role (e.g., Krall and Liewer, 1973)[6]. The model consists of a system of continuity, momentum and energy equations to be integrated numerically and including the effects of turbulence, selfconsistently considered through anomalous transport coefficients. The last ones depend on the unstable modes which can develop in time and space as the macroscopic parameters evolve; they are based on the quasilinear theory in conjunction with nonlinear bounds as obtained in computer simulation experiments. This method has been successfully used for the investigation of theta pinches, for example.

In the above models only, the case of collisionless plasmas was treated. That is, the transport coefficients were purely anomalous, as no particle-particle interactions were considered.

For the solar wind case in which collisions can play a non-negligible role over a significant part of the interplanetary range, a generalization of the hybrid-model is indicated. Thus, the higher order moment equations described in Section II (in which particle-particle interactions are considered) should be complemented by anomalous contributions to the collisional transport coefficients. The last ones should be obtained from quasilinear theories bounded by the results of computer simulation experiments as described in Section III.

V. NOTES

1. Actually the first six moments, if the radial and tangential random kinetic energies are considered to be different moments.

2. This electric field is responsible for the maintenance of both the equality in the particle fluxes (electrons and positive ions) in the solar wind, and the charge neutrality of the sun.

3. Here, as in the previous equations, kT is used to denote the mean random kinetic energy, rather than "thermal" energy (which is only defined in the Maxwellian case).

4. For comparison, for the case a=e, b=p, from (9) one recovers the following particular expression used by Braginskii (1965)

$$\tau_{ep} = 3m_e^{1/2}(KT_e)^{3/2}/4(2\pi)^{1/2}e^4 n_p \, \ell n\Lambda \, .$$

5. Particle-particle interactions can be also considered simultaneously in computer simulation experiments.

6. Sometimes, hybrid models consist of "fluid" electrons and "particle" ions (e.g., Hamasaki et al., 1977).

ACKNOWLEDGEMENT. The author would like to thank M. Dryer for helpful discussions and reading of the manuscript.

REFERENCES

Abraham-Schrauner, B., Asbridge, J. R., Bame, S. J., and Feldman, W. C.: 1979, J. Geophys. Res., 84, pp. 553-559.
Barnes, A.: 1975, Rev. Geophys. Space Phys., 13, pp. 1049-1053.
Braginskii, S. I.: 1965, Reviews of Plasma Physics (New York: Consultant Bureau).

Brandt, J. C.: 1970, Introduction to the Solar Wind, Freeman, San Francisco, Calif.

Burlaga, L. F.: 1975, Space Sci. Rev., 17, pp. 327-351.

Cuperman, S.: 1977, in M. A. Shea et al. (eds.), Study of Travelling Interplanetary Medium, D. Reidel Publishing Co., pp. 165-194.

Cuperman, S.: 1979, in H. Rosenbauer (ed.): Solar Wind 4 Conference Springer-Verlag, (in press).

Cuperman, S., Gomberoff, L. and Sternleib, A.: 1975, J. Plasma Phys., 13, pp. 259-272.

Cuperman, S., and Landau, R., 1974: J. Geophys. Res., 79, pp. 128-134.

Cuperman, S., and Salu, Y.: 1972, Proc. 5th European Conf. on Controlled Fusion and Plasma Phys., pp. 140-144.

Cuperman, S., and Sternlieb, A.: 1975, Plasma Physics, pp. 699-705.

Cuperman, S., Sternlieb, A., and Williams, D. J.: 1977, J. Plasma Physics, 16, pp. 57-72.

Cuperman, S., Roth, I., and Bernstein, W.: 1976, J. Plasma Physics, 15, 309-324.

Cuperman, S., Weiss, I., and Dryer, M.: 1979, Communicated.

Dessler, A. J.: 1967, Rev. Geophys., 5, pp. 1-41.

Dryer, M.: 1975, Space Sci. Rev., 17, pp. 277-325.

Dryer, M.: 1979, in Rosenbauer (ed.) Solar Wind 4 Conference, Springer-Verlag, (in press).

Dryer, M. and Cuperman, S.: 1972, in P. S. McIntosh and M. Dryer (eds.), Solar Activity Observations and Predictions, pp. 197-229.

Feldman, W. C., Abraham-Shrauner, B., Asbridge, J. R., and Bame, S.J.: 1976, in D. J. Williams (ed.), Physics of Solar Planetary Environments, Amer. Geophys. Union, Washington, D.C., pp. 413-427.

Feldman, W. C.: 1977, in C. F. Kennel, L. J. Lanzerotti and E. N. Parker (eds.), Solar System Plasma Physics, 20th Anniversary Review, North Holland, pp. 767-806.

Haber, I., Wagner, C. E., Boris, J. P., and Dawson, J. M.: 1970, Proc. Fourth Conf. Numer. Simulation Plasmas, p. 126.

Hamasaki, S., Krall, N. A., Wagner, C. E., and Byrne, R. N.: 1977, Phys. Fluids, 20, pp. 65-71.

Hollweg, J. V.: 1975, Rev. Geophys. Space Phys., 13, pp. 263-289.

Holzer, T. E.: 1976, in D. J. Williams (ed.), Physics of Solar Planetary Environments, American Geophys. Union, Washington, D. C., pp. 366-412.

Holzer, T. E., and Axford, W. I.: 1970, Ann. Rev. Astron. Astrophys., 8, pp. 30-60.

Hundhausen, A. J.: 1972, Coronal Expansion and Solar Wind, Springer-Verlag, New York.

Kennel, C. F., and Petschek, H. E.: 1966, J. Geophys. Res., 71, pp. 1-28.

Lemons, D. S., Asbridge, J. R., Bame, S. J., Feldman, W. C., Gary, S. P., and Gosling, J. T.: 1979, J. Geophys. Res., 84, pp. 2135-2138.

Liewer, P. C., and Krall, N. A.: 1973, Phys. Fluids, 16, pp. 1953-1963.

Morse, R. L., and Nielson, C. W.: 1971, Phys. Fluids, 14, pp. 830-840.

Parker, E. N.: 1958, Astrophys. J., 128, pp. 664-685.

Parker, E. N.:   1963, Interplanetary Dynamical Processes, Interscience,
    New York.
Parker, E. N.:   1969, Space Sci. Rev., 9, pp. 325-360.
Scarf, F. L.:   1970, Space Sci. Rev., 11, pp. 234-270.
Suess, S. T.:   1979, Space Sci. Rev., 23, pp. 159-200.
Völk, H. J.:   1975, Space Sci. Rev., 17, pp. 255-276.

THE CROSS SECTIONAL MAGNETIC PROFILE OF A CORONAL TRANSIENT

M.K. Bird and H. Volland
Radioastronomisches Institut, Universitaet Bonn, Bonn F.R.G.

B.L. Seidel and C.T. Stelzried
Jet Propulsion Laboratory, Pasadena, California 91103, U.S.A.

The role of the magnetic field in a coronal mass ejection event has not been unequivocally defined, and may in fact be quite variable in view of the large variety of shapes and sizes of coronal transients. Measurements of the magnetic field associated with these events have thus far been inferred from simultaneously observed radio bursts, which provide no information on the direction of the field and are limited in spatial resolution. Substantial improvement in these two areas could be achieved by continuous monitoring of the Faraday rotation of a linearly polarized spacecraft signal during solar occultation. A coronal event traversing the line-of-sight would yield a characteristic profile in cross section, which would be of value for discriminating between the various models of coronal transients.

I. Coronal Transients in White Light and Faraday Rotation

The coronal mass ejection event has been a widely studied solar phenomenon since the Skylab missions, and many important characteristics such as their plasma composition (Hildner et al., 1975), their speeds (Gosling et al., 1976), their frequency of occurrence (Hildner et al., 1976), and their associations with other solar activity (Munro et al., 1979) have been documented. Supplementary measurements of solar radio activity were made during the events of 14-15 September 1973 (Dulk et al., 1976) and 21 August 1973 (Cergely et al., 1979). If the broadband emission recorded could be interpreted as gyrosynchrotron radiation, then the magnetic field strengths associated with the enhanced density loops were of the order of a few gauss at a distance $r \simeq 2-3\ R_S (R_S = $ solar radius).

Dynamic coronal events were also observed during the 1968 solar occultation of Pioneer 6 (Levy et al., 1969) and even during solar minimum at subsequent occultations of Helios 1 and 2 (Bird et al., 1977). The Faraday rotation of the linearly polarized telemetry signal of these spacecraft was seen to abruptly deviate by tens of degrees before returning to its presumed baseline after about 2 hours. These deviations have been attributed to density enhancements (Pintér, 1973), or to moving mag-

M. Dryer and E. Tandberg-Hanssen (eds.), Solar and Interplanetary Dynamics, 475-481.

netic bottles (Schatten, 1970). The separation of magnetic field effects
from electron density effects is impossible without additional information such as dispersion measurements or white light observations. The
Faraday rotation due to the disturbance can be written

$$\Omega_t = K \int_t [N_t \vec{B_t} - N_o \vec{B_o}] \cdot d\vec{s} \qquad deg \qquad (1)$$

where $K = 2.58 \times 10^{-13}$ in gaussian units
          N = electron density in cm^{-3}
          $\vec{B}$ = magnetic field in gauss
          $d\vec{s}$ = path element in cm

The subscripts "t" and "o" refer to the transient and ambient values
respectively. The second term of (1) can be safely neglected for transient magnetic fields and electron densities well above the ambient (Dulk
et al., 1976). Furthermore, it will be shown below that for only a
small component of B along the Earth spacecraft line-of-sight one would
expect very large contributions to Ω_t from a typical coronal mass ejection
event. Combined with white light data, the magnitude and signature of
the time profile of the Faraday rotation could then be utilized to
determine the strength and orientation of the magnetic field in the
density enhancement.

Models of transient propagation have been designed to explain the
most conspicuous form of coronal event, the loop transient. These models
can be loosely subdivided according to the proposed driving force required to propel the observed density enhancement outwards through the
corona. The numerical simulations described by Nakagawa et al. (1978)
and Wu et al. (1978) in the solar equatorial plane or by Steinolfson et
al. (1978) and Dryer et al. (1979) in a solar meridional plane do not
require an enhanced magnetic field associated with the disturbance. A
pressure pulse applied at the coronal base propagates into the outer
corona provided the pulse (flare) occurs under an open magnetic field
topology. The perturbation to the ambient field is only slight, particularly if the plasma β is low as expected in the regions of interest.

A different philosophy is espoused by Mouschovias and Poland (1978)
and by Anzer (1978), who consider the coronal loop to be driven by the
magnetic Lorentz force arising essentially from field gradients in the
helical magnetic field threading the enhanced density loop. The higher
magnetic and thermal pressures in the loop cause it to expand as observed
as it propagates radially outward. The model of Mouschovias and Poland
(1978) will be employed here as an example of the use of the Faraday
rotation technique for determination of the magnetic composition of a
coronal transient.

II.   Faraday Rotation from a Transient Flux Tube

Consider a section of a coronal transient flux tube, be it a loop
or a streamer of arbitrary curvature, which is probed by an Earth/space-

Figure 1. Geometry of the occultation of the Earth/spacecraft line-
of-sight by a transient flux tube
in the solar corona. The axis of
the tube is defined by the angles
(θ_0, ϕ_0). The helical magnetic field
of the flux tube can be broken
down into longitudinal and azimuthal components.

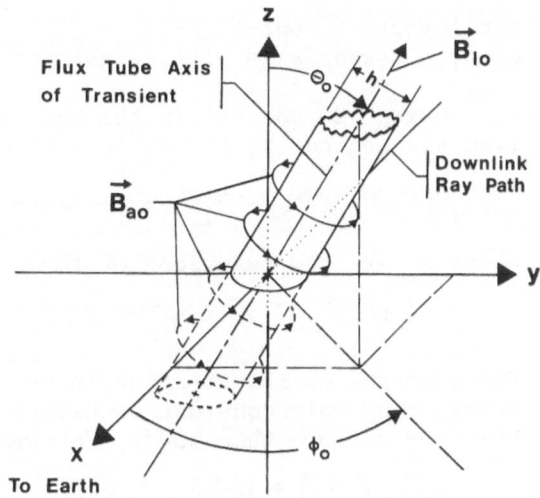

craft ray path in the z = 0 (ecliptic) plane. Figure 1 shows the orientation of the flux tube (thickness h), which is defined by the polar and
azimuthal angles (θ_0, ϕ_0). For west limb observations the Sun is located
at some point close to, but not necessarily right on the negative y-axis.
The angles (θ_0, ϕ_0) will be assumed uniquely determined in the following
from white light (with polarization) observations, although there is
likely to be some uncertainty in their exact values.

The flux tube is threaded by a constant longitudinal magnetic field
B_{lo} and a purely azimuthal field of constant magnitude B_{ao}. The x-comp-
onent of these two fields in the z = 0 plane, which is the only component
giving contributions to (1), is given by

$$B_x = B_{lo} \sin \theta_0 \cos \phi_0 - B_{ao} \frac{y}{r} \cos \theta_0 \qquad (2)$$

If the tube is slightly curved, the azimuthal field on the concave (CC)
side will exceed that on the convex (CV) side by an amount determined
from the following equality (e.g. Mouschovias and Poland, 1978)

$$B_a^2 (CC) - B_a^2 (CV) = \left[B_a^2 (CC) + B_a^2 (CV) \right] \cdot \frac{h}{R_c} \qquad (3)$$

where $B_a (CC)$ $=$ $B_{ao} + b$
          $B_a (CV)$ $=$ $B_{ao} - b$
          R_c $=$ Mean radius of curvature of tube

and   b is a small field contribution directed along $\vec{R_c} \times \vec{B_{lo}}$

Solving (3) for b, one obtains

$$b = \frac{h}{2R_c} B_{ao} \qquad (4)$$

which for typical transients at 5 R_S (h ≈ 0.6 R_S; $R_c \approx 1.6$ R_S), is a small

but important contribution to (1) if the field is directed primarily
along the x-axis as will be assumed here.

Using (2) and (4) in the expression for the transient Faraday rotation (1) neglecting the ambient term, one obtains

$$\Omega_t(y) = K \int_{x_-}^{x_+} N_t(x,y) \left[B_{1o} \sin\theta_o \cos\phi_o - \frac{B_{oa}}{r} y \cos\theta_o + \frac{h}{2R_c} B_{oa} \right] dx \qquad (5)$$

where $x_\pm$ are the solutions of the quadratic equation

$$(h/2)^2 = x^2 + y^2 - \sin^2\theta_o \left[x \cos\phi_o + y \sin\phi_o \right]^2 \qquad (6)$$

For purposes of illustration it is possible to simplify the analysis by
considering only constant density fluxtubes with small θ_o (i.e. θ_o less
than 30° or more than 150°), for which (5) reduces to

$$\Omega_t(Y) = K N_{to} h B_{lo} \left[T_1 + a T_2 + a T_3 \right] \qquad (7)$$

with N_{to} =   constant density in flux tube
          a = B_{ao} / B_{lo}

and T_1 = $\theta_o \cos\phi_o \sqrt{1 - Y^2}$

          T_2 = $-\frac{Y}{2} \ln\left[\frac{1 + \sqrt{1 - Y^2}}{1 - \sqrt{1 - Y^2}} \right]$

          T_3 = $\frac{h}{2R_c} \sqrt{1 - Y^2}$

          Y = $y / (h/2)$

Y, the normalized position of the ray path in the flux tube assumes the
value -1 (+1) on the CC (CV) side of the curved disturbance.

Taking reasonable values for N_{to} (1.2×10^6 cm^{-3}), h (0.6 R_S) and
B_{lo} (0.2 gauss) at a distance 5 R_S from the Sun, one may determine the
signature of $\Omega_t(Y)$ of a coronal transient flux tube as it is sampled
between $\pm$ h/2 by the Earth/spacecraft ray path.

The expected variations are shown in the three panels of Fig. 2,
which give examples when the longitudinal field of the transient is
directed toward (top), perpendicular to (middle) and away from (bottom)
the observer. The profiles are very sensitive to the parameter a, which
is a measure of the pitch angle of the helical transient field. Mouschovias and Poland (1978) argued that a should not be less than one nor
greater than 1.41. The upper bound is required by stability to the pinch
effect, and the lower bound is necessary to impart a positive magnetic
driving force to the loop from gradients in the azimuthal field. Values
of a over the range $|a| \leq 1.41$ were selected for this study. As a goes
to zero (vanishing azimuthal field), the Faraday profile loses its internal zero since the disturbance field is then unipolar. The maximum
excursions of Ω_t are a measure of the magnitude of B_{lo} provided θ_o is
not exactly zero. Higher values of the ratio h/R_c will shift the node

Figure 2. Faraday rotation profiles for Earth spacecraft line-of-
sight passing through a transient
flux tube of enhanced magnetic field
and electron density. The bottom
(concave) edge of the coronal loop
is at Y = -1, and the top (convex)
edge is at Y = +1. Curves are drawn
for various values of "a", the
ratio of azimuthal field to longitudinal field in the loop. The
three panels correspond to three
possible orientations of the flux
tube as defined by (θ_0, ϕ_0).

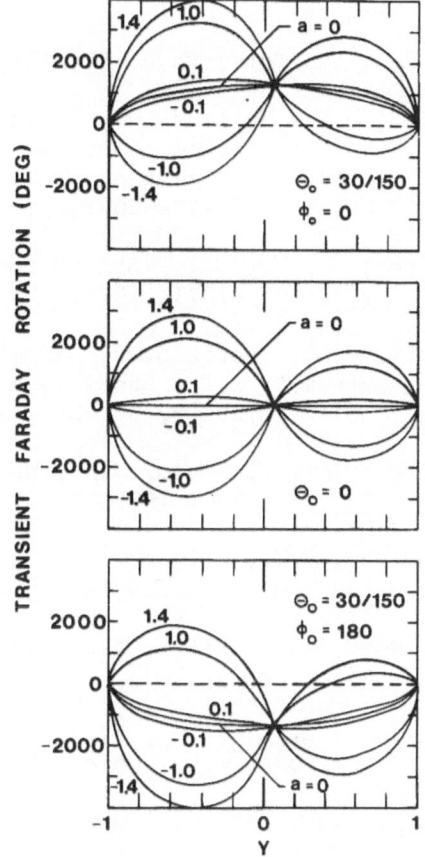

of the family of curves even farther
to the right of $Y = 0$. The more intense azimuthal magnetic field on
the concave side of the flux tube
is responsible for the higher maxima
of Ω_t at negative Y than at positive Y positions.

This model of Faraday rotation
expected from a transient flux tube
passing through an Earth/spacecraft
ray path could easily be extended to
the more complicated cases of arbitrary θ_0 and a variable electron density $N(x,y)$ as estimated from white
light data. In particular, the same
basic procedure could be developed for coronal streamers or any flux
tube containing greatly enhanced electron densities. Since the azimuthal field of approximately radial flux tubes is presumably small, one
would expect to see only one sign of anomalous Faraday rotation during
passage of the transient through the line-of-sight.

It should also be noted that a transient flux tube of width $h \simeq 0.6$
solar radii and velocity $v \simeq 400$ km s^{-1} needs only about 15 minutes to
traverse the ray path. The long duration and small excursions in Ω_t
seen at previous solar occultations indicate that these "Faraday rotation transients" were probably not isolated flux tubes as modelled here.
The Faraday profiles recorded then would be associated with much larger
and/or slowly moving coronal disturbances with much more modest field
and density enhancements than those estimated for a typical coronal
loop transient.

References

Anzer, U., Solar Phys. 57, 111-118, 1978.

Bird, M.K., Volland, H., Stelzried, C.T., Levy, G.S. and Seidel, B.L.,
    in Contributed papers to STIP Symposium, Tel Aviv, 1977, Eds. M.A.
    Shea, D.F. Smart and S.T. Wu (Also: Air Force Geophys. Lab. Report
    No. AFGL-TR-77-0309), pp 63-75, 1977.

Dryer, M., Wu, S.T., Steinolfson, R.S. and Wilson, R.M., Astrophys. J.
    227, 1059-1071, 1979.

Dulk, G.A., Smerd, S.F., MacQueen, R.M., Gosling, J.T., Magun, A.,
    Stewart, R.T., Sheridan, K.V., Robinson, R.D. and Jacques, S.,
    Solar Phys. 49, 369-394, 1976.

Gergely, T.E., Kundu, M.R., Munro, R.H. and Poland, A.I., Astrophys. J.
    230, 575-580, 1979.

Gosling, J.T., Hildner, E., MacQueen, R.M., Munro, R.H., Poland, A.I.
    and Ross, C.L., Solar Phys. 48, 389-397, 1976.

Hildner, E., Gosling, J.T., Hansen, R.T. and Bohlin, J.D., Solar Phys.
    45, 363-376, 1975.

Hildner, E., Gosling, J.T., MacQueen, R.M., Munro, R.H., Poland, A.I.
    and Ross, C.L., Solar Phys. 48, 127-135, 1976.

Levy, G.S., Sato, T., Seidel, B.L., Stelzried, C.T., Ohlson, J.E. and
    Rusch, W.V.T., Science 166, 596-598, 1969.

Mouschovias, T.C. and Poland, A.I., Astrophys. J. 220, 675-682, 1978.

Munro, R.H., Gosling, J.T., Hildner, E., MacQueen, R.M., Poland, A.I.
    and Ross, C.L., Solar Phys. 61, 201-215, 1979.

Nakagawa, Y., Wu, S.T. and Han, S.M., Astrophys. J. 219, 314-323, 1978.

Pintér, S., Bull. Astron. Inst. Czech. 24, 337-342, 1973.

Schatten, K.H., Solar Phys. 12, 484-491, 1970.

Steinolfson, R.S., Wu, S.T., Dryer, M. and Tandberg-Hanssen, E.,
    Astrophys. J. 225, 259-274, 1978.

Wu, S.T., Dryer, M., Nakagawa, Y. and Han, S.M., Astrophys. J. 219,
    324-335, 1978.

DISCUSSION

Anzer: What holds your loop together, if both the magnetic field and the density are much larger inside the loop than outside it?

Bird: The loops are seen to expand as they propagate radially outward indicating that $P + B^2/8\pi$ is higher inside than outside. I would therefore not expect a rigorous pressure balance across the loop boundary. It should be reiterated that radio observations indicate that the regions of high electron density and magnetic field are cospatial.

Stewart: The Faraday Rotation Transients last for 1 hr or so. Consequently, I do not think you are observing an isolated thin loop transient such as the model you described.

Bird: I agree. A coronal loop of thickness 0.5 R_s moving at a velocity 400 km s^{-1} would traverse the line-of-sight in 15 min. The coronal disturbances seen with previous Faraday rotation experiments lasted much longer and would therefore be attributed to larger and/or slower moving phenomena. There is no doubt, however, that the inferred enhancements in electron density and magnetic field in the isolated loops described here would produce an easily recognizable signal in Faraday rotation.

EVOLUTION OF CORONAL MAGNETIC STRUCTURES

R. S. Steinolfson and S. T. Wu
The University of Alabama in Huntsville
Huntsville, Alabama 35805, U.S.A.

INTRODUCTION

Several classes or types of coronal transients are believed to
originate near the solar surface at the base of pre-existing coronal
magnetic loops. These loops often lie beneath an overlying open-field
region in a typical coronal- or helmet-streamer configuration. Recent
experimental results indicate that these pre-existing magnetic loops
may be torn open by the explosive force of the solar phenomena
responsible for the transient. Numerical solutions of the time-
dependent, two-dimensional, dissipationless, MHD equations of motion
(the equations are discussed by Steinolfson et al. (1978)) are used to
examine the formation of a coronal streamer magnetic structure and the
evolution of the streamer following an explosive solar event in the
closed-field region.

CORONAL STREAMER

The coronal-streamer configuration is obtained by starting the
calculation with an initial state consisting of a closed dipole magnetic
field superimposed on a radial, hydrodynamic solution for the thermodynamic variables and velocity. This initial state, of course, does not
represent a steady-state solution to the complete two-dimensional
equations, so the time-dependent solution will evolve until it relaxes
to or approaches a steady-state solution (the coronal streamer).
Coronal-streamer configurations have been obtained previously by
Pneumann and Kopp (1971), who used a time-independent analysis, and by
Endler (1971) and Weber (1978). In all of these previous studies, the
temperature was assumed to be constant; that assumption is not made in
the present work. The initial values used in the simulation are as
follows: The temperature and density at 1 $R_\odot$ (solar radii) are taken
to be 1.8×10^6 K and 2.25×10^8 cm^{-3}, respectively, and the magnetic
field is 2.35 G at the solar surface at the equator which yields a
value for the plasma beta of 0.5. The polytropic index is 1.05. The
initial magnetic field lines are shown in Figure 1(a). The vertical

M. Dryer and E. Tandberg-Hanssen (eds.), Solar and Interplanetary Dynamics, 483-486.

axis represents the equator, the horizontal axis is the pole, and the
region shown is from the solar surface to 5 $R_\odot$. The solution is
symmetric about the equator. The dashed curves represent the radii at
which the sound speed Mach number (longer dashes) and the Alfvén speed Mach
number are equal to one. The initial state relaxes to approximately a
steady state after 16 hours and the resulting magnetic field lines are
shown in Figure 1(b). The relaxed configuration is that of a coronal
streamer with open field lines overlying and adjacent to the closed-field
region. The velocity is approximately zero in the closed-field region as
shown in Figure 2(a). The velocity at 5 $R_\odot$ at the equator is very nearly

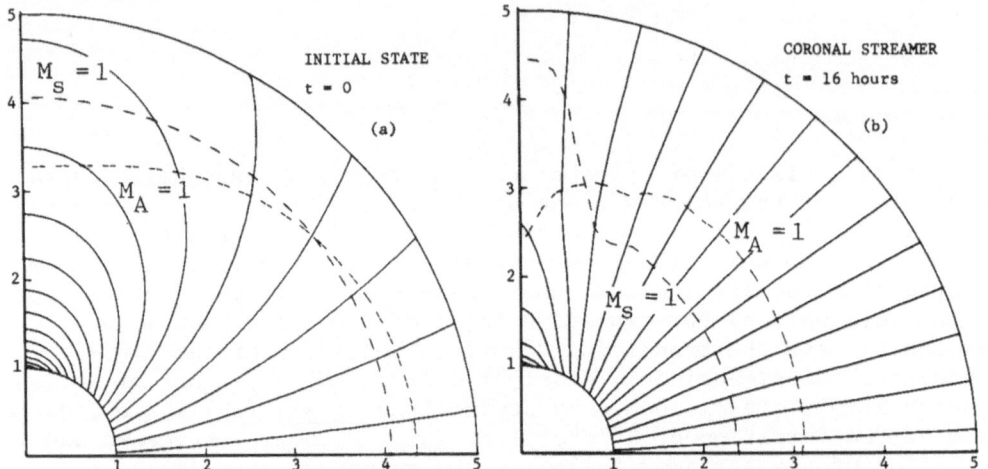

Figure 1.   Magnetic field lines (a) initially and (b) after the numerical
            solution has relaxed to a steady state coronal-streamer
            configuration.

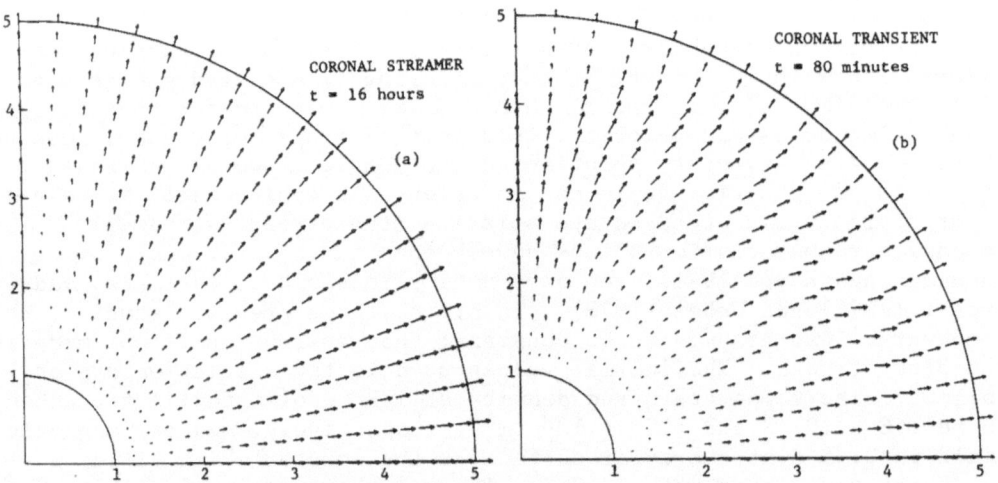

Figure 2.   Velocity vectors in (a) the coronal streamer and (b) the
            coronal transient where the time is now referenced to the intro-
            duction of the perturbation which simulates the solar event.

equal to the initial velocity there of 185 km s^{-1} while the velocity at the pole at 5 R$_\odot$ is increased to 417 km s^{-1}. The pressure and density in the closed-field region are increased over their initial values in this region, with the maximum increase at 1 R$_\odot$ being by factors of 4.8 and 4.2, respectively.

CORONAL TRANSIENT

   A simulated coronal transient is created by instantaneously in-creasing the pressure at 1 R$_\odot$ in the closed-field region by a factor of 10 over the initial value. The pressure is maintained at this value for the duration of the calculation. The velocity vectors and magnetic field lines in the resulting transient after 80 minutes are illustrated in Figures 2(b) and 3(a). The initially closed field lines are pushed outward both radially and azimuthally. An MHD shock is formed ahead of the transient at about 4.5 R$_\odot$. At later times the field becomes essentially open inside 5 R$_\odot$ as seen after 180 minutes in Figure 3(b).

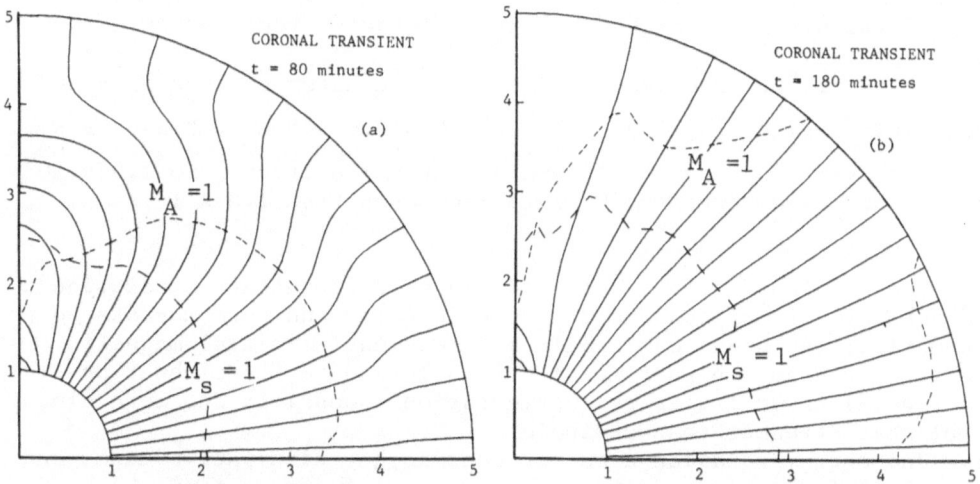

Figure 3.   Magnetic field lines in the coronal transient where the time is as for Figure 2(b).

ACKNOWLEDGMENTS

   The work of RSS was supported by NASA under Contract NAS8-33216 and by the National Oceanic and Atmospheric Administration under Contract NOAA/04-78-B01-6. The work of STW was supported by NASA under Contracts NAS8-33216 and NAS8-28097. Acknowledgment is made to the National Center for Atmospheric Research, which is sponsored by the National Science Foundation, for the use of its computing facilities.

REFERENCES

Endler, F.: 1971, Ph.D. Thesis, Gottinger University, Germany

Pneuman, G. W., and Kopp, R. A.: 1971, Solar Phys., 18, pp. 258-270.

Steinolfson, R. S., Wu, S. T., Dryer, M., Tandberg-Hanssen, E.: 1978,
    Astrophys. J., 225, pp. 259-274.

Weber, W. J.: 1978, Ph.D. Thesis, University of Utrecht, Holland.

DISCUSSION

   Levine: I am interested in the portion of your model which
represents the steady state (i.e., after your model streamer was found):
1. Can you describe the resulting temperature distribution, especially
on open vs. closed field lines?
2. Can your numerical technique handle other initial magnetic geometries,
as a dipole has the wrong strength distribution at the surface for a
realistic solar test case?
3. Would you please describe the energy equation you used in more
detail?
   Steinolfson: 1. The temperature at the equator decreases in the
closed-field region from 1.89×10^6K at 1 $R_\odot$ to 1.8×10^6K at 2.5 $R_\odot$,
and in the overlying open-field region further decreases to 1.6×10^6K
at 5 $R_\odot$. The temperature at the pole increases from 1.8×10^6K at
1$R_\odot$ to 2.82×10^6K at 5 $R_\odot$.
   2. The dipole field was selected since it represents a solution to
the equations for the absence of magnetic poles and zero Lorentz force.
The numerical technique can handle any reasonable initial magnetic
geometries such as a higher multi-pole solution to these equations (e.g.,
a quadrapole) or an arbitrary configuration - possibly one more closely
resembling a streamer than a dipole.
   3. The energy equation does not contain any dissipative terms and in
its simplest form can be written as

$$\frac{d}{dt} \left(\frac{p}{\rho^\gamma}\right) = 0,$$

where p is the pressure, ρ the density and γ the polytropic index. The
derivative is the total or Eulerian derivative, and the equation simply
states that the rate of change of the entropy of a fluid particle is
zero, or equivalently, that the changes of state of the fluid particle
are isentropic.

MAGNETICALLY DRIVEN MOTIONS IN SOLAR CORONA

B.V. Somov and S.I. Syrovatskii
Lebedev Physical Institute,
Academy of Sciences of the USSR, Moscow

Abstract. Solution of the nonlinear MHD problem of plasma flow in an increasing dipolar magnetic field is obtained in the approximation of a strong field. The distributions of plasma velocity, displacement, and density are calculated. The situation when the magnetic dipole is 'increased' by rapid process of magnetic reconnection or current sheet rupture is illustrated. Possible applications are discussed in connection with plasma ejections from chromosphere in corona.

In the upper chromosphere and low corona the conditions of a strong magnetic field are valid (see Syrovatskii and Somov, 1979). During flares and other nonsteady phenomena, magnetic field undergoes rapid local changes. Under the condition of high conduction these changes propagate at a rate close to the Alfven velocity and lead necesserily to plasma motion.

We consider, as example, plasma motions in the increasing dipolar magnetic field (Somov and Syrovatskii, 1972). The MHD equations for an axially symmetric flow in the approximation of a strong field and cold plasma take the following form in the spherical coordinates r, θ and φ:

$$\varepsilon^2 \, d\vec{v}/dt = K(r,\theta,t) \text{ grad } G, \tag{I}$$

$$dG/dt = 0, \tag{2}$$

$$d\varrho/dt = -\varrho \text{ div } v. \tag{3}$$

Here $G = G(r,\theta,t)$ is the dimensionless 'stream function' related to the single nonvanishing φ component of the vector potential $\vec{A}$ by

$$G(r,\theta,t) = A(r,\theta,t) \, r \sin \theta. \tag{4}$$

Let j be the φ component of the dimensionless current $\vec{j}$, then

$$K(r,\theta,t) = j(r,\theta,t)/(\varrho \, r \sin \theta). \tag{5}$$

In zeroth order in the small parameter ε^2, the Equation (I) yields

$$K^{(o)}(r,\theta,t) = 0 \quad \text{or} \quad j^{(o)}(r,\theta,t) = 0. \tag{6}$$

This corresponds to a time-dependent potential field described by the

M. Dryer and E. Tandberg-Hanssen (eds.), Solar and Interplanetary Dynamics, 487-489.

stream function $G^{(o)}(r,\theta,t)$. For the dipolar magnetic field

$$G^{(o)}(r,\theta,t) = m(t) \sin^2\theta/\ r. \qquad (7)$$

According to the Equation (2) this function is an integral of motion:

$$m(t) \sin^2\theta \ / \ r = m_o \sin^2\theta_o \ / \ r_o. \qquad (8)$$

The subscript 'o' designates the initial values of the Lagrange coordinates $r(r_o,\theta_o,t)$ and $\theta(r_o,\theta_o,t)$ of a 'fluid particle'.

From Equation (I) it follows that

$$d\vec{v}^{(o)}/dt = K^{(I)}(r,\theta,t) \ grad \ G^{(o)}. \qquad (9)$$

Dividing the radial component of (9) by the angular component eliminates $K^{(I)}(r,\theta,t)$. We also eliminate $r = r(r_o,\theta_o,t)$ by means of (8). The resulting ordinary differential equation for $\theta(\theta_o,\dot{\theta}_o,t)$ is of the form:

$$m(t)a(\theta)\ddot{\theta} + m(t)b(\theta)\dot{\theta}^2 + 2\dot{m}(t)a(\theta)\dot{\theta} + \ddot{m}(t)c(\theta) = 0, \qquad (I0)$$

where $a(\theta)$, $b(\theta)$ and $c(\theta)$ are the known functions.

The analytic solution for the linearized equation (Syrovatskii, I969) and the numerical solution of nonlinear problem for an increasing dipole (Somov and Syrovatskii, I972) show that the magnetic moment growth results in the 'raking-up' of plasma to the dense feature accelerated along the dipole axis. Platov et al. (I973) applied this mechanism to the formation of a surge in the initial exponential atmospere with the scale height h_{oo}. Figure Ia presents the calculated isodensity curves for m=5, I0 and 20. The dipole axis is inclined to the vertical at 30°. For comparison Figure Ib shows several successive optical contours for the growing surge observed on October 23, I970. However, to obtain observed velocities of surge a rapid change of the magnetic field is required. The needed characteristic time is of order I0 s (see Platov et al., I973).

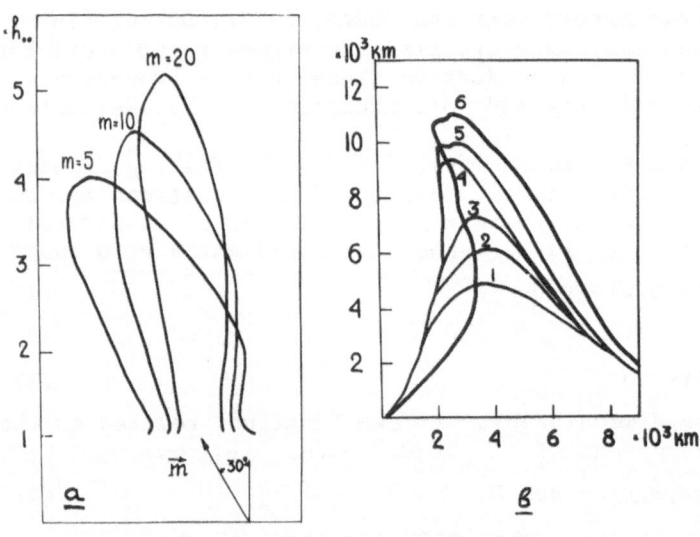

Figure I. (a) Isodensity contours for the inclined increasing dipole. (b) Constant brightness curves for some observed surge.

As yet, there are no data (see, however, Tanaka, I978) which can confirm or reject the possibility of so rapid changes in the photosphere. But these changes possibly can result from the reconnection in the corona.

Let us consider as example the picture of the magnetic field lines in the case when a new flux (N,S) emerges in the region between two sunspots N_I and N_2 (Figure 2).

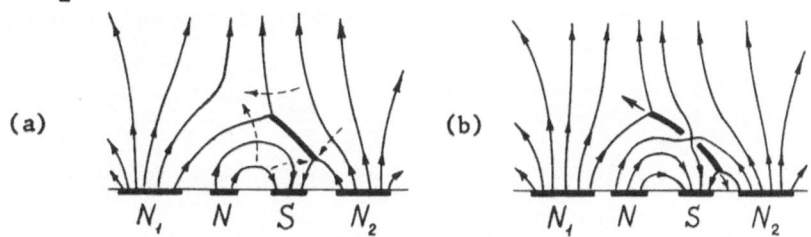

Figure 2. The straightening of field lines above the spot N_I. (a) Quasi-steady reconnection, (b) rapid field reconstruction at the current sheet rupture.

The quasi-steady magnetic reconnection in the current sheet provides some straightening of field lines above the spot N_I. The density of field lines is growing here. That corresponds in our model to the growth of the effective dipolar moment of this spot. If the current sheet is suddenly ruptured, this process is especially fast and is accompanied also by additional plasma ejection from the current sheet. Hence one should expect the plasma ejection (like a surge or spray) above the spot N_I. Note here that the model makes it possible to explain plasma ejections on the periphery of the developing spots and their connection with flarelike phenomena (sudden brightenings) in the chromosphere.

REFERENCES

Platov, Yu.V., Somov, B.V., and Syrovatskii, S.I.: I973, Solar Phys. 30, pp. I39–I47.
Somov, B.V. and Syrovatskii, S.I.: I972, Soviet Phys. JETP 34, pp. 332–335.
Syrovatskii, S.I.: I969, Astrophys. Space Sci. 4, pp. 246–25I.
Syrovatskii, S.I. and Somov, B.V.: I979, Physical driving forces and models of coronal responses (Invited paper presented at this symposium).
Tanaka, K.: I978, Solar Phys. 58, pp. I49–I63.

GASDYNAMICS OF IMPULSIVE HEATED SOLAR PLASMA

B.J.Sermulina[1], B.V.Somov[2], A.R.Spektor[1], S.I.Syrovatskii[2]
[1]Radioastrophysical Observatory of the Latvian Academy of Sciences, Riga, 226524, USSR
[2]P.N.Lebedev Physical Institute, Academy of Sciences of the USSR, Moscow

Abstract. Numerical solutions for problem of hydrodynamic response of the inhomogeneous (exponential) atmosphere on impulsive heating by energetic electrons or by very high-temperature thermal fluxes are discussed.

Impulsive heating of solar chromosphere during a single spike of hard X-rays (an "elementary flare burst" - EFB) can produce very fast motions. We consider them as two-temperature hydrodynamic plasma flows along a strong magnetic field. Let that be vertical one, for simplicity. Then the following equations are appropriate

$$\frac{\partial n}{\partial t} + n^2 \frac{\partial v}{\partial \xi} = 0, \tag{1}$$

$$\frac{\partial v}{\partial t} + \frac{1}{m_H} \frac{\partial}{\partial \xi}[nk(T_i + xT_e)] = \frac{4}{3} \frac{1}{m_H} \frac{\partial}{\partial \xi}\left[\eta_i n \frac{\partial v}{\partial \xi}\right] + g_0, \tag{2}$$

$$\frac{nk}{(\gamma-1)} \frac{\partial(xT_e)}{\partial t} - kT_e \frac{\partial n}{\partial t} + n\chi \frac{\partial x}{\partial t} =$$

$$= \mathscr{P}_c(n,T_e) + \mathscr{P}_e(n,\xi) - \mathscr{L}(n,T_e) - Q(n,T_e,T_i), \tag{3}$$

$$\frac{nk}{(\gamma-1)} \frac{\partial T_i}{\partial t} - kT_i \frac{\partial n}{\partial t} = \mathscr{P}_v(n,T_i,v) + Q(n,T_e,T_i). \tag{4}$$

Lagrange coordinate ξ is the plasma depth. Conductive heating and radiative cooling $\mathscr{L}$ are taken into account. Other designations, initial and boundary conditions etc. are in Somov et al. (1977, 1979) with one erratum there (the erroneous gravitational term in (4) must be omitted).

M. Dryer and E. Tandberg-Hanssen (eds.), Solar and Interplanetary Dynamics, 491-494.
Copyright © 1980 by the IAU.

Bearing in mind two possible interpretations of hard X-rays, we solve Eqs. (1)-(4) numerically for two models: nonthermal and thermal one. First of them assumes, as usually, rapid variation of the accelerated electron flux in accordance with the hard X-ray intensity. In our thermal model, electron temperature on the upper boundary of chromosphere follows that intensity.

For numerical treatment, the symmetrical EFB with a FWHM of 5 s is used. Note that Kostjuk and Pikel'ner (1974) solve the hydrodynamic problem for long-duration (100 s) heating by nonthermal electrons. In our treatment, very rapid ($t \ll 1$ s) change of plasma energy due to power radiative cooling is essential. This leads to the thermal instability (Somov and Syrovatskii, 1976).

Some results are shown in Fig. 1. For thermal model, the X-ray spectrum and emission measure (Fig. 2) are also calculated.

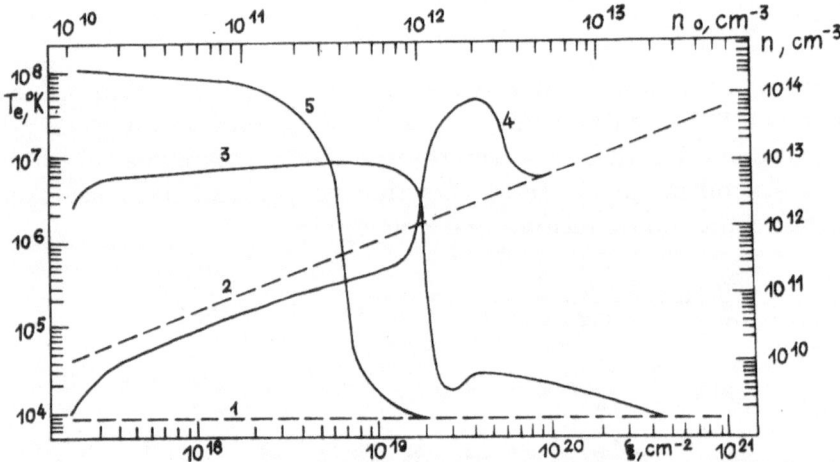

Fig. 1. Numerical solution at the time 5 s.(1) and (2) are initial temperature and density, respectively. (3) and (4) show the electron temperature and density in the nonthermal model. (5) is the electron temperature in the thermal model.

Conclusions are:
    (1) Impulsive heating can be produced by nonthermal electrons, as well as heat conductive fluxes.
    (2) In nonthermal model, the temperature in upper chromosphere rises to values of order 10^7 K. The heating in more dense layers is balanced by radiative cooling, that can give rise to short-lived EUV flash. As a result of radiative cooling and heat conduction, the thin flare transition layer (FTL) develops.
    (3) Maximum velocity of the heat front propagation in the chromosphere is of order 10^3 km/s before the FTL is formed according to the

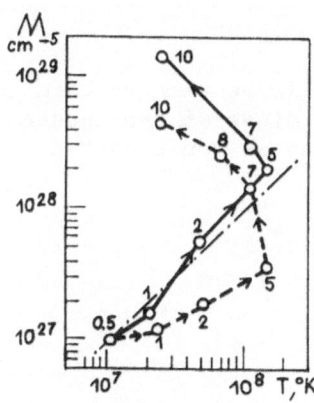

Fig.2. Relationship between emission measure M and effective temperature T_{eff}. The different lines correspond to different forms of limiting heat flux ($-\varkappa \sim T_e^{3/2}, ---\varkappa \sim T_e$). The time instants are indicated by figures (s). The dashed-dotted line corresponds to an adiabatic process.

thermal model.

(4) For both models, the thermal instability gives rise to cold condensation just below the FTL. That condensation moves downward at velocity exceeding the sonic one in the quiet chromosphere. The front shock fades gradually in denser layers.

(5) The heated high-temperature chromospheric plasma is ejected upward with velocity of order 1500 km/s. The plasma mass ejected during one EFB is of about 10^{-5} g per cm^2.

(6) In the impulsive ejected plasma, the ion temperature is more than two orders of magnitude less than the electron temperature.

(7) The thermal X-ray spectrum of the high-temperature plasma is approximated by the exponential law with some effective temperature T_{eff} well enough.

(8) The emission measure increases continuously even after the maximum of the T_{eff} (Fig.2). This differs the thermal model under consideration from the model with adiabatic compression and expansion (Matzler et al., 1978).

REFERENCES

Kostjuk, N.D. and Pikel'ner, S.B.: 1974, Astron.zh.51, pp.1002-1016.
Matzler, Chr., Bai, T., Crannel, C.J., and Frost, K.J.: 1978, Astrophys. J. 223, pp.1058-1071.
Somov, B.V. and Syrovatskii, S.I.: 1976, Soviet Phys.Uspekhi. 19, pp.813-853.
Somov, B.V., Spektor, A.R., and Syrovatskii, S.I.: 1977, Bull. Acad. Sci. USSR, Phys.series 41, pp.32-43.
Somov, B.V., Spektor, A.R., and Syrovatskii, S.I.: 1979, Proc. Lebedev Phys.Inst. 110, pp.73-94.

DISCUSSION

Nakagawa: (Comment) Current accepted idea is heating by hard and soft X-rays to the chromosphere. However, regardless of the mechanism of heating, the gasdynamic responses are similar as you reported.

DYNAMICS OF CORONAL TRANSIENTS: TWO-DIMENSIONAL NON-PLANE MHD MODELS

Y. Nakagawa
Space Sciences Laboratory/NASA
Marshall Space Flight Center, AL 35812, U.S.A.

S. T. Wu and S. M. Han
The University of Alabama in Huntsville
Huntsville, AL 35807, U. S. A.

ABSTRACT

Numerical results are obtained for non-plane MHD responses to a sudden energy release in a stratified model atmosphere. In agreement with observations, it is shown that after the energy release, the magnetic field affected by the energy release relaxes toward the potential configuration while the outer atmospheric fields increase its shear. Additional results suggest a new way of interpreting the energy storage and release in repeated flares.

INTRODUCTION

It was shown (Nakagawa et al., 1978a; Wu et al., 1978) that some observed characteristics of the coronal transients (Hildner et al., 1975) can be reproduced by two dimensional plane MHD models. However, the plane analyses (confined within a meridional plane) were limited to the initial configurations of magnetic field to be potential, also to the exclusion of transverse waves (often called the Alfven waves).

In the present non-plane analyses, transverse waves are included together with the out-of-meridional component of magnetic field. The initial magnetic field is represented by a global constant α force-free field (Nakagawa et al., 1978b), and numerical computations are initiated with an introduction of a short temperature pulse simulating the energy release by a flare.

RESULTS AND DISCUSSION

Two distinct initial magnetic configurations (shown in Figures 1a and 2a) are considered. These configurations correspond to the open and closed configurations of the previous study (Nakagawa et al., 1978a).

M. Dryer and E. Tandberg-Hanssen (eds.), Solar and Interplanetary Dynamics, 495-498.

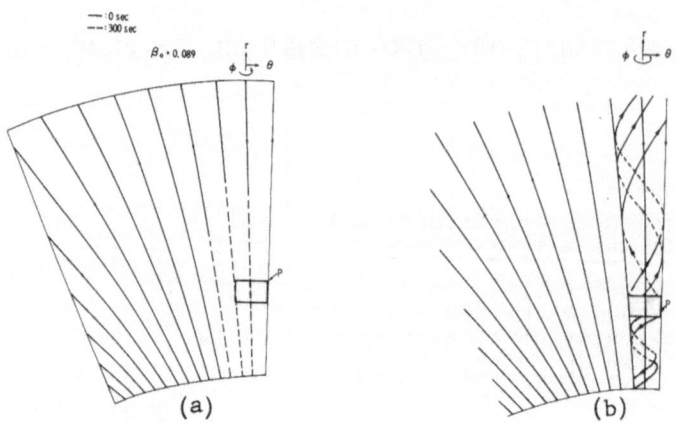

Figure 1. (a) The initial (solid) and disturbed (dashed) magnetic
              fields.
           (b) The resultant material motions, p the position of
              energy release.

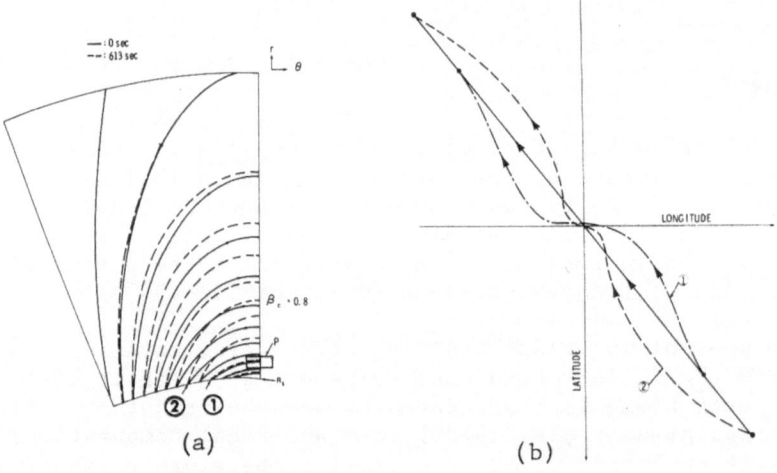

Figure 2. (a) The initial (solid) and disturbed (dashed) magnetic
              fields.
           (b) The schematic changes of θ-ϕ projection of field
              configuration after energy release.

In the open configuration, with a small value of $\beta = 0.089$ (where β is the ratio of the gas pressure to the magnetic pressure), it is found that the typical coronal response is a spiralling up or down motion confined around the axis as shown in Figure 1b: such spiralling motions are known with flare sprays (Tandberg-Hanssen et al., 1980).

With the values of β similar to those in previous studies (Nakagawa et al., 1978a; Wu et al., 1978), the general responses appear almost identical in the meridional plane projection. However, the velocities of expansion are reduced reflecting the fact that the presence of an additional degree of freedom, namely, the presence of transverse waves which carry away some of the energy released.

The most striking results are the difference in the responses of magnetic field in the closed configuration shown in Figure 2b. In the $\theta-\phi$ projection (where θ is the co-latitude and ϕ the longitude), the initial field lines are straight lines, and the foot-points of different field lines are located along this line at different distances from the center of symmetry (Nakagawa et al., 1978b). With the energy release, the field lines expand away from the point of energy release. Physically this expansion corresponds to the reduction of local shear, i.e., electric currents. Thus, the initial configuration relaxes toward the potential configuration. In other words, after the energy release, the initially sheared magnetic field appears to relax toward the potential configuration as shown by Tanaka and Nakagawa (1973).

In addition, the expanding field lines, particularly in the radially outward direction, increase the shear. This sheared field leads to the appearance of flat-top topology as shown with larger values of α by Nakagawa et al. (1978b) in interpreting the observed characteristics of coronal transient by Hildner et al. (1975).

Finally, by pursuing the computation long after the energy release, it is found that the heated atmosphere (in a form of a bubble) gradually moves out of the domain of computation yielding the field lines to return toward the initial configuration. This result implies that repeated flares are possible in a sheared magnetic field without apparent shearing motions at the photospheric level after the first flare. Further details of the analyses will be published elsewhere.

One of the authors (YN) is supported by the National Research Council Senior Research Associateship. STW and SMH are supported by a NSF Grant ATM77-22484.

REFERENCES

Hildner, E., Gosling, J. T., MacQueen, R. M., Munro, R. H., Poland, A.
    I., and Ross, C. L.: 1975, Solar Phys., 42, p. 163.
Nakagawa, Y., Wu, S. T., and Han, S. M.: 1978a, Astrophys. J., 219,
    p. 314.

Nakagawa, Y., Wu, S. T., and Tandberg-Hanssen, E.: 1978b, Astron.
    Astrophys., 69, p. 43.
Tanaka, K., and Nakagawa, Y.: 1973, Solar Phys., 33, p. 187.
Tandberg-Hanssen, E., Martin, S. F., and Hanssen, R. T.: 1980, Solar
    Phys. 65, 357.
Wu, S. T., Dryer, M., Nakagawa, Y., and Han, S. M.: 1978, Astrophys.
    J., 219, p. 324.

DISCUSSION

Stix: What is the reason for the helical motion in the open field
case?
Nakagawa: The helical motion is the consequence of inclusion of the
transverse waves in this non-plane study, which has never been treated
previously in one- or two-dimensional planar studies.

Kuperus: Why do you put your disturbance at some altitude above
the boundary instead of perturbing the field at the boundary?
Nakagawa: To introduce disturbance on the boundary leads to the MHD
initial-boundary value problem. Such a formulation is available now,
however, we have not tried it in this paper.

OBSERVATIONS OF INTERPLANETARY SCINTILLATION AND A THEORY OF HIGH-SPEED SOLAR WIND

H. Washimi, T. Kakinuma and M. Kojima
The Research Institute of Atmospherics, Nagoya University

1. IPS OBSERVATIONS OF HIGH-SPEED SOLAR WIND

It has been confirmed that the high-speed solar wind flows out of the coronal holes at low latitudes, where the magnetic fields open and the temperature is low (e.g., Krieger et al. 1973). But there has not been direct observation of the solar wind out of the polar regions of corona. We report here that the observations of interplanetary scintillation (IPS) show the existence of the high-speed flow of 800 km/s out of the polar coronal regions and the well-coincidence to the model of the coronal holes extending from the polar regions.

The observations of IPS of radio sources at 69 MHz have been made at three stations, Toyokawa, Fuji and Sugadaira. The velocity of the diffraction pattern moving across the earth can be derived by a cross-correlation analysis. The pattern speed, in most cases, is interpreted to be the solar wind speed at the point of closest approach of the line of sight to the sun, but we have to take into account the integration effect along the line of sight when there is latitude dependence of the solar wind velocity.

We have found that daily variations of pattern speed consist of two parts during the minimum phases of solar activity: the variation due to the latitude dependence of the velocity of the quiet solar wind which is represented by the lower envelope of the observations, and the variation due to a coronal high-speed stream.

We derived latitudinal velocity distribution of the quiet solar wind, which shows the wind velocity increases rapidly with latitude and is 800 km/s for latitudes higher than 45° (Kakinuma, 1977), that is, shows the existence of the polar high-speed region.

As seen in the observations of 3C48 in 1974, the amplitude of the speed enhancement due to a recurrent stream apparently decreases with increasing latitude, and the maximum speed remains about 700 km/s for several rotations. This stream can be interpreted to be by the equatorward extension of the above high-speed region at high latitudes. We have assumed two corotating streams in 1974, the extension from the north centered on 250° Carrington longitude and the extension from the south centered on 90° longitude. This model is consistent with the observations of K-corona

M. Dryer and E. Tandberg-Hanssen (eds.), Solar and Interplanetary Dynamics, 499-502.

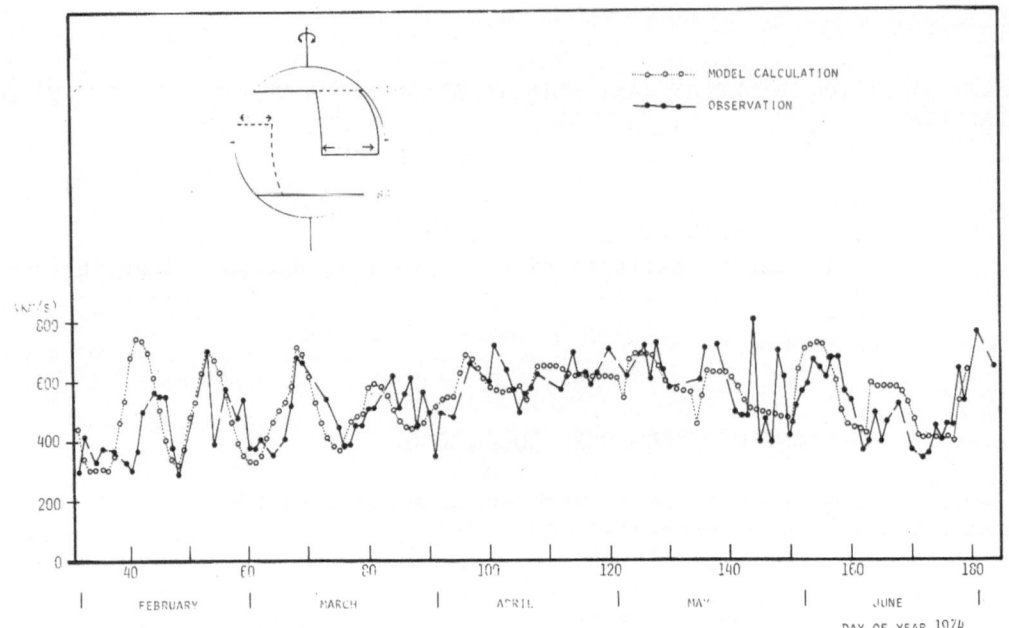

Figure 1. IPS velocity observations for 3C48
in 1974 and a model calculation.

(Hansen et. al., 1976). We have found that the observations in 1974 can
be accounted for by this velocity distribution, as shown in Figure 1.
IPS observations give evidence for the existence of high-speed streams
of 800 km/s extending from the polar regions.

2. A THEORY OF HIGH-SPEED SOLAR WIND

The super-sonic flow of the solar wind was first explained by Parker's
model in which the plasma kinetic pressure and the solar gravity are
considered to be basic forces which act on the solar wind near the sun.
But the high-speed solar wind from the open field regions where the
coronal temperature is low can not be explained by only two kinds of
forces.
We consider, as the third force, the ponderomotive force due to Alfvén
waves which propagate from the sun to the interplanetary space along the
magnetic fields. Theories of the ponderomotive force in magnetized plasma
(fluid model) have been given by one of the authors (Washimi, 1973) and by
Washimi and Karpman (1976). The expression of the force is reduced to

$$F = \frac{1}{8\pi} \left(\nabla_{\parallel} + \frac{2}{V_g} \frac{\partial}{\partial t} + 2\kappa \right) |\mathbb{B}|^2 , \tag{1}$$

where $\mathbb{B}$ is the magnetic field of Alfvén wave, V_g the group velocity
and κ the absorption coefficient of the wave.   Recently Hatori and

Washimi (1979) have derived the ponderomotive potentials by using Lie-operator approach for a single particle, in the oscillation-center coordinate,

$$\phi = - \frac{1}{2} (\xi \cdot E) \quad \text{and} \quad A = - \frac{1}{2} [\xi \times B] . \tag{2}$$

Here ξ means a excursion due to the electric and magnetic fields. They have found that the set of expressions in (2) can be rewritten to a covariant form and are consistent with (1).

It may be worth to note that the first term of the r.h.s. of (1) is not a repulsive force (solar wind acceleration) but an attractive force(deceleration), while the second and the third terms are for the acceleration. We consider the nonstationary Alfvén waves by which the effect due to the two terms overcome the one due to the first term.

We can see, in eq. (1), that the ponderomotive force due to nonstationary Alfvén waves shakes the element of the solar wind back and forth whenever wave-packets get ahead of the element, and on the average accelerates the element. In the course of the passage of wave-packets through the element, the heating mechanism which is a kind of stochastic process may also work. The heating rate is proportional to ion mass. This heating effect is expected to be so strong that the heating of the solar corona may be due to the mechanism (Washimi, 1979).

Our simple computational example shows that Alfvén waves of moderate intensity in the open fields first heat the low temperature plasma up to 10^6 K in the coronal regions and accelerate the solar corona. Subsequently they heat and accelerate the solar wind. Though the cooling effect due to the plasma expansion is dominant in the outer region of corona, the temperature of several times of 10^4 K can be still preserved near 1 AU while the velocity reaches to about 800 km/s.

Finally we suggest that the above acceleration and heating mechanisms may be expected in stellar atmospheres.

Some parts of IPS observations are indebted to Dr. T. Watanabe.

References

Hansen, R. T., Hansen, S. F. and Sawyer, C.: 1976, Planet. Space Sci. 24, 381.

Hatori, T. and Washimi, H.: 1979, in submission.

Krieger, A. S., Timothy, A. F. and Roelof, E. C.: 1973, Solar Phys. 29, 505.

Kakinuma, T.: 1976 'Proceedings of the L. D. de Feiter Memorial Symposium' (M. A. Shea, D. F. Smart and S. T. Wu, Eds.), D. Reidel, Dordrecht, p. 101

Washimi, H.: 1973, J. Phys. Soc. Japan 34, 1373.

Washimi, H.: 1979, in Proceedings of the Fourth Solar Wind Conference (H. Rosenbauer, Ed.), Springer-Verlag, Heidelburg, in press.

Washimi, H. and Karpman, V. I.: 1976 Soviet Phys. JETP 44, 528.

DISCUSSION

Couturier: I don't agree with the momentum equation you are using, particularly for the gradient of Alfven wave pressure. There is a lot of published papers which have treated that term (Belcher and Davis, Alazsaki and Couturier, Hollweg, Jacques..). As we have said to you last year in Solar Wind Conference 4, I suppose you miss the fact that Alfven waves are propagating radially outward from the sun and the gradient of Alfven wave pressure comes from the spherical geometry. The equations you are using are perhaps for plane geometry.

Washimi: Our expression includes the term due to the 'centrifugal wave effect' which you mentioned. In a weakly inhomogeneous plasma, such as in the interplanetary plasma, the Alfven wave is not purely a transverse one but is coupled weakly with longitudinal modes. In this case the parallel component of the Alfven wave, which I suppose you missed, can not be neglected. I am afraid all the papers which you referred to are incorrect.

Newkirk: You present velocity and temperature from a specific application of your model. What values of Alfven wave flux and absorption coefficient were used in that example?

Washimi: Alfven wave flux is given by $V_g \cdot \dfrac{|B|^2}{8\pi}$ which is about $3 \cdot 10^{-4}$ erg/cm^2 sec at 1 AU. The absorption of $|B|^2$ is assumed to be proportional to R^{-2} (R is the radius from the sun), so that $|B|^2 / |B_o|^2$ is a constant (= 0.04) for all regions.

GLOBAL MODELING OF DISTURBANCES IN THE CORONA-INTERPLANETARY SPACE

Tyan Yeh
Department of Mathematics, Metropolitan State College
Denver, Colorado 80204, U.S.A.

In the past few years, numerical studies of corona-interplanetary dynamics have been undertaken by many investigators. The results, as summarized by Wu et al. (1977), indicate that this approach seems very promising and worthy of further pursuit. So far, the calculations have been performed on two-dimensional codes, to simulate magneto-hydrodynamic flow in a meridional plane or in the equatorial plane. In order to fully account for the global features of the corona-interplanetary phenomenon, it is necessary to do three-dimensional calculations. Here we discuss some groundwork for the formulation of a time-dependent three-dimensional code.

To describe the physical phenomenon of the complicated disturbances in the corona-interplanetary space by mathematical solutions of an idealized initial-boundary value problem, it is necessary to use various approximations. If the solution is to be physically meaningful, the approximations used must be physically reasonable. This means a judicious choice of the governing equations and the initial-boundary conditions, with as few ad hoc parameters as possible. The governing equations represent the physical laws, according to them the corona-interplanetary medium undergoes changes from the state indicated by the initial condition, subject to the constraint indicated by the boundary condition. A part of the initial-boundary conditions specifies the source of disturbance. Of course, to find the solution explicitly, a stable computation scheme has to be devised, so that the numerical solution obtained is acceptably accurate.

The corona-interplanetary space has a spatial domain extending from the coronal base to the outreach of the heliosphere. In the length scale pertinent to the global phenomenon , the inner boundary can be represented by a spherical solar surface and the outer boundary by

M. Dryer and E. Tandberg-Hanssen (eds.), Solar and Interplanetary Dynamics, 503-506.
Copyright © 1980 by the IAU.

a heliocentric sphere of very large radius. Beneath the inner boundary
is the active subphotospheric sun, and beyond the outer boundary is
the passive interstellar medium. Effluxes of mass, energy, momentum,
and magnetic flux from the sun are transported by the corona-
interplanetary medium into the interstellar space.

A one-fluid description for the corona-interplanetary plasma is
appropriate for the large-scale global features. Thus, there are four
primary physical quantities, namely, mass density, temperature,
flow velocity,and magnetic field. Their temporal changes are governed
by conservation laws of mass, energy, and momentum and Faraday's
law of induction. These four physical laws can be written as partial
differential equations which relate the temporal derivatives to various
spatial derivatives. Other physical quantities, namely, plasma pressure,
specific thermal energy, current density, and electric field, are
secondary in the sense that they can be expressed in terms of the
primary quantities. The auxiliary equations, viz., equation of state,
a thermodynamic equation, Ampere's law with the displacement current
neglected, and Ohm's law with a scalar resistivity, involve no temporal
derivatives at all. The energy equation should include heat conduction
and ohmic dissipation. The former is an important process of energy
transport in the corona-interplanetary medium, and the latter is
necessary for the magnetic diffusion to allow the topological changes
between closed and open configurations in the temporal evolution of
the magnetic field. As to the equation of motion, the plasma motion
is driven by pressure gradient, solar gravitation, and magnetic force.
The equation of induction will preserve the solenoidality of the magnetic
field.

We envisage the disturbance as ensued from some sort of local
instability. Prior to the onset of the disturbance, the corona-
interplanetary medium is in a steady state, transporting the effluxes
from the sun to the interstellar medium. In this quiescent process,
somehow excessive amounts of mass, energy, momentum, or magnetic
flux are gradually built up in some localized region. If the region of
instability is located above the coronal base, the source of disturbance
is accounted for by the initial condition for that part of the corona-
interplanetary space. On the other hand, if the region of instability is
located below the coronal base, the source of disturbance is accounted
for by the boundary condition for a part of the inner boundary where
the excessive effluxes enter the corona-interplanetary space. Other
parts of the corona-interplanetary space have the quiescent initial
condition and other parts of the boundaries have the quiescent boundary
condition. The localized instability is only strong enough to affect a smal

part of the corona-interplanetary space at the initial instant or a part of the inner boundary for a short while. The unaffected parts remain at the quiescent state. The outer boundary is not affected at all till the arrival of the disturbance. In fact, if the outer boundary is idealized to be at infinity, the boundary condition at infinity is vanishing mass density, temperature, magnetic field and infinite flow velocity. The indefinite effluxes at this passive boundary are determined by the solar wind in the corona-interplanetary space. Therefore, we may pose two kinds of initial-boundary value problem. One has a disturbed initial condition and a quiescent boundary condition. The other has a disturbed boundary condition and a quiescent initial condition. Their mixture will have disturbed parts in both the initial and boundary conditions. With a static initial condition, as assumed in most of the previous work, the inner boundary would become passive . This means that some fluxes are allowed to enter through the coronal base from the corona-interplanetary space.

The quiescent initial-boundary conditions are in turn to be obtained by means of modeling which should utilize the observed solar data as much as possible. At the present time, solar magnetograms provide only the line-of-sight component of the magnetic field at the photospheric level. This is not sufficient to determine the current, hence the current-filled magnetic field, in the steady state corona-interplanetary space. We may use sheet-current approximation (Yeh and Pneuman, 1977), which requires only scalar data on the solar surface, to determine a piecewise current-free magnetic field partitioned by current sheets. The presence of the sheet currents means that the hydromagnetic interaction between the magnetic field and the plasma motion is not entirely ignored. Since small-scale details of the photospheric magnetograms have no bearing on the large-scale features of the corona-interplanetary space, only the large-scale photospheric data should be extrapolated as the data at the coronal base. The most important feature is the neutral lines. Their nested arrangement on the solar surface determines the magnetic topology of the corona-interplanetary magnetic field (Yeh, 1978). The H-alpha synoptic charts (McIntosh, 1979) are very useful in discerning the relevant neutral lines.

To solve the formulated initial-boundary value problem numerically by finite-difference method, the outer boundary is replaced by a heliocentric sphere of a suitable radius. The boundary condition to be specified at this ad hoc boundary is the quiescent values of the initial steady state. This replacement is valid only prior to the arrival of the disturbance. Hence, it is necessary to move the ad hoc outer boundary

farther out once in a while in the course of computation. Likewise,
ad hoc lateral boundaries may be used, as in the previous work. Again,
the boundary conditions on these ad hoc lateral boundaries are the
quiescent values of the initial steady state and the lateral boundaries
must be moved and eventually removed as the disturbance spreads
laterally. It is advantageous to use spherical coordinates. Special
care must be taken at the two polar lines. There the differential
operators for gradient, divergence, curl, and trajectory derivative,
written in the spherical coordinates, become indefinite in form. A
finite-difference approximation for the governing equations can be
devised by using the integral forms of these del operators at the
grid points located on the polar lines. For the grid points not on
the polar lines differential forms will provide the finite-difference
approximation. A scheme of third-order accuracy can be constructed
by the use of the two-step Lax-Wendroff treatment. The virtual grid
points for the intermediate steps may be chosen at the centers of the
boxes formed by eight neighboring real grid points. More important
than the accuracy is the stability of the scheme. The time step should
be less than the values of the local grid spacing divided by the fast
magnetoacoustic speed. The inclusion of dissipations in the governing
equations should enhance the numerical stability. Since a lot of grid
points are needed in such global modeling, a lattice of grid points
with uneven spacings will allow the allocation of more grid points to
regions where the physical quantities are expected to undergo large
variations. We are presently exploring the feasibility of implementation,
regarding the huge demand on the storage capacity and computation
time, for a three-dimensional code on NCAR's Cray-1 computer.

References

McIntosh, P. S.: 1979, Annotated Atlas of $H\alpha$ Synoptic Charts for
    Solar Cycle 20 (1964-1974), World Data Center A for Solar-
    Terrestrial Physics, NOAA, Boulder, Colorado.
Wu, S. T., Nakagawa, Y., and Dryer, M.: 1977, Study of Travelling
    Interplanetary Phenomena (ed. M. A. Shea, D. F. Smart and
    S. T. Wu), pp. 43-62, D. Reidel Publishing Co., Dordrecht-Holland.
Yeh, T.: 1978, Solar Phys. 56, pp. 439-447.
Yeh, T. and Pneuman, G. W.: 1977, Solar Phys. 54, pp. 419-430.

THE OPEN PROGRAM: AN EXAMPLE OF THE SCIENTIFIC RATIONALE FOR FUTURE SOLAR-TERRESTRIAL RESEARCH PROGRAMS

D. J. Williams
NOAA Space Environment Laboratory
Boulder, Colorado 80303

ABSTRACT

The field of solar-terrestrial physics has evolved to a point where quantitative theoretical modeling and cause-effect predictive techniques can be foreseen. Instrumentation now exists to quantitatively test these theories and lead to a significant improvement in our understanding of the solar-terrestrial environment. The proposed Origins of Plasma in the Earth's Neighborhood (OPEN) program will be used to trace this theoretical and experimental evolution and describe a solar-terrestrial research scenario for the 1980's which includes specific problems related to solar and interplanetary dynamics.

1.   INTRODUCTION

Exploration of the Earth's nearby space environment has revealed a dynamic and complex system of interacting plasmas, magnetic fields, and electrical currents surrounding our planet. This region, comprising the magnetized solar wind plasma plus the resulting perturbation in the heliosphere caused by the presence of the magnetic Earth, we call "geospace" (see Figure 1). Here plasma physics determines the behavior of matter on spatial and temporal scales vastly different from those that can be duplicated in earth-based laboratory devices. Geospace thus affords a unique and readily accessible laboratory for in-situ investigation of cosmic plasma processes. Through these plasma processes energy from matter expelled by the Sun is fed into the Earth's environment, constituting a small but highly significant part of our total solar energy budget.

The scientific thrust of the OPEN[1] program is:

1) to trace the flow of matter and energy through the system from input by the solar wind to ultimate deposition into the atmosphere;

2) to understand how the individual parts of the closely coupled, highly time-dependent geospace system work together;

M. Dryer and E. Tandberg-Hanssen (eds.), Solar and Interplanetary Dynamics, 507-522.

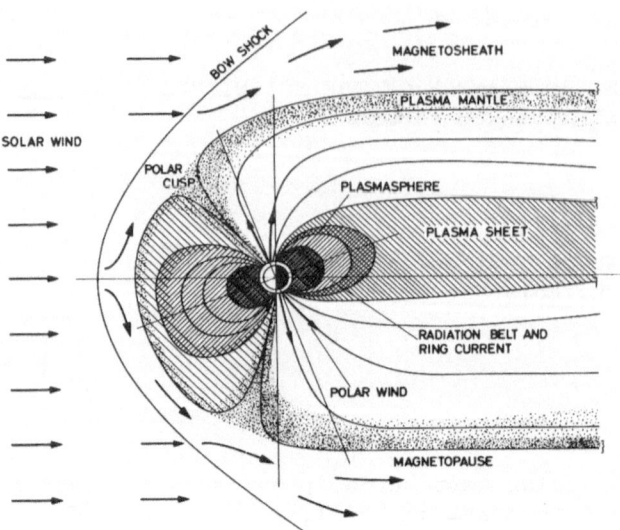

Figure 1.   The major plasma regimes of geospace.

     3)   to understand the physical processes controlling the origins,
entry, transport, storage, acceleration and loss of plasma in the
Earth's neighborhood; and,

     4)   to assess the importance to the terrestrial environment of
variations in atmospheric energy deposition caused by geospace plasma
processes.

     Previously, the near-Earth space environment has been explored and
studied primarily as a system of independent component parts--the interplanetary region, the magnetosphere, the ionosphere and the upper atmosphere.   From these earlier explorations, we now know that this environ-
ment--geospace--is a complex system composed of highly interactive parts
whose total behavior differs significantly from a simple linear sum of
the individual components.   While previous programs have advanced our
understanding of these geospace components individually, an understanding
of geospace as a whole requires a planned program of simultaneous observations and theoretical studies keyed to a global assessment of the
production, transfer, storage and dissipation of energy throughout this
system.   It is our current understanding of the various geospace components plus the long-awaited availability of required instrumentation
that allow us, for the first time, to define and plan a comprehensive
study of geospace as a whole.

     The past twenty-one years have brought us from the era of discovery,
through an era of exploration, to the beginning of an era of understanding of the complex physical processes in the geospace system.   Table 1
shows examples of this process.

Table 1.

EVOLUTION OF IN-SITU GEOSPACE RESEARCH

DISCOVERY — VAN ALLEN RADIATION BELTS
SOLAR WIND
MAGNETOSPHERE

EXPLORATION —		
BOW SHOCK	CUSP
MAGNETOSHEATH	MANTLE
MAGNETOPAUSE	AURORA
TAIL	PLASMASPHERE
NEUTRAL SHEET	RING CURRENT
PLASMA SHEET	SUBSTORMS

BEGINNINGS OF
AN UNDERSTANDING —

PARTICLE PRECIPITATION	—	EM CYCLOTRON INSTABILITIES
AURORAL PRECIPITATION	—	ES CYCLOTRON INSTABILITIES
		PARALLEL E FIELDS
PARALLEL E FIELDS	—	ANOMALOUS RESISTIVITY
		DOUBLE LAYERS
SUBSTORMS	—	TEARING MODE INSTABILITY
		MERGING

The discovery period was marked most notably by the discovery of the Van Allen radiation belts and the solar wind. The exploration period was characterized by the morphological delination of geospace components into localized substructures such as the bow shock, magnetosheath, plasmasphere, plasma sheet, neutral sheet, mantle, polar cusps, ring current, etc. The recent beginning of our understanding of specific geospace components and substructures has been characterized by the application of plasma physics to interpret observed phenomena such as particle precipitation via electromagnetic cyclotron instabilities, strong auroral precipitation via electrostatic instabilities, parallel electric fields via anomalous resistivity and double layers, substorms via the tearing-mode instability, and Alfvén waves in the solar wind.

It is clear that the scientific thrust of OPEN is intimately involved with a variety of questions in space plasma physics. In pursuing these questions OPEN will address directly the problems of magnetic field reconnection, the interaction of plasma turbulence and magnetic fields, the behavior of large-scale flows of plasma and their interaction with each other, acceleration of energetic particles, particle confinement and transport, and collisionless shocks.

2. THE OPEN OBSERVATIONAL LABORATORIES

Embedded in the geospace system are two major plasma sources--the solar wind and the terrestrial ionosphere--and two major storage regions --the geomagnetic tail and the near-Earth plasma sheet and ring current. These four basic geospace regions are interconnected by a complex network of transport processes which act to determine the highly interactive behavior of the system as a whole. This is illustrated by the topologically similar Figures 2 and 3. Consequently, it is necessary to

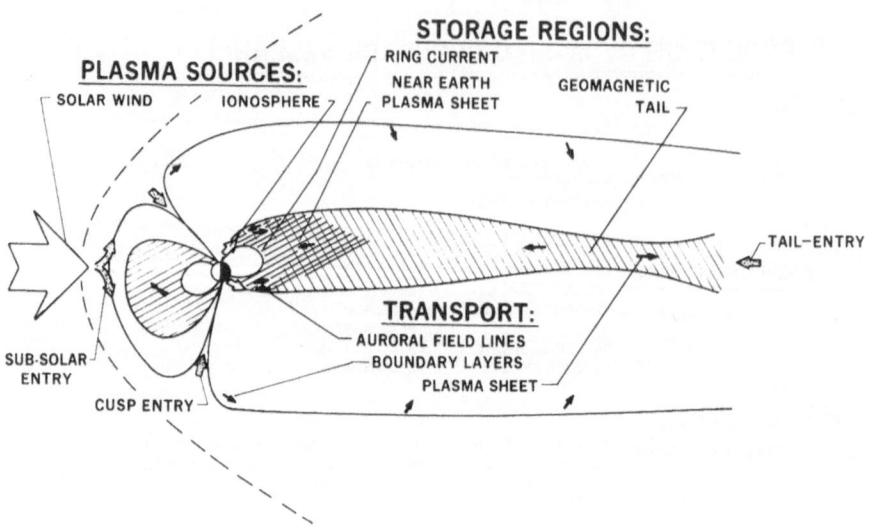

Figure 2. A schematic view of the entry of plasma from the solar wind
and ionosphere source regions, of the ring current and tail plasma sheet
storage regions, and of some of the transport paths which tie the system
together. Energization and dissipation can occur in all regions.

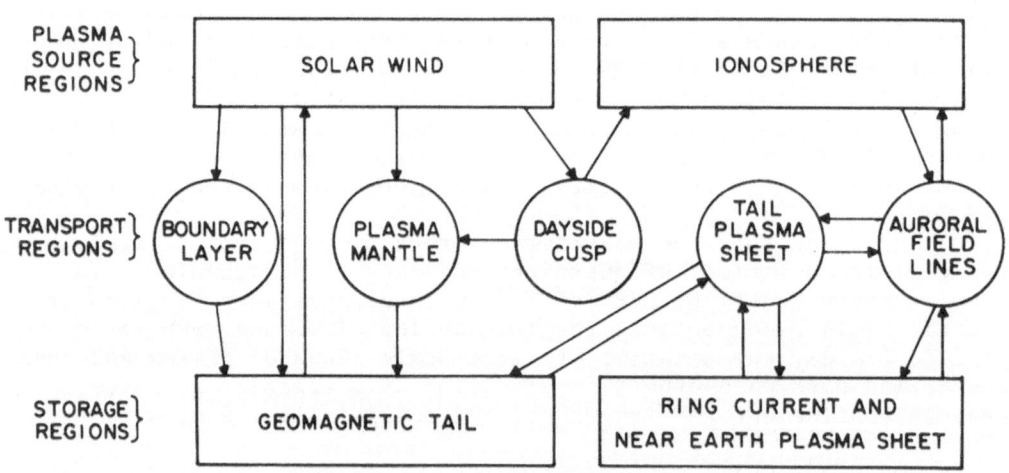

Figure 3. A simplified diagram of the major transport paths that couple
geospace source and storage regions. Since processes are highly time-
dependent, each of the key regions must be sampled simultaneously in
order to understand the behavior of the system as a whole. The OPEN
program will provide the first comprehensive attack on this problem.

obtain simultaneous, multipoint measurements in each of these key regions in order to meet the objective of a global view of the energy flow through the geospace system.

The OPEN Program described here is the minimum program capable of a global geospace study. In this program an observing laboratory with orbit-adjust capability is stationed to observe each of the two source and two storage regions. These laboratories are:

1)   Interplanetary Physics Laboratory (IPL)

The IPL will be stationed in the upstream solar wind near the sunward libration point. It will:

o   determine the characteristics of the solar wind source plasma,
o   provide the complete plasma, energetic particle and magnetic field input function for the complementary magnetospheric satellites, and
o   determine the magnetospheric input to interplanetary space.

(2)   Geomagnetic Tail Laboratory (GTL)

The GTL will utilize lunar swing-by orbit adjustments in order to maintain a distant apogee (80-250 R_E) in the magnetotail. The GTL will:

o   determine for the first time the characteristics of the distant geomagnetic tail,
o   with complementary magnetospheric satellites help determine the role of the distant tail in substorm phenomena and in the overall magnetospheric energy balance (energization, transport, storage and dissipation),
o   separate out the ionospheric and solar wind contribution to geomagnetic tail plasma, and
o   search for acceleration (reconnection, parallel electric fields, induction, etc.) processes.

(3)   Polar Plasma Laboratory (EPL)

The PPL will be placed in a highly eccentric polar orbit. An orbit adjust capability will be available to vary apogee radius in the 4-15 R_E range. The PPL will:

o   with complementary magnetospheric satellites help determine the role of the ionosphere in substorm phenomena and in the overall magnetospheric energy balance,
o   measure energy input through the dayside cusp and mantle regions,
o   determine characteristics of ionospheric plasma outflow from parallel electric field acceleration regions,

o   study characteristics of the auroral acceleration regions,
    and
o   provide global multispectral auroral images of the foot-
    print of the magnetospheric energy deposition into the
    ionosphere and atmosphere.

(4) Equatorial Magnetosphere Laboratory (EML)

The EML will be located in an equatorial 2 x 12 R_E orbit. It
will have an orbit-adjust capability to provide later a deep
tail orbit, thereby allowing (with GTL) simultaneous 2-point
samples of the distant tail. The EML will:

o   with complementary magnetospheric satellites help deter-
    mine the substorm trigger mechanism and the overall
    magnetospheric energy balance,
o   provide direct observations of the interactions of geo-
    magnetic tail and ionospheric plasmas in the equatorial
    magnetosphere,
o   measure the transport and storage of ionospheric and tail
    plasma in the near-Earth plasma sheet and ring current, and
o   measure the coupling of the solar wind to the magnetosphere
    at the subsolar magnetopause.

The orbit-adjust capability provides an observing flexibility
required by the comprehensive goals of the program. Figures 4 and 5
illustrate this concept of the OPEN Program by showing two of many
possible observational configurations which will be used to study the
global energy transfer problem. The figures schematically show the
major source and storage regions, locations of known and suspected
acceleration and dissipation, and the major transport avenues which
interconnect the basic geospace components.

Figure 4 shows the IPL measuring the input boundary conditions to
the magnetospheric system. The GTL, in the distant tail, will study
solar wind and ionospheric plasma entry and the effects of varying solar
wind input on entry, storage, acceleration, dissipation and transport
throughout these heretofore unexplored regions. Simultaneously the PPL
and the EML will perform complementary observations covering the high
latitude and subsolar magnetopause regions. In this case, not only will
the overall problem of entry be addressed, but also the connections
between the entry regions will be established by the simultaneous
observations from the OPEN laboratories.

In Figure 5, the IPL and GTL perform functions similar to those
described in Figure 4. However, the PPL has been placed in a low apogee
orbit to measure directly ionospheric acceleration and output to the
magnetosphere and to measure energy deposition into the atmosphere via
multispectral auroral imaging. To complete the picture of overall
energy flow, the EML will perform simultaneous observations of energy
flow, storage, acceleration and dissipation in the nightside equatorial

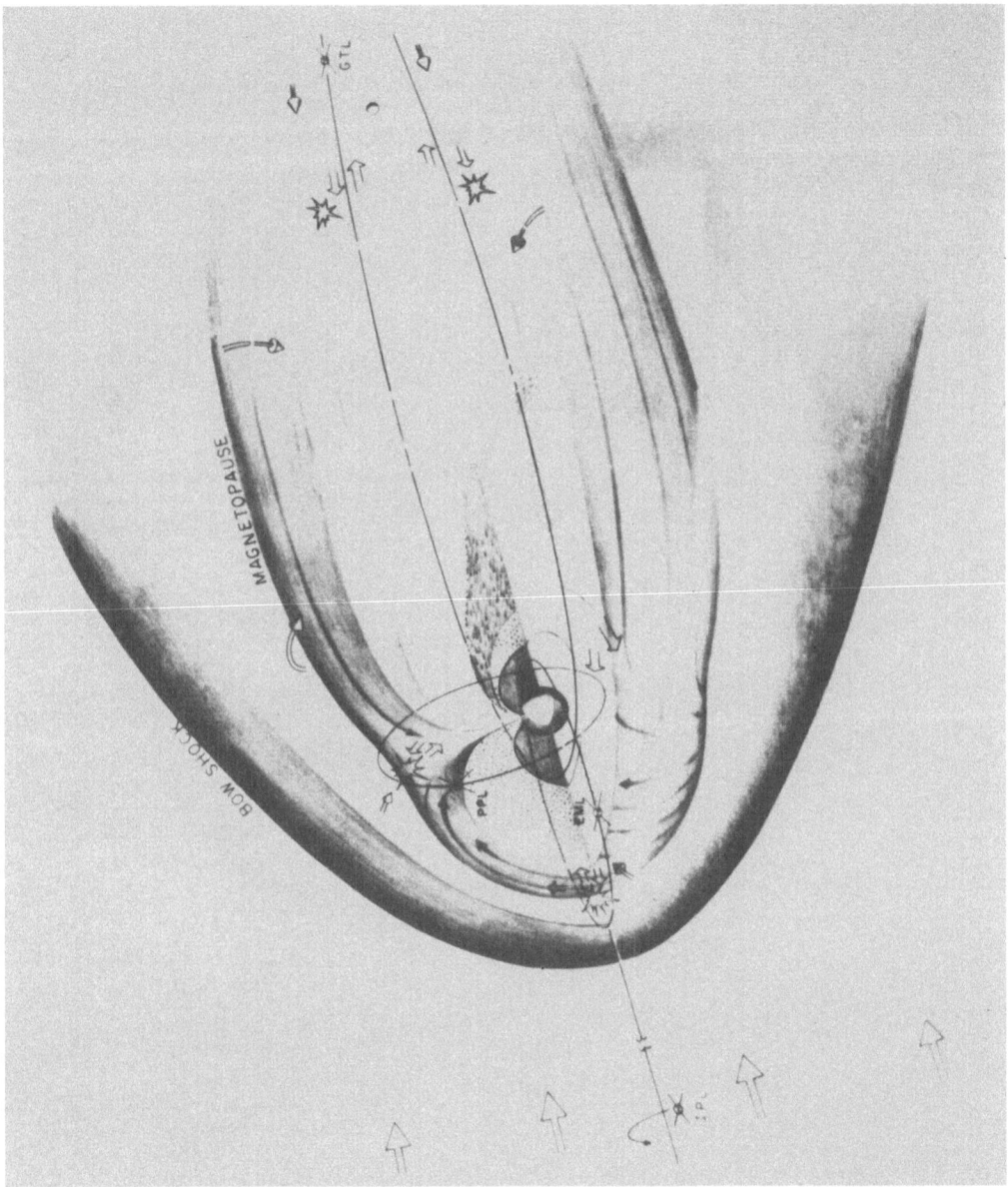

Figure 4. Example OPEN configuration for entry studies. Explosive symbols represent energization, three-dimensional arrows represent plasma entry into the magnetosphere, and open arrows represent transport. In this configuration the IPL measures the solar wind input; the PPL (in its high apogee orbit) and the EML measure entry, energization and transport spanning the dayside magnetopause from the subsolar point through the high latitude cusp regions; and the GTL measures entry, energization and transport in the distant, unexplored tail regions.

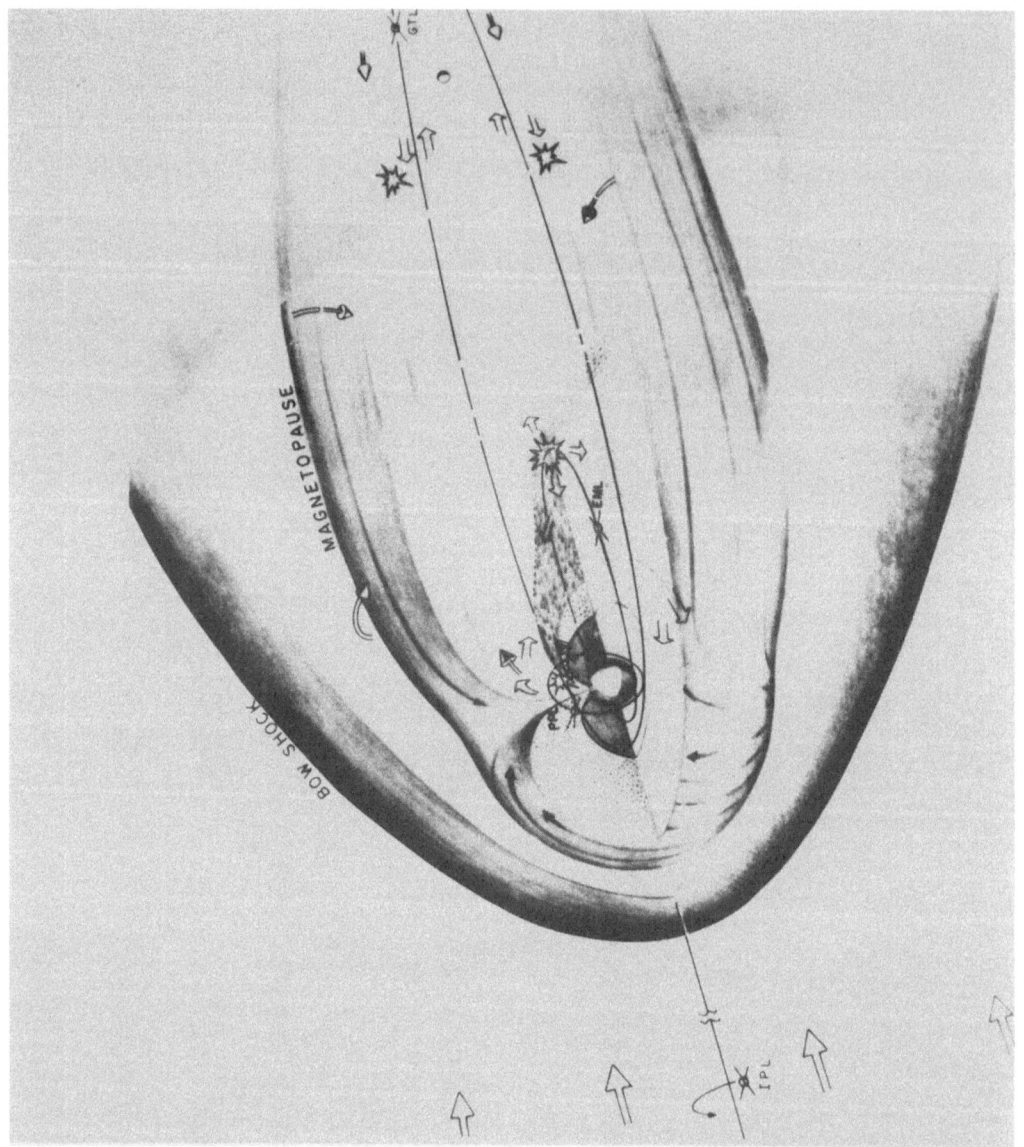

Figure 5.   Example OPEN configuration for internal magnetospheric plasma
flow studies.   Symbols are the same as in Fig. 4.   In this configuration
the IPL again measures the solar wind input.   However, the PPL, now in
its low apogee orbit, measures the ionospheric output to the magnetosphere and via multispectral imaging measures the atmospheric energy
deposition.   The EML (now in the nightside hemisphere), and the GTL
measure plasma mixing, energization, transport and storage in the
respective near-earth plasma sheet, ring current and geomagnetic tail
energy reservoirs.

magnetosphere--a region, largely unexplored, in which the ionospheric
and geomagnetic tail plasmas are thought to interact. The EML will
directly measure the amount of energy diverted from this region along
field lines to the atmosphere and will determine the relative amount of
energy diverted to flow across the field lines to the magnetopause.

The observational flexibility illustrated by Figures 4 and 5 is
required by the overall geospace energy flow problem. The orbit-adjust
capability also allows the transfer of a laboratory from one key region
of geospace to another and provides the capability for two-point tail
observations by injecting the EML into a tail orbit at the end of the
primary mission.

3.   THEORY

In addition to the observing laboratories discussed above, a vital
ingredient of the OPEN Program is the incorporation of a strong and
active theoretical studies and modeling program. The observational
goal of the OPEN Program is to provide an overall assessment of energy
flow throughout the geospace system. The physical goal of the program
is to use this observational data to construct physical models capable
of cause-effect predictions throughout the geospace system. Strong
theoretical and modeling input is envisioned from the inception of OPEN
to insure the accomplishment of this physical goal.

Theoretical progress in physics is usually closely linked to
observational discoveries and advances. In space physics, although
speculative predictions have occasionally preceeded observations,
quantitative theoretical models have developed only after detailed
satellite measurements have clarified the basic physical processes.
The reason theories of the earth's magnetosphere have usually lagged
observation is not hard to discern. The magnetosphere is such a complex
and dynamic plasma-physical system that most fundamental questions cannot
be resolved by theoretical reasoning alone. However, once the initial
observations were made, many theoretical models, especially in the area
of microscopic plasma dissipation were developed to a level of sophistication which exceeded that of the existing observations. Hence, the
future evolution of theoretical magnetospheric physics is directly
dependent on future satellite observations both to clarify those
physical processes still not understood and to challenge present theoretical models with complete plasma data for the first time.

The OPEN Program will settle several basic questions which have
seriously inhibited previous theoretical efforts. The first question
is what is the total mass, momentum, and energy budget to the magnetosphere? This bound represents the most basic constraint on all quantitative magnetospheric models and processes. The structure of the
distant tail, the closure of the plasma mantle flow within the tail,
the determination of the relative importance of reconnection and viscous
transport, the structure of the dayside equatorial and polar magneto-

spheric plasma, and the temporal morphology of magnetic storms and the ring current, substorms, and magnetospheric configuration changes are all theoretical issues which can be resolved only by the simultaneous observations from the four OPEN spacecraft. In these areas either the absence of information or major uncertainties in previously obtained data have stymied theoretical progress. Fairly detailed theoretical models have been developed to describe the microscopic plasma turbulent dissipation in the solar wind, the radiation belts, and in the auroral arc field-aligned current regions. For these problems, OPEN will obtain high resolution measurements of the complete ion, energy, and pitch-angle distributions which can then be directly compared with the theoretically expected distribution functions. Only when such microscopic comparisons are achieved can theory advance to quantitatively more accurate models.

As fundamental plasma dynamics are revealed and clarified by observations, theorists can begin to utilize the powerful computational tools of plasma particle simulation and hydromagnetic fluid codes in order to develop a complete understanding of solar wind and magnetospheric processes. Each microscopic dissipation mechanism should be studied until quantitative agreement with observations is achieved. Only then can these microscopic processes be incorporated into the development of global models of the entire geospace system. The ultimate goal of theory will be to construct global models that accurately predict the time-dependent response of this system. A first step toward this goal will be the global observational study of geospace by the OPEN mission.

4.   A SAMPLE PROBLEM

It is not only the present scientific maturity of geospace research that enables us to define and conduct a positive attack on the goal of assessing mass, momentum and energy flow throughout the geospace system. Without the evolution of spacecraft to their present weight, power, telemetry rate and orbit-adjust capabilities the measurements required could not be made. Spacecraft evolution and recent instrumentation breakthroughs now allow the measurement of a complete set of key physical parameters of the geospace plasma to be obtained. Critical instrumentation which has been developed recently or is being developed for approved missions includes ion composition and charge state measurements covering the few eV to several MeV energy range, three-dimensional plasma and energetic particle distributions, auroral imaging and active particle loss-cone scan instrumentation.

Here we give one example of how present scientific understanding couples with new instrumentation developments and makes the OPEN Program possible. Of the many examples available from geospace research, perhaps one of the clearest is the long standing problem of the earth's ring current. The ring current is one of the two major energy storage reservoirs in geospace and is responsible for the occurrence of worldwide geomagnetic storms. The major questions associated with this geospace energy reservoir are: (1) what is its source? (2) what are its constituent charged particles? (3) how does it develop? (4) how does it

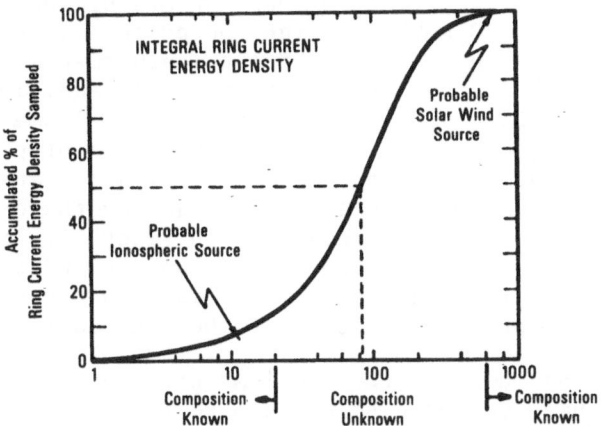

Figure 6(a). A major storage region--the earth's ring current. The accumulated percentage of ring current energy density is shown versus energy. Presently available direct composition measurements and inferred sources also are shown.

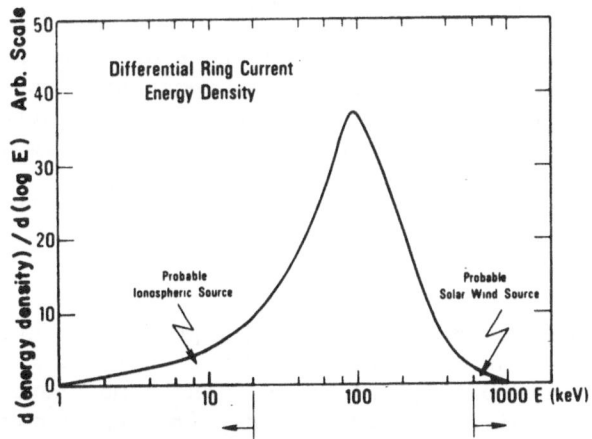

Figure 6(b). The differential d(Energy Density)/D(Log E) of curve (a) versus energy.

dissipate? and (5) how does it interact with the other components of geospace? Several years ago it was thought that the basic answers to these questions were known--the source of the ring current was assumed ultimately to be the solar wind, it was therefore made up of protons, it probably developed via convection of plasma from the geomagnetic tail, it dissipated via charge exchange with the neutral atmosphere, and it had no strong interaction with other geospace components.

The limited composition measurements that have become available in the past few years have completely changed our concept of this major geospace energy reservoir. Figure 6 summarizes what is known of the

ring current at present. Direct composition measurements exist for only
the low energy tail (< 20 keV) and high energy tail (> 600 keV) of the
ring current distribution. No composition measurements exist for 80-90%
of the energy density in this storage region. In addition, we also now
know that the interaction of the ring current hot plasma with the cold
plasma of the ionosphere results in the stimulation of plasma instabilities which cause significant ring current dissipation and ionospheric
heating. The most probable source identified from the existing
composition information also is indicated in Figure 6. Note that the
low energy ions have a probable ionospheric source and the high energy
tail has a probable solar wind source. No information exists as to how
these sources mix to produce the bulk of the ring current distribution.

Thus we see that none of the problems previously posed concerning
the ring current have yet been solved due to the present lack of required
measurements. However, our present scientific understanding allows us
to specify the problems quantitatively, and recent instrumentation advances now permit us to implement a program for their solution. Note
that for this specific example simultaneous measurements are required
of the ionospheric source, the solar wind source, the convecting geomagnetic tail plasma, and the equatorial mixing, acceleration, storage,
transport and dissipation of these plasmas.

A similar situation exists if we are to understand many other
important problems such as the role of the remarkable auroral magnetic
field-aligned acceleration regions in coupling the magnetospheric energy
storage regions to energy dissipation regions in the atmosphere, or the
global dynamics of the geospace system during substorm energy release
events, or the role of the distant geomagnetic tail in the solar-
terrestrial plasma chain.

5.   OPEN SUMMARY

To pursue the goals defined for OPEN a program of correlated
ground-based observations is required. While the direct energy input
represented by particle precipitation into the atmosphere will be measured by the PPL via multi-spectral auroral imaging, another intense
and complementary atmospheric energy input exists via Joule heating due
to large currents flowing at low altitudes. To assess this additional
heating it is necessary to have ground-based observations of electric
fields, conductivities, and currents. Thus an active program of radar,
magnetometer, photometer and riometer observations must be conducted
simultaneously with the OPEN spacecraft measurements. Figure 7 thus
summarizes the main elements of the OPEN Program.

6.   TIMELINESS AND RELATION TO SOLAR-TERRESTRIAL AND
     HELIOSPHERE STUDIES

1)   It is the next logical scientific thrust in geospace research.
Our understanding has evolved from an early exploratory phase through a
stage of detailed, localized investigations to our present capability
of undertaking a comprehensive, global program.

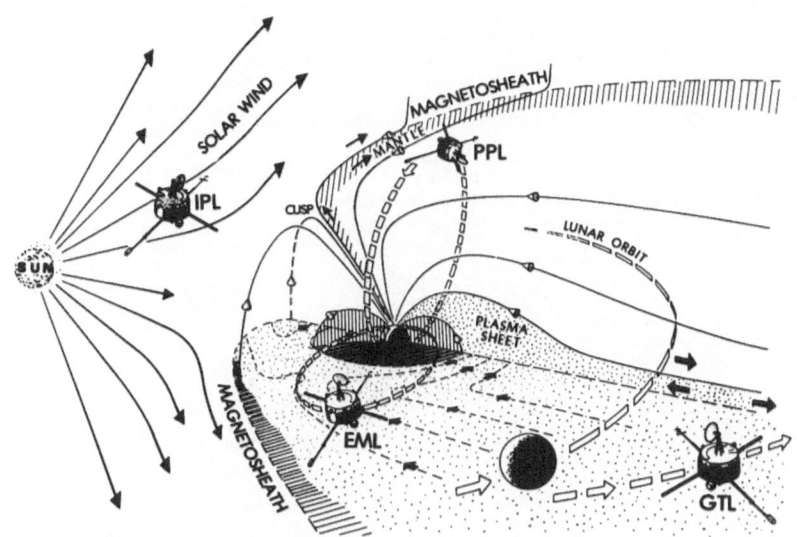

o Interplanetary Physics Laboratory (IPL), Sunward Libration Point
o Geomagnetic Tail Laboratory (GTL), Remain in Tail via Lunar Assist,
     80-250 R_E Apogee
o Polar Plasma Laboratory (PPL), Highly Eccentric Polar Orbit,
     4-15 R_E Apogee
o Equatorial Magnetosphere Laboratory (EML), Equatorial
     2 x 12 R_E, Orbit
o Theoretical Studies and Modeling Program, Centralized Data Base
o Ground Based Observations

Figure 7. A summary of the elements of the OPEN program. Each OPEN
spacecraft will have propulsion systems which provide substantial orbit
change capability so that a number of different orbit configrations can
be utilized during the course of the program.

    2) New instrumentation developments have also become available.
We can now, for the first time, scientifically and technically define
the program, and normal scheduling will implement the program in the
mid-1980's.

    3) OPEN is a strong, self-contained program in its own right.
However, it is also a key element in a larger, overall solar-terrestrial
program for the 1980's. Together with the Solar Polar Mission and the
Upper Atmosphere Research Satellite, OPEN will permit us to conduct a
coherent study of the transport of mass, momentum and energy from the
sun through the heliosphere, the magnetosphere, and the ionosphere and
into the atmosphere.

    4) Studies of the magnetosphere and heliosphere are complementary,
and simultaneous, coordinated investigations will be synergistic. In
both the heliosphere and magnetosphere there are bulk motions of magnetized plasma, irregular motions of energetic charged particles, and beams

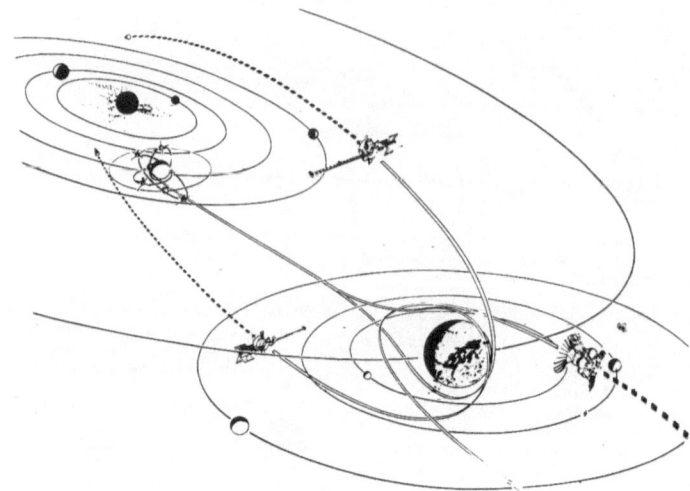

Figure 8. A possible solar terrestrial studies configuration in the
1980's. Shown are the OPEN laboratories, the Solar-Polar spacecraft,
and the Galileo orbiter. Voyagers 1 and 2 will be in the ecliptic at
distances of ∿ 8 AU and 15 AU.

of suprathermal ions and electrons. Similar types of plasma physics
problems exist in both regions: what are the sources of plasmas, magnetic
fields and energetic particles, how are thermal plasmas and energetic
particles accelerated to form directed flows and suprathermal populations,
and what are the energy transport and "loss" (conversion) processes?
Previous studies have already shown that an understanding of physical
processes in the magnetosphere can contribute to understanding processes
in the heliosphere, and vice versa. The 1980's provide a timely opportunity for an organized effort to study problems common to both fields.

    The OPEN study is part of an investigation of solar-terrestrial
dynamics (Figure 8). It will take place during an investigation of
global heliospheric dynamics by: (1) the International Solar Polar
Mission (which will move from near Jupiter at 5 AU and pass over the
sun's poles at a distance of 1 to 2 AU, (2) Voyager-1 and -2 (which will
be in the ecliptic, moving between ∿ 8 AU and 15 AU), and (3) Galileo
(in orbit around Jupiter). The IPL, stationed in the solar wind near
earth, is a vital part of both the heliospheric and geospace studies.
It will serve as a baseline for heliospheric investigations and a remote
sensor of solar conditions. It will identify the processes between the
Sun and Earth which link solar and terrestrial activity, and it will
provide input functions for magnetospheric studies at Earth and, to
some extent, at Jupiter.

    5)   Since geospace-like systems appear to be common elements
throughout the universe (Figure 9) the knowledge gained from the OPEN

Program can be applied directly to various planetary and astrophysical studies. While the boundary conditions and scale sizes vary significantly from one system to another, we expect that a firm understanding of the overall geospace system will contribute fundamentally to studies of other cosmic plasma systems.[2,3]

6)   A recent comprehensive study of space plasma physics[2] has shown that this field is of fundamental significance and that future progress in the field will depend on a balanced program of theoretical and observational studies that can systematically address basic scientific problems. OPEN is directly responsive to the specific recommendations of that study.

7)   Knowledge gained of the geospace system has already been used to provide routine and specialized services to societal systems affected by geospace perturbations.[2,4,5] For example, alerts, warnings and forecasts of conditions throughout the geospace system are supplied to national defense systems and to communication, power and oil exploration industries as an aid to their day-to-day operations.[5] Variations in the geospace environment also lead to significant variations in energy deposition in the atmosphere and may have important consequences in the Earth's weather and climate. Thus an understanding of the geospace system as a whole is expected to have practical benefits. OPEN can develop the physical basis for assessing solar wind particle and field perturbations to atmospheric processes and for routinely monitoring those space plasma processes that affect satellite communications, satellite orbit lifetimes, and terrestrial power and exploration geophysics systems.

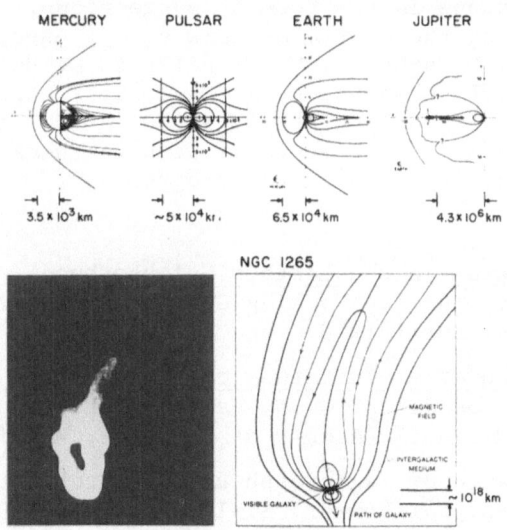

Figure 9.   Geospace-like magnetosphere systems are probably common throughout the universe even though their scale sizes vary over an enormous range. For example, the subsolar magnetopause distance for Mercury is 3.5x10³ km; for the radio galaxy NGC 1265 the analagous distance is roughly 10^{18}!

ACKNOWLEDGEMENT

I gratefully acknowledge the two years of exceptional effort given
by the Science Definition Working Group, and especially my Co-Chairman,
J. K. Alexander. Members of the group were:

Co-Chairman	J. K. Alexander	Goddard Space Flight Center
	H. S. Bridge	Massachusetts Institute of Technology
	J. L. Burch	Southwest Research Institute
	L. F. Burlaga	Goddard Space Flight Center
	D. P. Cauffman	Lockheed Palo Alto Research Laboratory
	F. V. Coroniti	University of California, Los Angeles
	W. C. Feldman	Los Alamos Scientific Laboratory
	L. A. Frank	University of Iowa
	T. A. Fritz	NOAA Space Environment Laboratory
	J. Geiss	University of Bern, Switzerland
	G. Gloeckler	University of Maryland
	G. Haerendel	Max-Planck Institute, Astrophysics
	R. E. Hartle	Goddard Space Flight Center
	R. G. Johnson	Lockheed Palo Alto Research Laboratory
	S. M. Krimigis	JHU, Applied Physics Laboratory
	L. R. Lyons	NOAA Space Environment Laboratory
	A. Nishida	University of Tokyo
	B. Whalen	National Research Council of Canada
Co-Chairman	D. J. Williams	NOAA Space Environment Laboratory

I also wish to acknowledge the many individuals who provided
valuable contributions in the form of suggestions, ideas, information
and hard work during the course of this study, especially S.-I. Akasofu,
R. W. Farquhar, W. D. Hibbard, T. W. Hill, B. C. Holland, D. Hovestadt,
C. E. McIlwain, D. P. Muhonen, N. F. Ness, G. P. Newton, G. A. Paulikas,
G. T. Roach, E. R. Schmerling, D. P. Stern, A. F. Timothy, D. D. Wallis
and J. D. Winningham. I particularly wish to thank K. Papadopoulos for
his invaluable contributions to the program.

REFERENCES

1. Origin of Plasmas in the Earth's Neighborhood. Final Report of
the NASA Science Definition Working Group, Co-Chairmen J. K. Alexander
and D. J. Williams: April 1979.

2. Space Plasma Physics: The Study of Solar-System Plasmas. Vols. I &
II; Space Science Board. A. G. W. Cameron, Chairman (S. A. Colgate,
Study Chairman): National Academy of Sciences, Washington, D.C.: 1978.

3. "The Significance of Magnetospheric Research for Progress in
Astrophysics"; C.-G. Fälthammar. S.-I. Akasofu and H. Alfvén, Nature,
275, 185: 1978.

4. Measures for Progress: A History of the National Bureau of
Standards. Rexmond C. Cochrane, U. S. Department of Commerce: 1966.

5. Watch Upon A Star: A Description of Space Environment Services.
NOAA PA 77002, U. S. Department of Commerce: 1977.

A PROGRAM FOR THE OBSERVATIONS OF
THE SUN AND HELIOSPHERE FROM SPACE 1980-1995

J. David Bohlin and Eric G. Chipman
Headquarters
National Aeronautics and Space Administration
Washington, DC 20546

ABSTRACT

Recent, fundamental discoveries of the phenomena of the Sun
and of interplanetary space have led to a far broader
definition of the term "solar physics" than was generally
perceived a decade ago. The implications of this broadened
definition of solar and heliospheric physics will be studied
by essentially every solar space mission either now
approved, or in the planning stage, for the period of 1980
to 1995. These missions include traditional Earth-orbiting
satellites; Shuttle/Spacelab sortie missions, free flyers
that transit the solar polar caps and probe the innermost
corona (both frontier regions of the heliosphere), and
finally possible semi-permanent orbiting platforms for
advanced solar/heliosphere observations.

1. INTRODUCTION

The decade of the 1970's saw the discovery of a wide variety
of phenomena in both solar and interplanetary physics using
space as well as ground-based techniques. Among the
observations made from space was the discovery of solar
coronal holes and the identification of these holes as the
source of the high speed solar wind streams that give rise
to recurrent geomagnetic storms. A second observation was
that regular expulsions of large quantities of mass from the
Sun - called coronal transients - accompany nearly every
form of energetic solar event. A third major discovery was
that the large-scale, extended structure of the solar wind
can be described by a so-called "warped current sheet"
model, at least during the declining phase of the solar
cycle, which in turn can be related to the large-scale
pattern of the Sun's photospheric fields. These

M. Dryer and E. Tandberg-Hanssen (eds.), Solar and Interplanetary Dynamics, 523-539.
Copyright © 1980 by the IAU.

developments have led to an intense reawakening of interest
in the relationship of solar activity to interplanetary
phenomena, so much so that the space about the Sun is now
increasingly referred to as the heliosphere.

A second class of discoveries relate to the interior of the
Sun. Among these are that the observed neutrino flux is
only about one-third as much as predicted by standard models
of a solar-type star; that the solar cycle apparently
disappeared for the 70 year period from 1645 to 1715 (the
Maunder minimum); that the five minute p-mode oscillations
of the Sun can be used to probe the interior structure and
rotation of the Sun (sometimes referred to as "solar
seismology"); and that the total solar irradiance (the
"solar constant") may vary at a level of a few tenths of one
percent on a long time scale, if not with the solar cycle
itself.

In order to pursue all of the implications of these discoveries, a number of the major space programs now started
or being considered for the period 1980 to 1995 emphasize
both the relationship of solar to heliospheric phenomena and
solar interior dynamics in addition to the more traditional
aspects of solar physics. The rest of this paper will
briefly discuss each of these missions to the extent that
the current level of planning allows.

2.   THE SOLAR MAXIMUM MISSION

The Solar Maximum Mission (SMM), has as its primary goals
the study of solar flares, activity-associated coronal
phenomena, and the long term monitoring of the total solar
irradiance. The phenomenon of the solar flare is one of the
oldest problems of solar physics, and whereas the Skylab
missions in 1973-74 provided some of our most incisive
observations yet of these events, the flare still lacks a
satisfactory explanation. Since the hottest parts of the
flare, and thus perhaps the key to the mechanism itself,
appear in the XUV, X-ray, and gamma-ray portions of the
spectrum, the majority of the payload of the SMM consists of
these types of experiments (Table 1). Another one of the
major discoveries of the Skylab/ATM solar experiments was
that virtually every energetic solar event (flare, eruptive
prominence, or surge) is accompanied by a coronal transient.
Thus, the SMM also carries a coronagraph to monitor the
location, frequency, intensity, and polarization of these
spectacular events.

Table 1.
Payload of the Solar Maximum Mission

SMM Experiment	Spectral Range	Spatial Resolution
• Gamma-Ray Spectrometer	0.3-17 Mev	Full Sun
• Hard X-Ray Spectrometer	20-300 Kev	Full Sun
• Hard X-Ray Imaging Spectrometer	3.5-30 Kev	8" x 8"
• Soft X-Ray Polychromator	1.4-22.4 Å	10" x 10"
• UV Spectrometer and Polarimeter	1100-3000 Å	4" x 4"
• Coronagraph/Polarimeter	4435-6583 Å	6.4" x 6.4" or 12.8" x 12.8"
• Solar Constant Monitoring Package	UV-IR	Full Sun

One of the major benefits of this time phasing of the SMM will be the ability to correlate the coronal observations with in situ solar wind measurements (composition, ion and electron velocity distributions, magnetic fields, and plasma waves) made by the ISEE-3. This spacecraft is the third of the trio of the International Sun-Earth Explorer series, and is located in a "halo orbit" about the Sun-Earth gravitational libration point located about 0.01 AU toward the Sun along the Earth-Sun line. Thus, the ISEE-3, launched in August, 1978, is always outside the Earth's magnetopause and continuously immersed in the solar wind. (The ISEE-1 & 2 spacecraft orbit the Earth such that they pass through the magnetosphere). Thus, this combination of the SMM coronagraph and ISEE-3 will provide a major set of solar activity/solar wind data at the time of the maximum of the Sun's cycle, to complement the Skylab data taken during the declining phase of the last solar cycle.

A combined group of SMM Principal Investigators, Co-Investigators and Guest Investigators have been meeting regularly for the past two years to plan the SMM mission operations. These operations will be carried out from an elaborate Experiment Operations Facility (EOF) at the Goddard Space Flight Center, which has project management of

the mission. Extensive plans for near real-time evaluation
and comparison of data will also be carried out at the EOF,
thus enabling the rapid updating of observing modes to
capitalize on the continuously evolving patterns of solar
phenomena. The SMM will operate for a minimum of two years
with an option for a third, depending on spacecraft health,
value of returned data, and availability of funds.

3.   THE THREE DIMENSIONAL STRUCTURE OF THE HELIOSPHERE

3.1.   The International Solar Polar Mission

One of the few remaining "frontiers" of the heliosphere is
the third dimension well above and below the ecliptic. The
new effort which will explore this region of space is the
International Solar Polar Mission (ISPM), sponsored jointly
by the NASA and the European Space Agency (ESA). The
mission architecture of the ISPM will be to send two
similarly instrumented spacecraft, one built by the ESA and
one by the NASA, out of the ecliptic plane and over the
poles of the Sun itself (Figure 1).

There are three main categories of science objectives for
the ISPM:

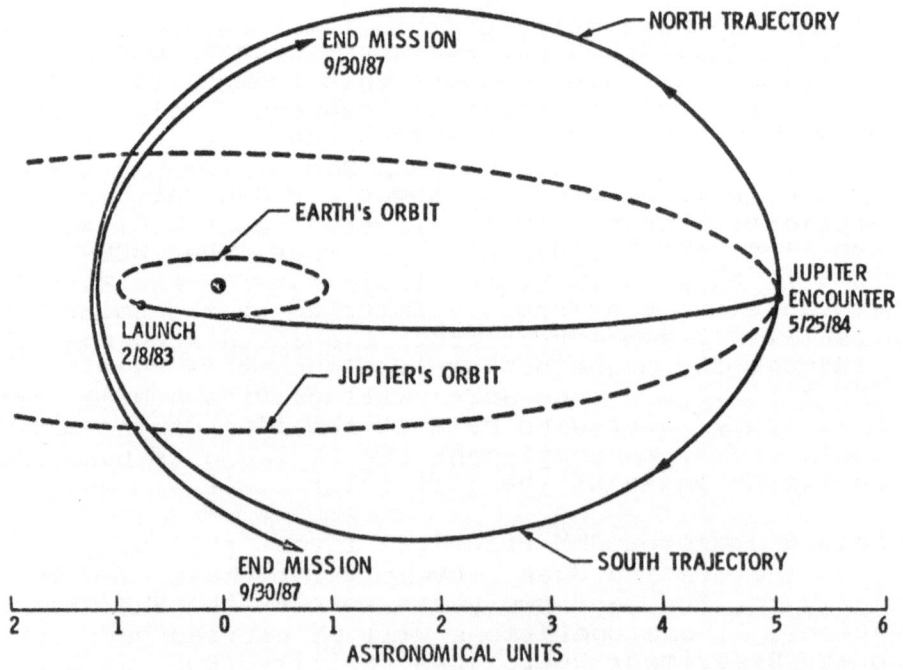

Fig. 1 Schematic of trajectories of the two ISPM spacecraft.

o First, both theory and observations give strong
reason to believe that the solar wind from the Sun's polar
caps should have high velocity (500-800 km/s) owing to the
presence of the coronal holes usually present at the solar
poles. This polar solar wind should be far more homogeneous
and radial in structure than the wind from the lower,
equatorial "active" solar latitudes.

o Second, the polar view afforded by the ISPM
trajectory is ideal for observing in an unambiguous manner
the longitudinal structure of the inner solar corona (out to
10 solar radii or so). This longitudinal structure is basic
to understanding how coronal holes and streamers relate to
the solar wind farther out in the heliosphere.

o Third, the ISPM will have the opportunity to observe
the nearly unmodulated spectrum of the cosmic rays, since
these particles will only have to traverse the nearly
radial, weak polar magnetic field lines rather than the
tangled, stronger solar wind fields toward the ecliptic.

The ISPM is a joint venture between the NASA and the ESA,
with each agency providing one spacecraft. The management
centers for the two spacecraft are the Jet Propulsion
Laboratory and the European Space Research and Technology
Center, respectively. The payloads of the two spacecraft
are similar and divided roughly equally between US and
European investigators (Table 2). The major difference
between the two spacecraft is the presence of a coronal
imaging investigation, consisting of a coronagraph and soft
X-ray telescope, on a Sun-pointing gimbal on the NASA
spacecraft. This experiment will thus allow imaging of both
the inner and outer corona from the unique perspective of
the solar poles.

Both spacecraft are to be launched simultaneously by the
Shuttle and then boosted by an auxiliary rocket on a transit
trajectory in the ecliptic to Jupiter (Figure 1). At
Jupiter, about 1.3 years after launch, both spacecraft
undergo gravitational swing-bys, one directed north-bound
and one south, out of the ecliptic plane and back toward the
Sun in "mirror" images of each other. The two spacecraft
pass over the north and south solar polar caps approximately
3.8 years after launch, at a distance that will not exceed 2
AU, then swing down to perihelion at 1 AU on their way to
respective passages over the other solar poles. Thus, both
spacecraft will pass over both solar poles in the period
from mid-1986 to early 1987. The total nominal mission
duration from launch to shortly after the second polar
passages is five years. For the February, 1983, launch

window, it is anticipated that each spacecraft will spend
about 50 days above 70º heliographic latitude. This
duration compares favorably to the rotation period of about
35-40 days of the polar caps, over which significant
evolution of the solar polar conditions might occur.

3.2. In-Ecliptic Observations with the ISPM

Even though the ISPM will make fundamental and exciting

Table 2.
Payloads of the International Solar Polar Mission
(* U.S. PI; ** European PI)

NASA SPACECRAFT	ESA SPACECRAFT
SOLAR PHYSICS	
• White light/X-ray Coronal Telescopes*	
• X-ray Burst Monitor*	• X-ray Burst Monitor**
• Energetic Particle Detectors*	• Energetic Particle Detectors*
• Radio Burst Antenna*	• Radio Burst Antenna*
SOLAR WIND PHYSICS	
• Magnetometers*	• Magnetometers**
• Plasma Velocity/Density Exp.**	• Plasma Velocity/Density Exp.*
	• Plasma Ion Composition* &**
	• Plasma Wave Detector*
COSMIC ASTROPHYSICS	
• Cosmic Ray Spectrometer*	• Cosmic Ray Spectrometers*; **
• Gamma-Ray Burst Exp.*	• Gamma-Ray Burst Exp.**
INTERPLANETARY MEDIUM	
• Zodiacal Light Photometer**	• Interplanetary Dust Exp.**
• Interplanetary Gas Exp.**	
RADIO SCIENCE & GRAVITY WAVES	
3 U.S. PI's; 2 European PI's	
THEORETICAL & INTERDISCIPLINARY PHYSICS	
5 U.S. PI's; 3 European PI's	

observations by itself alone, its observations will be considerably enhanced by in-ecliptic space measurements of the solar wind plasma and the solar corona. Actually, owing to the well-known "convergence" effect of the high-latitude corona towards the solar equator, such simultaneous in-ecliptic measurements with the out-of-ecliptic ISPM may reveal that a substantial portion of the solar wind flow at the Earth may originate from the edge of the solar polar holes themselves.

The mechanism for accomplishing these in-ecliptic wind measurements is the Interplanetary Plasma Laboratory (IPL) of the space program called the Origin of Plasmas in the Earth's Neighborhood (OPEN). This OPEN program is designed to study the plasma sources, sinks, and processes within the Earth's magnetosphere and ionosphere using a system of four satellites: the IPL measures the "input" solar plasma, whereas the other three spacecraft are located at various places within the magnetosphere (and thus shielded from direct access to the solar wind). Thus the IPL, which carries a relatively standard complement of solar wind detectors, will serve the double function of monitoring the in-ecliptic wind for both the OPEN program and the ISPM. The plan is to launch the IPL in 1985, in time to be in its halo orbit about the Sun-Earth libration point when the ISPM spacecraft begin their out-of-ecliptic ascent following Jupiter swing-bys.

The in-ecliptic observations of the visible light and X-ray corona would be accomplished by a Solar Cycle and Corona Mission, which is part of a Solar Cycle and Dynamics program discussed in detail in Section 5 below.

4. THE SOLAR SHUTTLE/SPACELAB MISSIONS

4.1. The Shuttle/Spacelab Transportation System

The Space Shuttle/Spacelab System is considered a keystone to the NASA's solar physics programs of the 1980's through its ability to carry payloads of experiments for discipline-dedicated sortie missions of up to 30 days. The advantages of these Shuttle missions for solar physics, in spite of their relatively short length, are several:

o The Shuttle can carry much larger experiments than can the typical free flying spacecraft;

o The pointing and operation modes can be

under direct, real-time control of on-board crew
members (Payload and/or Mission Specialists);

o Sub-arc second pointing will be possible through
   a variety of ESA and NASA pointing gimbal
   systems; and

o The instruments can be retrieved for
   refurbishment, evolutionary upgrading, and/or
   changing film canisters.

The expected mode of operation for the Shuttle science
missions is through the Spacelab system, which is an array
of pressurized modules (for manned operations consoles)
and/or unpressurized pallets (for experiments) that are
mounted in the Shuttle bay (Figure 2). These Spacelab
systems are being built by the ESA as part of a joint
US-European venture. The general plan for Shuttle/Spacelab
sortie missions is to mount preintegrated experiment pallets
and the module (if required) into the Shuttle bay at the
Kennedy Space Center. After launch and in-orbit checkout,
the instruments are operated by the on-board crew working in
close collaboration with scientists at the Payload
Operations Control Center (POCC) at the Johnson Space
Center.

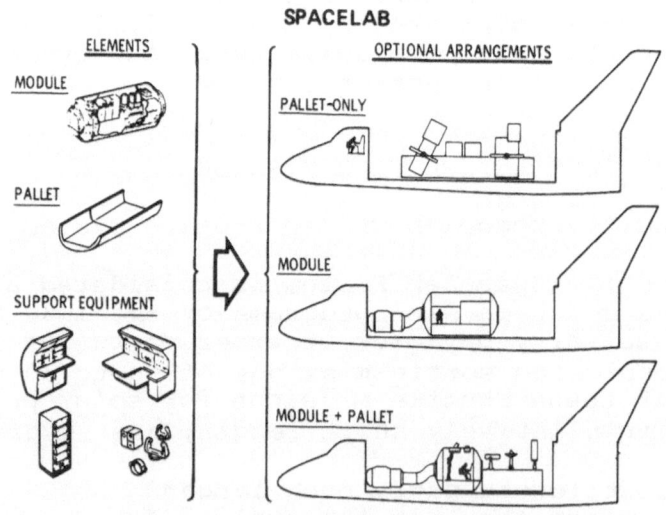

Fig. 2 Schematic of Shuttle with different Spacelab
        module and pallet configurations.

4.2. Shuttle/Spacelab Experiment Categories

Two classes of experiments will be flown: First are the
Principal Investigator (PI)-class experiments for which
Announcements of Opportunity (AO's) have been released.
Several fundamental aspects are envisaged for the on-going
Spacelab opportunity which differ from the usual PI flight
opportunities (Figure 3). First, the AO is not for a given,

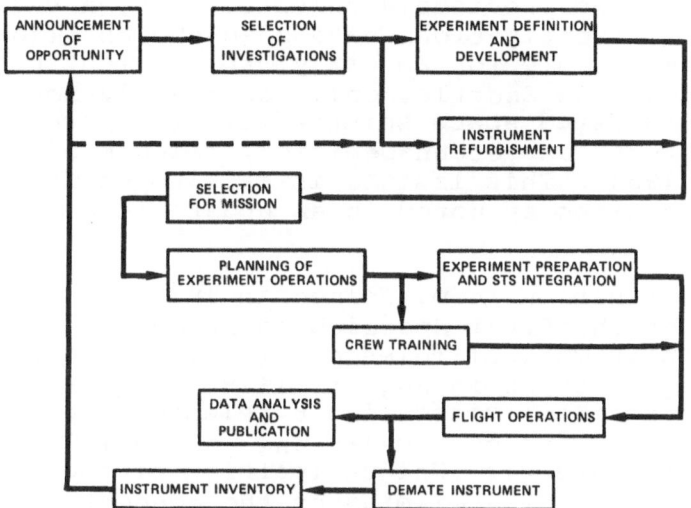

Fig. 3 Evolutionary schematic of Principal Investigator
        Class Spacelab experiments.

single flight. Rather, selected experiments are to be
developed in parallel to a state of flight readiness, and
then a second review will choose from among those
experiments for a designated Shuttle flight. Second, the PI
will be solely responsible for delivering a functioning,
documented instrument; the NASA will exercise only general
control to insure compliance with negotiated cost,
interface, and safety requirements. Third, all developed
instruments will be held in inventory and are expected to be
reflown a number of times, either by the original PI or by
other investigators who propose to use them.

The second class of Shuttle experiments is that of facility
or Multi-User Instruments (MUI's), whose cost and
versatility for the science community puts them in a class
apart from the lower cost, more single-purpose PI
instruments. An MUI is developed by a Facility Definition
Team (FDT) chosen by the NASA and is developed as a NASA

project through one of the NASA centers. MUI's have a much
higher cost than PI experimens and work only in conjunction
with other focal plane and/or auxiliary instruments.   In
general, these auxiliary pieces of equipment will be
solicited by AO from the science community. Once developed
and tested, they become part of the facility system and thus
available for a variety of Guest Investigations.

4.3. Solar Physics Spacelab Investigations in the 1980's

As a result of two previously released AO's, a variety of PI
solar physics investigations have been chosen (Table 3).
These experiments are for an early Orbital Test Flight
(mainly to test the Shuttle, only secondarily for science)
in 1981; a dedicated space science flight, Spacelab 2, in
1982; and future, as yet unspecified, flights of Spacelab
starting in 1983. This last AO is still open for a second
round of selections as noted in an update issued in June,
1979.

A Facility Definition Team, formed in 1975, resulted in the
development of the first Spacelab solar physics facility,
called the Solar Optical Telescope (SOT), which was approved
as a new project start in August, 1979. The basic SOT is a
1.25 m, diffraction-limited, f/3.6 telescope designed to
accommodate a variety of focal-plane and auxiliary PI-class
instruments (Figure 4). Unique features of the SOT are its
Gregorian design, which eliminates the concentrated heat
load on the secondary mirror (at the expense of
field-of-view); a fine focus/alignment system using six
linear drives to translate the primary mirror through all

Table 3.
Solar Physics Investigations Selected for Shuttle/Spacelab

Orbital Test Flight 4
- Solar UV Spectral Irradiance Monitor
- Solar Flare X-Ray Polarimeter

Spacelab 2
- Coronal Helium Abundance Experiment
- Solar UV High Resolution Telescope & Spectrograph
- Solar Magnetic & Velocity Field Measurement System

Future Spacelab Flights
- Solar Optical Telescope Facility (including baseline
   of two focal plane instruments)
- Lyman-Alpha & White Light Coronagraphs
- X-Ray Telescope & Spectrometer System
- Solar EUV Telescope & Spectrograph

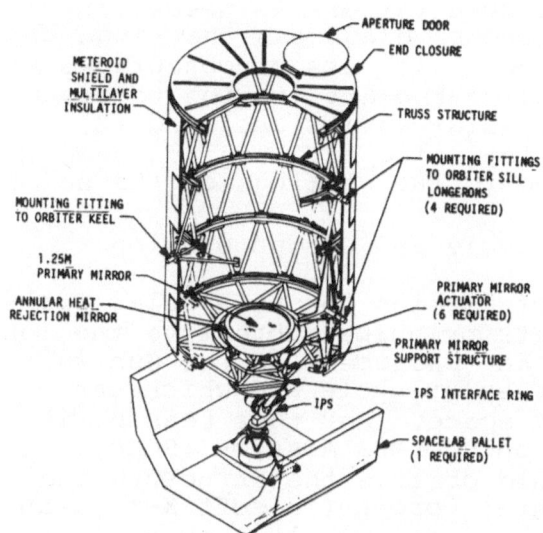

Fig. 4   Cutaway of the basic Solar Optical Telescope.

six degrees of freedom (thereby accomplishing focus and
alignment continuously in-orbit); and a large (3.8 m dia)
truss structure that can accommodate up to six focal plane
instruments and three, 1 m diameter, PI-class or other
facility instruments simultaneously. Thus, the fully
configured SOT alone offers the possibility of up to nine
instruments all co-aligned and co-mounted on a single
pointing system within the Shuttle bay.

5.   THE SOLAR CYCLE AND DYNAMICS PROGRAM

5.1.   Background

The cyclic nature of solar activity has been known since the
1840's and recorded with relatively high accuracy back to
the early 1700's. However, in the last decade, several
facts have emerged which indicate that the Sun's cycle and
thus, by implication, its structure are not nearly as
regular or as well understood as had been thought (Section
1).

These findings, combined with the fact that the solar dynamo
itself is not understood, led to a recent study to identify
the key questions of solar variability (to the extent they
can currently be identified) and to determine which of these
questions would benefit from study by space techniques. The
conclusion of this Solar Cycle and Dynamics study was that a

program of solar observations is needed involving probably
two missions in order to address the range of problems.
Briefly, these two basic sets of objectives would be met by
(1) a fully state-of-the-art mission to measure certain
known aspects of solar variations; and (2) a far more
exploratory mission using advanced instruments to measure
phenomena related to the workings of the solar dynamo itself.

5.2.   The Solar Cycle and Corona Mission

The first type of mission for this Solar Cycle and Dynamics
Program has as its principal objective the long term
observations of key phenomena of the Sun which are now known
to vary with the solar cycle and which can be only, or best,
carried out from space.   For this reason, it has been called
the Solar Cycle and Corona Mission (SCCM).   In particular,
this mission would observe the structure and evolution of
the inner and outer corona (in soft X-rays and white light,
respectively), the resonance line corona (to infer outflow
velocities of the solar wind), the total and spectral solar
irradiances, and perhaps the photospheric velocity and
magnetic fields.

The technology for all of these instruments is currently in
hand, and little if any new development work is required.
Ideally the mission would be launched in early 1986 for a
nominal three-year lifetime, in a standard, low-inclination
Earth orbit.   As such, the SCCM is also an ideal complement
to the ISPM and the IPL experiment discussed in Section 3.
In particular, the SCCM would allow a 90⁰ "stereo" view of
the corona in conjunction with the ISPM, and would thus
fully exploit those out-of-ecliptic observations.

5.3.   The Solar Dynamics Mission

The second part of this Solar Cycle and Dynamics Program
would be far more ambitious than the SCCM in that the
required instrument technologies are not all currently
available.   The thrust of this second mission is to
investigate the interior workings of the solar dynamo
itself, and so it is aptly named the Solar Dynamics Mission.
To this end, the payload should include an ultra-high
precision photospheric velocity analyzer (with a sensitivity
of 1 m/sec in radial velocity on the Sun's surface); a solar
diameter and global oscillations telescope to detect
possible oblateness and/or global oscillations of the Sun;
and an ultra-high precision differential intensity analyzer
to measure extremely small differences in temperature over
large-scale solar convection cells.   The first instrument
would also serve to measure p-mode oscillations, which act
as a probe of the solar interior.

This mission should have a polar Earth orbit to provide
continuous sunlight and will require sub-arc second pointing
stability. It is reasonable that such instruments could be
proof-tested on Shuttle sortie missions before committing to
a free-flying satellite. A final mission duration of the
order of several years is required.

6.   AN ADVANCED SOLAR OBSERVATORY

A natural consequence of an agressive program of Shuttle
flights in solar physics in the 1980's will be a rather
complete inventory of sophisticated, large solar
experiments. A strong case can be made for flying a subset
of these experiments on some type of permanent orbiting
Space Platform in the time frame of the early 1990's for the
following reasons.

First, the Shuttle sortie flights will have necessarily been
of rather short duration (30 days or less), and thus any
solar phenomena having a long evolutionary period or low
event rate will not have been satisfactorily observed.
Second, the next maximum of solar activity should occur in
the early 1990's, and observation of flares with these new
high resolution instruments requires both long orbit-stay
times and large free-flyer capabilities.

Combined, these requirements call for an Advanced Solar
Observatory (ASO), visited periodically by the Shuttle for
servicing, which can make full, continuing use of the
experiments and facilities (including the SOT) previously
developed. (In fact, it is conceivable that the Solar
Dynamics Mission would constitute one subset of the
instruments on an ASO.) The basic concept would be for the
Shuttle to mount pre-integrated Spacelab pallets of
experiments to a basic orbiting bus that provides power,
telemetry, and orientation services. Such platform concepts
are currently under study for feasibility, and could be
orbited by the late 1980's.

7.   THE SOLAR PROBE

Perhaps the most audacious mission under consideration is
that of the Solar Probe, the intent of which is to send a
spacecraft to within three solar radii of the Sun's surface
(Figure 5). The science objectives for this mission were
discussed at the Jet Propulsion Laboratory in a symposium in

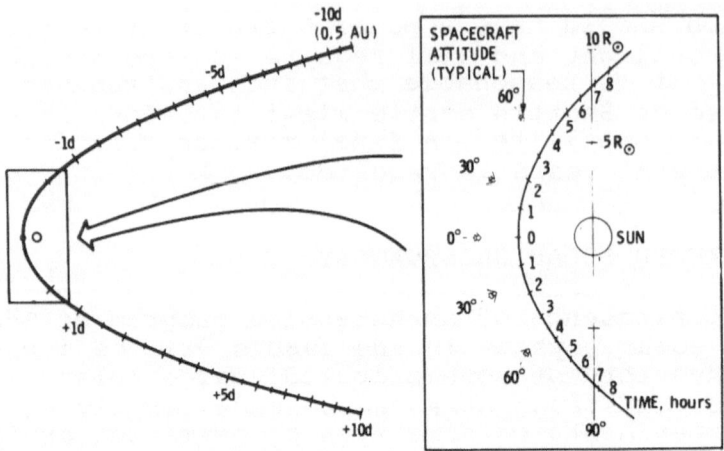

Fig. 5 Schematic of Solar Probe trajectory near perihelion.

May, 1978. It is currently envisaged that only a Solar
Probe can address certain fundamental aspects of the Sun and
heliosphere that the other missions previously discussed
above will leave unanswered:

> o the mechanism(s) for the acceleration of the solar
>   wind;
>
> o the acceleration and/or storage of energetic
>   paticles in the corona;
>
> o the distribution of mass and angular momentum
>   in the solar interior;
>
> o the three dimensional structure of photospheric and
>   chromospheric fine structure at the limits of the
>   scale heights of those atmospheres; and
>
> o the plasma temperature, density, and magnetic field
>   strength as a function of radius in the inner
>   corona.

In addition, several important other objectives can be
studied, such as the distribution of interplanetary dust
(and its ultimate thermalization) in the inner heliosphere
and the possible existence of gravity waves as measured
during the cruise phase of the mission.

Several mission architectures have been considered, but the
one currently most in favor is called the Delta
Velocity-Earth Jupiter Gravity Assist (Delta V-EJGA)

trajectory. This scenario calls for a launch in the late 1980's into an initially elliptical orbit about the Sun; near aphelion (about 2 AU), a rocket burn retargets the spacecraft to achieve a grazing encounter with the Earth (about 1.8 years after launch), the gravity assist from which sends the spacecraft out to Jupiter. A retrograde gravity swing-by around Jupiter results in the Probe essentially free-falling back to the Sun for a perihelion encounter 4.9 years after launch. In this mission version, the Solar Probe spends 20 days inside 0.5 AU and 16 hours inside about 9 solar radii (Figure 5).

Several severe technological problems faced by the Probe are under intensive study. First is the thermal shield which must keep the payload to the order of "room" temperature during perihelion, while itself sustaining a photon flux load of about 3000 "Suns." Second is a communications system for dependable telemetry during perihelion passage through the corona and a large on-board storage device for data. Third is a "drag-free" control system to reduce the non-gravitational accelerations on the spacecraft to the order of 10^{-10} g, for solar mass and oblateness determinations and the gravity wave search during the mission's long cruise phase.

7. SUMMARY

The period from 1980 through 1995 could prove to be a definitive epoch of solar and heliospheric studies by the NASA space program. The array of missions begins with the Solar Maximum Mission, which will be launched by February of 1980 and operate for at least two years. The International Solar Polar Mission will begin its five year sojourn through the heliosphere in 1983, culminating in passages over the solar poles in 1986-87. In this same time frame, the Solar Cycle and Corona Mission should begin a series of basic observations of key phenomena of the solar cycle and, in concert with the Interplanetary Plasma Laboratory, take critically important complementary observations with the ISPM.

Ideally the Solar Probe will be launched in 1987 or 1988 to begin a five year trip through the solar system, finally having its perihelion at only four solar radii in order to measure in situ the inner corona. Meanwhile, a Solar Dynamics Mission should be returning definitive data for the study of the workings of the solar cycle dynamo itself. Finally, a series of Shuttle/Spacelab flights throughout the 1980's will allow a variety of PI-class experiments to be

developed in conjunction with the Solar Optical Telescope
facility to achieve a breakthrough in the temporal and
spatial resolution limits of solar observations. These
instruments will measure phenomena important to the solar
dynamo, to the composition and structure of the Sun's
surface layers, and to the still enduring problem of the
origin of the solar wind. The end evolution of many of
these Spacelab experiments could be their inclusion on an
orbiting platform called the Advanced Solar Observatory by
1990, in time for the next maximum of solar activity.

DISCUSSION

Moore: What are the possible durations of SMM, Solar Polar and
Solar Synoptic Mission, and the probable relative timing of Solar
Synoptic with respect to Solar Polar?

Bohlin: The SMM will be launched in very early 1980, and nominally
will have a two year lifetime. Extended operations will depend on the
health of the spacecraft, the expected science return by the longer
operations period and the availability of funds. The International Solar
Polar Mission (ISPM) will be launched in February 1983, and nominally
will end in September 1987 (the end of the fiscal year 1987), after
the second solar passage of each spacecraft. Additional operations
depend on the same factors as for SMM. Our plan is to launch the Solar
Synoptic Mission (or SCADM) by mid-1986, in order to provide in-ecliptic
coverage of the corona to complement the ISPM polar passage period from
mid-86 through mid-87. Nominally I propose at least a three year lifetime, with the option for in-orbit refurbishment.

McIntosh: Ground-based solar synoptic programs are presently in
dire straits. Is NASA actively concerned about revitalizing these
programs as a necessary complement to the long-term NASA solar programs?

Bohlin: Yes, the NASA is deeply concerned about the status of
adequate ground-based observations as they relate to the support of its
solar programs. To the extent that financial resources allow, we are
trying to keep the most important elements active and healthy.

Petelski: Is there a chance for direct, in situ measurements of
interplanetary <u>neutral</u> gas to be performed by one of the probes you
mentioned?

Bohlin: Yes. The Interplanetary Plasma Laboratory (IPL) of the
OPEN program does not have all of its spacecraft resources (mass,
telemetry, power) by the strawman payload needed for the prime science
objectives. These residual resources should be available for secondary
objectives, of which the neutral gas is a good example. Moreover, the
location of the IPL permanently in the solar wind flow is an ideal
location for a neutral gas experiment.

Callahan: What is the difference between IPL and Solar Synoptic observatory?

Bohlin: IPL is to make <u>in situ</u> solar wind measurements with no direct solar observations, while the synoptic observations will be just the opposite.

PROPOSAL FOR AN INTERPLANETARY MISSION TO SOUND THE OUTER REGIONS OF THE SOLAR CORONA

H.Porsche[1], H. Volland[2], K. Bird[2] and P. Edenhofer[3]

1 Deutsche Forschungs- und Versuchsanstalt für Luft- & Raumfahrt e.V.
  Forschungszentrum Oberpfaffenhofen, D 8031 Wessling
2 Institut für Radioastronomie der Universität Bonn, D 5300 Bonn
3 Institut für Hoch- und Höchstfrequenztechnik der Ruhr-Universität
  D 4630 Bochum
  Federal Republic of Germany

The mission of HELIOS had been started in order to investigate in situ the innermost regions of the interplanetary space. The two spacecraft achieved a perihelion of about 0.3 AU solar distance. Fig. 1 is a sketch of the two orbits. The orbital periods are 190 resp. 186 d.

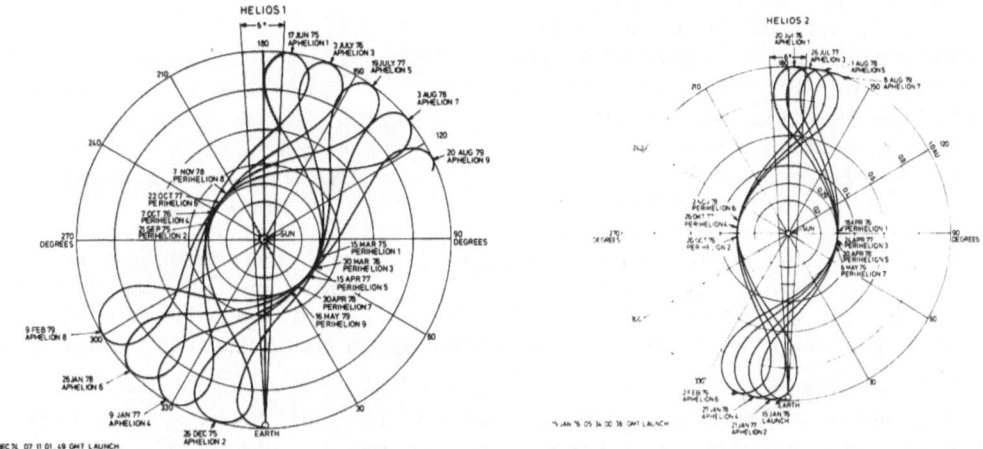

Fig. 1: Orbits of HELIOS 1 and HELIOS 2

This is pretty close to a half-year orbit (182.6 d). Therefore, the spacecraft are standing in and near outer conjunction for a relatively long time. Even now, almost five years after launch of HE1, each of the two spacecraft is running through a blackout in outer conjunction once a year. This is a reason of undesired data loss. On the other hand, however, this constellation gives the opportunity to investigate Sun's corona by sounding using radio science means, i.e. evaluation of polarization, of signal propagation time, of Doppler shift, of spectral line width, etc. (1,2,3,4). It is possible thus to measure the electron content, from which

normally the electron density distribution can be deduced, the velocity
and the size of transients travelling through the field of view (5), the
magnetic field distribution, density waves, etc., i.e. many characteristics
of the corona which define the source conditions of the interplanetary
plasma farther out. As an example Fig. 2 gives the frequency spectrum of
a pass, as it has been deduced by
Edenhofer et al. (3).

Fig. 2: Example of a power density
spectrum of the electron density
fluctuations during DRVID-Pass
abscissa: frequency (mHz)

The HELIOS corona sounding experiment did unfortunately not perform as we had desired. In general
the ground stations are overloaded
with projects. Therefore the number
of long passes of big ground stations was small, in spite of relatively good coverage. Short passes have
turned out to be of little value for
part of these investigations. Moreover after 18 months of operation of
HELIOS 1 the transponder failed. The
analysis of the failure resulted in
suspecting a certain IC to operate
insuffifiently. As a consequence the
transponder on HELIOS 2 was also
turned down in order not to jeopard-
ize the spacecraft. This all together
may be considered as only a partial
success of the corona sounding experiment on one side; on the other

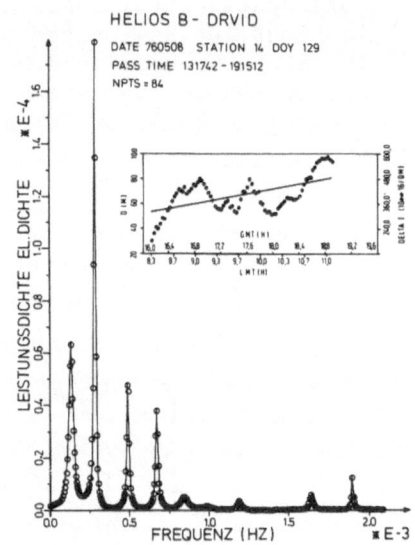

side it showed that radio science methods are well suitable and powerful
tools to investigate Sun's corona, especially those regions which neither
can be observed optically nor by in situ experiments out to say 20 R_O.
Data of the plasma behaviour out of that region are urgently needed,
however.

A proposal is given here, how to overcome this difficulty. For a detailed and thorough investigation of the corona by sounding, a spacecraft
is needed which is standing near outer conjunction for a long time. This
condition is fulfilled for a spacecraft on earth-orbit, but orbiting the
Sun near the opposite side of the sun, i.e. near outer conjunction. However such a spacecraft would either never be seen from Earth, if it is
standing exactly in conjunction, or it would stand at the same point
forever, if it is shifted a little bit to the west or east of the Sun.
I disregard orbit perturbations (etc) which obviously would lead to orbit
changes.

Consider now a spacecraft on an orbit with major axis equal to that
of the Earth, but at a slightly larger eccentricity near outer conjunction. This spacecraft would oscillate between two points west and east of
the Sun. The period would be exactly one year. The amplitude of the
oscillation depends on the eccentricity of the orbit. Thus one would

be able to sound the corona in or near the ecliptic plane. The blackout periods would be short compared to the whole mission time, i.e. of the order of a few days each.

The next step would be the extension of the mission to high solar latitudes by giving the orbit not only an eccentricity but also a small inclination. Depending on the phase angle one can end up in an orbit virtually surrounding the Sun in a short angular distance. Fig. 3 is a plot of such an orbit.

The real period as well as the virtual period is exactly one year, i.e. it takes one year to travel from the north pole of the Sun to the south and back. Thus not only equatorial regions of the corona can be sounded, but also high solar latitudes including the polar regions.

Fig. 3: Suggestion for an orbit of a Coronal Sounding Probe

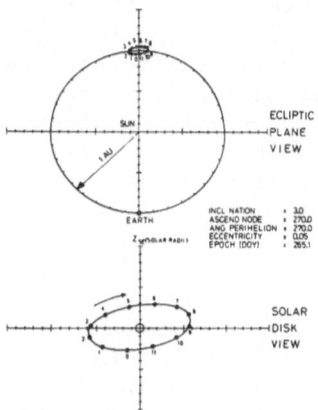

This might be of a special interest during that period, when ISPM is in its primary mission. One of the two ISPM-probes will run from north to south, the other one vice versa. The travel time from the first polar passing to the next will be about 200 d. This is in fairly good agreement with the 182.6 d of the coronal sounding mission. Thus the proposed mission could be understood as directly correlating with ISPM. A linear or a near-linear alignment would occur at the polar crossings. In between ISPM and Corona Sounding Mission are almost on the same latitude but shifted a few degrees in longitude.

Unfortunately it is not possible to arrive at the desired orbit without maneuvering. There are a few transfer orbit options. The transfer time is either about half a year or about 1.5 or 2.5 years (Hohmann transfer). The 2.5 y transfer needs least maneuvering power. This would fit very well into a time frame in accordance with ISPM, having a transfer time of about 3.5 y between launch and first solar polar passing. At least 1 y between the launch of ISPM and Corona Sounding Mission would be left for the preparation of the latter.

The scientific objective of the Corona Sounding Mission is of course the investigation of Sun's corona.
1. Faraday rotation: Linear polarization of S-band and X-band links allow the determination of $\varphi \approx \int \vec{B} \cdot n_e \cdot d\vec{s}$ to an accuracy of better than 1 angular degree and a time resolution of better than 1 second. Typical coronal oscillations of the order of some mHz may be expected. Therefore the consecutive pass length should be of the order of 20 h, preferably continuous over periods of several days. Special attention must be given to observe transients in those regions which correlate with ISPM and where the boundaries between fast and slow solar wind regimes may be expected.

2. Ranging: Determination of the signal propagation time between spacecraft and ground station in S-band and X-band. Plasma influences as well
as gravity effect delays have to be considered. The plasma-generated time
delay is a measure of the electron content between transmitter and receiver. The achievable time resolution is of the order of some 10 min.
By comparison of two frequencies plasma- and gravity contributions to the
time delay can be separated.
3. Range rate (Doppler): In addition to the determination of the relative
velocity of transmitter and receiver the Doppler shift of the transmitted frequencies depends on the electron density between transmitter and
receiver. It influences the phase velocity of the wave. Therefore the
Doppler shift is a sensitive indicator for changes in electron density
anywhere along the ray path of the wave, especially in the corona at high
time resolution (better than 1 sec).
4. DRVID (Differentiated Ranging Versus Integrated Doppler): This is a
combination of phase- and group-velocity measurement. It determines the
electron content between transmitter and receiver.
5. Linewidth determination: The line width of the received signals
comprises information about the dynamical state of the matter between
transmitter and receiver, i.e. about turbulences etc.

All those data include local information of the observed transients.
It can be extracted by correlation methods. Thus such a mission is able
to give a comprehensive review of the state and of the dynamical behaviour of the corona. It is obvious that the spacecraft can additionally
be equipped with in situ experiments or with experiments to observe the
Sun optically (spectroheliograph, coronagraph, etc.) to become a solar-
backside probe.

References:

1 H. Volland, M.K. Bird, G.S. Levy, T.C. Stelzried and B.L. Seidel
  HELIOS 1 Faraday Rotation Experiment: Results and Interpretations
  of the Solar Occultation in 1975; J.Geophys. 42, 659, 1977

2 P. Edenhofer, P.B. Esposito, R.T. Hansen, E. Lüneburg, W.L. Martin and
  A.I. Zygielbaum, Time Delay Occultation of Data of the HELIOS
  Spacecrafts and Preliminary Analysis for Probing the Solar Corona;
  J.Geophys. 42, 673, 1977

3 P. Edenhofer, M.K. Bird and H. Volland, Comparison of Time Delay and
  Faraday Rotation Measurements from HELIOS Spacecraft; Kleinheu-
  bacher Ber. 21, 305-312, 1977 and H. Süss, Privat Communication

4 R. Woo, Radial Dependence of Solar Wind Properties Deduced from
  HELIOS 1 and 2 and PIONEER 10/11 Radio Scattering Observations,
  Astroph. J. 219, 727, 1978

5 E. Lüneburg and P.B. Esposito, A Method for the Evaluation of Solar
  Coronal Plasma Propagation Speeds by Radio Occultation of Space
  Probes; Nat. Radio Sci. Meeting, U of Washington, Seattle, Wash.
  18 - 22 June 1979, p. 284 and JPL - Engineering Memo 315 - 90,
  1979

DISCUSSION

Bratendal: Do we need Venus or some other clever scheme to inject into earth orbit in opposition?

Porsche: The spacecraft needs a motor for orbital corrections, but no swingby maneuver at a planet to achieve the desired orbit and position in superior conjunction.

SOLAR AND INTERPLANETARY DYNAMICS
(Symposium Summary)

M. Kuperus
Astronomical Institute of the University of Utrecht

Solar and interplanetary dynamics comprises dynamic and plasma-
physical phenomena in the solar atmosphere, the corona and the interplanetary medium in the broadest sense. In this symposium, however, one
has essentially tried to restrict the subject matter to the study of
the propagation of a disturbance, produced in the solar atmosphere,
through the corona and the interplanetary medium. In studying solar and
interplanetary dynamical phenomena we find ourselves in the unique
position, with respect to other astrophysical disciplines, to be able
to relate solar observations obtained with the highest possible spectral,
spatial and time resolution with in situ measurements made in the interplanetary medium. It has now turned out that the two fundamental
questions to be answered are:

a) How does the medium in between the sun and the earth and beyond the
   earth's orbit, the socalled *heliosphere*, look like? Does a basic undisturbed heliosphere actually exist, and is one able to model its
   observed magnetic structures and plasma motions with their spatial
   and temporal variations?

b) How and where in the solar atmosphere are the disturbances generated
   and what are the characteristic time scales, geometries and energies
   involved?

   In trying to answer the above mentioned complex of basic questions
one is invariably confronted with three outstanding problems in solar
physics:

 i) What causes solar magnetic fields to appear at the solar surface
    and how do they extend into the overlying atmosphere?

 ii) How is the strongly magnetically structured corona produced and
    heated?

iii) What is the mechanism of solar eruptions such as flares and
    erupting prominences?

M. Dryer and E. Tandberg-Hanssen (eds.), Solar and Interplanetary Dynamics, 547-552.

1. CORONAL AND INTERPLANETARY STRUCTURE

As to the first problem the magnetic configuration in the photosphere is largely determined by the circulation in the subphotospheric
layers. Conversely, the magnetic structures and their motion relative
to the ambient photosphere can be used to get some insight in the
rotation of the deeper layers. Once the magnetic structures, presumably
present in the subphotospheric layers as flux tubes, have emerged, they
must rather quickly evolve into one or the other of the two basic configurations: the open field structures and the closed field structures.

The open magnetic configurations, whose footpoints are grouped in
large regions of essentially one polarity, are directly associated with
the coronal holes, observed as persistent depressions in soft X-ray
emission.

The coronal holes are the seats of the high speed solar wind. They
are born in association with active regions and they grow as new active
regions emerge and old regions disperse so as to form a new magnetic
cell pattern. However, the lifetime of the open field structures is
much longer than the lifetime of the individual active regions, which
suggests that the active regions only produce minor changes in the
large scale magnetic field pattern.

The closed magnetic configurations are observed in great concentration inside the active regions as loop structures visible in
X-ray emission, the green coronal line, UV lines and in Hα. In between
different active regions one frequently finds the interconnecting loops
as convincing evidence that large scale magnetic reconnection must have
taken place after the birth of active regions.

The closed magnetic regions bottle up part of the energy received
as they show bright in X-ray emission, while the open regions dispose
of their energy received by a high speed solar wind. It should be
stressed here that no satisfactory theory of the heating of the corona
for the holes as well as for the loop structures exist at this moment,
though considerable progress is made on the propagation of Alfvén waves
through an inhomogeneous corona. Only Alfvén waves seem to be able to
transfer sufficient energy and momentum to great heights. In the lower
transition layer and chromosphere acoustic shock waves are still undoubtedly the most likely source of heating.

The magnetic structure of the corona can be modelled in various
ways. The simplest way is to assume a potential field with the photospheric field distribution as the boundary. Such a potential field
appears indeed to be a reasonable approximation in many cases.

A still better approximation is a force free field distribution.
Since β in the corona is much smaller than unity, this must be a good
approximation for most of the coronal magnetic field. However, two
critical remarks are appropriate:

Firstly: It is not at all certain that a force free field solution
can be used as an equilibrium solution in the case that gas pressure is

not negligible. Stable equilibrium solutions seem to be much more restricted than those of a zero pressure plasma ($\beta = 0$).

Secondly, the corona is not a static atmosphere and not even stationary. The lower boundary is in a state of continual motion while drastic field changes occur when new flux emerges. Moreover, several solar radii above the surface the magnetic Reynolds number is so large that all magnetic fields are stretched by the solar wind into interplanetary space. A force free field solution should be a bad approximation in this case and undoubtedly current sheets of different scales and geometries are present in the corona above an active region.

In the interplanetary medium the sector structure, with its sharp boundaries, is one of the most well established corner stones of the outer magnetic configuration.

Yet, it appears difficult to trace these sector boundaries back into the photospheric magnetic field or in any other solar phenomena. This is probably the result of the fact that the strong fields, which are the most pronounced, close rather low in the corona, while small though large-scale fields, which may remain unnoticed in the photosphere, are stretched outwards and thus constitute the important boundary condition for the magnetic structure of the "more or less undisturbed" outer interplanetary medium.

The sector structure is certainly related to the pattern of coronal holes of one polarity, which can frequently be interpreted as the extension of the polar coronal holes. The neutral sheet separating the north and south polar coronal holes encircles the sun like a "ballerina skirt". The large scale pattern of magnetic fields, clearly outlined by filaments, is only weakly affected by solar activity. However, when large scale magnetic fields of one polarity merge into that of the opposite polarity the filaments become distorted and disrupt.

Filaments always occur at neutral lines and are the inevitable evidence that reconnection of magnetic fields does take place on a large scale in the corona, thus giving rise to numerous kinds of disturbances.

The process of large scale reconnection in the corona is intimately associated with the formation of prominences and loop type structures. The whole dynamics of the corona and the interplanetary medium must be seen in the light of these facts: Large-scale long lasting open unipolar magnetic regions with, at their borders, the shorter-living and smaller bipolar active regions. At many places reconnection occurs at the borders of the unipolar magnetic regions as well as inside the bipolar regions.

2. CORONAL TRANSIENTS

A great deal of work has been reported on the study of the coronal and interplanetary responses to disturbances at various time scales. Long time scale variations of the heliosphere follow the evolution of

solar activity and its influence on the large scale magnetic structure.
Short time scale variations which are observed as coronal transients
are associated with eruptive solar phenomena of various kinds in
particular with flares and prominence eruptions.

Coronal transients have been extensively studied in white light,
X, radio and, in most cases, they can be associated with mass ejections.
Some of the transients are nothing else but the remnants of the transient producing agent. In particular transients observed in white light
are mass ejection transients which are associated with Hα flares, X-ray
events, eruptive prominences, sprays, surges and metric type II and
type IV bursts. The mass ejections that have been identified have often
a loop type appearance. They are strongly associated with flares and
filament disappearances (Disparitions Brusques) and often occur when
the surface fields are strong and complex. The correlation with flares
and erupting filaments is particularly strong if one restricts attention to the large flares and filament eruptions only, evidently the
energetic events. In these cases the correlation with soft X-ray
events and metric radiobursts is very good. The correlation of coronal
disturbances with erupting prominences is certainly better than with
flares. The important conclusion that can be drawn from all this is
that eruptive prominences and disparitions brusques must be the central
phenomenon that produces a coronal transient. There is increasing
evidence that a prominence eruption in an active region is to a large
extent the solar flare. It is therefore of paramount importance to
study the relation between prominence eruptions and solar flares and
the cause of prominence eruptions that do not seem to be related to any
flare. Detailed studies of preflare emissions and disturbances can
possibly solve this outstanding problem in solar physics. The radio
transients are in general associated with shock waves and particle
acceleration except for the moving type IV which is direct evidence for
a moving plasma cloud. Since particle acceleration is likely to be
associated with magnetic reconnection , any form of magnetic reconnection may give rise to a radio transient. In the very powerful cases mass
ejection may well result from reconnection and so-called current sheet
rupture. It is of interest to note that metric type III bursts increase
in number prior to mass ejection transients. Their peak seems to
correspond with the forerunners.

Theoretical work on the origin and nature of coronal transient
falls in two categories. First there are the model calculations of
coronal disturbances where the mechanism causing these disturbances has
been unspecified but where one particularly concentrates on modelling
the spreading and propagation of a magnetohydrodynamic perturbation.
Such a perturbation can be either a temperature or pressure pulse
caused by an as-yet unspecified impulsive heating mechanism in the lower atmosphere or a well defined magnetic perturbation at the lower
boundary. Impulsive heating can give rise to mass ejections such as
surges and spicules. To give full credit to all possible wave modes and
interactions these calculations ought to be made in three dimensions
using kinetic theory instead of the fluid approximation. Notwithstanding
the restrictions one is forced to make, these numerical calculations

give promising results, although the large variety of velocities of coronal transients already indicates that a pure wave mechanism certainly cannot explain the majority of white light transients. Instead real mass ejection has to be of central importance.

Secondly there are the models of evolving magnetic structures such as sheared fields, helical structures, loop type structures on the one hand and on the other hand the models of emerging magnetic fields creating large scale current sheets, where the process of magnetic reconnection is the major cause of magnetic reorientation and its dynamical consequences.

Both types of field configuration are used as model fields in flare theory and both types of field may actually occur in a flaring active region or in prominences. The extreme cases that are studied are the force free fields and the neutral sheets. I repeat the conjecture expressed earlier that hopefully force free fields can be used in a low β plasma when the boundary conditions change very slowly but it seems unrealistic to model the field this way when a fast disturbance occurs. The dynamic evolution of coronal flux tubes and arcades has recently come off the ground but none of the studies reported have considered the complicated boundary value and initial value problem in a fully consistent way.

Any magnetic configuration in a low β plasma that is in equilibrium must have the major component of the current density parallel to the magnetic field. Hence, the magnetic field is thus likely to be subject to MHD instabilities of the kink-mode as well as of the tearing mode type. On the resistive instabilities no work has been reported during this conference, while kink type structures in 3-dimensional non-linear calculations have been shown to occur under certain conditions during force free field evolution.

Large perpendicular current densities can only occur in neutral sheets, where $\beta \gg 1$ or they may be generated because of short time scale disturbances, e.g. at the boundaries. In that case the interchange instability may develop, increasing the rate of heating and possibly giving rise to detached plasma clouds. It should be emphasized that plasma turbulence may play an important role in the heating and dynamics of neutral sheets as well as in other configurations where magnetic reconnection occurs. Several attempts have been made to find the observable signatures of plasmaturbulence especially in radio waves. Radar measurements indicate that a high level of low frequency plasma waves is present above active regions, possibly giving a clue to the mechanism of the heating of the active solar corona. However, no satisfactory theory has been presented for the onset and maintenance of plasmaturbulence in coronal and interplanetary structures.

The presence of plasmaturbulence in the corona and heliosphere clearly indicates that the study of the propagation of a disturbance through the outer layers can only be treated correctly when a microscopic description is used. Only the microscopic description treats properly the processes of dissipation and wave-particle interactions.

The corona is not a fluid but a collisionless multicomponent plasma subject to perturbations of various time scales that sometimes can be surprisingly well approximated by a fluid.

I have tried to summarize the present state of knowledge and the trend of research in this field as it has come across during this symposium in a general sense without a discussion of any detailed work that has been presented.

SUBJECT INDEX